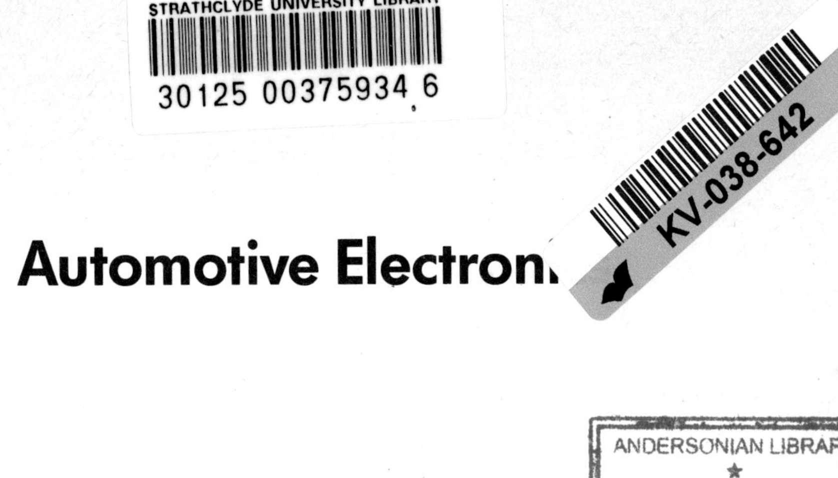

Automotive Electron

Conference Planning Panel

P C Capon (Chairman)
Leyland DAF
Leyland, Preston

J P Cousins
AB Automotive Electronics Limited
Whitchurch, Cardiff

W Gibbons
MIRA
Nuneaton, Warwickshire

T J Goldrick
Hereford

S Hutson
Rolls Royce Motor Cars Limited
Crewe

P Kay
Leyland Technical Centre
Leyland
Preston

A Noble
Ricardo Consulting Engineers plc
Shoreham-by-Sea

C Tapsfield, BSc(Eng), AMIEE
Texas Instruments
Bedford

J D Turner, BSc, PhD, CEng, MIEE
Department of Mechanical Engineering
University of Southampton

A C N Wilkinson, CEng, MIEE
Ford Motor Company
Basildon, Essex

M Williams, BSc, CEng, FIEE
Jaguar Cars
Coventry

J Wood
RAC
London

Proceedings of the Institution of Mechanical Engineers

Seventh International Conference

Automotive Electronics

9–13 October 1989
Olympia Conference Centre
London

Organized by
Automobile Division of the
Institution of Mechanical Engineers
Computing and Control Division of the
Institution of Electrical Engineers

In association with
Associazione Elettrotecnica ed Elettronica Italiana
Institute of Physics
Japan Society of Automotive Engineers
Society of Automotive Engineers
Svenska Mekanisters Riksförening
Verein Deutscher Ingenieure

With the patronage of
FISITA (Fédération Internationale des Sociétés d'Ingénieurs des Techniques de l'Automobile)

IMechE Conference 1989–11

MEP

Published for the Institution of Mechanical Engineers by
Mechanical Engineering Publications Limited

First Published 1989

ISBN 0 85298 697 1

A CIP catalogue record for this book is available from the British Library.

Printed by Waveney Print Services Ltd, Beccles, Suffolk

Contents

C391/KN1

The state of automotive electronics in the year 2000: a perspective of the North American marketplace

J J PAULSEN, BS, MS
Ford Motor Company, Dearborn, Michigan

INTRODUCTION

The emergence of electronics has changed the character of motor vehicles in a myriad of ways. From engine control to suspension leveling, from instrumentation to entertainment, the "electronics revolution" has made a significant impact and promises more of the same. This paper has been prepared to help interested observers and auto industry participants understand the future direction of automotive electronics in North America. It's hoped that both the general and technical reader will enjoy our vision for the year 2000 automobile.

HISTORICAL PERSPECTIVE

North American automotive electronics has evolved in three distinct but overlapping stages. Each stage influenced the industry's practices and the nature and application of electronic systems in vehicles for about 15 years. Figure 1 shows the stages of evolution that cover the periods from the mid '60s to the late '70s; from the mid '70s to the late '80s; and from the mid '80s to about the year 2000.

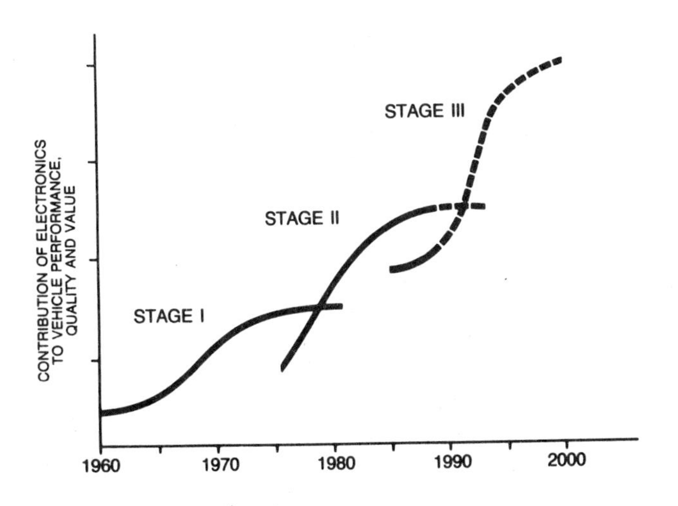

Fig 1 Automotive electronics technology evolution

The three evolutionary stages can be distinguished by the type of semiconductor devices in use in automobiles during each period (Figure 2). The first stage saw the application of diodes, discrete transistors and analog IC's. The second stage was brought about by the availability of digital IC's followed by four, eight, and sixteen bit micro-processors. The third stage, which has already begun, will see increasing use of smart power devices, smart sensors and massive memories.

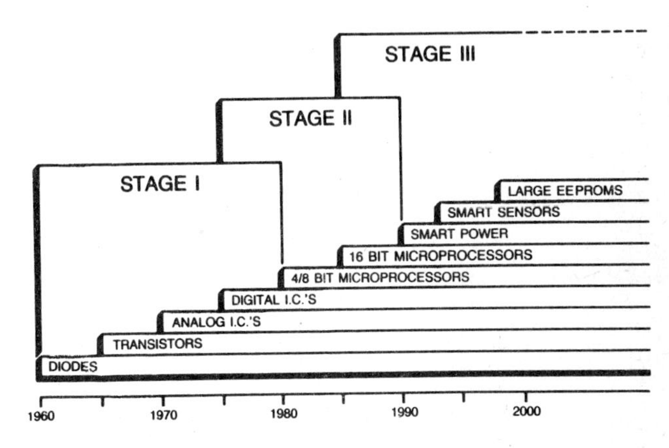

Fig 2 Semiconductor evolution during the three stages of automotive electronics in North America

STAGE ONE - Solid state electronic devices were first widely used for motor vehicles in the 1960s. Over the following fifteen years, diodes, discrete transistors and analog integrated circuits were used to solve problems in stand-alone electronic components (Figure 3). The earliest installation in Ford vehicles of an all solid state radio occurred in 1961. Solid state ignitions were first incorporated in Ford's production in 1973 (Figure 4). A stand-alone electronic clock appeared in the instrument panel in 1974. These early applications provided a learning experience with the emerging electronic technology.

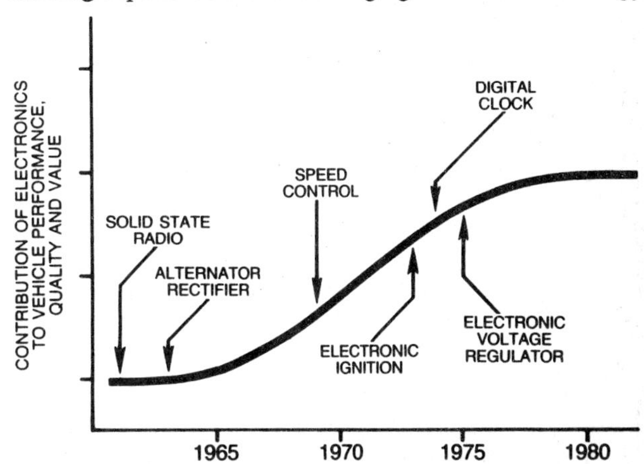

Fig 3 Stage One was characterized by stand-alone electronic products – illustrated with Ford experience

More importantly, they demonstrated that electronics could achieve the levels of reliability required by the automotive industry, while providing dependable and affordable service to the customer. For example, new electronic devices increased the dependability of many automotive components. Radio failures became much less frequent than in the old vacuum tube

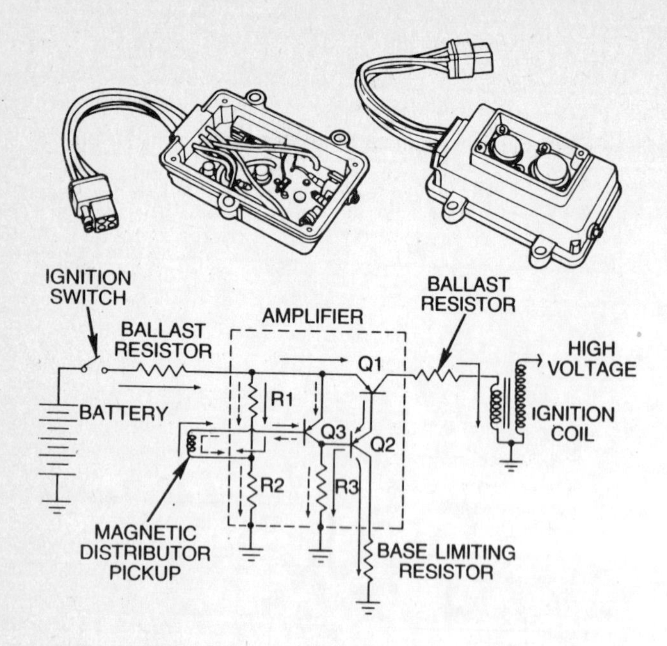

Fig 4 The first Ford solid state ignition module and circuit diagram

days. Car clocks became more reliable timepieces. Electronic ignition eliminated the routine maintenance associated with changing the breaker points.

STAGE TWO - In the mid 1970s, new electronic capabilities became available to the North American automotive industry. Their applications were different in significant enough ways to define another stage in the evolution of automotive electronics.

The primary source of new capability was the microprocessor. It came along at a time when the American industry was experiencing upheaval. Gas lines, CAFE standards, emissions controls - created unprecedented challenges. The microprocessor led to many solutions.

Stage Two was characterized by a shift from independent components to increasingly sophisticated systems, which link components together (Figure 5). These systems were first used for engine controls. Ford, for example, introduced the EEC-I in 1978. Several sensors were linked with a computer which, in turn, was linked to various output devices such as the ignition module. Similarly, multiple electrical and electronic components were tied together in driver information and entertainment applications. Electronics have been used to integrate functions which previously had been separate.

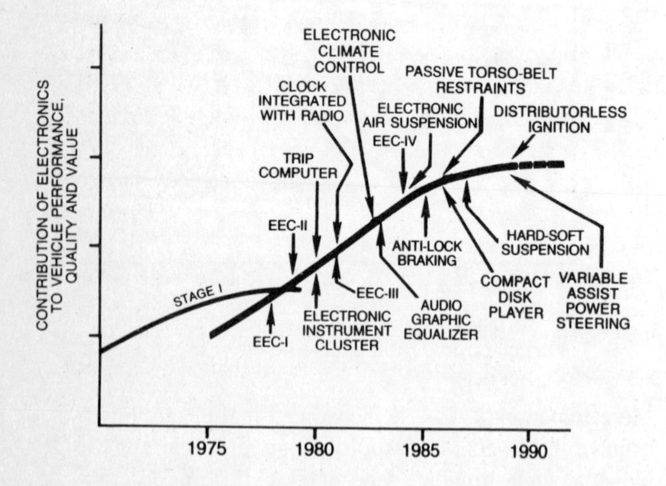

Fig 5 Stage Two has seen electronics linking components in systems – illustrated with Ford experience

Late Stage Two developments will make more widespread use of advanced packaging techniques such as surface mounted devices, increased processor speed and capacity, increased memory and refined I/O methods to obtain optimal functional performance from the auto's special purpose sub-systems.

Powertrain control will be adapted to a greater range of conditions, and there will be more integration between engine and transmission. Anti-lock braking will be more common in North America (Figure 6). Improved displays will offer drivers more choices of format and content. Cellular telephone communication will be increasingly common in North America. All of these improvements should create vehicles that are more responsive, useful and reliable.

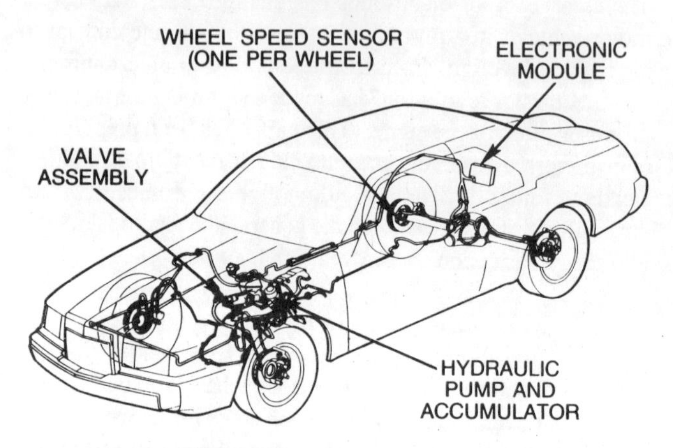

Fig 6 Ford four-wheel anti-lock braking system

As the vehicle's sub-systems evolve over the next few years, they will form a total vehicle network. Sensors, processors and actuators will be interconnected, with power and control signals distributed in a highly efficient manner. This functional integration will lead to Stage Three.

STAGE THREE - This phase of automotive electronic development will be characterized by the emergence of a totally integrated vehicle electrical and electronic system. Designers will escape from the mechanical function replacement and "add-on" approaches that have characterized Stages One and Two. They will seek to optimize the performance of the total vehicle through electronics. The total system will have great flexibility and adaptability, with extensive software control of multi-function features. This will offer customers unprecedented opportunities to customize their vehicle. Vehicle characteristics such as ride quality, handling properties, steering effort feedback, brake "feel," information display format, and even engine power versus economy tradeoffs will be controllable by the driver.

Operating as an information-based system, the automobile's on-board electronics will use extensive computing capacity, multiplexed circuit technology, and amounts of program memory that would be considered extremely large by today's standards.

Some early third stage features are already beginning to appear (Figure 7). These include speed control integrated with engine control -- and transmission controls integrated with engine

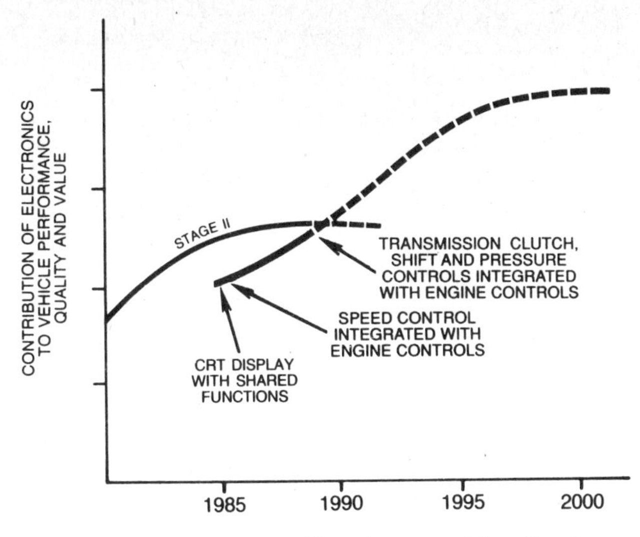

Fig 7 Stage Three will evolve around functional integration of electronics — illustrated with Ford experience

controls (Figure 8). These examples represent only the leading edge in the systematic integration of functions, a transition that will completely change the role of electronics in the automobile by the year 2000.

Stage Three will bring:

- Torque-demand powertrain control which fully integrates the response of the engine and transmission.
- Vehicle dynamics which integrates braking, steering, and suspension.

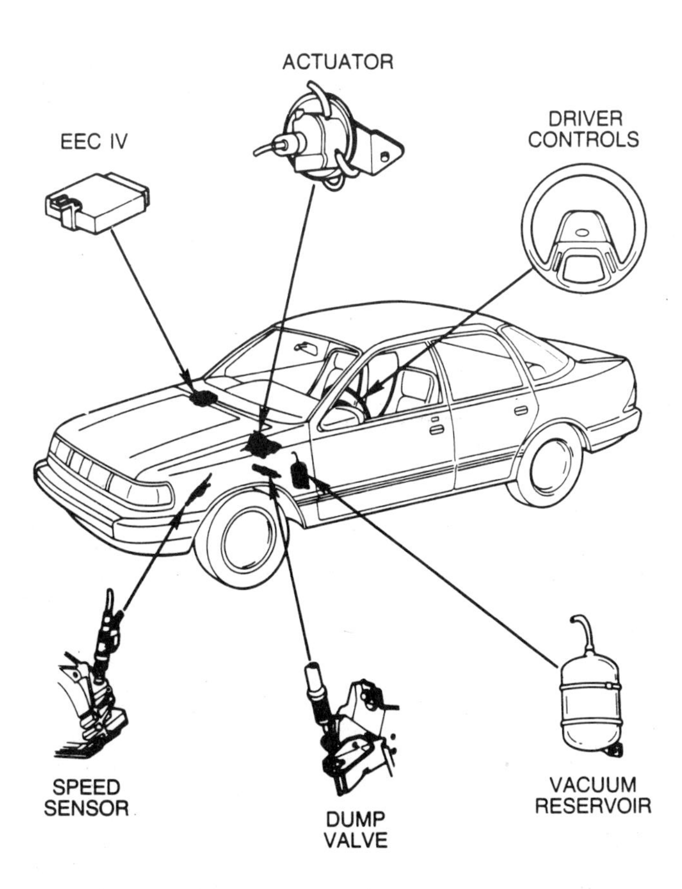

Fig 8 Ford speed control system with electronics integrated into EEC-IV

- Electric power management based on new power generating components and sophisticated load management controls.
- Traction control derived from full powertrain control coupled with anti-lock braking.
- Multi-purpose soft switches and shared displays for driver information, climate control and entertainment functions.

The Third Stage has begun - it will be fully evolved when the American cars of the 2000 model year roll out of the assembly plants. By the year 2000 the value of electronics for the average North American automobile will reach $2000 (Figure 9). Let's explore some of the features that such a car should bring to the market.

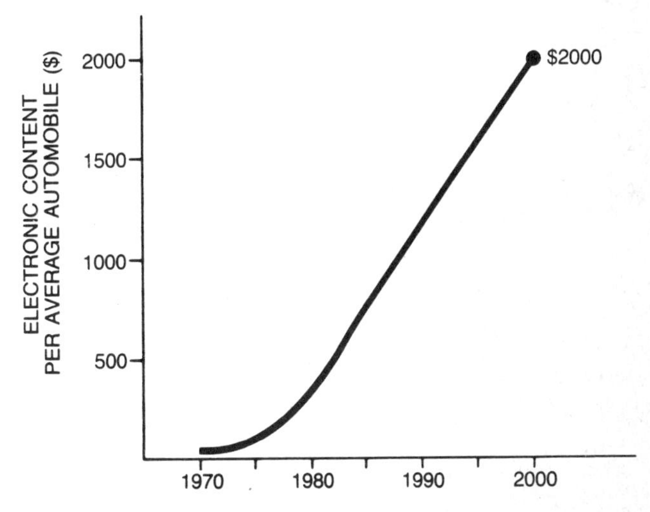

Fig 9 Automotive electronics content in North America will reach $2000 by the year 2000

THE NORTH AMERICAN FORD AUTOMOBILE IN THE YEAR 2000

The emergence of Stage Three in electronics will allow Ford to introduce many new electronic features on its motor vehicles. A window sticker of the year 2000 (Figure 10) should have a good array of electronics. Generally, the following electronic systems are expected to be available.

POWERTRAIN - The engine compartment of the year 2000 Ford automobile may contain a lightweight, supercharged or turbocharged 4 cylinder or 6 cylinder, multi-valve engine of 1.5 to 2.5 liters displacement (Figure 11). It will probably be equipped with multipoint electronic fuel injection and distributed ignition which is distributorless, and has a high-voltage coil at each spark plug. The system will control the engine on a cylinder-by-cylinder basis.

The powertrain will be electronically controlled by a highly advanced system. Engine operating parameters will be adaptively controlled over the full range of torque and RPM, and will allow the driver to select for performance or economy. Variables under active and continuous electronic control will include manifold boost pressure, fuel mixture, spark timing, valve timing, and variable intake manifold geometry.

The information needed to manage the engine control will come from a small number of high-performance sensors.

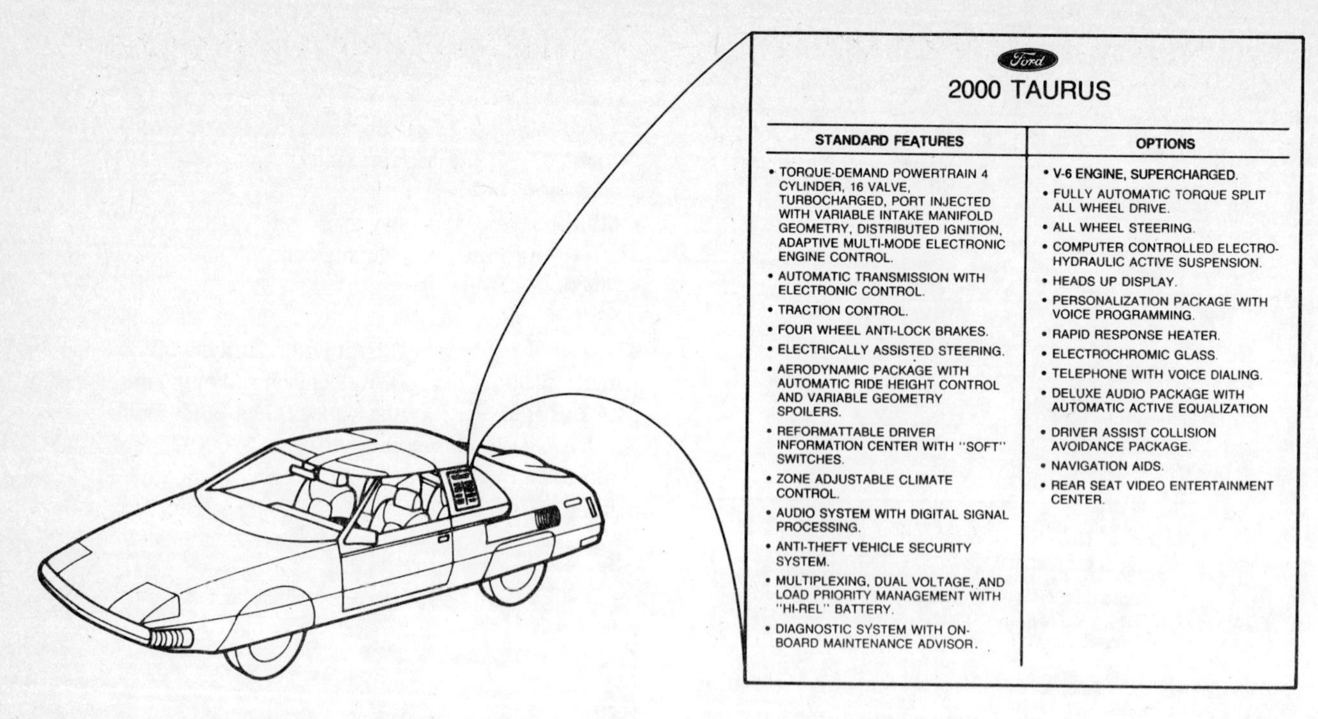

Fig 10 The window sticker on the year 2000 Taurus with Stage Three electronics

These sensors will monitor, analyze and transmit data on fundamental engine performance parameters. Improved sensors will be necessary before this is possible. Primary data will include combustion chamber conditions and exhaust gas chemistry. This data will be compared by the engine controller to a performance algorithm of much greater sophistication than today. Then, all controlled variables will be adjusted to optimize vehicle performance according to the algorithm. This will be a great improvement over current systems which measure secondary parameters such as inlet charge temperature and barometric pressure.

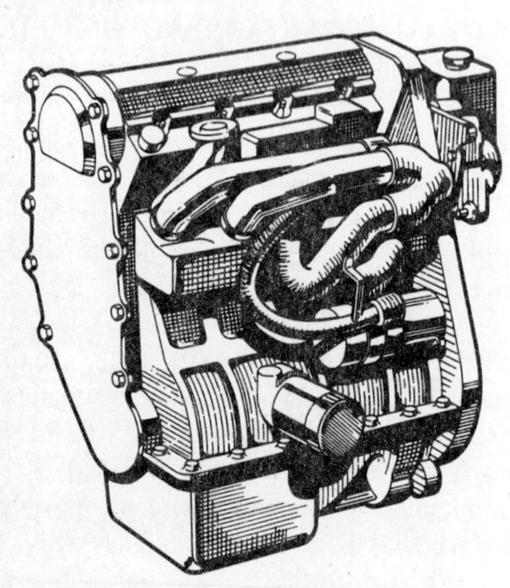

Fig 11 Four-cylinder engine performance will improve with Stage Three electronics

The engine will be closely coupled with a transmission of advanced design. The transmission will either be continuously variable (Figure 12) or will have a highly adaptive shifting capability. Transmission and engine will be electronically controlled as a unit in response to the driver's demand for more power. Responding to engine speed, vehicle speed and command input from the driver, the powertrain controller will decide whether to supply torque by increasing engine output, altering the drive ratio or both. Figure 13 contrasts the degree

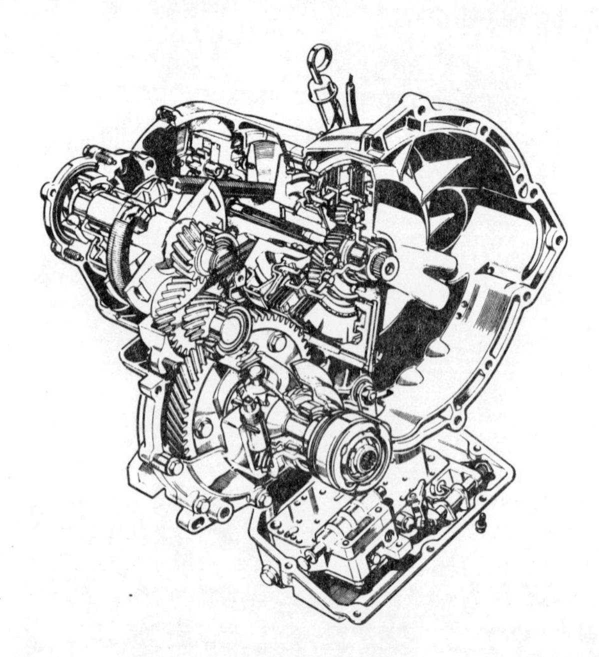

Fig 12 Continuously variable transmission (CVT)

of powertrain integration typical of Stages Two and Three. In some cases, the transmission will drive all four wheels and utilize a sophisticated electronic control scheme.

Drivers in the year 2000 will find that the powertrain will perform smoothly under all conditions. Changes in transmission ratio and adjustments in engine speed will seem nearly imperceptible compared to those in today's cars. The powertrain management system will enhance the car's safety and improve its performance in ways that are meaningful to all users - by keeping the driver in control.

CHASSIS - Perhaps the most revolutionary new features and performance enhancements resulting from Stage Three systems integration will occur in the area of the chassis system consisting of steering, brakes and suspension. Drivers will experience a new level of performance from these chassis

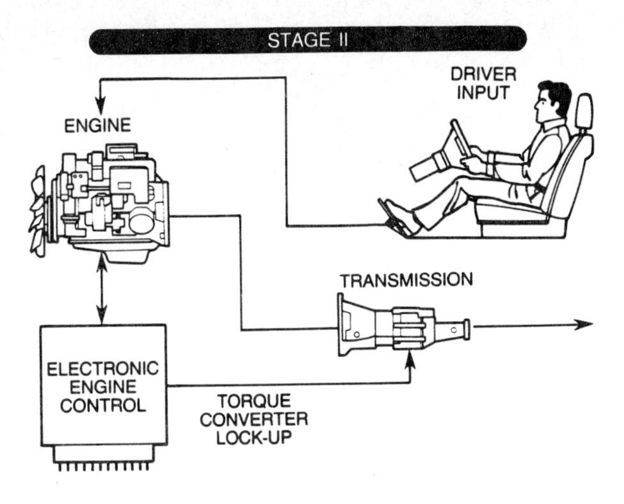

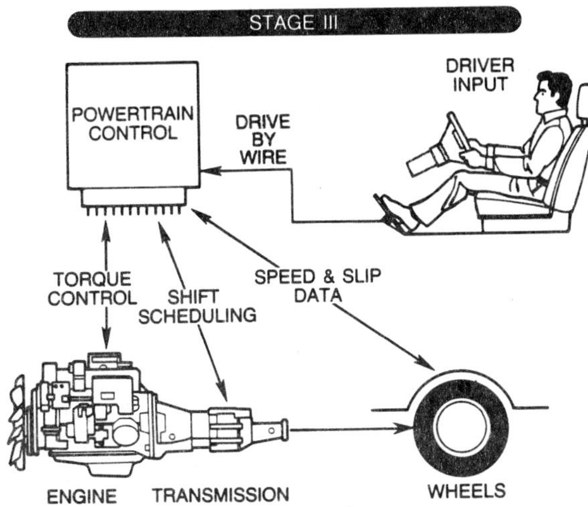

Fig 13 Electronic powertrain systems for Stages Two and Three

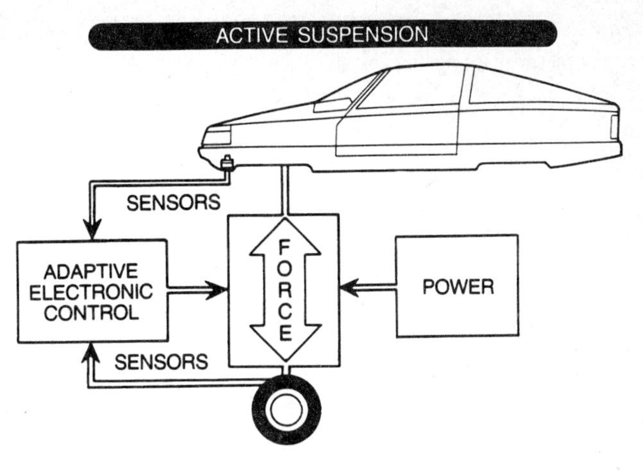

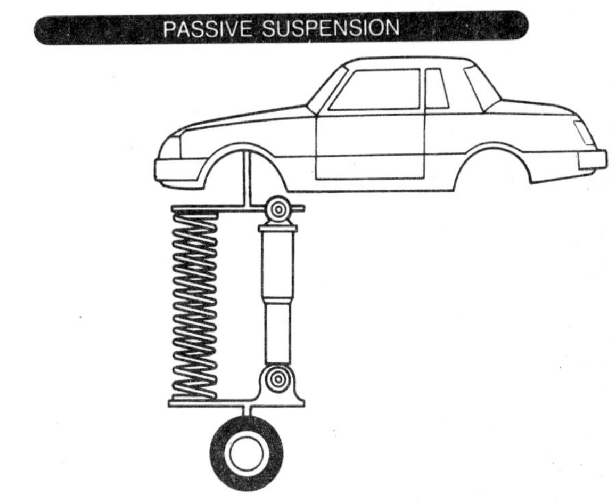

Fig 15 Suspension evolution to active control

systems through their synergistic interaction achieved through an integrated electronic network as depicted in Figure 14. To better examine this interactive chassis system, let's examine the suspension, steering and braking elements.

The growing capability of electronics in Stage Three will allow adaptive control of the suspension - springs, shock absorbers and suspension geometry - which were restricted to a passive response in Stages One and Two (Figure 15). In order to control ride height, aerodynamic angle of attack and dynamic response of the body, an electronic system will sense displacements and accelerations in the suspension system and will control spring rate and damping independently at each wheel. A first step in this direction was Ford's introduction of ride height control (Figure 16) in 1984.

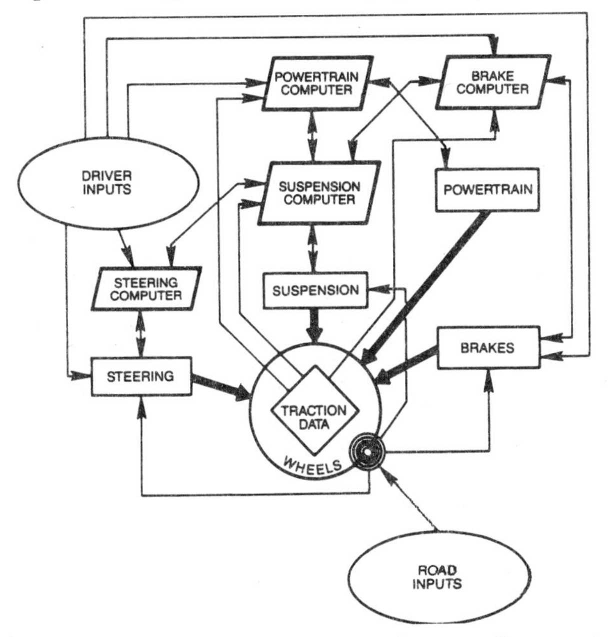

Fig 14 Stage Three chassis control system diagram

In most year 2000 Ford cars, the suspension control will be "semi-active." There will be continuous modulation of devices

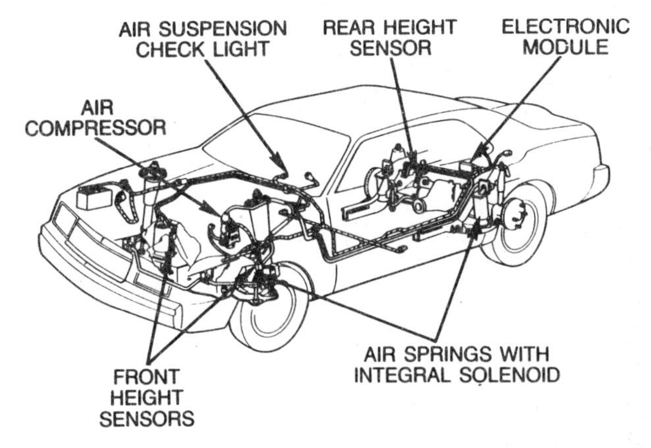

Fig 16 Ford air suspension/ride height control

like valves to control shock absorber damping but with no external power input. In some high-performance vehicle applications, however, fully active suspensions (Figure 15) will incorporate controlled energy input from a dedicated power source. While the ultimate evolution of both semi-active and active systems will probably be electro-mechanical, those systems in use in the year 2000 will employ electro-hydraulic control units. In either case the driver will experience a marked improvement in handling quality and ride comfort.

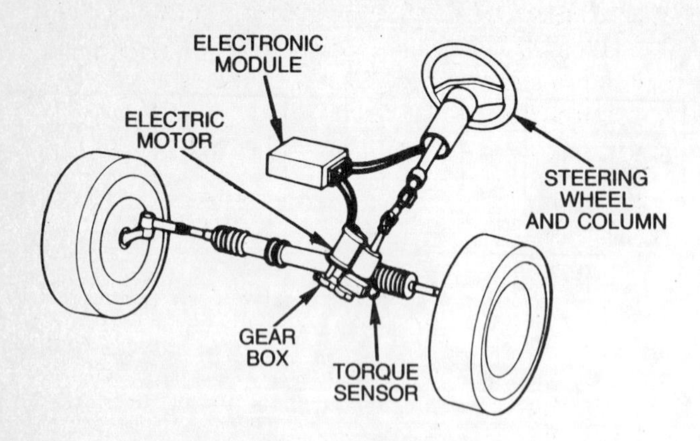

Fig 17 Electric power steering

Electrically actuated front wheel steering (Figure 17) will be found on most cars by the year 2000. Among the advantages of these systems will be compactness, energy efficiency and adaptability.

Braking performance will improve steadily as it evolves from the anti-lock braking systems that are just now becoming widely available. By the year 2000, traction control systems will fully integrate braking with the powertrain. The functional flexibility of these systems will be far greater than the simple "anti-lock" capability that's presently available. Conditions which could cause slipping will be monitored during both acceleration and deceleration. The system will modulate torque and braking inputs to provide both maximum acceleration and minimum stopping distance. The driver will be unable to break the wheels loose from the road under anv normal driving conditions.

An enhanced traction control system will improve the vehicle's ability to avoid collisions. In some applications, the space all around the vehicle will be monitored for the presence of collision risks using some combination of sensing technologies such as radar, laser, visual, infrared and ultrasonic. Not only will the area in front of the vehicle be scanned to detect a rapidly closing interval, but the "blind spots" on the rear quarters will be monitored to assure safe lane changes (Figure 18). The sensors' output will be analyzed by artificial intelligence software that will direct controllers to reduce acceleration, or in extreme cases, to apply the brakes and tighten seat belts if an accident is impending.

For some time now, the handling capability of an automobile has been beyond the average driver's skills. Advances in steering, braking and suspension technology in Stage Three will allow the average driver to employ the full performance potential of the vehicle in exceptional situations (like avoiding accidents), without subsequent loss of control. The subtle and rapid corrections needed to deal with the complex dynamic transients will be handled automatically.

DRIVER INFORMATION AND PERSONALIZATION - By the year 2000, human factor design for driver information displays and controls will be greatly advanced. "Cockpit workload" will be much reduced. Essential information such as vehicle speed will be provided continuously by a holographic heads-up display (Figure 19), similar to what is now being used in aircraft. Other information will be displayed on a reformattable multifunction display panel (Figure 20). It will use one of several technologies such as liquid crystal, vacuum fluorescent or light emitting diode.

The system will display performance data whenever it senses something unusual. Also, the driver will be able to select a particular array of information and the style in which it is presented. For example, the driver will be able to request a complete display of all engine operating parameters such as RPM, oil pressure, coolant temperature, fuel pressure and so

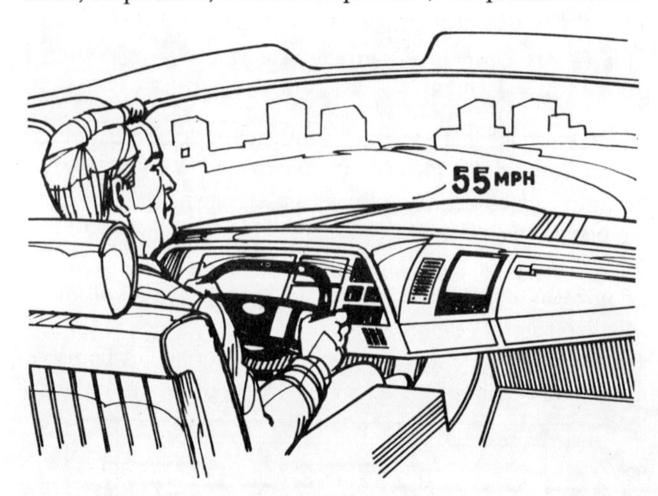

Fig 19 Heads-up display of vehicle speed

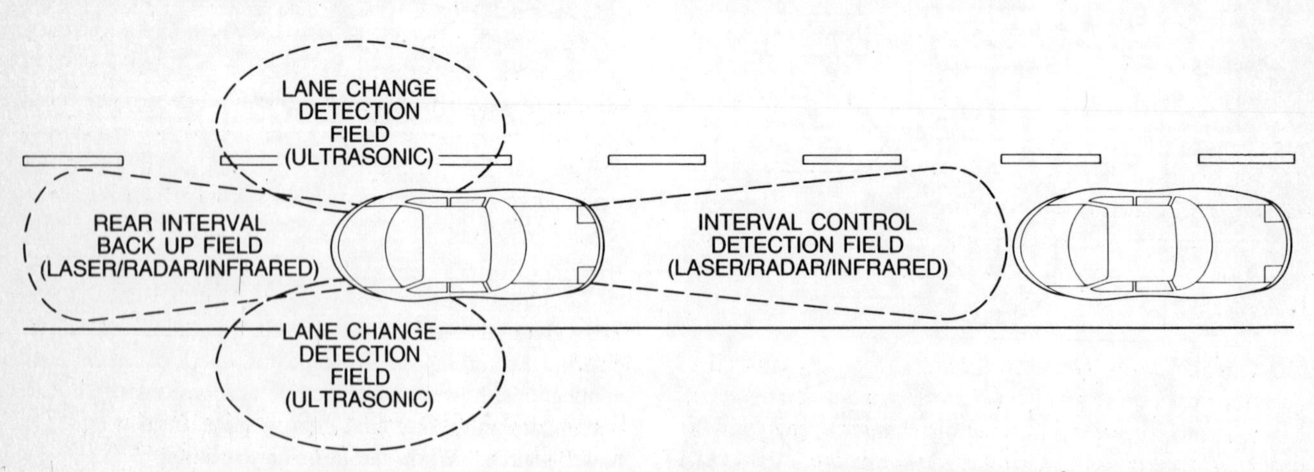

Fig 18 Collision avoidance aids require sensors to monitor the space around the car

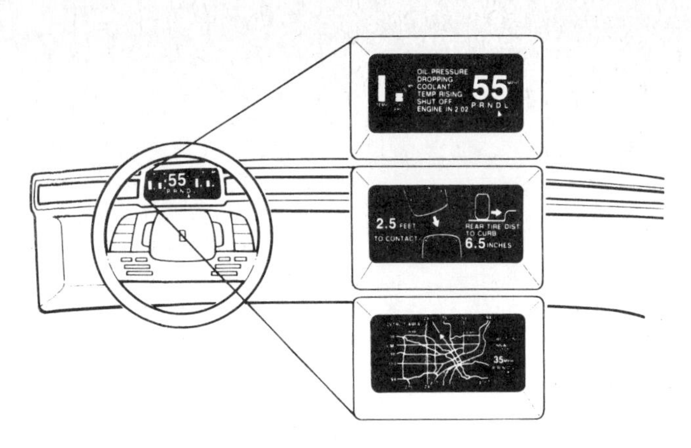

Fig 20 Reformattable driver information centre

forth. He or she will also be able to select this data in digital format or as an analog display. Further, he or she will be able to call up displays of maintenance related data such as advice on when to change engine oil and brake linings.

Notification of emergency and alarm conditions or warnings from the collision avoidance aids will be transmitted to the driver by a variety of audible and visual signals. These will communicate clearly the nature of the problem and the recommended corrective action.

Voice recognition (Figure 21) will be used for functions such as entertainment system control, driver information display mode selections, and telephone dialing. Many other control inputs will be made using programmable multi-function "soft" switches.

Fig 21 Voice recognition for dialling cellular telephone

There will be systems which are capable of detecting impairment or loss of alertness on the part of the driver. They will focus on actions that are fundamental to the driver's safe operation of the vehicle, such as appropriate steering and braking behavior.

The climate control system will be electronically controlled and electrically powered. It will be able to adjust temperature, humidity and air quality. The air distribution system will be designed to provide unique temperature variations in different zones of the passenger compartment to insure comfort of all passengers. Supplemental electric heat will beused to provide faster warmup during cold engine starts (Figure 22). Electrochromic glazing (Figure 23) will moderate sun loading by darkening in hot weather but remaining transparent when the outside temperature is low. Sensors mounted on the windows will automatically control the defroster to provide clear vision under all conditions. Air quality will be maintained by mechanical filters and possibly electrostatic precipatators.

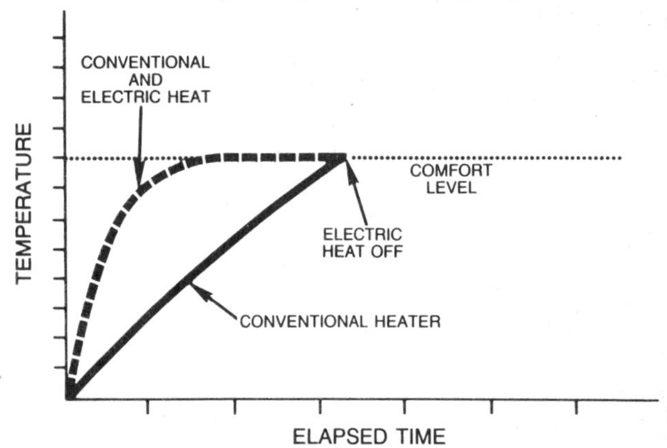

Fig 22 The heater achieves comfort level faster with addition of electric heat

Fig 23 Electrochromic glass

Seats in the vehicle will be able to change their configuration in many ways. Passengers will be able to change the position of the seats with voice commands. Figure 24 shows a concept version of such a seat.

AUDIO - Throughout Stages One and Two, audio system performance and features have been increasingly important to the customer. Figure 25 illustrates that the AM radio has declined in cost, thus offering the owner a better dollar value. Additionally, Figure 25 shows that a second market has emerged for the premium audio system which offers the consumer an ever-increasing feature content such as FM, stereo, AM stereo, cassette player, compact disc player, more speakers and graphic equalizer. In the year 2000, the automobile's entertainment system will still be primarily an audio system, although there should be some rear seat video applications.

The year 2000 premium systems will automatically deliver high quality sound to all seats in the car regardless of the number of passengers present. A fully automatic equalizing

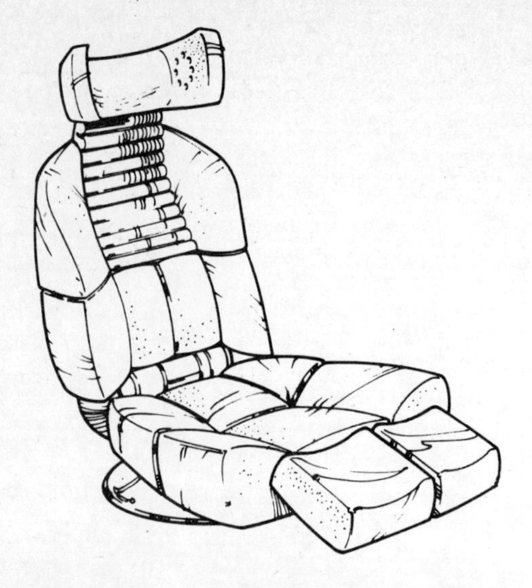

Fig 24 'Smart' seat concept

system will optimize the quality of the sound delivered to each occupied passenger zone based on input from distributed audio sensors. Digital technology will predominate in components. Radios, for instance, will have digital signal processing.

In another case of Stage Three integration, the entertainment system's displays and controls will be shared with the displays and controls of other kinds of information such as climate controls and instrumentation.

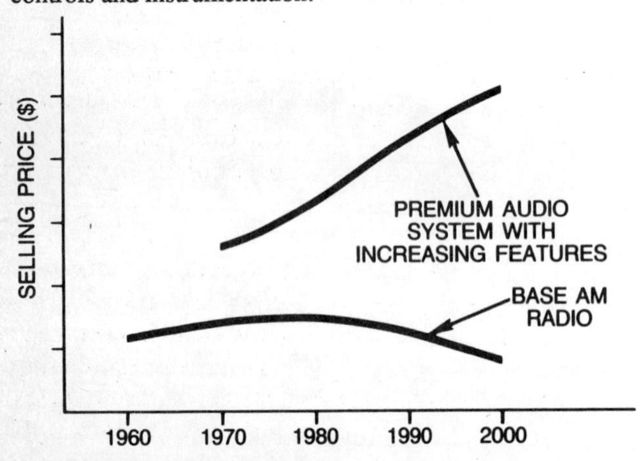

Fig 25 Audio system price trends (constant dollars)

COMMUNICATION AND NAVIGATION - Cellular telephones now represent the leading edge of a new area for automotive electronics (Figure 26). Many cars in the year 2000 will be equipped with telephones which will serve as the communication centerpiece of the mobile office.

Highly capable navigation aids will be especially useful in commercial delivery, service, and rental vehicles. These navigation systems will operate on the combined principles of dead-reckoning and map matching, utilizing a wide area digital map that is stored on a CD-ROM. The CD-ROM will also provide the user with a wealth of additional "yellow pages" information. By the year 2000, global positioning satellites will be employed in order to track a vehicle's geographic location. These systems will also utilize an external datalink that will greatly enhance the on-vehicle navigation system by providing the driver with current traffic information. There will also be real time satellite updates on new roads, road construction and new landmarks.

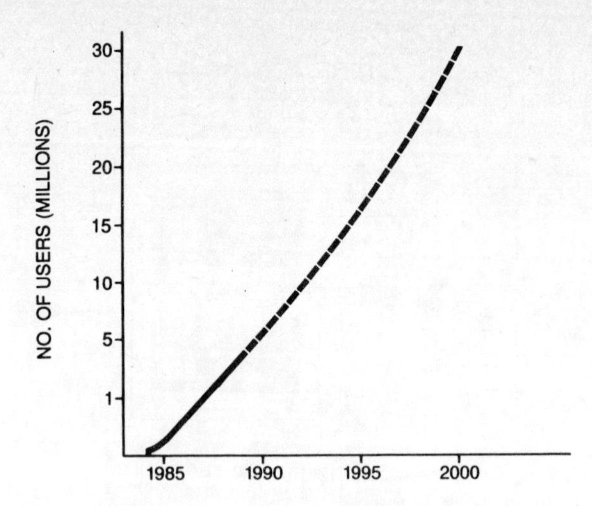

Fig 26 Cellular telephone usage trend in U.S.

Other information will be exchanged via datalink. For example, information on malfunctions will be transmitted to maintenance systems. This information will be particularly useful in the case of intermittent failures that are difficult to isolate in the service bay. Also, business, personal, security and emergency information will be handled.

Virtually all of the new features and performance enhancements discussed so far could be accomplished today - if cost were no object. For these benefits to represent high value to our customers, however, significant advances in several key technologies will be required.

STAGE THREE TECHNOLOGIES

In order to make the automobile of the year 2000 a reality, it will be necessary to develop new technologies that take full advantage of the flexibility of electronics (Figure 27).

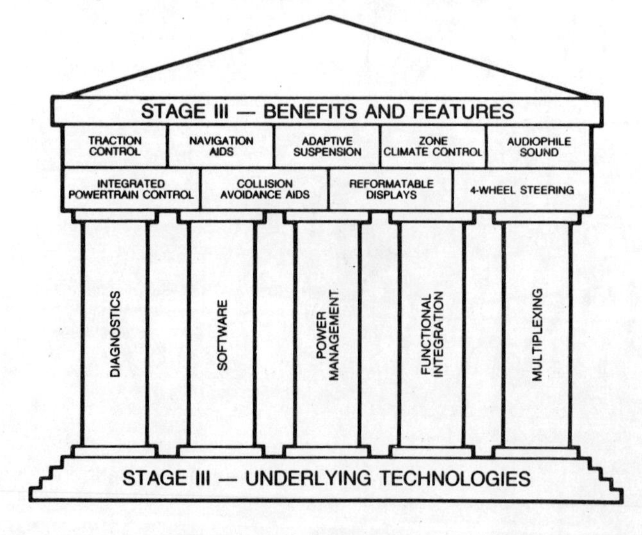

Fig 27 Stage Three automotive electronics technologies and features

OVERALL APPROACHES - Integration of different vehicle functions will be the strongest electronic trend in Stage Three. An example is traction control which is illustrated in Figure 28.

To succeed in achieving the overall functional integration that will characterize Stage Three, electronic developments must focus on the following key technical strategies:

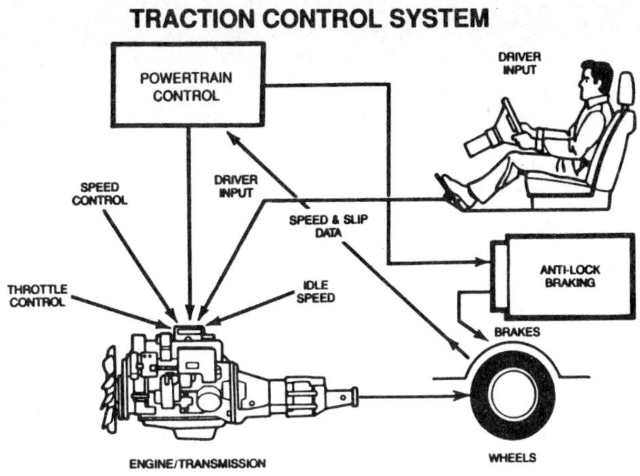

TRACTION CONTROL SYSTEM

Fig 28 Traction control achieved with only added software and the functional integration of powertrain control and anti-lock brake control

- Power management
- Multiplexing
- Software
- Diagnostics

Power Management - The requirement for electrical power will rise (Figure 29) from today's 1.8 kilowatts to 2.5-4.5 kilowatts. This increase may lead to a dual voltage power distribution system (Figure 30). The "low" voltage supply (12 volts) will power most of the vehicle's electrical and electronic products, including the lights. The "high" voltage supply (at a voltage between 24 and 48 volts) will power heavy loads such as cooling fans, steering motors, chassis hydraulic power sources and the motor-driven air conditioning compressors.

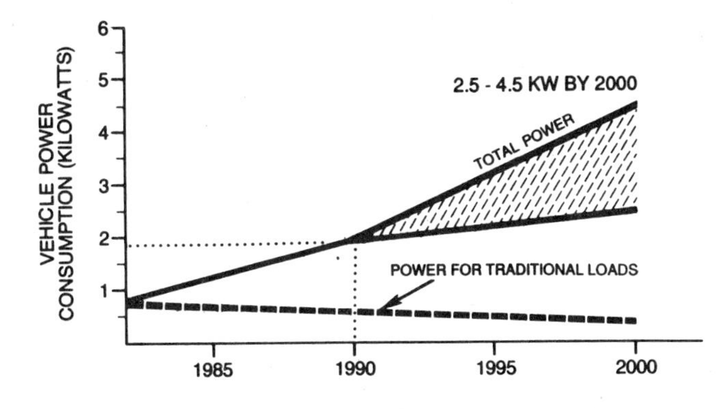

Fig 29 Electrical power consumption trends

The feasibility of a dual voltage system will depend to a great extent on the development of suitable batteries. These batteries must provide high energy density, flat discharge curves, tolerance for a variety of duty cycles, and high reliability. Battery reliability will be increasingly vital as the battery will assume the role of the backup power source for critical vehicle functions such as steering.

Battery charging circuits will be electronically controlled. This will require new types of sensors to accurately measure the battery's state of charge. The system will include active load management to unload the engine on startup or at idle. Under such conditions, load shedding will be common and loads will be assigned different priorities as environmental conditions vary.

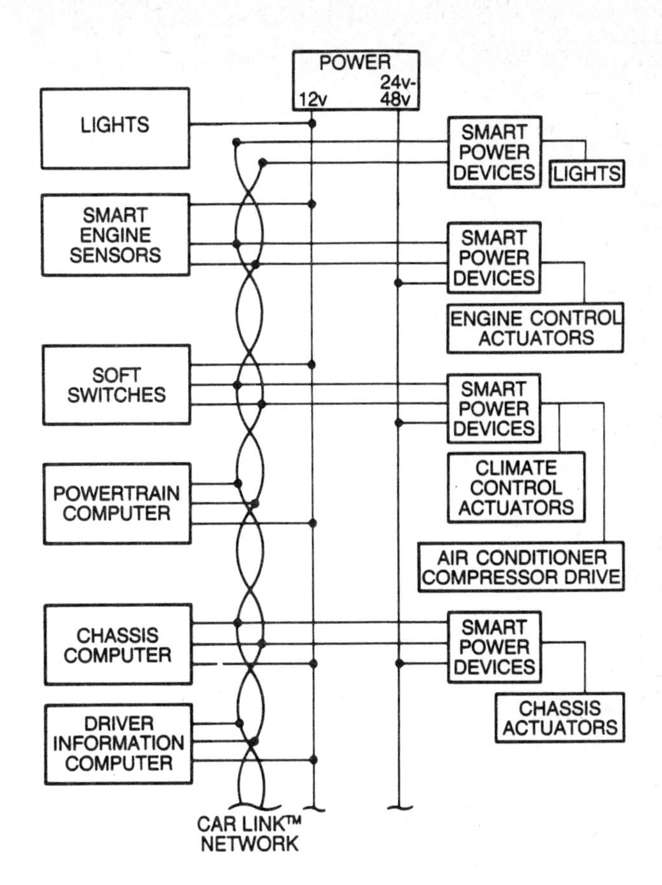

Fig 30 Partial diagram of a multiplexed system and dual voltage power supply

Multiplexing - A multiplexed system architecture based on the SAE J1850 CarLink™* network, will be adopted to make system integration more effective. Figure 30 shows how components could be linked by a data-carrying signal network and a power distribution system for each of the system's two operating voltages.

The CarLink™ will create a network to share information with the vehicle's central computers. Each computer will place signals on the network where they will be available to other components. The processors will also receive signals from other components via the network. Sensors and actuators will have "smart" interfaces so that they can communicate with the multiplex system. In the event of failure, firmware will provide "limp-home" capability.

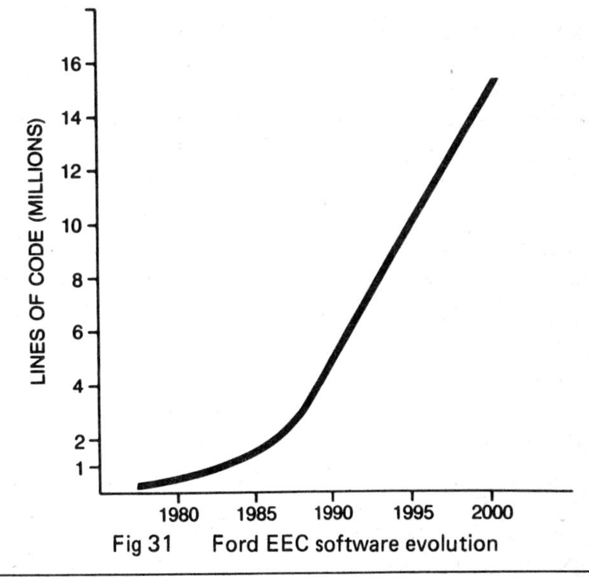

Fig 31 Ford EEC software evolution

*CarLink™ is the property of the Society of Automotive Engineers (SAE).

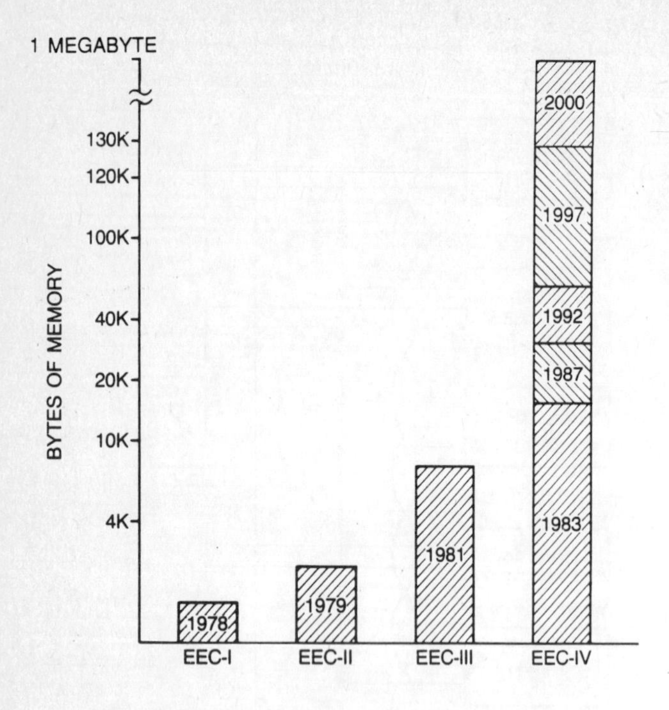

Fig 32 Ford EEC memory evolution

Software - Total on-board processing capability of the year 2000 automobile will be approaching that of today's typical minicomputer. The car's multiplexed network will be heavily dependent on software. Figures 31 through 33 show succeeding generations of Ford engine controls which have used growing amounts of software to accomplish increasingly sophisticated tasks. As Stage Three brings increased networking among computers, even more software will be required.

Models of powertrain and chassis performance spanning the extremes of vehicle operating conditions will be a vital part of software development programs. Accurate and precise control systems will require the development of more sophisticated control algorithms.

Fig 33 Ford EEC module evolution

Prognostics - Multiplexed systems together with massive memories will significantly improve service diagnostics by allowing capture and storage of large databases. Prognostics, the projection of component life expectancy, is achieved through the accumulation of statistical data on vehicle performance (Figure 34). This will greatly benefit the consumer by providing for optimally programmed maintenance which in turn should make vehicle failures a rarity. Component fault prognostics will enable alternate control activation, allowing the fault to be by-passed until maintenance can be performed.

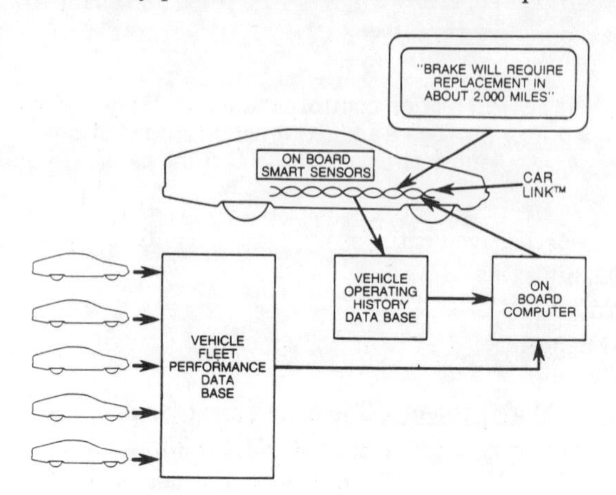

Fig 34 Vehicle databases will improve service diagnostics and prognostics

SOURCES OF TECHNOLOGY - New Stage Three automotive electronics technology will be adopted from several sources (Figure 35).

- Software and data management techniques will be derived from the data processing industry, with adaptations to fit the unique aspects of the automobile.

- Systems engineering technology and multiplexed systems approaches will come from the aerospace industry.

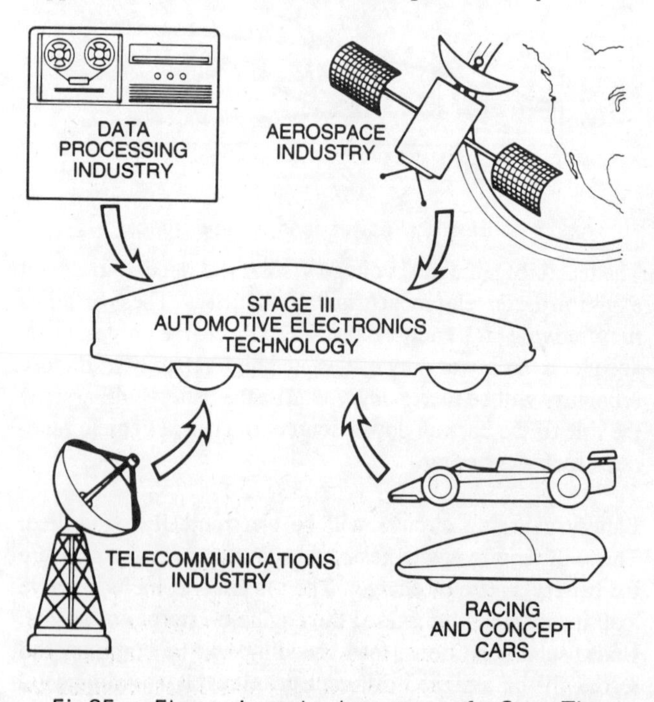

Fig 35 Electronics technology sources for Stage Three

- External communications technology will be adopted from the telecommunications industry.

Adaptations of such technologies will be developed and refined in the traditional automotive laboratories, the race track and concept cars.

DEVICE CONFIGURATIONS - Stage Three semiconductors will evolve from the ones in use today. Except for a few special items, they will be silicon devices. They will be present in three major components:

- Computer modules - incorporating microprocessors, I/O ports and several types of memory (DRAM and EEPROM).

- Smart sensors - hybrids of electrical or electronic sensing elements integrated with digital electronics.

- Smart actuators - hybrids of electromechanical actuating devices, power switching semiconductors and digital electronics.

CHALLENGES TO GET FROM HERE TO THERE

These projected features and underlying technical strategies depend on the maturing of several electronic technologies. It is the North American industry's challenge to make this happen, and particularly, to make steady progress in the following areas.

SEMICONDUCTOR DEVICES - While progress is expected across a broad front in semiconductor devices, several areas are worthy of being singled out as exceptionally important.

<u>Solid State "Smart" Power Devices</u> - These have been mentioned often in the past as a key to future automotive electronics technology. Such devices are an absolute requirement for taking full advantage of multiplexed systems.

The first devices are available (Figure 36). Continuing improvement in the on-resistance of power devices (see Figure 37) has produced components which can handle the switching of many automotive loads with "on" state power dissipation of less than 1 watt. These devices must also become available at significantly reduced cost before they can be widely employed in automobiles (Figure 38).

Fig 36 'Smart' power semiconductor

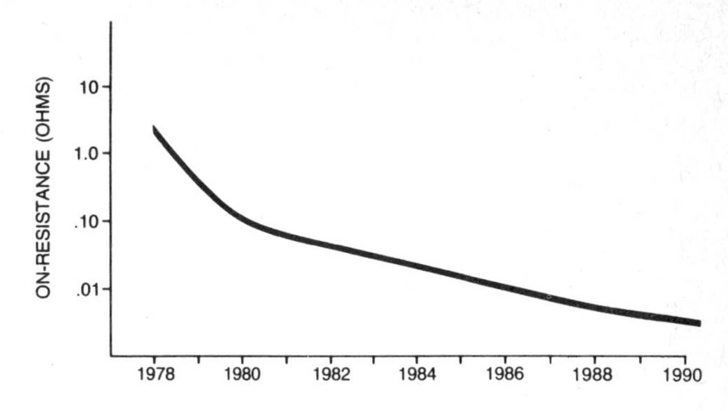

Fig 37 MOSFET's on-resistance trend

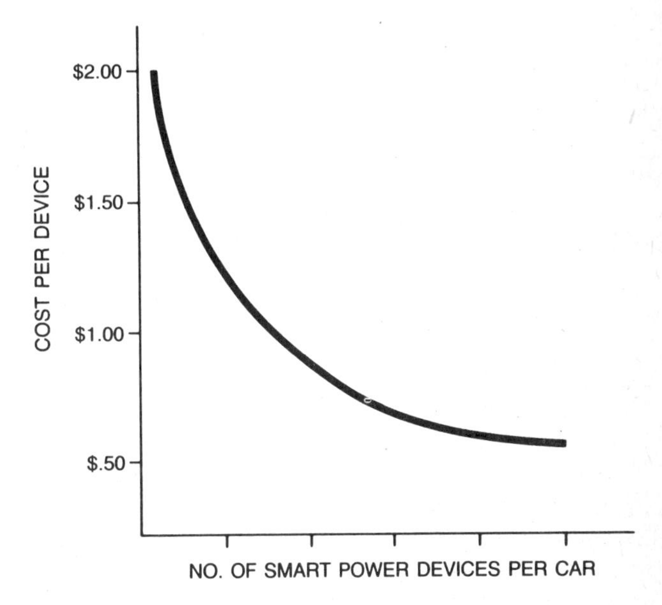

Fig 38 'Smart' power usage will increase significantly

<u>Digital Signal Processing</u> - DSP integrated circuits with supporting software will be needed to realize some of the data-intensive capabilities outlined earlier. Speech processing for voice actuated controls, interpretation of collision avoidance data, real-time diagnostics, interaction of navigation systems with external positioning aids and digital audio entertainment systems (Figure 39) will all make use of digital signal processing.

Of these, the most challenging requirement is for a practical speech recognition system. Currently available systems rely on several techniques to isolate speech elements and match them against recorded templates. Presently, such systems have about 90% accuracy with up to 70 decible environmental sound level, but they do not handle "continuous speech." Clearly, there is substantial progress yet to be made in the area of speech recognition.

Digital Signal Processing reduces hardware complexity but results in increased software complexity. It allows faster prototyping and easier modification and upgrading of system capability. DSP is an area where the software development challenge is at least as formidable as the need for specialized processing devices. Both are important, neither can be neglected.

<u>**Memory**</u> - The databases that will be required in the year 2000

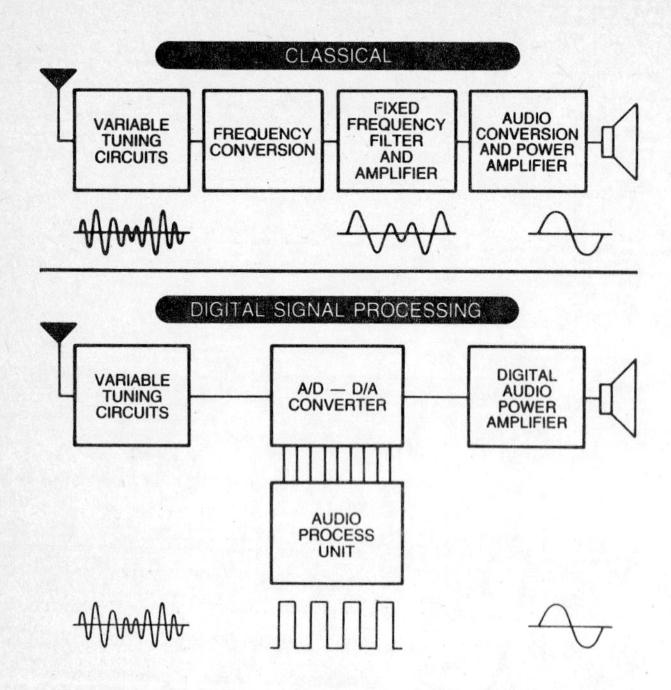

Fig 39 Radio block diagrams comparing the classical analogue approach with digital signal processing

make megabyte-scale electrically eraseable memory essential. It is likely that this scale of memory can be achieved with improved silicon technology which can offer sub-micron resolution, less than 3 volt operation and extremely efficient chip architecture. For archiving applications (such as the accumulation of vehicle history for maintenance diagnostics) writable CD-ROM will offer 500 to 1000 megabytes - more than enough to record every significant event that occurs in the life of the vehicle.

Packaging - As the use of surface mounted devices (Figure 40) increases, a complete set of industry standards must be established. Packages should conform to standard patterns, mounting tape should be of standard type and size and tape reels should mount on any machine. In addition, testing methods for SMD's should be standardized.

Fig 40 Ford electronic search radios contain hundreds of surface mounted devices

Environment - Since Stage One, the semiconductor industry has made remarkable progress to produce cost effective and reliable devices which operate in the harsh automotive environment. Two of the more notable achievements are environmentally-rugged plastic packages and built-in electrical transient protection.

However, higher operating temperature semiconductors for underhood applications is one area where further progress is required. Underhood temperatures have increased due to more aerodynamic vehicle shapes (Figure 41). Ambient temperatures exceeding 125°C are present at some potential electronics locations in the engine compartment.

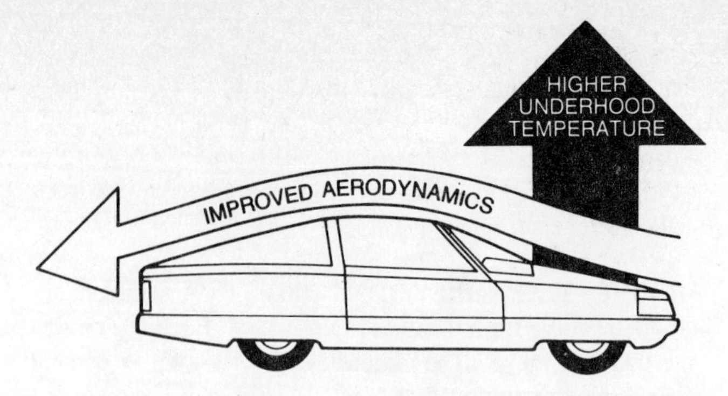

Fig 41 Improved car aerodynamics have caused higher underhood temperatures

Underhood electronics applications will continue to increase since it is very desirable to move powertrain and chassis electronics underhood. The benefits include smaller wiring harnesses and reduced system complexity through integrating more electronics with sensors and actuators.

A partial solution is the semiconductor industry's current transition from NMOS to CMOS which helps by reducing the power requirements. Additional changes must be considered. They could include larger die sizes and unique design rules.

The North American automotive industry will help by designing for better underhood airflow management and improved engine cooling.

"SMART" SENSORS - Typically, these devices contain micromechanics (physical structures generally produced in silicon by semiconductor processing techniques) and interfacing electronics. The "smart" sensors which will be used in Stage Three will likely process signals from analog to digital and will be capable of performing some logic functions (Figure 42). For example, smart sensors will be able to perform self-diagnosis and will be remotely recalibratable by calling-up correction factors stored in internal memory.

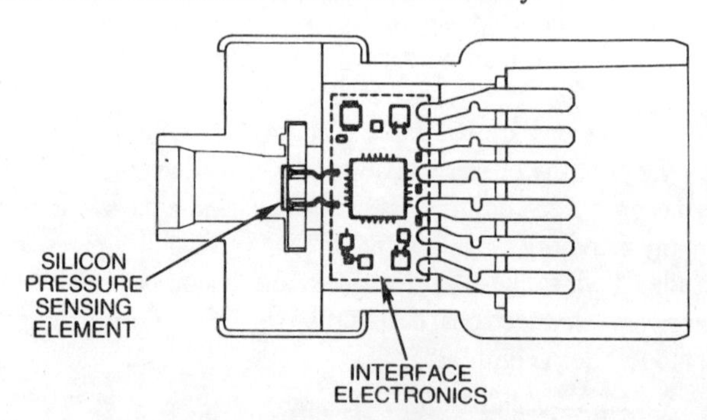

Fig 42 'Smart' pressure sensor

Micromechanics will reduce system cost and improve reliability in much the same way the VLSI IC's reduced the cost and improved the reliability of complex electronic functions.

To date, most sensors have been supplied by specialty suppliers who concentrate on devices which measure one or several of the following parameters (all of which will be required for Stage Three electronic systems):

- Acceleration
- Pressure/force
- Displacement
- Fluid flow

- Gas chemistry
- Fluid level
- Temperature

In the future the semiconductor suppliers will also become major suppliers of smart sensors due to the silicon sensing elements and the related in-sensor electronics.

The optimum utilization of smart sensors within a total vehicle electronic system requires careful partitioning of functions between the sensor and its central computer. This is necessary to achieve maximum system performance and cost efficiency. The automotive system engineers who are responsible for total vehicle electronic systems will need to work with the sensor suppliers at all stages of the design work, starting at the very earliest periods, to assure that such optimum system partitioning is achieved.

"SMART" ACTUATORS - Just as incorporation of electronic logic in sensors is needed to fully achieve the benefits of multiplexed circuits, so also are hybrid actuators needed. Linear and rotary actuators with integral electronics will be needed in a wide range of displacement and force outputs. Figure 43 shows such a unit.

The electronic portion of these actuators must be able to linearize and standardize the output (self correcting zero-point), to compensate for environmental variables (especially temperature), to protect themselves from power transients and to directly interface with the vehicle's central computers.

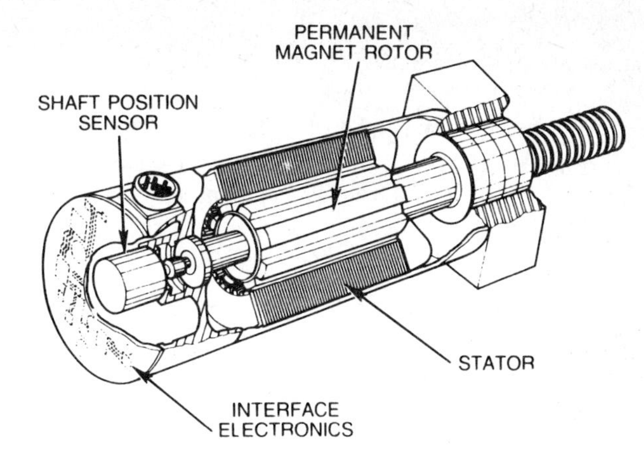

Fig 43 'Smart' motor

The actuators themselves are likely to be advanced versions of currently available components such as solenoids, motors and valves. Some new applications, however, might lead to actuator configurations so radically different from today's components that they seem to result from breakthroughs in actuator technology. For example, new electrically controlled actuators will be required to drive variable valve timing as part of the engine control.

In short, challenges which the industry must meet include dependable electromechanical actuation combined with repeatable electronic control.

SENSACTOR - The sharp distinctions between sensors as independent eyes and actuators as separate muscles could

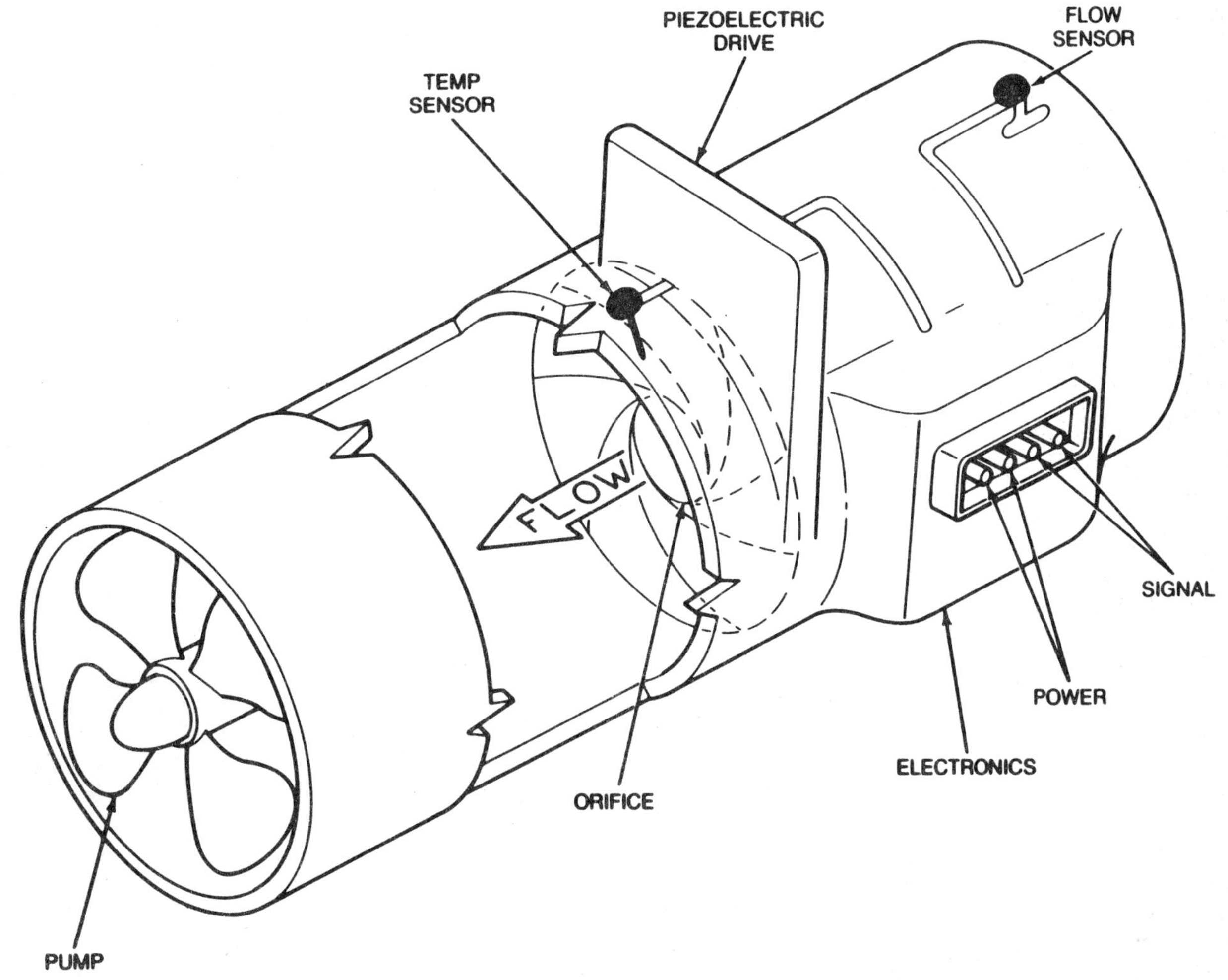

Fig 44 'Sensactor' concept

begin to blur somewhat. As sensor, microprocessor, and actuator become integrated (perhaps even upon the same silicon chip), we may find ourselves speaking and writing of "sensactors." Unique combinations or systems of smart sensors and actuators could distribute the intelligence now centralized in the microprocessor. Perhaps phrases such as "functional block" or "control sub-system" capture the essence of the sensor and actuator vision for the future.

Figure 44 illustrates a "sensactor" concept. It's a small, self-contained component that controls four functions (two sensors and two actuators) with its on-board electronics. In communication with the central control module, the "sensactor" monitors fluid temperature and flow and is capable of altering an orifice/valve with its piezo-electric drive - - or by activating a pump.

SOFTWARE - The development of increasingly sophisticated microprocessor based controls is likely to be limited by software, rather than hardware development. New and more efficient software tools are required, including new languages and new protocols for communication on-board the vehicle within and among processor modules. Equally important will be models of automotive system behavior. Powertrain, braking systems, steering systems, suspensions and interior environments will need to be modelled with previously unheard-of thoroughness.

The potential of artifical intelligence techniques in automotive applications will be thoroughly explored in Stage Three. It will not be a panacea, but it could certainly be useful in several complex areas such as on-board diagnostics and prediction (alerting driver to impending service), intelligent control, smart features (user friendly), and navigation (Figure 45). It is necessary to identify these areas and to adapt the theories of AI to practical automotive needs.

Perhaps most importantly, a larger group of software professionals will have to be trained. They are men and women who are skilled in the intricacies of software engineering and at the same time are intimately familiar with the unique requirements of the automobile and the automotive industry.

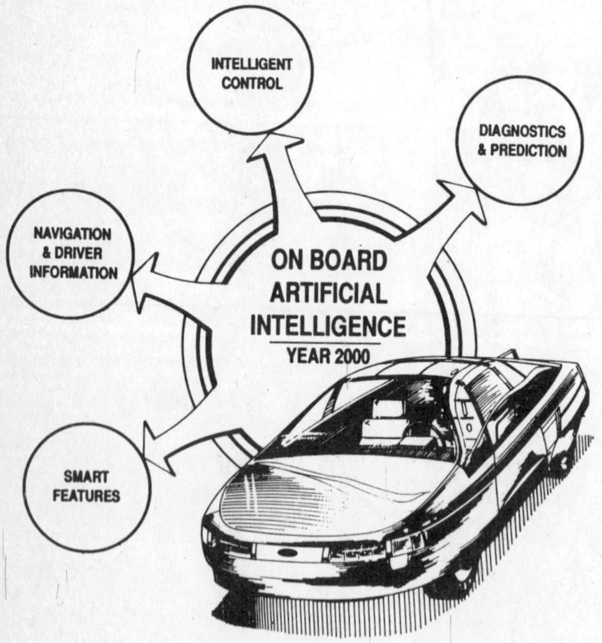

Fig 45 Potential automotive applications of artificial intelligence

INDUSTRY STANDARDS - Industry cooperation is essential to realize the potential of automotive electronics in certain areas. The European joint research project, Prometheus, is one example of such cooperation. This eight year project involves thirteen European automotive manufacturers and has a principal objective of enhancing safety through an integrated systems approach involving standardized vehicle-to-vehicle communications.

There's also a similar Japanese project among various automakers and electronics manufacturers. They have agreed to the joint development of navigation and travel control systems that will use electronics and telecommunications technologies to guide automobiles.

In the United States, the CALTRANS research project in California is testing the feasibility of linking traffic information with an in-vehicle, TV-type computerized mapping device.

In the state of Michigan, twenty partners, including the National Highway Traffic Safety Administration, are supporting a program to bring about Intelligent Vehicle Highway Systems (IVHS) on a national basis.

One of the most critical standardization issues for Stage Three development is the need for an industry-wide multiplexing protocol. The SAE, with representatives from the automotive and electronics industries, has initiated an effort to establish a vehicle multiplexing standard. Good progress has been made with the SAE J1850 CarLink™ network. As the process continues, the participants should expand its scope so that a world-wide standard is achieved.

EARLY COOPERATION BETWEEN AUTOMAKERS AND SUPPLIERS - It is very important for suppliers of electronic devices and components to become involved with automobile programs earlier in their development than they have in the past (Figure 46).

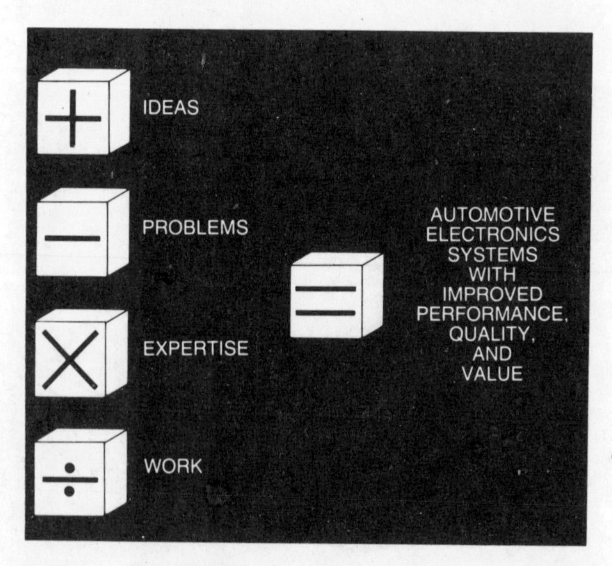

Fig 46 The 'arithmetic' of earlier co-operation between automakers and suppliers

Historically, the automakers and suppliers have started programs two to three years before automobile production. This has resulted in designs being "frozen" before all parties have been able to exchange information on their ideas, needs, problems and solutions. Automakers and suppliers should be working together at least five years into the future. The vision of the industry should be extending in a cooperative way, out to the introduction of the 1995 model year cars and beyond. All parties should be applying their unique expertise to projects aimed at that market.

CONCLUSION

The automotive electronics industry is at a transition point where one level of technology has been achieved and the thrust toward a new level is beginning. Stage Three - characterized by extensive functional integration of systems - will bring great improvements to automobile performance and reliability. The year 2000 car - because of Stage Three electronics - will have an exceptional quality level.

Is the automotive electronics industry on the verge of developing the ultimate automobile through the use of electronics? Certainly not. There is no reason to believe that Stage Three will be the last stage in the evolution of automotive electronics.

Looking into the future - beyond the range where precise speculations can be made - will be the advent of Stage Four. It will begin to emerge late in the 1990's. It's character will be defined by the integration of the automobile - itself by then a fully integrated system - into its external environment.

As Figure 47 shows, expect to see direct communication between car and satellite, or between car and cellular network. The techniques of artificial intelligence will be widely used. Collision avoidance aids will become true collision avoidance systems with computers reacting many times faster than the most skillful human driver. Navigation will include guidance, beginning with "guidewire" control on major highways.

Fig 47 Vehicle-to-vehicle communication

Later, an "expert" system will act as "copilot/navigator" and, sometime in the Fourth Stage the car that drives itself will be a reality.

Lest this vision seem too close to science fiction, Figure 48 shows an experimental driverless vehicle built by Martin Marietta. It is equipped with a guidance system developed by AI experts at Carnegie-Mellon University who, under contract to the Defense Advanced Research Project Agency (DARPA), are working to advance the technology still further.

And perhaps the reduction to practice of optical computers will herald the arrival of Stage Five! Bell Labs has already demonstrated optical devices which portend a single computer capable of handling the traffic of the Earth's entire population in simultaneous conversation.

The long range vision is exciting, but it is also exciting to be a part of Stage Three. There are many challenges inherent in the vision described in this paper of automotive electronics in the year 2000. With dedication and persistence, the automotive electronics industry will meet these goals in a timely manner.

Fig 48 Driverless vehicle developed by Martin Marietta

C391/KN3

Automotive electronics in Japan—present and next five years

T INUI, BA, MSAEJ, MIEEJ
Hitachi Limited, Tokyo, Japan

SYNOPSIS Electronics has now become quite popular among people within the automotive industry. Many microcomputers, along with sensors and actuators, are used to control the engine, brakes, etc. However, from a social point of view, new requirements have recently appeared. These include cleaner exhaust gas, better fuel economy, improved safety, and a more harmonious co-existence of cars with our society. By more precise investigation, fuel injection can be improved to bring about better fuel economy and less exhaust emission, and the number of traffic accidents can be reduced. Systematic organization of cars and stations will become beneficial to both driver and administrators.

1 INTRODUCTION

Already 12 years have passed since the first international conference of automotive electronics was held in London. During this time the application of electronics to the automotive field has become more and more popular and is now widely accepted in the automotive industry.

Recently, electronics has given us very fruitful improvements in exhaust emission control, maintenance, safety and ease of driving.

Even though reliability problems caused by inadequate design or production occurred in some cases, generally speaking, automotive electronics will continue to be a powerful means to pioneer the future.

However, recently we have encountered a new problem. More stringent regulations on car exhaust emission are now being discussed all over the world. Furthermore, carbon dioxide which has not been specified as pollution gas, is now claimed to cause, so called, global warming. Eventually more improvement in fuel economy will be required.

Anti-skid braking and traction control are being appreciated for their safety aspects.

Because of their convenience, car telephones are becoming popular, and navigation systems are being introduced.

In order to meet these new requirements, electronics is expected to be indispensable. I will report about the present and the next five years in automotive electronics in Japan and cover several important items, which will be explained in more detail.

2 POWERTRAIN

2.1 Fuel control

In Japan diesel engines are mainly used for trucks. Here I will discuss gasoline engines for passengers cars, which have 2 to 8 cylinders, 550 to 4000cc displacement, and OHC type structures. In the market various types of engines, like DOHC with 4 valves, turbocharged, supercharged, or with variable camshaft, are supplied to attract new demands.

However, very common and fundamental concerns about these engines are related to control intake air-fuel mixture, ignition, and EGR (Exhaust Gas Recirculation), in order to improve their exhaust emission and fuel economy without subsequent loss of performance.

Similarly to the US and Europe, fuel injection systems, shown in Fig 1, are rapidly penetrating the passenger cars engine market, even small car engines, while carburetors are drastically decreasing.

Each injected fuel, whether it is multipoint or single point injection, is basically controlled to have the relation of equation (1) with intake air. In order to approach the maximum mechanical efficiency of engine at any state, this calculation is continuously done by the microcomputer.

$$\Delta m = K \times \Delta M \;/\; 14.7 \qquad (1)$$

Δm: mass of each injected fuel per pulse
K: proportional coefficient
ΔM: mass of intake air for each cylinder

But under the various running conditions, including transient, the computer also has closed loop control, making use of the signal from the oxygen sensor, so that we can keep the air/fuel ratio close to the stoichiometric condition. This is because the three-way catalyst, which is widely adopted as the exhaust gas processing system even for small cars in Japan, has the maximum conversion efficiency at this condition (1). Moreover, against the change in ΔM under various conditions, for example, change in volumentric efficiency caused by a change in valve clearance with engine age; we apply the adaptive control, in which EPROM (Electrically Programmable Read Only Memory) in the control unit is rewritten to approach the real value of ΔM according to the criteria given by the oxygen sensor. This then has less effect on the engine performance and exhaust emission.

2.2 Mass air flow sensor

Clearly ΔM is very important in controlling the fuel injection system. Originally ΔM was found by an indirect method, that was, sensing of pressure in the intake manifold, and calculated as follows.

$$\Delta M = K_1 \times p \times \eta \times V \quad (2)$$

p: intake manifold pressure
V: cylinder displacement
η: volumetric efficiency
K_1: proportional coefficient

Then, mass air flow sensor was developed, which sensed air flow into intake manifold directly, and gave us ΔM by the equation (3), as follows.

$$\Delta M = K_2 \times Q \mid N \quad (3)$$

Q: mass air flow
N: engine revolution speed
K_2: proportional coefficient

These two methods have been called *speed density method* and *mass air flow method*, respectively.

Though there was a lot of discussion about the comparison between the *speed density method* and *mass air flow sensing method*, many test results reported mass air flow sensing revealing more accurate control even by simple fuel calibration work (2)(3). The foreseeable stricter regulation on fuel economy and exhaust emission all over the world and tougher developments in the market will encourage *mass air flow sensing method.*

An example of a mass air flow sensor is shown in Fig 2. It has a tiny, but precisely wound platinum wire with a ceramic bobbin as the sensing element. This gives us enough accuracy, response speed and durability. The integrated electronic module has its own heating circuit which keeps the element at about 120 degree C. This heating current corresponds to mass air flow (4)(5). The sensor is so compact that it can be integrated into the throttle body.

2.3 Injector

The injection system is appreciated for its better fuel distribution, air/fuel ratio accuracy, consistency in its performance, and less fuel calibration work compared to carburetors. One of the main components in the injection system is the injector. The injector measures, transports, and atomizes the fuel to each of the cylinders. But in some cases, we know that multipoint injection gives us lower combustion temperatures and less stable idling than carburetors.

In Japan it is becoming evident, even with multipoint injection systems, that a significant amount of injected fuel makes wall flow in the manifold and combustion chamber. This plays a certain part in diffusion burning which generates heat and exhaust hydrocarbons (6). It is said that less than 50 per cent of the injected fuel is introduced into the cylinder at that injected cycle, and remained fuel becomes wall flow as shown in Fig 3. In this regard, with a large droplet size and fast droplet motion, the majority of the fuel collides on the wall and cannot pass through the narrow passage around the intake valve. In other words, we should make use of this to convert it into mechanical energy. This depends partly on the improvement in injector which is currently being tried.

An example of the injector is shown in Fig 4. It has quite good atomization because of the fuel swirler integrated within it and gives us much better cold starting performance, 3 per cent less fuel consumption and 8 per cent less exhaust hydrocarbon compared to our conventional pintle type injector.

With regard to dual intake valve engines, the single fuel beam injector, from the multipoint injection system, cannot be suitably applied either because significant wall flow is generated on the branch wall of the intake port by symmetrical installation of the injector, or because a less uniform mixture is formed in the cylinder by the introduction of the fuel through single intake valve. Eventually, two injectors, or one injector with two outlet holes, are applied. The above mentioned injector with a fuel swirler can be modified to have two fuel beams which gives good atomization.

2.4 Wide range air fuel ratio sensor

The zirconia oxygen sensor has played an extremely important part in the air/fuel ratio monitoring for engine management. However, recently a wide range air/fuel ratio sensor has been requested so that we can have more accurate closed loop control even about the acceleration or at larger throttle openings. In these conditions, richer mixture than stoichiometric is required, and we cannot depend on the oxygen sensor. Also we want a quick response sensor to control air/fuel ratio quickly. In order to improve these, a new wide range air/fuel ratio sensor, shown in Fig 5, has been developed, which has almost linear output from rich state (air/fuel ratio = 12) to lean state (air/fuel ration = 23). The principle of this sensor is to measure the pumping current of oxygen ions through zirconia electrolyte, as well as cell voltage (7)(8).

For future applications, optical combustion flame sensors are under development. A glass fiber penetrating the center of the spark plug electrode transfers the combustion light which is then analyzed for its intensity and spectrum by an optical process to find the initiation timing, knocking, and even air/fuel ratio.

2.5 Methanol-fueling

In order to reduce NO_x emission and secure new fuel sources, methanol-fueling is under development. By simply mixing methanol with gasoline, exhaust NO_x present in the engine is almost halved compared to gasoline-fueled engines, even if the emission of aldehyde, and unburnt methanol still has to be dealt with (9). Also the air/fuel ratio and ignition timing should be optimized according to the data supplied by the methanol sensor.

2.6 Ignition control

Ignition also affects decisively on output torque (fuel economy), exhaust emission etc. Basic ignition timing is determined by using table lookup method, where the advance angle to each combination of air flow and engine speed is tabulated by an experimental process in the microcomputer. Furthermore, the knocking control function is added to the above basic timing. At first, the control unit sets the ignition timing to advance until the engine begins to knock. When the output of the knock sensor is fed back to it, the computer starts to make the ignition timing retard, until the knocking ceases. These operations are repeated, and on average, ignition timing is kept closely near its maximum limit.

The conventional canned ignition coil is replaced by a resin mold coil, which can be contained in the distributor. This coil is more reliable and easy to handle.

Simultaneously, owing to the development of the mold coil, we can produce the distributorless ignition system, where one coil is attached to each spark plug, or one coil is connected to two spark plugs and is then mounted on the engine. High voltage is supplied to each spark plug by electronic switching.

Much development has been done to decrease the electromagnetic noise about the rotating distributor switch and the distributorless system removes this completely. Also, by this method, restriction of advance angle by mechanical distribution disappears, and ignition timing can be decided more freely in wider range (10).

2.7 Intake-valve control

Clearly it is ideal to change valve timing and lift for high revolution speed, or for low (especially idling) speed. Recently such systems have been produced by Japanese car manufacturers and are controlled by electronic and hydraulic mechanisms. One type uses a timing pulley with a helical motion (11). The other type uses two combinations of the rocker arm and cam assigned to low and high speeds, with only the low speed combination connected to the intake valve stem. When the system is energized to convert low speed to high speed, the high speed rocker arm which was following the high cam (without valve movement) is united with the low speed rocker arm by connecting pins. The intake valve is then lifted by the high cam through the high and low speed rocker arms (12). With these systems, better performance for a broader speed range is obtainable.

2.8 Powertrain control

The engine management system, which conventionally controls fuel and ignition, is now processed by a single chip, or a two chip microprocessor (13), and about the former the composition of each function block is shown in Fig 6. Recently, engine management systems are co-operating with transmission control. In other words, one microprocessor controls them using a single software, with the engine and transmission controllers exchanging data continuously with each other until it finds the best selection of gear-train thus avoiding useless fuel injection and eventually, exhaust emission (14).

Instead of the gear-train and torque converter combination, the CVT (Continuously Variable Transmission), shown in Fig 7, is being produced for small cars with 550 - 1000cc engines. This is combined with an electromagnetic clutch to separate the engine from the drive mechanism during idling. This system is controlled by an electronic circuit which controls the hydraulic mechanism to change the effective diameters of the pulleys, according to the drive ratio (15).

3 VEHICLE CONTROL AND SAFETY

3.1 Vehicle control

In Japan 4 per cent of total cars are equipped with an anti-skid brake system, which is either mechanically or electronically controlled. However, this figure is gradually increasing. Electronic control depends on a conventional principle which decides the hydraulic circuit operation by a signal corresponding to the rapid deceleration of wheel speed.

Combined with the above anti-skid brake system, the traction control system was recently introduced to some luxury cars. During acceleration on a slippery road, free from any strong pedal pressure, the secondary throttle valve with motor actuator limits intake air via a signal from the control unit so as not to increase the engine torque and thus avoiding wheel spin.

Suspension systems are controlled electronically to improve driveability in some cars.

Four wheel drive (4WD) systems, which were either part time or full time types were introduced to cars, used for off road driving. Four wheel steering (4WS) systems have also been recently introduced and it may be useful in some situations, for example, parking in narrow spaces. Some 4WD and 4WS systems use, to a certain extent, electronic control.

3.2 Safety

For the passive restraint, the air bag system is equipped in some cars. Air is introduced through a valve which is activated by the trigger signal from a collision sensor. The sensor and control circuit are still being developed by electronic technology, so that its reliability may be improved.

On the other hand, accidents related to, so called, sudden acceleration in automatic transmission cars have raised a lot of discussion recently. After the analysis of 1108 accident cases, the Institute of Traffic Safety and Pollution Control in the Japanese Ministry of Transportation reported its results. As a conclusion, it recommended several items to be observed, regarding the design of automatic cruise and idle control and the quality of electronic devices. Along with this, from 1988, the installation of a shift lock mechanism, which does not allow the gear to be shifted from the park position without the brake pedal being pressed, became mandatory in Japan (16).

4 COMMUNICATION AND TRAFFIC CONTROL

4.1 Communication

The assembly of the conventional wire harness in the car body has reached its peak complexity. The multiplex system has long been developed to meet it, even though it is not yet applied to the production car. However, very recent requirements to meet the coming diagnosis or other data exchanges may lead to new concrete development.

Car telephones are becoming popular day by day. For safety requirements, some of them are modified to the handfree type. In this case the microphone mounted near the operator receives both the voice of operator and the sound from loud speaker simultaneously and these interfere with each other. To overcome this problem, several developments have taken place. One is to switch the function, with the microphone not being able to catch the output of loud speaker. The other, called echocanceller type, is to apply just the opposite signal of the loud speaker to the microphone, so that the echo from the loud speaker is cancelled.

Electronic displays are gradually being introduced to the instrument panel. Vacuum fluorescent tubes or liquid crystal tubes are used in the display device. Information, like speed, temperature, etc. is given by a fixed pattern electrode display, and some additional items, like diagnosis, are given by the so called dot matrix type display. Liquid crystal devices must be combined with a light source and a small fluorescent lamp is developed for this purpose.

The temperature limitation of the liquid crystal becomes high enough to be applied to car, and now it can be used in production car systems.

4.2 Navigation and traffic control

While on the road. we often need some information through either visual displays or acoustic means. Navigation systems, by which the route to our destination is automatically displayed on the screen, were then developed. Road maps drawn to a scale of 1 to 25,000 are digitally stored on optical discs, installed in car and displayed on the 6 inch CRT (Cathode Ray Tube) screen. The car position is determined by the signal from the direction and speed sensors. If the direction sensor, which is now depending on terrestrial magnetism, becomes more accurate by depending on, for example, glass fiber gyroscopes, the system may become more useful, as the delivery vehicle driver can find his destination more easily on Japanese complicated urban streets.

Several traffic management government projects are being tried for the future. One of them is called AMTICS (Advanced Mobile Traffic Information and Communication System) in which teleterminal, distributed in the range of 3km from central facility, sends to each car its exact position, where the car is, or the traffic condition (Fig 8). Then the driver can find his position on his display and find his way to reach his destination earlier and more easily (17). Part of these projects seems soon to begin practical service in the metropolitan area, and then all over Japan.

5 CONCLUSION

In Japan automotive electronics has substantially grown both in product variety and production quantity. These support the advancement of passenger cars by the improvement in affinity with people, performance as transportation, and the relation to the environment.

Electronic devices in cars have flourished independently. Recently discussions were held to organize networks, not only in the inside of cars, but also with outside stations. This will bring to us new possibilities in improving both individual functions and mutual cooperation, so that we can develop new functions and improve reliability.

Present anxiety concerning our environment and natural resources requests us to study various kinds of subjects, whether these are past or new fields of development. However, I think electronics will be closely related to them and continue to be the most dependable means.

REFERENCES

(1) HEYWOOD, J.B. *Internal combustion engine fundamentals*, 1988, 655-657, (McGraw Hill).

(2) FUJISAWA, H. Trend of gasoline injection system for conventional Otto-engines (in Japanese). *JSAE Review*, 1976, 5.

(3) INOUE, E. Electronic engine control based on high performance micro-computer (in Japanese). *JSAE Review*, 1983, 2.

(4) OHYAMA, Y. Hot wire air flow meters for engine control system. Proceeding of ISATA, 1981.

(5) SASAYAMA, T. A new engine control system using a hot wire air flow sensor. *SAE 820323*.

(6) OHYAMA, Y. Effect of mixture formation on idling performance for gasoline engine (in Japanese). Preprint of 1st JSAE conference, 1989, 891007.

(7) SUZUKI, S. Air-fuel ratio sensor for rich stoichiometric and lean ranges. *SAE 860408*.

(8) UENO, S. Wide-range air fuel ratio sensor. *SAE 860409*.

(9) ITO, K. Exhaust emission from methanol-fueled engines (in Japanese). *JSAE Review*, 1986, 9.

(10) MINORIKAWA, H. Direct ignition system (DIS). 3, 145 *Hitachi Review*, 1986, 35.

(11) MASE, Y. Development of valve-timing control system (in Japanese). *JSAE Review*, 1987, 9.

(12) NAGAHIRO, K. A high power, wide torque range, efficient engine with a new variable-valve-timing and -lift mechanism (in Japanese). Preprint of 1st JSAE conference, 1989, 891004.

(13) SHIDA, M. Development of engine controller using high performance microprocessor H8. Proceeding of ISATA, 1989, 89160.

(14) OHYAMA, Y. A totally integrated electronic control system for engine and driveline. Proceeding of 7th International Conference on Automotive Electronics, England, IMechE, 1989.

(15) SAKAI, Y. Subaru ECVT (in Japanese). Preprint of 2nd JSAE conference, 1987, 872094.

(16) Investigation report on accidents of sudden starting and acceleration about automatic transmission cars (in Japanese). Institute of Traffic Safety and Pollution Control, Ministry of Transportation of Japan, 1989.

(17) OKAMOTO, H. An overview of AMTICS. Proceeding of International Congress of Transportation Electronics, 1988.

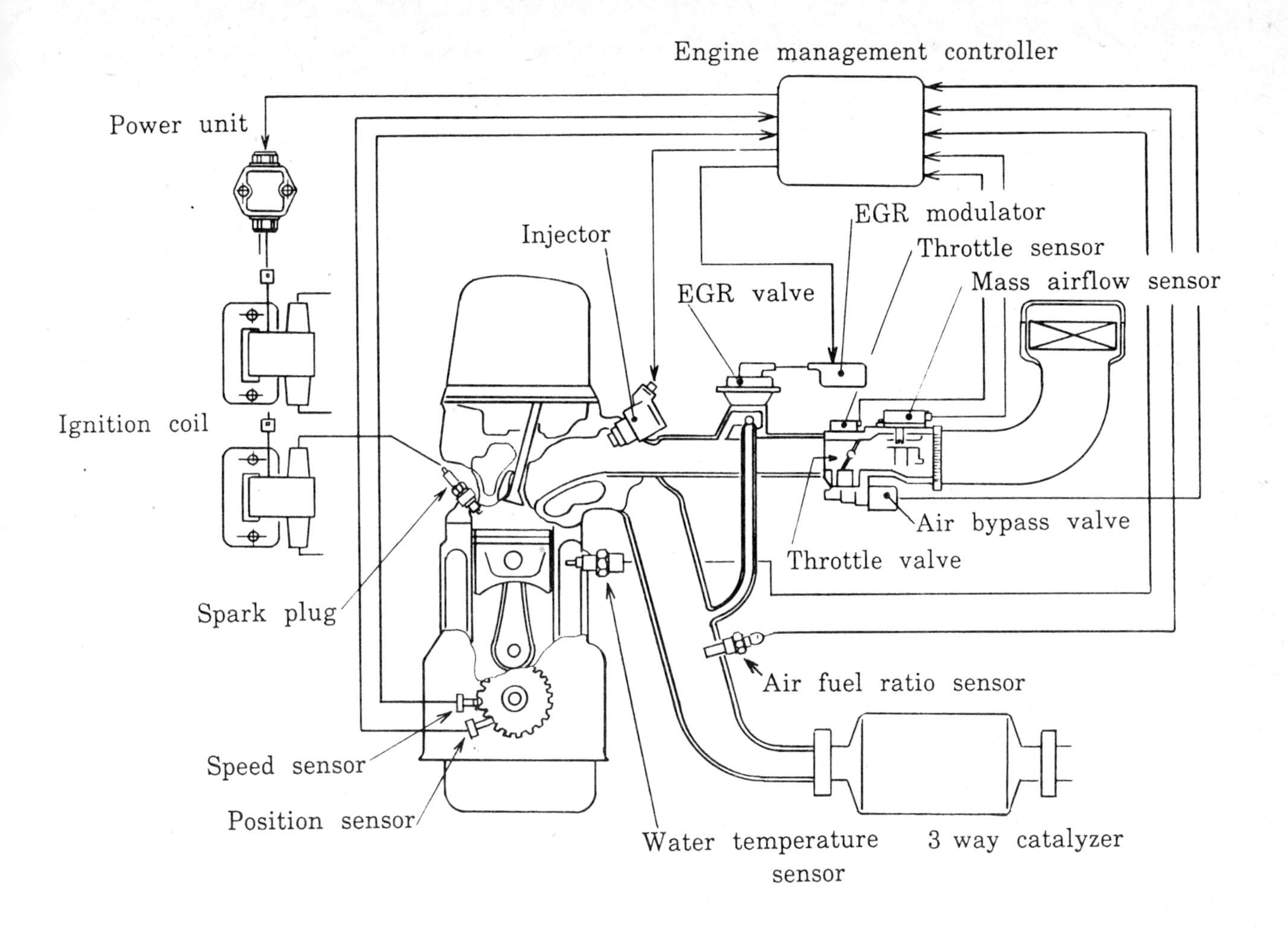

Fig 1 Engine management system

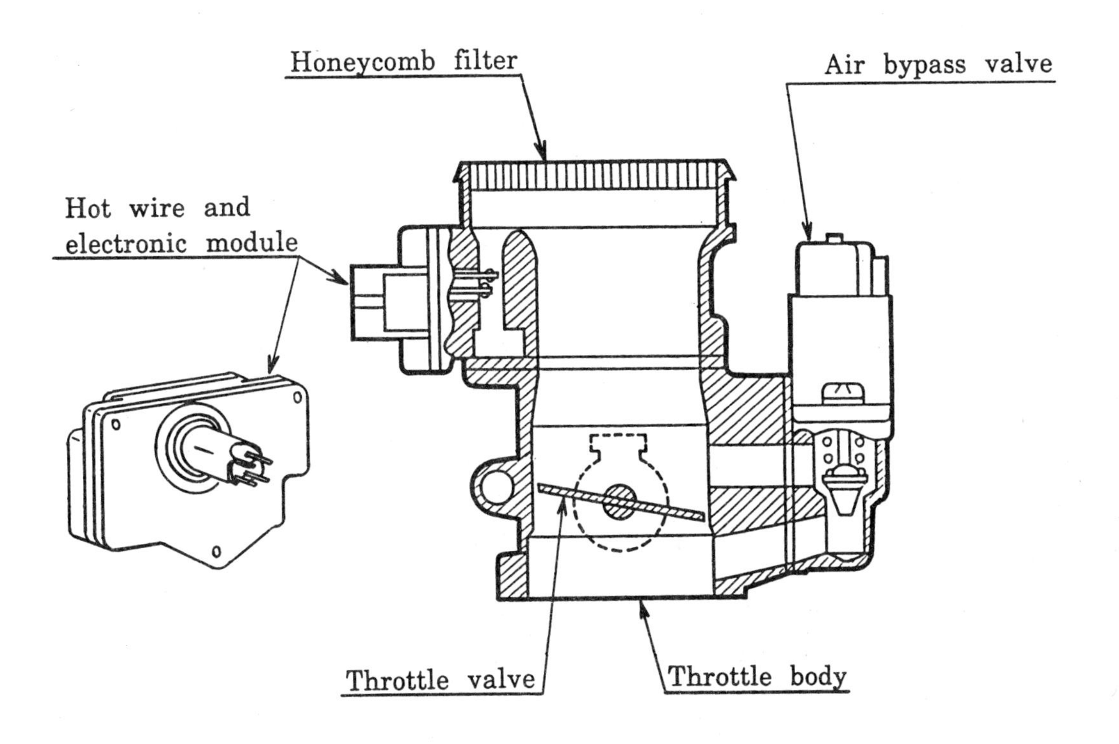

Fig 2 Hotwire mass air flow sensor integrated into throttle body

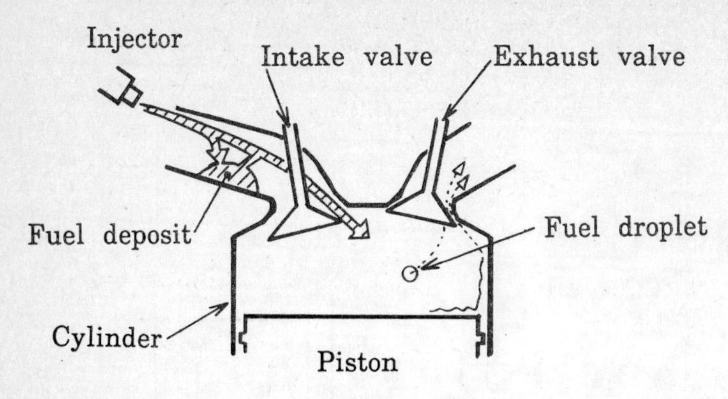

Fig 3 Charging of fuel into cylinder

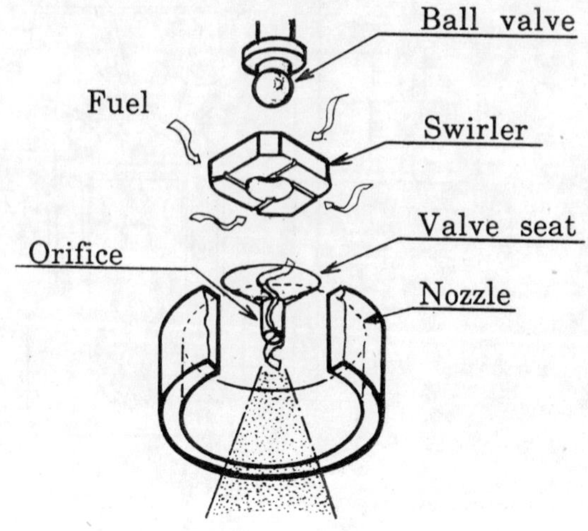

Fig 4 Fuel injector with swirler

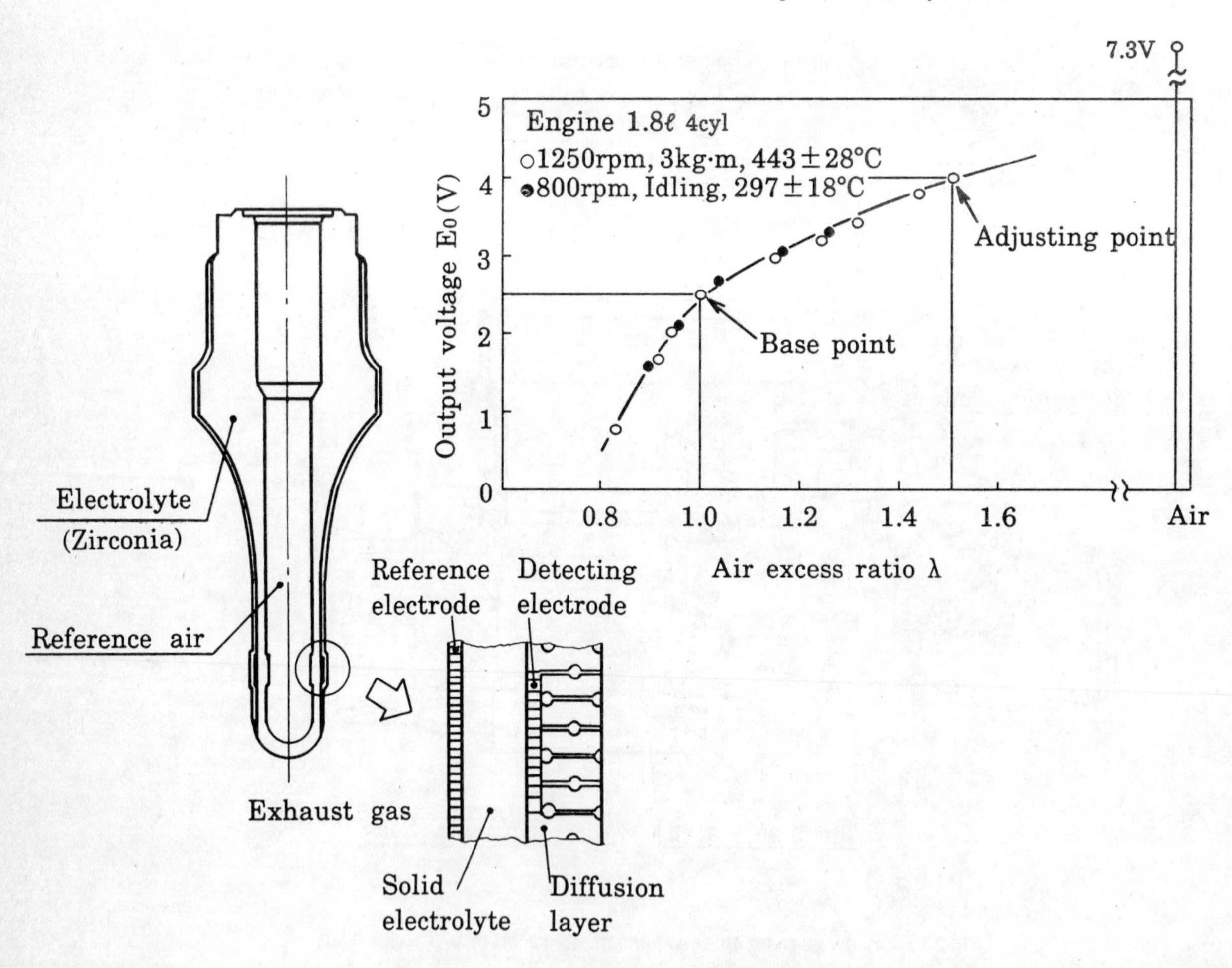

Fig 5 Wide range air fuel ratio sensor

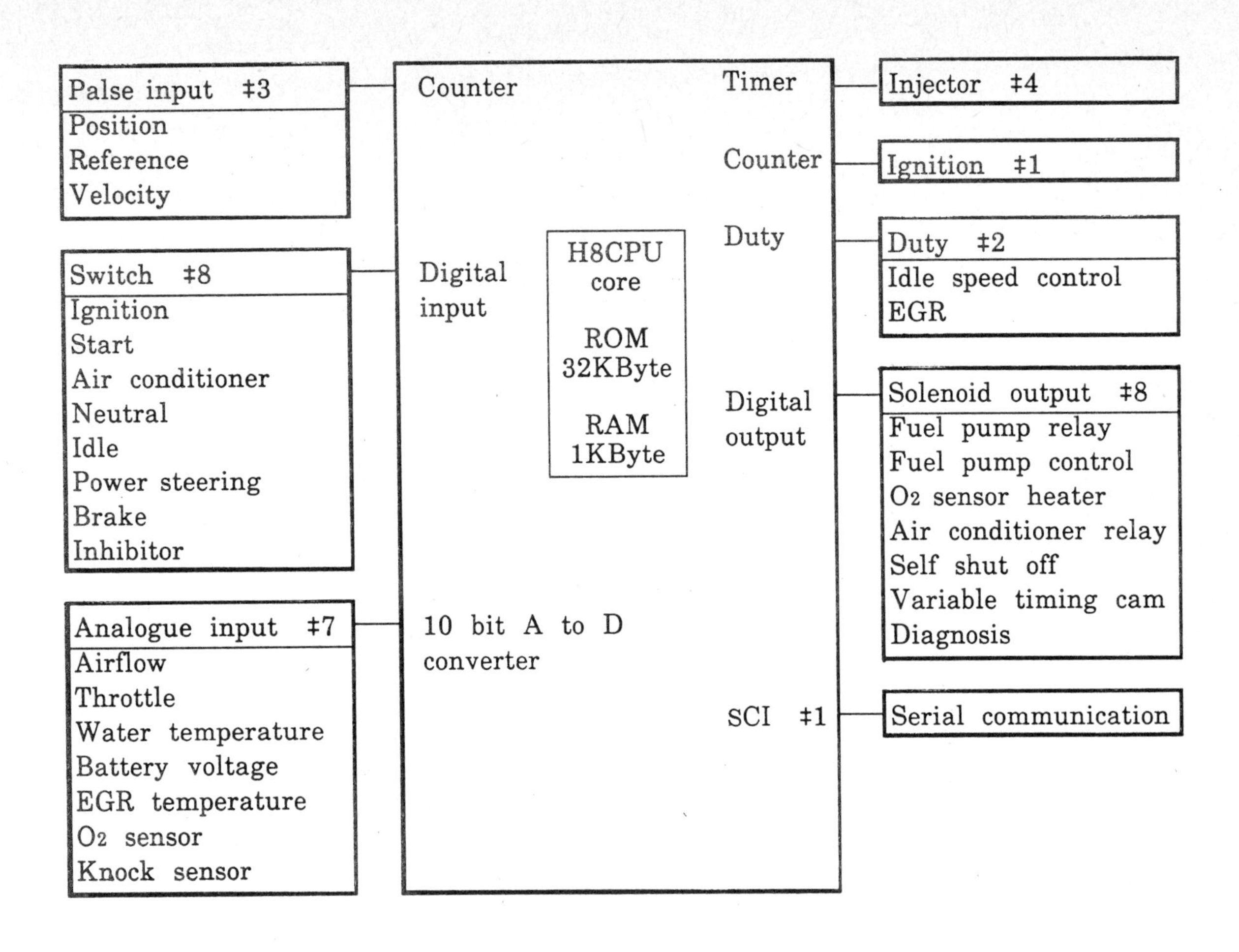

Fig 6 Single chip microprocessor for complete engine management

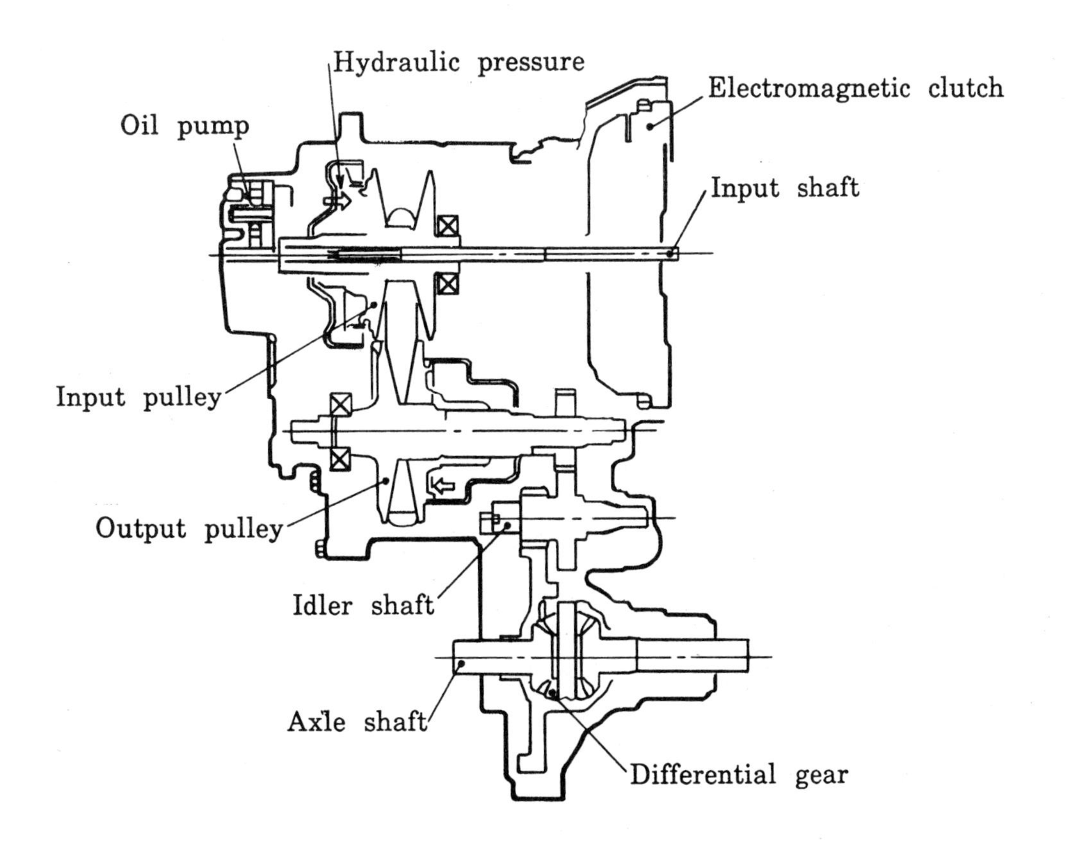

Fig 7 Mechanism of ECVT (15)

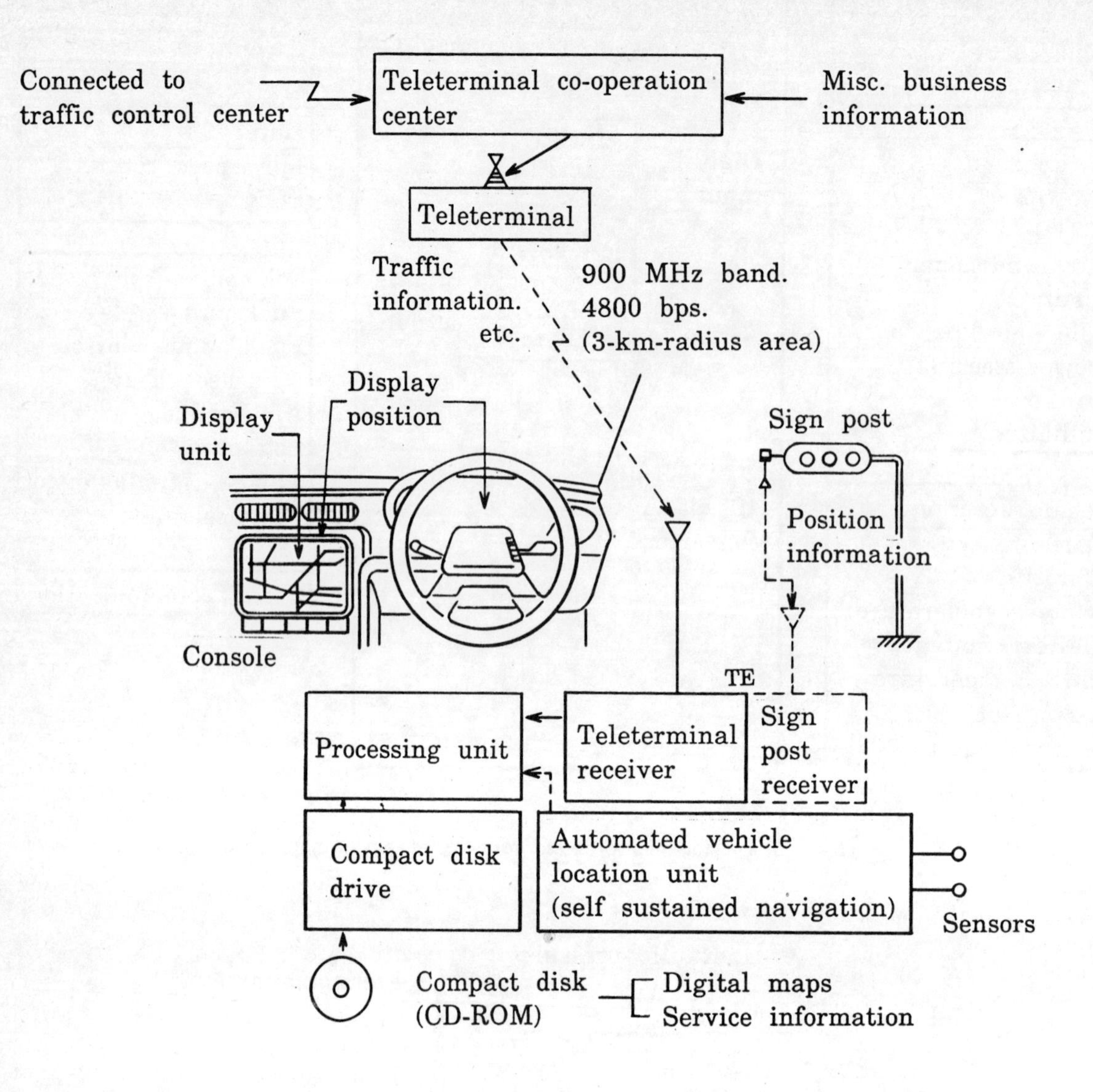

Fig 8 Schematic diagram of AMTICS (17)

C391/088

The operator's view of automotive electronics

G SPOONER, BTech, CEng, MIMechE, MIRTE, MILDM, MCIT
Spooner and Spooner, Woodford Green, Essex
A PARKER
TNT Express (UK) Limited, Atherstone, Warwickshire

SYNOPSIS

The factors that affect the fleet operator by the increasing use of automotive electronics are reviewed and discussed. Although commercial vehicle manufacturers are cautious about introducing electronic systems, many have developed systems. Training in the understanding and acceptance of electronic principles is an important factor as is the objective assessment by operators of problems associated with electronic systems. The advantages and problems of diagnostic testers are discussed. Vehicle technology has to be cost effective to the operator although cost advantages are quite possible. The fragmented road transport industry is recommended to coordinate efforts to overcome general or specific problems.

1 INTRODUCTION

The aim of this paper is to highlight factors that directly affect the commercial fleet operator in the rapidly increasing use of electronics in vehicles.

Any advance in technology must meet a real need. That need could be to satisfy a driver's ego by supplying a company car which impresses both the driver and the people with whom he or she comes into contact. But this paper addresses the commercial fleet operator, where real needs cost real money and needs to be justified.

With downtime becoming more costly, skilled mechanics becoming more scarce and supervisory skills of those controlling vehicles becoming more demanding, vehicle system reliability and cost effectiveness is essential if an operator is to be able to retain any commercial advantages of operation. There is no doubt that much of the technology and systems now available can be cost effective.

But:

- are the circumstances under which they will be operating too demanding?
- is the technology developed sufficiently?
- are the players involved ready for such a revolution?
- what are the critical factors and how can operators improve their own situation?
- is the operator really in a position to influence the situation?

This paper will not review all the component areas of trucks in which the operator will be interested. This is excellently covered in the 1988 SAE/IMechE Exchange Lecture paper by Aravosis (1) and by Bergqvist (2) to which those interested should refer. This paper will concentrate on the operational problems affecting those who run vehicles commercially.

2 TRADITIONAL PROBLEMS

Vehicle maintenance has traditionally incurred high maintenance costs in electrically related items. Batteries, generators, and starter motors have historically caused more problems than many other single items on vehicles. The numerous switches and connectors in vehicles inherently incur reliability problems.

All manufacturers know of the potential problems of electrical systems not only because of the nature of electronics but more importantly because of the vast number of individual components in electrical systems and the scarcity of trained staff to diagnose and repair any faults. Each component is a potential failure point. There are also the many components and assemblies returned under warranty that have been disallowed because no fault was subsequently found even though the replacement of the component was said to have cured the reported fault. While many faults could be more effectively cured using the correct diagnostic procedures by mechanics trained in the relevant systems, the innate susceptibility of electronic systems to show intermittent faults makes cure by replacement of components one after another an effective alternative method although it is suspected that many of these faults may be caused by poor contacts which are cured by the removal of connectors and their reconnecting.

The industry has overcome many of these problems in the past. But new and more complex vehicle systems are being introduced giving new problems. Operators are not so optimistic that the new equipment is going to give them a net advantage. Lighter materials, thinner wires, more emphasis on cost as well as weight reduction has heightened fears that reliability is affected, even though modern developments have increased material and production quality.

However which ever way it is looked at, the example of the now universal use of alternators surely demonstrates how improvements can be made both technically and operationally. Why cannot other forms of electronic equipment in vehicles become equally advantageous, commonplace and accepted?

3 ATTITUDES

Many manufacturers of the larger trucks believe that operators do not want sophisticated electronics. Their attitudes in fitting complex systems inevitably range from the conservative to the cautiously exploratory. In 1988 during an investigation by Spooner into electrical and electronic reliability, truck engineers visibly winced at the mention even of wiring and connectors. (3) Engineers are being cautious about implementing electronics because they have experienced enough problems with conventional systems. The two founding companies of Leyland DAF have between them been involved for many years in the development of advanced technology. The Daf 95 series premium tractor is the first of the new company to have sophisticated electronic systems but the model is feared by some operators because of the reputation for unreliability that has not yet been overcome. One driver has innocently asked how long will the fault diagnosis display last before it has a fault itself?

The engineering director of a specialist British truck manufacturer said earlier this year in response to questions about use of vehicle on-board computers that his company was 'against complexity for complexity's sake'. Manufacturers are also guided by their marketing departments. Are customers asking for electronic systems? Are they asking for the positive advantages which only electronics can give them? The answer, like many in this complex age, is not clear although generally it is apparent that while vehicle users do not want electronics per se, many look forward to reaping the advantages of better operational control that electronics can give. These problems do not only apply to electronics. The same criticisms can be made of air brake systems which have become increasingly complex mainly due to the demands of legislation and now represent a significant proportion of heavy vehicle maintenance costs. Operators do cope with these systems but they accept them grudgingly because they have little choice and no-one would compromise the safety that such systems are intended to provide.

4 TRIALS

If an operator is to be convinced that a new vehicle design is going to be reliable and suitable in service, he can either try one or more under his own operational conditions or listen to other operators who have experience of them. It is to the industry's credit that there are many operators who will take the lead. Demonstration vehicles may be offered free by manufacturers or dealers but operators will still be concerned about the disruption caused if anything goes wrong. Few operators have the resources to monitor vehicles on trial and compare them against their standard operating vehicles let alone do a statistically correct study. For instance, few 'average' fuel consumption figures quoted by operators would stand up to statistical interrogation. Proper evaluation requires thought put into the trial right from the very start. What is being evaluated? What is a successful conclusion? What constitutes a failure of the system? Is downtime being measured and, if so, does downtime include the time that the vehicle is ready after repair but the operator is not ready to use it because a hired vehicle is still being used? There are many different factors that need to be taken into account. By the time these are all analysed, the operator could well have forgotten the objective.

How long is a trial? If failures are expected on average, say, not less than every 30,000 km, should the vehicle have to run twice this distance to ensure that a fair trial is carried out? Even on a vehicle working a triple shift (with resultant problems from using different drivers on the same vehicle), this can take six months to complete. With many service vehicles doing less than 25,000 km a year, the technology will have changed before any conclusion is obtained.

Many of the decisions that an operator makes, assuming that there is a choice, are therefore made with too little data. No wonder that decisions are often based on rumours or stories that are heard from other operators.

5 TRAINING

Training is an often ignored requirement. But it is essential with electrical applications where many otherwise excellent mechanics or mechanical engineers often publicly admit that they do not understand the subject. With technology now intimately integrating mechanical and electrical principles into mechatronics, engineers and technicians will no longer be able to specialise in one narrow discipline within engineering. The role of the autoelectrician will be usurped by technicians trained in all aspects of engineering.

There are three factors for believing that this will happen.

The first is that the basic principles of mechanical engineering such as power, energy, and work can easily be compared to the basic principles of electrical engineering, electronics and of course pneumatics or hydraulics. Force and rate of movement in mechanical structures are analogous to pressure and flow rates in hydraulics or pneumatics and to voltage and current in electrical engineering. Power assisted lever systems, air and hydraulic valves and transistors are all capable of multiplying work input - the principles are the same. With the right approach to training, the principles of electronics can be more easily understood. Understanding how something works is contrary to the current trend towards vocational qualifications but is essential for our workforce of tomorrow to be able to analyse changing situations.

The second factor is that the use of electronics in many situations does not need those maintaining or repairing them to understand semi-conductors or electronic circuits. The prime requirement is for logical analysis of what is happening in a failure mode although troubleshooting manuals and diagnostic equipment are essential in this case. With the cost of microchips reducing rapidly, repair by modules replacement may be economic. The two major problems of electronics compared with electrical installations are caused by electromagnetic interference and low signal currents in a poorly protected environment. These need basic understanding by repairers and optional equipment installers as well as correct design by manufacturers.

The third factor is that diagnostic testers will enable even semi-skilled mechanics to diagnose faults using simple troubleshooting procedures. However if a different diagnostic box is required for each make of vehicle or even for different models with a range of vehicles from the same manufacturer then the cost or even the logistics of accommodating different equipment in a workshop environment may be prohibitive. The fleet operator currently has three options:

1 Pay the extra cost for a diagnostic box, assuming that one is available to repairers other than franchised dealers.

2 Use a franchised dealer for those repairs that need the use of a diagnostic unit.

3 Buy from another manufacturer which provides vehicles that do not require the use of a diagnostic unit to troubleshoot problems.

A fourth option is not yet available but should be. And that is for some enterprising supplier to provide a universal diagnostic unit that can be connected to all makes of vehicles. However it is hard to see manufacturers like Jaguar, BMW or Leyland DAF who all have developed diagnostic testers for their dealers to use on their own vehicles agreeing to supply information on diagnosing problems in their cars and trucks to a third party manufacturer. They are generally committed to ensuring that their dealer network has first option on service and repair work on their vehicles. Connecting into different systems is likely to involve a range of connecting cables and adaptors similar to the problems of connecting engine diagnostic and tuning equipment. Even with the SAE standards being proposed in the USA, this problem is unlikely to be reduced to a practical working level. (1) And there is also the problems of updating procedures as vehicle specifications change and develop.

Vehicle manufacturers need to:

- keep things simple. This does not mean that innovation should not be encouraged but that the consequences should be simplified.

- avoid change for the sake of it.

- be cautious in introducing changes into production without ensuring that the service aspects, including training, have been adequately planned.

- beware of introducing new parts or technology because of the requirements of a marketing department or pressure from competitors.

In summary, the needs should be justified by the ultimate advantages to customers.

6 COSTS

The effective cost of new equipment can be very high. There is not only the component cost which often can be low with modern electronics but also the cost of learning by operators, drivers and repair staff even if formal training is not required. Much of the cost of electronic systems is in the development which needs to be recovered on each unit. High volumes will dramatically reduce unit costs.

The residual value of vehicles fitted with sophisticated equipment must be affected. Many owner-drivers or small hauliers will buy second-hand trucks because they are good value having lost a high percentage of their original cost with their first owner. Unfortunately older vehicles need more maintenance and even by choosing a well looked-after vehicle, there is a high chance of electrical failures through corrosion, disconnection, or worn contacts. That is a risk that some operators will take. Even these can be diagnosed and repaired by the roadside. But a vehicle with unnecessarily sophisticated equipment which will need specialised attention if something goes wrong will never be top of the list for choice.

Downtime can be the biggest cost to an operator and with the current emphasis on quality assurance, with tighter timescales to deliver goods within a specified delivery window at many locations, this is the biggest risk. Companies in such critical operations will not want to chance a reputation and even worse a valuable contract in order to try a new piece of equipment even if there may seem to be an immediate direct advantage to them?

7 BENEFITS

So far the negative aspects have been predominant.

But there are definite benefits in using electronics in vehicles. Electronic control systems are lighter and give superior response and performance characteristics to mechanical systems which often are not even capable of performing the prime functions available through electronics. Electronic systems are the only effective way to record data for subsequent display and analysis. Compare the tachograph and its chart to an electronic data recorder.

Consider the savings possible by being able to use this data which can be analysed immediately the driver's shift has finished or even before by remote data transfer from vehicle to base. The route mileage, delays, and fuel consumption can all be available to the transport supervisor for discussion with the driver before clocking off.

Engine management systems can give considerably more precise control of fuel delivery especially on petrol engines where the ideal ignition map can be very complex. Electronic memory can react much faster with better control than a vacuum and centrifugal weight control system used on conventional spark ignition engines. These concepts are now increasingly being used on diesel engines. Fuel savings will give an immediate direct increase in profits.

Electronic controls can monitor many factors of engine and drive line performance, constantly analysing data looking for indications of even the slightest deviations from what should be the normal condition. Problems can be spotted before they actually occur. It may still need a trained ear and eye to make the final diagnosis but the potential problem will have been spotted before expensive damage occurs.

Condition based maintenance will involve monitoring many factors and processing them for comparison against set parameters. Currently the sensors and the environment in which they work are costly for retrofit applications but with the on-board processing power now available, operators will soon develop ways to use this information to obtain optimum service and major unit overhaul intervals.

8 OPERATOR REACTION

What will operators do in this situation? The less optimistic ones can refuse to buy vehicles which they believe may not be reliable. They can refuse to repair components that are not essential to the vehicle's operation. They can complain that the manufacturer is not listening to their criticisms. They can make bald derogatory statements to anyone who will listen that certain vehicles that they operate are unreliable and convince themselves and others that all such vehicles are a waste of money. This will not advance the operators' cause.

Positive efforts based on fact need to be made. Operators generally are not well enough organised to be able to operate a unified effort to communicate with manufacturers. With over 100,000 companies running some 3 million business vehicles in the UK, it has proved in the past impossible to organise a large numbers of these into coordinated action let alone logical action. The large number of institutes, trade journals and trade organisations that operate in the UK, often representing the same companies, illustrates this. Due to the varied and fragmented nature of the industry there are far too many different points of view. There is also still a lot of emotion and subjectivity in vehicle purchasing and maintenance decisions. An immediate reaction in refusing to buy vehicles with electronics is self-defeating in the long-run. Electronics are often intended to give the operator a long term cost advantage. Why throw this away?

9 CONCLUSION

There is no doubt that electronic components and assemblies have been developed to work in the extreme ranges of conditions in which vehicles operate. Further advances in design and manufacturing techniques will ensure that this improvement is sustained.

The technology is available to carry out many of the functions that have until recently only been postulated although in many cases the cost of doing so is not yet economic. Lower cost systems need to be developed although if built-in as standard on every vehicle, the cost will reduce through high volume. Integrated systems will bring down the overall cost but must be proved to be reliable and easily and cheaply repaired if there is a fault.

However it is believed that many transport managers and engineers are not yet ready to use the vast amount of information that is now becoming available; many are still too involved in basic day-to-day problems because of the lack of management capability of transport staff in general although the extremely fragmented nature of transport generally does not make it easy to overcome these problems.

The critical factors are believed to be psychological rather than actual. The systems should be seen as logic systems based on common sense rather than electronic technology. Should operators be more understanding of manufacturers' problems when a component or assembly failure can provoke major disruption in transport arrangements? An impartial analysis of the cause of problems must be the first and only objective, not wild conjecture and scaremongering. Training must play a more important role. Not only the formal training available from vehicle manufacturers and local educational establishments but on-going informal training through in-company supervision.

Most of the solutions are known and have been discussed but modern society places pressures on many of us to outperform competitors and even colleagues. It is unlikely that companies or individuals succeed by being complacent but the costs to others in their aggressive efforts for success are immeasurable. What is being suggested here is that it is perhaps the socio-economic pressures that are the prime influence in providing equipment for which we are not mentally or physically prepared. However by insistent coordinated pressure by the industry on operator companies and vehicle and equipment suppliers to provide the training and skills necessary to overcome the problems facing us today, we are convinced that the gap between the technology and the ability to supply and use it will rapidly converge to enable the science and art of electronics to be mastered.

REFERENCES

1 **ARAVOSIS G.D.** 1988 SAE/IMechE Exchange Lecture: Twenty-first century truck electronics - today's global challenge. Proceedings of the Institution of Mechanical Engineers, Journal of Automobile Engineering, 1989, Vol 203 No D1, pp 1-9

2 **BERGQVIST, L.** Electronics in Trucks. Scania/ Commercial Motor Technical Conference - 'Moving with the Micro', England, June 1988, pp 4-8

3 **SPOONER, G.** Prospects for Reliable Electrics. Transport Engineer, November 1988.

C391/076

Software safety and reliability—achievement and assessment

M BARNES
AEA Technology, Warrington, Cheshire

ABSTRACT:

The increasing use and sophistication of computers in vehicle control and monitoring heralds a new era in motoring. Significant advantages can be gained by the use of Programmable Electronic Systems (PES) to control braking, steering, engine management, etc.

Such advantages are not limited to control systems; the scope of vehicle information systems has been greatly increased by the advent of PES. In the past, primitive systems were limited to warnings (usually late) to the driver of such events as low oil pressure, increased coolant temperature, or low fuel. When PES are applied, early warnings can be generated relatively easily; the scope of the warnings is very large, for example, adverse weather conditions, such as frost or fog, proximity information to warn of safe driving distances, or on-board optimum route planning.

When all this is supplemented by the flexibility of software, in that it is relatively quick and easy to change functionality or implement extra functions, it is easy to see why we can expect vehicles of the future to have many computer-based systems. However, one problem remains to be solved: when the software of these systems "goes wrong", the systems can fail catastrophically, sometimes leading to loss of life (in the case of control systems), but certainly always leading to financial loss, either through bad publicity, recall and modification, or product liability.

This paper discusses the errors that can occur in the software life-cycle, what can be done to reduce them, and an effective startegy to assess the software safety and reliability.

INTRODUCTION

In recent years, many kinds of engineering and other systems have become increasingly dependent upon Programmable Electronic Systems (PES) for their reliable operation. The automobile industry provides perhaps some of the most innovative examples of the use of PES:

- engine management systems provide enhanced and economical performance, and can be kinder to the environment

- advanced braking systems can provide enhanced safety

- sophisticated ergonomic indication and warning systems to aid the driver in determining correct vehicle system status

The increasing use and sophistication of computers in vehicle control and monitoring heralds a new era in motoring. There are signficant benefits to be expected from the use of Programmable Electronic Systems (PES) - ie computer systems - in automobile applications. The complex nature of systems such as engine management, safe steering and braking make them candidates for the application of computer control, since complex control algorithms are relatively easy to implement via computers. Other features of the computer, such as response time, small size, relatively low cost, and most important of all - its flexibility to modify its transfer function - all contribute to its potential in autombile control applications.

The potential does not end with control systems; there is also significant benefits to be obtained from using PES in vehicle information and alarm systems. The early trend of individual parametric information presentation, eg:

- low oil pressure
- high coolant temperature
- low fuel

was limited to primitive warnings (usually visibly annunciated only, and all-too-often recognised after the event) to the driver. This trend is now disappearing, and in most vehicles these individual systems will soon be replaced by integrated systems processed by PES, including Built In Test Equipment (BITE). The application of PES in this area has not only increased the sophistication, but also the scope of application. The scope of the warnings is very large, for example:

- environmental information, eg adverse weather conditions, such as frost or fog

- proximity information to warn of safe driving distances between vehicles

- navigation systems, eg on-board optimum route planning, with local and global direction information, traffic bottlenecks, etc

This sophistication can be applied to very large number of useful applications, and therefore are an extremely attractive selling point to a potential customer, either on the basis of safety, comfort, entertainment, or even novelty value.

FINANCIAL LOSS

It is easy to see why we can expect proposals for vehicles of the future to be fitted with an increasing number of computer systems. The standardisation of hardware and flexibility afforded by the software means that it is relatively quick and easy to change the existing functionality or implement extra functions.

However, one problem remains to be solved: when the software of these systems "goes wrong", the systems can fail catastrophically, resulting in a whole spectrum of problems, eg:

- the need to redesign the software from scratch
- loss of life
- bad publicity
- vehicle recall and modification

All of these problems result in the manufacturer's financial loss, either directly or indirectly. It is this fact which is providing the driving force to get the software "right first time".

The automotive industry will wish to ensure that the use of PES does not fall into disrepute. The public at large are slow to forgive unreliable features of vehicles, and the manufacturer's reputation can be tarnished for a very long time, even on something trivial which merely creates a nuisance value. When safety features become unreliable, the public's reaction can be adversely fuelled by the media.

There is therefore an acknowledged need to get the safety and reliability of PES correct at the start of a project, so that these features are catered for early in the development lifecycle, and engineered into the product. Capital invested in these activities is not usually considered by senior management as a visible asset, but it has been shown to pay handsome dividends when considered against financial losses caused by the modified/aborted designs that often occur. Companies will also wish not to fall foul of any liability law with respect to their software.

THE SOFTWARE PROBLEM

Since there is a widely acknowledged need to assess the reliability of a variety of systems which incorporate PES, there is in consequence a need to address the computer software elements of these systems in order to produce a comprehensive assessment.

The enormous benefits that can be obtained from the use of PES in real-time control systems are often overshadowed by hidden defects in the applications software that provides the transfer function of the system (not to mention the vast costs that can be incurred in getting the software "right"). These software defects consistently escape detection during development and testing, but repeatedly demonstrate the validity of Murphy's Law by manifesting themselves during operation.

Techniques for the assessment of hard-wired electronic systems are well-established and widely practiced; unfortunately this is not the case for PES. A particular difficulty arises for those individuals or companies which have made a sudden transition to PES, without gradually aquiring expertise in software issues.

The burning question, therefore, is "how can the reliability of software be assessed"? The answer to this question is that there are many techniques and methods that can be used to assess software, and currently no single method will suffice by itself. This paper highlights the more important of these techniques and discusses a framework to integrate them. The framework is not unique; others are possible which could approach the task from a different angle. The paper mainly focuses on "state of the practice" techniques available.

WHAT IS MEANT BY SOFTWARE RELIABILITY ?

In order to be able to assess "software reliability", it is first necessary to define what is meant by this term. For the purposes of this paper, "software" means computer programs and the data they process.

Currently there are many qualitative and quantitative interpretations of "software reliability"; for example:

a) the number of failures (ie non-conformance to specification) observed in a time interval

b) the time to the next software failure (this is estimated from previous failure history)

c) the number of faults remaining in the software (this is estimated from previous failure history)

d) the software has been well-produced, therefore it should be reliable

Whilst the above examples are concerned with software reliability, each is incomplete; there are also flaws in all of the above interpretations. In order to understand this, it is necessary to examine the differences between software faults and hardware faults:

> Electronic hardware is subject to random failure mechanisms whereby a component can be functioning to specification one minute, and be in a failed state the next. Software (ie computer programs) is different; software either contains faults or it is fault-free (this overstates the obvious, but is a point which is often missed). Thus unlike hardware, it does not deteriorate in service; however its <u>performance</u> might fluctuate in service, because software can only fail when its inputs cause execution of the areas of the code which contain the fault(s). Thus software reliability is a function of both the faults in the

software, and the sets of input data which exercise the software paths on which the faults lie.

Examples a), b), and c) will be affected by the distribution of inputs which is used to test the software; thus if quantified operational reliability is required, it is important that the software is tested with data representative of the operational profile. It should be noted that the operational profile is not always easy to define.

The number of faults remaining in a piece of software is not considered to be a good indicator of software reliability. If there are numerous faults in parts of the software which are infrequently or never exercised, then the software might exhibit a low failure rate, in spite of the many faults. Conversely, if software has a single fault which lies on a common path, it will be exercised frequently, and exhibit a high failure rate.

The argument put forward in example d) does not hold. High quality in the software production processes does not automatically result in highly reliable software; however it should be noted that high quality in the software production processes should be regarded as a necessary pre-cursor to obtaining highly reliable software.

In defining PES system reliability we are concerned with a PROBABILITY of the SYSTEM to function for a specified period of TIME, under specified CONDITIONS. When addressing software reliability, it is this aspect and this aspect alone that is of interest, ie how software failures will affect the overall systems reliability. Thus we now have a firm definition which provides us with basis for assessment:

> "SOFTWARE RELIABILITY IS THE PROBABILITY THAT SOFTWARE WILL NOT CAUSE A FAILURE (to be defined) OF A SYSTEM FOR A SPECIFIED PERIOD OF TIME UNDER SPECIFIED CONDITIONS".

The above description therefore embraces the number of faults in the software, as well as the distribution of inputs to the software.

It would appear that with such a firm basis, the task of assessment is now straightforward. This is untrue for a variety of reasons, the main one being that the quantification of software reliability is difficult to determine with any precision for high reliability applications. Another problem is in the definition of "system failure"; if it is to be determined which of the software faults can lead to a specified system failure of interest, then a lot of detailed analysis could be involved. A short-cut approach could be to assume that every software failure will result in the system failure of interest.

THE APPROACH TO SOFTWARE ASSESSMENT

It should be realised that software should never be assessed in isolation only; software is just one element of a system, and therefore its effects within a system need to be considered. The configuration of the system is a very important element with respect to system reliability, and this will strongly influence the amount of effort required to assess the software; this aspect is discussed fully in the HSE PES Guidelines (see /HSE1-87/ & /HSE2-87/). For example, the amount of assessment required for a software-based system might vary enormously if there was a diverse hardwired back-up system available. Therefore system aspects should be considered as well as the assessment of the software itself.

As stated previously, software reliability depends not only on the number of faults in the software, but also on the input data profile. This explains why:

- software that contains many faults can still work satisfactorily over a given range of inputs

- a program which has been running satisfactorily for many years can suddenly fail to produce correct answers, or even "crash"

Thus the focus of attention should be directed towards a thorough understanding of the processes whereby faults can be introduced into software, and determining the countermeasures required to remove these faults, or rendering them impotent.

When assessing software as a stand-alone product, the assessor is faced with three fundamental questions;

1) what good software engineering practices have been invoked to avoid software faults originating?

2) given that the human being is imperfect and will still make errors, what has been done to detect and remove the faults which have not been avoided?

3) given that nothing has changed, and that the human being is still imperfect and may fail to detect all the faults, what has been done to tolerate the faults not detected?

Although this three-pronged approach to the achievement of software reliability is simple in concept, it hides a multitude of software engineering issues. Nevertheless, it provides a framework for the assessment of software.

The first prong of the approach (FAULT AVOIDANCE) addresses the assessment of the software PROCESS. This is concerned with issues of quality, and the best software engineering and "housekeeping" practices available to reduce the chances that errors will be made during the software development processes. As stated previously, it should be noted that even the very best effort invested in Fault Avoidance techniques is no guarantee that the software will be fault-free; however such techniques are a necessary prerequesite for high-integrity software. Therefore because Fault Avoidance techniques do not guarantee the reliability of the produced software, it is also necessary to carry out an assessment of the software product itself.

The second prong of the approach (FAULT REMOVAL) addresses the assessment of the software PRODUCT, ie the delivered code. It is concerned with issues of testing and analysis, in order to detect and remove faults which escaped the Fault Avoidance "filters" used during the software development. The assessment of the product could also be extended to include the specification and design.

The third prong of the approach (FAULT TOLERANCE) addresses the assessment of reliability PERFORMANCE. Fault tolerance is also product-related, and is concerned with the issues of:

- measurement/prediction of the reliability, to determine if the current reliability can be tolerated, and/or
- analysis of features of the software or system which have been incorporated to reduce the effects of software faults to an acceptable level, for example system configuration, software diversity.

CONTINUOUS versus RETROSPECTIVE ASSESSMENT

An assessment can be continuous (ie the assessment starts as software lifecycle starts, and continues through the software lifecycle), or retrospective.

Continuous assessment: is considered vital for high reliability projects, since it is desirable that software faults do not propagate from one life cycle to another. Such fault propagation results not only in greatly increased cost and time penalties, but also increases the probability that faults will escape detection, or repair-induced faults will be created.

Retrospective assessment: may be unavoidable, for example in cases whereby a software product is already completed, or the software is a commercial package over which the purchaser has had no development control. In such cases, an assessment cannot influence the quality of the delivered software, because the software is already complete; however, it might provide an indication of the quality of the software, as discussed below.

Retrospective assessment suffers from the distinct disadvantage that the vital documented information necessary for the assessment may not be available, or may be difficult to obtain / decipher at so late a stage. Also members of the software team may have dispersed at such a late stage, and so answers to pertinent questions may be difficult to obtain. These disadvantages reinforce the fact that the approach to assessment of high-integrity software should be continuous throughout the software life cycle.

For high integrity applications, software assessment should also be independent from the "software-house". This means that an independent "third-party" (ie external to the organisation) should carry out the assessment of the software.

Some examples of the more important issues relating to the three-pronged approach of FAULT AVOIDANCE, FAULT REMOVAL, and FAULT TOLERANCE are now discussed in further detail.

1) Fault Avoidance

This is concerned with ensuring that a great deal of effort is put into the best possible software engineering practices to produce high quality software (for a good overview of the disciplines of Software Engineering see /Smi 87/). It embraces all of the activities within the software life-cycle, and it is therefore essential that a quality management system, such as BS5750, is invoked and executed in order to enforce the engineering disciplines that will result in a high-quality product. (A recent report in a technical periodical has indicated that IBM has recently installed such a quality management system and as a result has reduced its software fault rate by a factor of 4 on its application software packages - so the message is "get it right at the first attempt"). The assessor should check that an adequate Quality Management System has been invoked, is functioning satisfactorily, and is auditable (and audited) at suitable intervals.

One of the most important QA activities is the construction of a "Software Quality Plan", which addresses the development techniques, procedures and standards which are to be applied throughout the software life-cycle in order to avoid faults. Such a quality plan should be provided for each software project, and should be based upon a recognised standard, for example, see ref /AQAP-13/. The assessor should ensure that the Software Quality Plan addresses all the relevant activities and items for the task.

Throughout the software lifecycle, tools should be employed to aid in the software construction processes and to reduce the likelihood of faults. The assessor should be satisfied that such tools (see the STARTS guide, ref /Sta 87/) are adequate for the task, and are of good pedigree (ie have been successfully used in other similar projects). The assessor should also be aware of any limitations of such tools, and analyse the effects that such limitations could have on the software produced.

The starting point for software is with a requirements analysis; this is an iterative procedure between the customer and the software producer, whereby the customer tries to convey the software requirements to the producer (usually in plain English) so that the producer may express these requirements more formally in a requirements specification. For further reading on requirements specifications, see /Dah 88/. Formality is aided by the use of "Formal Methods" (eg VDM, Z) which are a mathematical approach to software development, and which lend themselves to "proof" of the software. This includes the aim of achieving unambiguous, consistent, and complete specifications, - although it is widely recognised that formal specification techniques do not fully address issues of performance specification (eg timing and accuracy).

Formal methods will be made mandatory in all MoD Safety-critical projects, according to a new MoD software standard (00-55) which is due for release around summer-autumn 1988.

An essential quality which aids in the avoidance of faults is the visibility of the software, such that it can be read and easily understood, both by the originator or other team members. So the assessor should consider the choice of structured development techniques (eg Jackson, Yourdon), and of the choice of programming language and its constructs. Currently, safe sub-sets of ADA and PASCAL are the favoured languages because of their good structure, and "english-like" visibility which makes them self-documenting, and the fact that such languages lend themselves to formal analysis and proofs by automated tools such as LOGISCOPE, MALPAS, and SPADE. Overviews of these tools are provided in refs /Mee 88/, /Web 87/, and /Car 86/, respectively.

An assessor should possess a thorough (and sometimes suspicious) mind. Good assessment involves reviewing ALL activities of the software throughout its development life-cycle in order to determine where faults can occur. To illustrate this point, consider a piece of software which has been produced to the highest quality standards, undergone formal proofs & static analysis, and has been thoroughly tested. It is of little use investing such effort if "bad housekeeping" results in a previous (faulty) issue of the software being installed in the system, and escaping detection (see /Law 87/ for a practical discussion of these aspects). Thus configuration control of the software processes is vital to aquiring correct software.

Similar comments apply to the security of software. An article in a recent technical journal claims that the software for the next shuttle mission has been subjected to low security control for five months, and a number of unauthorised changes were made to the software. Thus security is important, whether or not the unauthorised changes are malicious (sabotage) or due to good (but damaging) intentions.

Summarising for the fault-avoidance activities, the assessor should determine that:

- a quality management system has been invoked and implemented,

- a good software QA plan has been produced and implemented, and that an appropriate level of Quality Assurance (QA) has been specified and implemented.

- a requirements specification has been produced which is complete, consistent, and unambiguous

- that a visible, understandable, top-down design approach has been implemented.

- that best structured coding practices have been observed and implemented.

- strict configuration control is defined and enforced

2) Fault Detection and Removal

Once the software construction is complete, the assessor is then faced with the task of determining what has been done to detect and remove software faults which have not been avoided. There are many methods that can be used here, and they generally fall into two main categories:

- testing
- analysis.

The STEM project addressed issues of software testing and evaluation methods, and details can be obtained from ref /Bis 87/. Testing can be both "black box" , ie based upon the specified functions of the software with no knowledge of the structure, and "glass box", ie based upon the structure of the software code with no knowledge of the specification (see /Oul 86/ for a thorough review of testing techniques). Both types of testing have their strengths, and should be considered as inseparable for high reliability applications. Also the efficiency of test data can be measured by observing the statements or paths encountered by various test data sets. The code needs to be "instrumented" for such tests, and this is best achieved by automated tools, such as LOGISCOPE, or the Liverpool Data Research (LDRA) automated test bed; further information on the LDRA test bed can be found in the user documentation, ref /LDR 81/.

It should be noted that for all practical purposes, testing can never prove that the software is correct - it can only prove that the software is incorrect. This is because of the large number of premutations of digital inputs that can occur in software, even with relatively few inputs. For example, a control system processing just 5 analogue inputs, each digitised to 12 bits, would take of the order of 36 million years to test exhaustively, assuming 1 microsecond per test. Of course, not all of the input permutations would be possible in practice, and of those that were possible, only a subset would be in the region of interest. Nevertheless, the quantity requiring exhaustive testing would be far too great for practical purposes.

Clearly testing has to be sample-based, and needs to be supplemented by other means of fault detection. One of the most powerful techniques for detection of faults is static analysis. This involves examination of the source code of the software (without execution of the code), to determine certain qualities of the code, eg:

- if the code is well constructed (Control Flow analysis)
- if there are anomalies in data usage (Data Flow Analysis)
- if there are anomalies in the use of information (Information Flow Analysis)

The analysis can be carried out manually; however automated tools are commercially available (LOGISCOPE, MALPAS, SPADE) which speed up the analysis, removes the tedium, and hence reduce human error from the process.

Tools such as MALPAS these have more powerful features (such as Semantic Analysis, Conformance Analysis / Proof Checking) than those mentioned above, and should be used not only to validate the code, but throughout the software lifecycle to verify each stage of software development.

Static analyses can reveal areas which warrant closer investigation, and has been shown to lead to detection of very subtle faults, which would be difficult to detect by testing alone.

3) Fault Tolerance

What can be done about software that still contains faults after testing and analysis has been carried out? Most software of any size contains residual faults, and in cases of large complex programs, the repair of such faults can create other subtle faults; in some instances the fault content of a program has increased after repair.

One way of handling software with residual faults is to determine whether or not the measured or predicted reliability (expressed as a failure rate or mean time to failure) of the software can be tolerated. For example, a mean time to failure of 3 months might be considered acceptable for an industrial control system which fills metal cans with soup, the greatest consequence here being the overflow or short measure of soup. Hence reliability modelling could be used to predict the reliability of the software; for a review of software reliability modelling, past, present, and future, see /Dal 87/.

One major drawback in the use of some reliability models is that the models require a minimum number of faults before a reasonable prediction is obtainable. For very high-integrity software, such a large number of faults during testing would indicate that the software is unsuitable for its application. Thus reliability models are more suited to large complex applications where large numbers of faults are originally generated, such as in wordprocessing packages or operating systems, ie where the reliability growth can be demonstrated.

Another approach to Fault Tolerance is one that is traditionally applied to electronic hardware in order to achieve high system reliability, ie "the achievement of a function by at least two different and independent means". The hardware technique of redundancy ie "the achievement of a function by at least two identical but independent means" cannot be implemented in software, because if two identical pieces of software are produced, they will behave in exactly the same manner for the same set of inputs, and thus exhibit complete dependency.

For software, two or more different (diverse) versions of a program can be implemented to overcome some of the dependencies. Experiments (see the PODS project, /Bis 86/) have shown that diverse software (N-version) can provide significant benefits over single software applications. Diversity can be applied at all levels of software development, and to varying degrees; eg diverse software specification methodologies, design methodologies, diverse languages, diverse teams,

The opponents of software diversity argue that the requirements specification (where most faults are traditionally found) is common to all diverse versions, and therefore reduces the benefits of software diversity. This of course is also true for electronic hardware, yet diversity is the favourite mechanism in high-reliability hardware applications.

Other means of Fault Tolerance exist, eg :

- recovery block analysis, whereby the output from critical modules is subjected to an "acceptance test" (ie sensibility checks), and if it fails, the input is redirected to another module which carries out the same task, but by a diverse algorithm.

- inverse algorithms, whereby for critical modules, a number of intermediate outputs are generated together with the main output, and from these, the input is calculated (using an "inverse" algorithm), and compared with the original input.

- serial diversity, in which one program performs the same task n times, using a different algorithm each time. The advantage of this is that if such software is used in redundant hardware systems, the software in each chain is identical, and therefore problems of maintenance are reduced. The disadvantage is that the software in each redundant system is still identical, and suffers from dependencies.

THE NATIONAL CENTRE OF SYSTEMS RELIABILITY (NCSR)

NCSR is currently producing a guidance document called the "Guidance for the Reliability And Safety assessment of Software (GRASS)" document. This document will provide a framework for the assessment of software for a wide range of applications. It will address the important issues necessary to ensure reliability of software, and the current techniques and tools available to undertake an assessment.

It is intended that the GRASS document will be applicable to a range of users, including those involved in the assessment, purchase, production, or licensing of software. It will be valuable both as a guide to the non-specialist who seeks to understand the issues involved in the assessment of software, and as a useful reference document for those who have greater expertise. With this in mind, it will be produced in a well-structured and user-friendly format.

The implementation of the techniques discussed here to specific application is beyond the scope of this paper, however, the GRASS guide will provide a pragmatic framework for the assessment of specific categories of application, including, for example, engine management systems.

The GRASS guide will provide practical advice on the limitations, advantages, and disadvantages of the tools/techniques.

REFERENCES

/AQAP-13/
NATO Software Quality Control System Requirements

/Bis 86/
"PODS - a Project on Diverse Software" by P G Bishop et al. IEEE Transactions on Software Engineering Vol SE-12, Number 9, ISBN 0098-5589.

/Bis 87/
"STEM - A project on Software Test and Evaluation Methods" by P G Bishop et al, (presented at SARS '87, and published in "Achieving Safety and Reliability with Computer Systems" edited by B K Daniels, Elsevier Applied Science, ISBN 1 85166 167 0)

/Car 86/
"SPADE - the Southampton Program Analysis and Development Environment" by B Carre, C Debenny, I o'Neil, D Clutterbuck. Published in "Software Engineering Environment" (IEE computing, Series 7, 1986), edited by I Sommerville.

/Dal 87/
"The development of techniques for Safety and Reliability Assessment: Past, Present, and Future" by C J Dale, S Foster (NCSR). First published in "Achieving Safety and Reliability with Computer Systems" - proceedings of SARRS'87. Elsevier Applied Science, ISBN-1-85166-167-0.

/Dah 88/
"Software Specifications - The Requirement", by G Dahll, OECD Halden
Reactor Project. Presented at the ESRRDA Seminar on Software Reliability, Brussels, 14 September 1988.

/HSE1-87/
HSE Guidelines on Programmable Electronic Control Systems in Safety-related Applications, PART 1 ", ISBN 0 11 883906. Her Majestys Stationary Office, PO Box 276, London SW8 5DT.

/HSE2-87/
HSE Guidelines on Programmable Electronic Control Systems in Safety-related Applications, PART 2 "General Technical Guidelines", ISBN 0 11 883906. Her Majestys Stationary Office, PO Box 276, London SW8 5DT.

/IEE 87/
Guidelines for the Documentation of Software in Industrial Computer systems, the Institution of Electrical Engineers.

/Law 87/
"Configuration Identification and Control of Software for Microprocessors" by A Lawrence, CEGB. Presented at the 4th Annual Symposium on Microprocessor-based Protection Systems (for the Institute of Measurement and Control), 10 December 1987.

/LDR 81/
LDRA Software Test Bed user documentation, from Liverpool Data Research Associates, Victoria Buildings, Liverpool University, Liverpool, Merseyside.

/Mee 88/
"LOGISCOPE - A Tool for Maintenance", by J Meekel & M Viala. To be published in the Proceedings of IEEE Conference on Software Maintenance, Pheonix, Arizona, October 1988

/STA 87/
The STARTS (Software Tools for Application to Real Time Systems) Guide DTI and NCC. Available from the National Computing Centre, Oxford Rd, Manchester. ISBN 0 85012 619 3.

/Oul 86/
"Testing in Software Development", monograph from the British Computer Society, edited by Martyn A Ould and Charles Unwin; from University Cambridge Press. ISBN - 0 - 521 - 33786 - 0

/Smi 87/
"Engineering Quality Software" by D J Smith & K B Wood, Elsevier Applied Science. ISBN-1-85166-074-7

/Web 87/
"Static Analysis - a Technique for Software Verification" by J T Webb. Published in "Computer Techniques" July - August 1987

C391/016

Software need not be a risk

P A BENNETT, PhD, FIEE, FBCS, FIQA
CSE Limited, Flixborough, Scunthorpe, South Humberside

SYNOPSIS Increasingly motor manufacturers, and their suppliers, are turning to embedding software within the micro-electronics so as to achieve the complex control algorithms required of todays vehicles. The use of such software introduces a new set of problems for the design engineer: design, development, quality assurance and safety.

The acknowledgement that software can have a significant effect upon the safety criticality of a system is a recent development which responsible engineers cannot ignore. Similarly engineers cannot ignore the responsibility for the liability attaching to such systems.

This paper will discuss and demonstrate that the development of safety critical software and systems need additional levels of skill to other types of systems. There are international standards being developed to aid the developer and there are cost-effective methods of testing and assessing the safety without the recourse to formal mathematical methods in every case. Both these developments will be explored. One approach is called RISCS - Reliability, Integrity and Safety Critical Studies which accords with the standards being developed.

1 INTRODUCTION

Safety critical systems have been around in nuclear power plants for more than 30 years and in major chemical plants for a longer period of time, so what's new? What's is new is the growing use of computers to control such processes especially in the motor vehicle. Initially, they performed similar functions to the earlier relay logic but as their power has increased, it has become possible to operate plants with much faster decision-making by computer. This has brought economic benefits but also the need for a more critical appraisal of the whole process from the specification of what the system is required to do, through to the system design phase which identifies how the system may be structured to its implementation in software. As the system becomes more complex, each of these phases need codes of practice and, in some cases, regulations to ensure safe operation of the system over its life.

In the past, we have relied on the education, training, experience and wisdom of engineers to ensure that our systems are safe. In the future, engineers will need to be brought up to date with the ever-increasing armoury of tools available to aid the design process involving software. The existence of a pool of experts will be of value in conducting peer reviews of the competence of practising engineers.

As the costs associated with the development of computer-based systems continues to fall Industry worldwide is increasingly coming to depend upon new technology to maintain a competitive edge whilst being able to operate ever more complex control systems that require a high degree of reliability, integrity, dependability and safety. Such system are generically referred to as being "safety-critical".

Engineers and managers have begun to realise that with this new technology comes flexibility and additional safety considerations. These considerations include:

i) the knowledge that the correct and safe operation of the business, capital equipment or control system is dependent upon such technology

ii) the risk of causing a loss of human life, economic loss or environmental damage is outside their normal engineering judgment

iii) that the conventional safety assessments are insufficient and fail to identify that there are weaknesses in the safe operation of the system.

What is now needed is a structural approach for the assessment of Safety Critical Systems to which the experienced engineer can turn. In addition to being sufficiently flexible to encompass the latest technique as well as ones which are more mature the scheme must also be cost-effective.

Till now the only assurance of correct operation of a safety-critical system has been by conventional software testing techniques based on the functional requirements. However such testing is not without flaws and cannot locate all faults. Faults often only materialise many months or years after a system has been put into operation, possibly leading to very expensive recall procedures or "retro-fits" on site, it not dangerous incidents. Most industries have an urgent requirement to gain confidence in the correct and safe operation of their computer-based systems this is no less a requirement in the automotive industry.

The concerns of software safety began to surface early in the 1980's [1] and prompted a number of "standards" activities. Many readers will have seen much in the press during 1989 on the new Defence Standard (Def Stan 00-55) which places emphasis on the use of formal mathematical methods for defence procured software. Surprisingly though there has been little press interest in the work of the professional bodies, like the IEE and I.Gas.E., in advising their members on this important subject.

A recent survey of practices in safety critical systems has shown that dependence by the engineer for safe operation is placed on the use of traditional quality assurance methods, such as BS 5750 and ISO 9000. This is understandable as there are few standards for such systems and those that there are have only just come about.

In June 1987 the Health and Safety Executive published a document commonly known as the "PES Guidelines" [2]. This document, as its name implies, established guidelines and procedures for the implementation of where are called Programmable Electronic Systems (PES) used in industrial control, instrumentation and consumer products. Similarly, guidelines have been, or are being developed by trade associations, professional bodies and other UK standards organisations. Similar activity exists in a number of other countries, including France, West Germany, the Scandinavian countries and the USA.

In parallel with all this activity the International Electrotechnical Commission (IEC) established two working groups in 1984 with the objective of producing draft international standards in 1989. One working group (SC65A WG9) is concerned with software for safety related applications and is expected to present its work to Geneva in June 1989 whilst the other, concerned with generic safety aspects (SC65A WG10) expects to submit its material sometimes in the summer of 1989.

The significance of these two IEC activities must not be overlooked. As international standards they have the advantages over national standards of being a reference source across national boundaries and therefore common basis for determining working practice. Also both working groups were charged with preparing generic standards which can be tailored for specific industries within the broad scope of being "civil non-nuclear". Clearly the IEC work has a real significance to the automobile industry.

The draft standard from WG9 is extensive and addresses the full lifecycle of software development. To precis the standard is beyond the capability of this paper but as chairman of WG9 the author is able to give some insight into the standard.

There are three significant points to the standard from WG9. Firstly, the standard advances a five-point scale for the levels of safety integrity so as to direct the engineer towards a line of thought that safety is not a two-state condition but one of degree of risk. Secondly, the standard does not place emphasis on the use of formal methods for developing and assessing software unlike Def Stan 00-55. Thirdly, the standard is in two parts; part one contains the normative standard whilst part two contains an informative annex to expand upon part one so as to aid the developer of systems. Part two contains, amongst other details, tables to cross-reference particular techniques and the integrity levels. Clearly the work of the IEC groups is going to open up a discussion on the subject internationally whilst in itself is going to advance the cause for safety in software.

2 EXPERIENCE WITH SAFETY CRITICAL SYSTEMS

As the engineer becomes more aware of needs (contractual, legal and insurance) to assess the computer-based systems on the vehicle for safety he is going to become aware of the enormous costs that can arise from using formal methods alone on the software. Additionally the other parts of the system will need to be assessed as well as the software.this problem based on research it has conducted on the subject. The result is a methodology called **RISCS** - Reliability, Integrity and Safety Critical Studies.

The **RISCS** approach was developed by drawing on extensive experience in:

a) developing and applying these methods to evaluate large complex computer systems in many spheres of industry.

b) of these systems have had a safety-critical or safety-related requirement and have been successfully evaluated.

The combination of this expertise in using advanced verification and validation techniques coupled with its detailed knowledge of Industry has enabled **RISCS** to be developed and applied successfully to many systems.

The methodology permits differing hazards of the system to be assessed at differing levels, making use of the most powerful assessment techniques available. In this way a very high level of assessment can be achieved for dependable systems in a cost-effective way as the most powerful techniques are only used where necessary.

The methodology can be applied on existing systems or on ones being developed. Though it is clearly more cost-effective to apply this standard approach throughout development stages, it can also be successfully applied to systems already implemented.

RISCS assesses the reliability, system integrity and safety in the context of the complete system and not just the software, though this is an important part of the assessment. What is more the client is able to decide on the level of assessment required.

The RISCS approach consists of a number of consecutive stages.

1) hazard criteria selection
2) legal/economic consideration
3) determination of assessment level
4) selection of assessment
5) evaluation

3 HAZARD CRITERIA SELECTION

The Hazard Criteria are established in discussion with the client. These are the criteria by which the system is expected to conform for it to be judged operable. Examples could be:

a) no loss of power assisted braking

b) no loss of control on the engine management system such that stability is lost

c) the system shall be capable of fail-safe operation under definable operating parameters

The Hazard Criteria differ for each system, industry sector and application because of the way in which the system is to be used, given legal, economic and environmental considerations.

4 ASSESSMENT LEVELS

For each Hazard Criteria a specific Assessment Level is established after discussion with the client. Since there is no universal technique for making this type of assessment the criticality of the system, with respect to the clients application, is determined and from this the appropriate technique chosen. By this consultation the client is assured that the correct emphasis is placed on the application.

5 EVALUATION

Each Hazard Criteria is evaluated using the selected technique, in terms of the hardware, the software and as a complete system.

For each Hazard Criteria, the system will be assessed to conform to some recognisable standard of safe/reliable operation for the agreed Assessment Level or will be found to be deficient in some way. If the system is found to conform a Conformance Report will be issued. Alternatively, if the system is found to be deficient a Deficiency Report will be issued suggesting ways in which the system should be changed in order to conform.

6 THE TECHNIQUES

The methodology developed is an advanced method for assessing the Hazard Criteria for systems requiring a high degree of reliability, integrity, dependence or safety, generically referred to as being "safety-critical". There is no universally applicable method for making an assessment of safety-critical systems as the criticality of the application, the user industry and the technological culture of the developer all have an influence on the set of techniques that can be used. Therefore RISCS is a method of determining the appropriate technique to use given the Hazard Criteria and the Assessment Level.

RISCS has five levels of assessment, increasing in rigour:

6.1 Level 0 - System Overview

Level 0, System Overview, is characterised by the use of Design Reviews, often referred to as Formal Design Reviews. A design review is a formal, documented, comprehensive and systematic examination of a design to evaluate the design requirements and the capability of the design to meet those requirements.

A Formal Design Review is an advisory activity whose primary purpose is to provide verification of the work of the development team. In general, the review should ensure that a product meets its specified requirements, that it satisfies customer needs, and optimises costs. The secondary purpose of the design review is to provide input to the creative process, although care must be taken to ensure that the review team do not interfere with the work being undertaken by the design team. In short, a design review should identify problems and propose solutions, where the solutions are likely to be of a general, rather than detailed, nature. Within the context of RISCS, the formal design review should assist in revealing areas where safety critical features have been overlooked and in highlighting areas where the system can be improved.

6.2 Level 1 - System Structure Analysis

Level 1, System Structure Analysis, is characterised by the use of Checklists, and two types of formal design review, applicable to software development, that is Fagan Inspections and Structured Walkthroughs. Note that these techniques are not necessarily exclusive as Fagan Inspections also incorporate the use of checklists.

Checklists are a useful tool to aid assessment of reliability and safety in high integrity computer systems. They contain a structured set of questions which help the assessor make a qualitative judgment of the system, and also help to focus the assessment by identifying areas requiring further investigation.

The need to assess safety and reliability of any computer based system may occur at any stage of the project, where the actual point of assessment will be dependent on the particular project. Furthermore, it is not reasonable to expect that a rigid set of questions can be applicable to every application. Therefore RISCS allows checklists to be applied at every stage of software development, but expects them to be applied in an intelligent manner, adapting them to the project under consideration.

To accommodate wide variations in software and systems being validated, these checklists contain questions which are applicable to many types of system. Nevertheless, in some cases it may be desirable to supplement the standard checklist with questions specifically directed at the system being validated.

Structured Walkthroughs are a type of technical design review, applicable to software development. During a structured walkthrough, a developer's work, for example program design, program code or documentation, is reviewed by fellow project members. In essence, structured walkthroughs differ little from the formal design reviews described previously. The main difference concerns the duties and responsibilities of the participants.

Structured walkthroughs are arranged and scheduled by the author, that is the person developing the product to be reviewed, rather than by an appointed chairperson. The author selects the list of reviewers. Furthermore, it is usual for the author to chair the walkthrough meeting, and to compile an action list of all errors, discrepancies, and inconsistencies uncovered during the walkthrough.

Software inspections are a form of quality review. Based on well established review techniques, Fagan Inspections provide a more formal and rigorous method of performing technical reviews at the end of development phases.

Because it is a creative activity, the development of software is a process which is prone to error. Testing is the traditional way of reducing software errors, which, if well planned, can be reasonably successful at detecting certain types of error, for example coding errors. This approach has two distinct disadvantages. Testing can rarely be considered completely comprehensive as checking every path, through even a simple program, can be extremely time consuming and therefore expensive. Secondly, the cost of correcting an error increases as the project life cycle unfolds, that is the later in the life cycle the error is detected. Testing inevitably detects errors during the later stages of development.

There is therefore a need for a review which can detect errors in the earlier stages of development, and which may improve the quality of the final product by detecting the type of errors which are not easily detected by testing. Fagan's original paper, comparing a team using inspections against a team using less formal reviews, reported a 23% increase in productivity with 38% fewer errors.

6.3 Level 2 - System Hazard Analysis

The goal of system safety is to incorporate acceptable levels of safety into the system design prior to production and operation. One of the first steps in any attempt to achieve this goal is to perform System Hazard Analysis. Hazard analysis involves the use of procedures to identify, and assess the criticality levels of, hazards within a design. This facilitates the elimination, from the design, of those hazards which pose an unacceptable level of risk. In some cases elimination may not be possible, in which case the associated risk should be reduced to an acceptable level.

Within RISCS, hazard analysis is characterised by a variety of techniques intended to complement the use of Design Reviews, Walkthroughs and Checklists which are associated with the previous level, that is System Structure Analysis. The particular techniques associated with system hazard analysis can be split into two categories, corresponding to two of the main modes of human reasoning, that is induction and deduction. These two generic analytical methods may be described as follows.

One of the most effective, and most commonly used, techniques for performing deductive analysis is Fault Tree Analysis (FTA). FTA is widely used within the nuclear industry as a systematic method for acquiring information about a system. FTA therefore merits further mention. Many approaches to the inductive style of systems analysis have been developed, and two approaches are considered to be the most widely used. They are Failure Modes, Effect and Criticality Analysis (FMECA) and Reliability Block Diagrams (RBD).

Fault Tree Analysis (FTA) is concerned with identification and analysis of conditions and factors which cause or contribute to the occurrence of a defined undesirable event, usually one which significantly affects system performance, economy, safety or other required characteristics.

The fault tree itself is a graphical representation of the conditions and factors contributing to the occurrence of the undesirable event, often referred to as the top event. Examples of factors which could affect the reliability of the system include, amongst others, component failure modes, operator errors, environmental conditions, and software errors. Furthermore, fault trees facilitate the identification of common events, affecting more than one functional component, which could cancel the benefits of specific redundancies.

6.4 Level 3 - Rigorous Analysis

This level is characterised by the use of rigorous methods, for example Finite State Machines and Time Petri Nets, which do not require a high degree of mathematical ability. Nevertheless these methods are generally recognised as part of a wider design approach, commonly referred to as Formal Methods. Formal methods is a philosophy that places great emphasis on rigour, in particular the use of mathematical modelling, throughout all phases of system development. The following discussion, of finite state machines and Petri nets, demonstrates that these methods are actually a subset of formal methods, termed Dynamic System Specifications. This compares with the methods associated with Level 4, Formal Mathematical Modelling, as described in the following chapter. These are also considered to be a subset of formal methods, termed Model Based Specifications.

Finite state machines provide a method of modelling systems which react to changes in their immediate environment, such that the reaction is recognised by changes in the internal states of the system. A simple finite state machine consists of a finite set of internal states, a finite set of input symbols, and a next state function. At any time, the system will be in one of the internal states. The input symbols represent the environmental influences on the system. The next state function maps the Cartesian product of the set of internal states and the set of input symbols onto the set of internal states.

A more complex finite state machine may be required to include some form of system output in the model, and will thus include a finite set of output symbols and an output function which maps the Cartesian product of the set of internal states and the set of input symbols onto the set of output symbols.

The Petri net, a tool for system modelling, has been developed from the original work of C.A. Petri, who developed a new model of information flow in systems. The model was based on the concepts of asynchronous and concurrent operation by the parts of a system, and the realisation that the relationships between the parts could be represented by a graph or net. Since that time, research has continued into the use of Petri nets, and a number of modifications and enhancements have greatly increased their modelling power.

One of the major advantages of using a formal specification methodology is that it can sometimes be used in conjunction with an automatic program generator. For example, in the telecommunications industry, there are systems which will generate code in a variety of languages, such as Pascal or Coral. These systems require as input a specification written in Structured Design Language (SDL), which incorporates mathematical representations of both Petri nets and finite state machines.

6.5 Level 4 - Formal Mathematical Methods

Level 4 is the most rigorous level of assessment within RISCS, and is characterised by the use of specification methods such as the Vienna Development Method (VDM) and Z. These methods provide a mathematical language in which the developer may describe the system data types and the operations to be performed upon them. The developer is therefore able to build a mathematical model of the system which is exact, that is free from ambiguity, and should therefore be internationally appreciated.

Having formulated a design specification, the developer is able to proceed with the various stages of system development, and at each stage check that the realised design still meets its specifications. Under certain circumstances, it is possible to formally verify, using formal logical deduction, that a given program is correct, that is it satisfies its specifications. In most cases however, it is sufficient to apply less stringent validation techniques, in combination with standard testing methods, to ensure that the design satisfies its specification.

Automated software testing tools such as SPADE and MALPAS may be used to assist in the analysis.

7 EXPERIENCE OF USING RISCS

The RISCS methodology has been deployed on many complex embedded control systems including emergency shutdown systems for oil rigs, fault control for nuclear power and in the automotive industry such applications as anti-lock braking and engine management. In all such cases the client has been an integral part of the assessment; defining the hazard criteria and in the determination of assessment level and technique. This has resulted in the client being fully aware of the developing assessment and in the rationale for the outcome. The assessment has in each case been shown to be cost-effective for the client and efficient in staff usage.

With the arrival of international standards many engineers are faced with the need to have assessments conducted on their safety critical or safety related systems. As might be expected in a commercial environment the assessment needs to be conducted in a cost-effective way, with the constraints of developing standards, RISCS is one approach which has been demonstrated to meet these conflicting requirements.

REFERENCES

[1] BENNETT, P.A., The Safety of Industrially-based Controllers Incorporating Software, PhD Thesis, 1984

[2] THE HEALTH & SAFETY EXECUTIVE, The Safety of Programmable Electronic Systems in Safety Applications, H.M.S.O., 1987

C391/037

Achieving high quality software

D P YOULL, BSc, PhD
Cranfield Information Technology Institute, Milton Keynes

Introduction

The achievement of fault free software is not attainable for any practical systems today. In fact, the actual quality of any software we do produce is very hard to quantify, thus making the software development process very difficult to manage. Since software is being used increasingly in all sorts of automotive applications where safety and reliability are of concern (e.g. automatic transmission control, anti-lock braking, cruise control, engine management), procedures for the development of high quality software are required.

Over the last twenty, or so, years the development of software has been moving away from a concentration on programming to a full lifecycle engineering approach. In other words, to produce complex software effectively one needs skills in requirements analysis, design, testing strategies, project management, et al. This has greatly improved the quality of software and computing systems in a large range of products. However, in high quality systems where reliability is critical, yet more rigorous approaches to the development of software are required (e.g. mathematical analysis, reliability models). This paper reviews the current state of the art in software development and considers the strengths and weaknesses of some of the available techniques. The effects of the recent rapid development in computing hardware and future trends are also considered.

The Software Development Process

Software development can be a very error prone process making high quality difficult to achieve. Control of the problems in software development can only be achieved through a thorough understanding of the software development process. A simple summary of this process is shown in Figure 1. The special problems faced in high quality software require different actions in each phase of the process. These problems and current techniques for tackling them are reviewed in the following sections which cover the major phases of the process.

Requirements Specification

For many years systems developers have liaised with their clients and recorded their discussions in a natural language description (e.g. English) of what the client would wish a system to perform. This specification often turns into a contractual document, defining the customer's view as to what is to be delivered.

This method of working will no doubt continue for a long time because developers and their clients need a common mode of communication. However, the use of natural language can lead to many problems for the software developers. It leads to documents which are:

(1) verbose - documentation, in order to be readable, often repeats information in order that a new point can be understood without a need to read the rest of the document. This can lead to confusion and error due to unintentional rephrasing. Maintenance becomes much more expensive and error prone due to the need to change all descriptions of any requirement which is changed;

(2) ambiguous - natural language can communicate difficult concepts very quickly, but it requires great care when it is used to define precise requirements. Unless such care is taken, the resultant documentation can be vague and ambiguous, possibly leading to a system which does not meet the client's true needs;

(3) inconsistent and incomplete - any analysis of a natural language specification of a large system requires a good memory! Since automated tools do not exist that can *understand* natural language, an analysis has to be left to manual methods. It is not surprising, therefore, that requirements specifications are often incomplete, or are even inconsistent.

One's first thoughts when reviewing these problems is that a natural language is not an appropriate means for specifying the requirements. However, this is normally the only technique that a system's developer has to communicate with their client. Hence, two specifications may be appropriate when developing high quality systems. One specification is developed using advanced techniques for specification and analysis (see below) and a second, natural language version, is derived from it. This enables the client to be provided with an understandable statement of the system's requirements, together with the confidence that it is based on a firm, analysed foundation specification. Any changes proposed will be implemented in the advanced specification technique and then reflected back into the natural language specification if they pass analysis (McDermid, 1987).

The review of the problems with natural language specifications tells us what to look for in advanced specification techniques i.e. preciseness, conciseness, machine analysability. Such techniques started appearing in the early 1970's based on a few simple mechanisms e.g. the *actions* that a system must perform and the *data flows* that pass between the actions. It was found that a large part of a system's requirements could be defined using these simple mechanisms. Specification techniques have been developed over the years, together with tools to support their use. There is still no perfect technique for use with all application types (e.g. database systems, communication systems, real-time process control) so a technique has to be chosen that is appropriate to the application. Examples of techniques currently in use include Yourdon Real-Time (Ward and Mellor, 1985), Jackson's System Development (Jackson, 1983), Specification and Description Language (SDL) (CCITT, 1983) and CORE (Systems Designers, 1981). These techniques are generally referred to as *structured* techniques due to their structuring of a system's requirements into discrete actions.

In recent years it has been realised that the structured techniques themselves are imperfect and can leave hidden errors in a system. Mathematicians realised that the only way to achieve perfect specifications is to base them on mathematical principles (McDermid,1987). Hence, they have developed notations for systems specifications which permit analysis by mathematical proof. This process still requires some mathematical skill so, until the techniques have been improved and provided with effective automated support, they are not generally applicable to most systems today. However, in high quality systems where safety is of paramount importance, there are critical elements where these techniques are already being used to benefit.

The application of techniques based on mathematical principles is generally referred to as the application of Formal Methods. Formal Methods currently in practical use include the Calculus of Communicating Systems (Milner, 1980), Z (Sufrin, 1986) and the Vienna Development Method (Jones, 1986).

Once a specification of the system requirements has been completed and analysed for internal consistency and correctness it can be put through further analysis to check for feasibility and "appropriateness" for the client. This can be done by creating a simulation or model of the requirements which can be executed. This of course takes time and requires some assumptions to be made about the design, but it has the advantage that it provides valuable feedback to the client and the designer on key system features e.g. user interface and performance bottlenecks. It can also highlight areas of incompleteness in the requirements which no amount of analysis could detect.

Design Specifications

For a long time Design Specifications were, like Requirements Specifications, recorded in a natural language (e.g. English). This led to the same problems as described above, so designers also developed new specification techniques in order to record their design. Design specifications have similar characteristics to requirements specifications, so often the same notation is used for both needs. This has the advantage that verification of a design against its requirements is simplified, but experience shows that it may also lead to a requirements specification being corrupted because the systems analysts include design within the requirements specification (this is due to it often being easier to specify a particular requirement in the form of a design solution). When this occurs designers can be misled into adopting a particular solution because it is defined as a requirement.

The design of ultra high quality systems requires a different approach to that adopted for traditional systems. After deciding on the level of reliability required (what is safe enough?), the designer must choose a combination of verification and validation techniques, together with the development of diverse (independent) designs (Avizienis, 1985) and fault tolerant control techniques (Anderson et al, 1985) in order to achieve the desired quality. The objectives for the designer being:

(1) try to avoid introducing an error into the design

(2) check thoroughly to find any errors that were introduced

(3) assume that errors were introduced and use techniques which will enable the system to continue operation after a software fault occurs (fault tolerance)

(4) do not forget to design the system such that a failure in a hardware component will not cause the software and the rest of the system to fail (Halbert, 1987).

As with requirements specifications, there is no perfect specification notation for use in designing all types of software applications. Examples include Mascot (Mascot Suppliers Association,1980), Yourdon Real-Time and SDL for use in systems where a structured technique is of benefit. An example of a Mascot Design Diagram is given in Figure 2. Formal Methods can also be used for design of systems where safety or security is of paramount concern. In order to achieve a formal proof of correctness, the same notation should be used for the design as for the requirements specification (see above) although care must be taken to avoid the problems identified above.

So often today we find that a new system requires all of its software to be developed from scratch. However, designers and their managers are beginning to realise the benefits of reusing software from previous projects. This can be especially relevant on an high quality development because reused software has often been thoroughly tested and is of known quality. Care must be taken, however, to ensure that the reuse of software does not force a complex design and checks must be made to ensure that the new use of the software does not lead to previously undetected errors.

The increasing computing power available to software developers (including the trend towards massively parallel systems) opens up new approaches for the designer to consider, e.g.

(1) simulation and rapid prototyping of high level specifications is made feasible where many of today's techniques would lead to very slow operation of the prototype

(2) higher level implementation languages can be supported

(3) the ability to handle hardware failure and to design gracefully degrading systems is simplified by having multiply redundant CPUs

(4) the computational power made available to designers will enhance the automated verification and validation they can perform on their design.

Software Code

Many languages have been used over the years for writing software code, all of which have their strengths and weaknesses. In the 1970s, the U S Department of Defence (DoD) decided they would define a new language to replace the wide variety used to develop their systems. They initiated this process partly because maintenance of their systems was expensive, and partly because they believed better languages could be developed. The result was the language Ada.

Ada is not perfect, but many software users and developers found that, for the same reasons as the DoD, they would also benefit by adopting the Ada language. Examples of the benefits that the Ada language can bring include: improved static (compiler) analysis; real-time constructs; improved software structuring constructs; re-useable code. A major repercussion of the adoption of Ada by many users as a standard is that the developers of software tools have been given a huge market for their Ada products. Hence, even more software developers are adopting Ada as a standard simply because they find that there is a better choice of tools than for any other language. The large number of users, plus the validation tests of the Department of Defence in the USA, increases the probability that compilers are very thoroughly tested and of high quality.

Ada might have many advantages for general use, but it has two major drawbacks for use by the automotive industry: its complexity and size make it difficult to analyse for formal correctness and there are few, if any, compilers for the microprocessors used in vehicles. Unfortunately, these factors and performance requirements forces many software developers to use assembler code, or a low level language and an unvalidated compiler, to develop their application. This approach, based on a "needs must" requirement, leads developers into producing software of dubious quality. The software is difficult if not impossible to verify with any confidence and the tools themselves may contain errors. With recent changes in National and European laws, engineers should view such an approach with great skepticism.

What then is the alternative if microprocessors are to be used in vehicles? Well, one solution is to say that the technology is not ready for such applications and microprocessors should not be used in vehicle control systems when safety is of concern. Another approach is to implement the software using simple, analysable languages which have certified compilers. Such languages don't have to be new; "safe" subsets exist for several commonly used languages today (e.g. SPADE Pascal). If a compiler does not exist for a particular microprocessor, then system manufacturers should collaborate together with the microprocessor supplier to develop the compiler, or simply choose another microprocessor. These solutions might appear too simplistic, but the repercussions of a failure in a safety critical system should be a motivating force for the industry to make them realistic.

What is meant by a "safe" subset of a language. A subset is chosen on the basis of our concern that certain operations are liable to erroneous use, and our current ability to reason (in a mathematical sense) about the correct operation of the implementation. As an additional benefit, this restriction facilitates our ability to verify formally the implementation against its design and requirements specifications (if all the specifications and implementation have a consistent basis). Tools are now available (e.g. Bramson, 1984) which that will automatically analyse source code (e.g. in Pascal or Coral) for, inter alia:

control flow- identifying unreachable code or code from which there are no exits

data flow- identifying undefined variables, unused variable definitions

information flow- identifying use of undefined variables in constructing export data, ineffective imported data

Such analysis is almost indispensable in systems where quality is critical because they have been shown to be capable of detecting many types of obscure errors at the code level.

In systems where it is not cost effective to limit the staff to what is usually a very small safe subset of a language, it is usually prudent to at least restrict the programmers from using some of the more dangerous or complex features of the chosen language (e.g. untyped pointers, dynamic storage allocation). Also, the use of interrupts presents serious difficulties with current techniques for static analysis, makes software more complex and is difficult to test dynamically in a systematic way. Hence, they are usually avoided in high quality applications.

Some designers have chosen not to use any high order languages in high quality systems because the compilers required are large and complex, and hence may not always create object code correctly. If they choose not to use a subset of a high order language (which requires a small, simple compiler) then they must use manual techniques to create the assembly code. This is notoriously laborious and error-prone and, even though the resultant code can be manually or automatically analysed, is unlikely to lead to better quality systems. Also, the use of a compiler for a high order language (or a subset) will lead to consistent results over a long period of time and, with many users of this language worldwide, any faults in the compiler will be identified leading to improved quality software.

Software Building and Testing

The applications software will be controlled by an underlying operating system. Operating systems today are very complex and ill understood so their use in ultra high quality systems require special care. Usually, they are reduced to the minimum functionality required (e.g. the use of direct addressing for all memory with no paging or need for base registers; see also restrictions on implementation language above) and the key elements are included in a "kernel" which is subjected to the most thorough analysis. Such operating systems should be used on microprocessors which have been designed with safety in mind rather than the more powerful, but complex processors that are widely used for more general applications. A processor that has been specially developed for this purpose is the Viper Processor (Halbert, 1987).

In a high quality system more time is spent in the earlier phases of software development to stop errors being introduced into the software because it is much more expensive to locate and remove them later. Nevertheless, it is still necessary to build and test systems to ensure that nothing has been overlooked. The use of some high order languages (e.g. Ada) as an implementation language assists the building process because more checks can be performed by the system builder than could be performed with other languages e.g. interfaces are more thoroughly specified.

Execution of the software to verify it will meet the client's needs will always be necessary. However, it can be demonstrated that a typical system would take hundreds of years of computer time to test, if all paths through the code were to be exercised. This leads to the question, "if we cannot test all of the paths through the code, then which ones should we test?". Unfortunately, there is no general purpose answer to that question, so the system developers have to make careful analysis of their system in order to ensure that they are using effective tests. This problem can be reduced by the use of system test coverage analysers (Hennell et al,1983). These are tools which monitor the execution of tests and provide the test team with reports on what has been tested and, more importantly, what has not been tested. This enables a test team to improve the thoroughness of testing and to ensure that key features of the system receive greater attention.

The operation of a system can be very hard to monitor, which leads to problems if the system has to be tuned in order to increase its performance. Performance analysis tools are not generally available due to the wide variety of features that a system can have.

Since the software can not be tested exhaustively, there is still a possibility that an error still remains in the software when it is released. In order to assist managers to decide when software can be released several software "Reliability Models" have been developed (e.g. Abdel Ghaly, 1986). Since there is no universally best model, users must take great care when choosing a model and using it to predict reliability.

Project Management

The sections above have shown how the use of advanced methods and tools can improve product quality. But what is there to help the project manager?

A project manager's major role on a project is to direct and coordinate a team of people in order to produce a high quality product on schedule, within budget. The project manager must therefore be able to plan the development, monitor progress and take effective action if, and when, problems occur. Monitoring has traditionally been achieved by informal reports that project tasks have been completed and milestones have been achieved, but experience has shown that this technique is ineffective because staff have no hard measure of achievement.

The use of structured notations for recording specifications, together with tools that analyse them, enables this situation to be changed. Staff can be given concrete targets to achieve in, say, the amount of specification to be performed. Achievement of a target can be accurately monitored by defining an analysis that has to be performed by one of the support tools.

This approach gives a project manager true visibility of progress, but there is still another step to be taken. That is to integrate the set of tools used by the technical staff with that of the project manager i.e. the actions and results of each tool in the toolset must be communicated automatically between all tools. This is typically achieved by requiring that all tools update a central database with the results of their actions. In an ultra high quality software development it is critical that this database supports Configuration Management so that changes to the system can be carefully monitored and controlled (Wilson and Youll, 1986).

In the early 1980s there were attempts to create integrated toolsets by taking existing tools and patching them to a central database. This approach turned out to be very cumbersome, and sometimes impossible if a tool supplier would not provide an interface to their tool. Purpose built Integrated Project Support Environments (IPSE) now exist to overcome this problem (e.g. ISTAR (Dowsan, 1986), see Figure 3).

An IPSE provides most of the tools a software developer requires and a standard interface which can be used if more are required. The toolset includes tools for estimating, planning, resource scheduling, progress monitoring, quality control, configuration management, requirements specification, design specification, coding and system building. A central database is provided to store information provided by the tools, together with communications facilities that will support a distributed project team.

Conclusions

Some of the key problems facing developers of high quality software have been discussed, together with the advanced techniques that can be adopted today in order to overcome them. However, the adoption of new techniques can only be successfully achieved by thorough preparation. In some areas there are no off-the-shelf solutions to the problems so a Company will benefit from the support of a Research and Development team in order to ensure that effective practices are being used throughout the development process.

References

Abdel Ghaly, A. A., Chan, P. Y., Littlewood, B. (1986) Evaluation of Competing Software Reliability Predictions. *Transactions on Software Engineering*, 12, no 9, IEEE.

Anderson, T. et al. (1985) Software Fault Tolerance: An evaluation. *Transactions on Software Engineering*, 11, no 12, IEEE.

Avizienis, A. (1985) The N version approach to Fault Tolerant Software. *Transactions on Software Engineering*, 11, no 12, IEEE.

Bramson, B D. (1984) Malvern's Program Analysers, *RSRE Research Report*

CCITT Z100-105 (1983) "SDL"

Dowson, M. (1986) ISTAR- An Integrated Project Support Environment. *Second SIGPLAN/SIGSOFT Symposium on Practical Software Development Environments*

Halbert, M P. (1987) A self checking computer module based on the Viper microprocessor- A building block for reliable systems. *Conference Proceedings on Microprocessor Based Protection Systems*. Institute of Measurement and Control.

Hennell, M. A., Hedley, D., Riddell, I. J. (1983) The LDRA Testbeds: Their Roles and Capabilities. SOFT-SAIR, Arlington, Virginia.

Jackson, M. A. (1983) *System Development*. Prentice Hall.

Jones, C. B. (1986) *Systematic Software Development using VDM*. Prentice Hall.

Mascot Suppliers Association (1980) *Official Definition of MASCOT*, UK

McDermid, J. (1987) The role of Formal Methods in Software Development, *JIT* 2,3.

Milner, A. J. R. G. (1980) A Calculus of Communicating Systems. *Lecture Notes in computer Science*, 92, Springer Verlag.

Suffrin, B. (1986) *Z Handbook*, Oxford University Computing Laboratory

Systems Designers Ltd. (1981), CORE , Seminar Manual

Ward, P. T. and Mellor, S. J. (1985) *Structured Development for Real-Time Systems*, Yourdon Press.

Wilson, G. and Youll D. P. (1986) The Implications of Ada for Configuration Management. In *Proceedings of the Ada Europe Conference*, Cambridge University Press

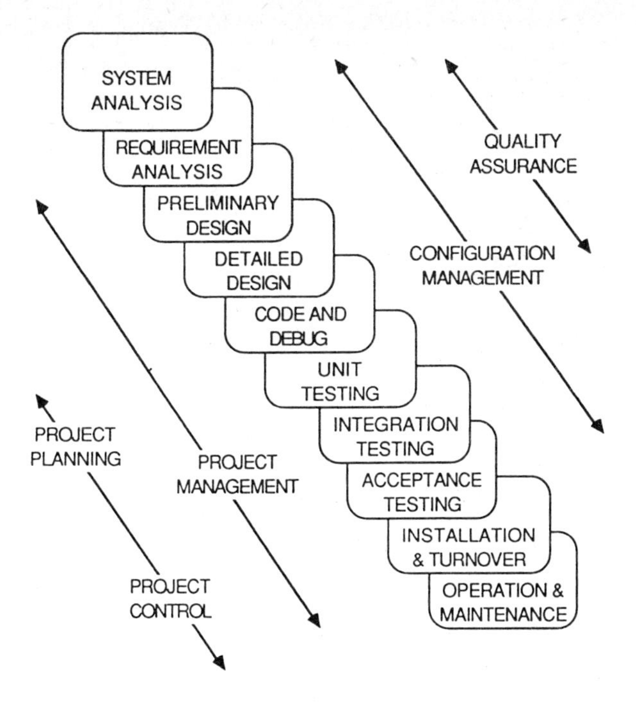

Fig 1 Software development process

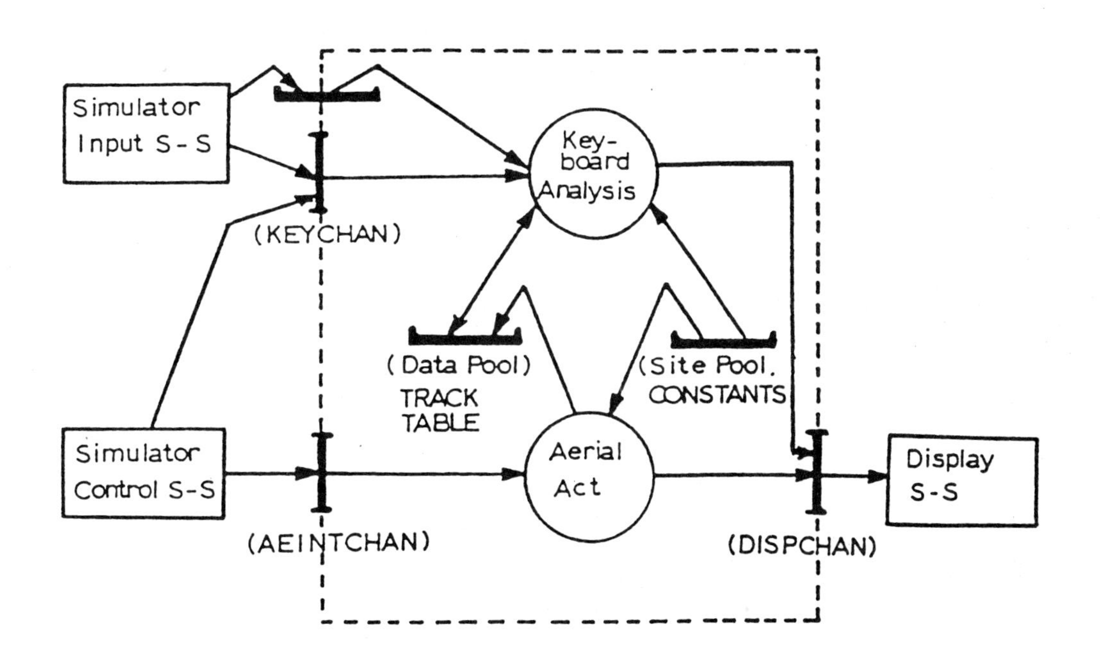

Fig 2 A Mascot design diagram

ISTAR FRAMEWORK

ISTAR TOOLSET

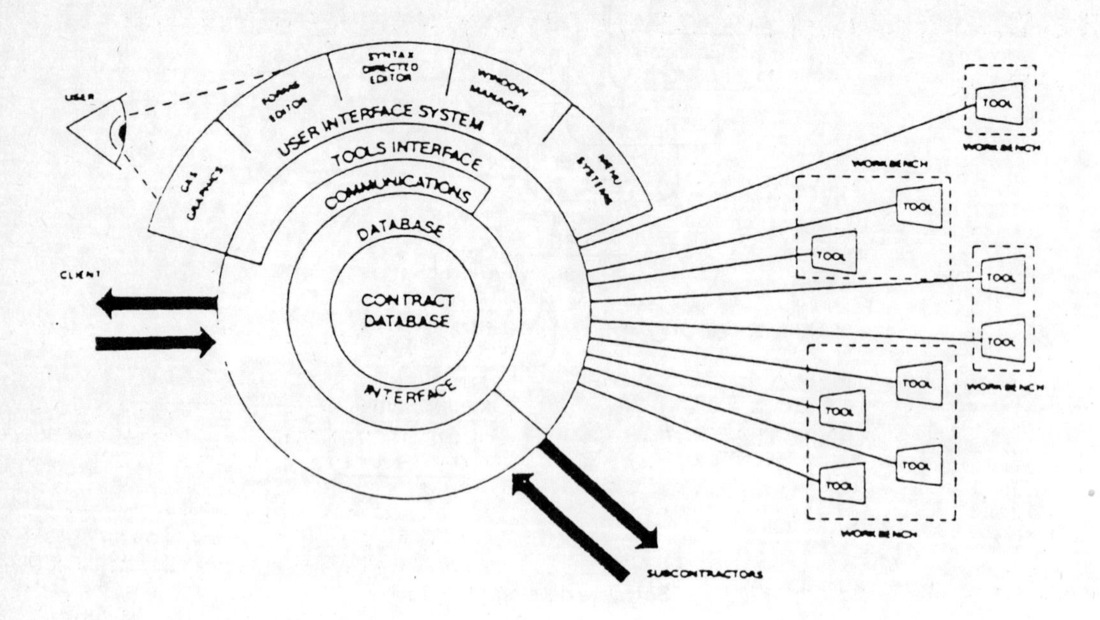

Fig 3 Istar

C391/093

Considerations about the use of surface mount technology in automotive electronic applications

L CERGEL
Motorola Inc – European Semiconductor Group, Geneva, Switzerland

The increased complexity of electronic systems in automotive applications makes the use of Surface Mount Technology necessary. Since the automotive electronics is characterized by the very high reliability requirements and at the same time has to be cheap in high volume production, it is necessary to study the implications of Surface Mount Technology on the overall reliability of the system. Surface Mount Technology imposes severe strains on plastic leaded chip carriers. Susceptibility to package cracking was observed in plastic packages after vapour phase or infrared solder reflow processes. The loss of package integrity is a potential concern for all surface mount components.
Recent investigations into the effect of surface mounting using vapor phase, IR and wave solder of plastic IC packages have led to different reliability requirements for surface mount packages when compared to through-hole mounting processes. Some of these requirements for the new surface mount packages, such as SOIC's, SOJ, PLCC's,
QFP's and PQFP's, are defined.
The solder joint provides the electrical, thermal and mechanical functions so its reliability is of major importance for correct functioning of the electrical systems. Flex Test method as specified by IEEE is presented for the solder joint reliability investigation and some results are listed.
Future development in packaging materials and Multichip Modules are mentioned which will have applications in automotive electronics. Since the solder joint and the package lead is the weakest point in the electronic systems their elimination by the extensive use of ASIC's will lead to the overall reliability improvement.

INTRODUCTION

The continous increase of integrated circuits in the car electronics, which has been recorded in the past will not only continue but will accelerate in the future. According to Prometeus study (1), the automobile of 1995 will have in the order of 100 intelligent sensors, 80 intelligent actuators, 45 motors, 5 displays, 2 imagers and 1000 integrated circuits of 1987 complexity, (2-2.5µ technology). It is believed that there will be further increase of the electronic complexity in the cars by the year 2000. This increase in electronic value in cars will go from 4 - 5% of today to 15 - 20 % in the year 2000. There will not only be a simple increase of the electronic contents in the car, but the electronic system in automotive applications will also change from the autonomous systems of today to a more integrated approach, where many electronic systems send and receive information from or to other systems. This situation will require higher operating frequencies and the use of multiplexed wiring.

It is fully understood today that the implementation of the electronic systems for the future cars, of such complexity as mentioned above, can be done only by means of Surface Mount Technology. In order to produce the systems of required reliability it is very important to study the implication of Surface Mount Technology on the overall system reliability.

The mounting of plastic IC packages directly onto the surface of the printed wiring boards has created different reliability requirements for this new technology. The Surface Mount process exposes the component to significantly higher stresses because the whole component is heated, often rapidly, to solder melting temperatures. These stresses open up a whole new area that needs to be explored to ensure the long-term reliability of components. The entire package is subject to higher temperatures than the through hole mounting processes. The nominal surface mount soldering process temperatures are 220 °C, for vapour phase, IR is slightly higher and wave soldering is apprx. 260 °C. (Figure 1).

Exposure to the temperatures and times associated with these soldering processes have be seen to degrade the moisture resistance of the package and in some cases causing the bulk of the plastic to crack. Sometimes the crack propagates to the outside of the package. This has resulted in modifications to the package designs, materials and development of mold compounds that can withstand the surface mount processes. The loss of package integrity is a potential concern for all surface mount components. It presents a challenge to component and material designers in order that surface mount technology can reach its full potential.

To investigate the influence of Temperature Cycling on Solder Joint reliability is of major importance for the automotive electronic applications. But it is a lengthy process. The IEEE proposed Flex Test for accelerated Solder Joint Reliability evaluation. It is believed that the acceleration is by factor 10, in relation to standard thermocycling test. This test is described with some results obtained with this method are presented.

"J" leaded PLCC packages have been registered in JEDEC and accepted as standard surface mount packages in recent years throughout electronic industries. One of the key features with this lead form is the leads coplanarity. The coplanarity was found to have a direct impact on board mounting yields as well as the reliability of lead solder joints during the thermo cycling of the system. Especially at higher lead counts this becomes a very important yield factor, so there is a great deal of investigation going into means of coplanarity improvement. One possible way might be to replace the J lead shape with Butt lead shape or Gull Wing shape, and introduce the concept of Molded Carrier Ring.

RELIABILITY REQUIREMENTS OF AUTOMOTIVE ELECTRONICS

The reliability demands on electronic systems in future automotive applications automatically set-up the reliability conditions for the electronic elements and the entire electronic system. Some electronic systems will be not allowed to fail during the life time of the car, or will not be allowed to determine the lifetime of the car. This means that the lifetime of those systems has to be longer than that of the mechanical parts. Today the "Lifetime of a car" is understood to be (2):

- 15 - 17 years of operation or 300.000 km.
- During this time there will be 12-13.000 temperature cycles between low and high temperature. Low can be -30 degC to +20 degC, and high +90 degC to +100 degC.
- Non-operation for 150.000 hours in humid atmosphere.
- Humid and salty atmosphere during an operating time of about 5000 hours.
- High vibration levels during the operation.

The reliability requirements as mentioned above are similar to aerospace electronics. But the very high reliability of automotive electronics has to be obtained at a very low price in high volume production.

In order to satisfy the reliability requirements for automotive electronic applications it is neseccary to study not only the reliability of the semiconductor elements itself, but to examine the overall production and design activities and modify them in such a way that the reliability required will be obtained. E.g.: What will be the effect of a particular SMT soldering method on the selected IC's? What will be the effect of the selected solder paste and IC's package leads type in thermo cycling test? Which type of the package leads would suite best to thermo cycling ?

PACKAGE CRACKING - TESTING AND PACKAGE PERFORMANCE

The performance level of semiconductor packaging is becoming more demanding. The need for high density board mounting has driven a trend toward thin surface mount components, along with the higher integration of semiconductor chips resulting in large die. This has made the packages more sensitive to failure due to thermal stress.

The surface mounted devices must withstand a high temperature in the board mounting process. At the board mounting process the entire semiconductor device is exposed to temperatures of 210 to 260 °C, while the PDIP mounting process, only the leads of the device are exposed to the molten solder temperature (260 °C).

After the initial discovery of the package cracking problem it has been quickly determined that the problem is an "Industry Problem"; not a problem of a particular vendor. The reason why it came as a suprise after the "Package Cracking" had been discovered was that the Plastic Packages have been often qualified for surface mount processes based upon the same type of reliability testing used for through-hole mountable components. The SMT processes expose the component to significantly higher stresses because the whole component is heated to solder melting temperatures. That is why this problem is present only when infrared, vapour phase, hot air ovens or hot plates soldering methods are used. HAND SOLDERING, HOT AIR PENCILS, and IRON STAMPING DO NOT CAUSE ANY problem because they localise the heat to the leads only. But these soldering methods do not appear as attractive production solutions at present.

Figure 2, shows a Cross-Section of a cracked package. Usually the origin of the crack lies at the die attach flag and is directed to the bottom of the package. The crack presents a major cause for enviromental contamination, e.g. moisture, to travel directly into the immediate vicinity of the silicon. It also appears that the cracks propagate from the high stress area caused by the sharp edge of the lead frame. Externally the cracks appear to be slightly bulged, basically circular bubbles centered on the lower side of the package directly under the die flag. They are not easily visible. External inspection of the package may reveal cracks, generally on the back of the package. Cross sectioning may reveal internal cracks both from the bottom of the flag edge and from the upper corners of the die. In some cases with very large die in small packages, cracks may appear on the mold line. A Scanning Laser Acoustical Microscope (SLAM) is another method to determine the presence of package cracks. The crack will grow larger after thermal cycle stressing and can extend completely across the package.

Fukuzawa et al (3) first proposed that the mechanism of cracking was due to moisture absorption in the package. Moisture condenses between the plastic and bottom side of the flag. During the reflow process, the moisture is converted to steam, which exerts high preasure on the bottom plastic. As a result of the high steam preasure, cracking of the bottom plastic was observed.

In order to examine the packages to what extend they are succeptible to damage during the production processes and at the same time to have a tool for packages evaluation during the development stage, a test standard has been developed. This test standard is called "PRECONDITIONING" and simulates high risk cracking conditions. It has been developed in close colaboration with Motorola major customers of different fields of IC's applications. This test involves saturating the surface mount package with moisture to simulate worst case storage conditions and then simulating the surface mount process to test package integrity. MOTOROLA proposed standard is showen in Figure 3.

The package crack test has been performed after the packages have been soaked. They were subjected to heat cycles similar to the vapor phase surface mount process, but without the preheat. The full sequence for the test is:

* Dry package in oven at 150 degC for 16 hours
* Weigh and record the dry weight
* Place in 85 degC/85 % RH for 144 hours
* Weigh and record the saturated weight
* Subject the package to Vapour Phase heat for times within one hour
  after removing from the 85 degC/ 85 % RH soak
* Inspect for cracks

Using the above described test method it has been found that the susceptibility of a package to cracking depends on:

- Reflow process temperature
- Amount of moisture absorbed
- Flag size
- Plastic thickness
- Lead frame design (stress concentration factor)
- Strength of plastic (molding conditions)

It has been generally understood that the cracking is due to the decrease of the mechanical integrity of the package. The development of packages for surface mount goes towards smaller, more dense, and thinner packages. While the die sizes become larger. Since the moisture content in the plastic is considered as a most important factor, the investigations were directed on moisture characteristics of plastics used in packages. The plastic is actually a single component epoxy which is highly filled, typically with granular silica, often about 80 % of the total volume. This fill rate is used to limit the thermal expansion characteristics of the end material, to improve strength and to assist in the moldability. The apparent critical parameter is the percentage of water content by weight. Figure 4 shows the moisture saturation test of different packages, and Figure 5 showes the moisture loss during bake. The results from our tests for absorption and desorption are equivalent to the values published by different authors, e.g. (4). It has been proven that the component will crack if it was last in a high humidity environment for a significant length of time or was in normal ambient for a longer time.

Another very important conclusion is that the component will not crack if it was last baked for a sufficiently long period of time. We believe that the 0.1 % of humidity content is a sufficiently low value to prevent the cracking of the component.

PACKAGE PRECONDITIONING AND AUTOCLAVE PERFORMANCE

This test is done in a similar manner to the package crack test but followed by an autoclave test. The autoclave test is adding preasure to humidity test. Although preasure testing under extremes of temperature and humidity is not directly comparable to test conducted at the standard 85°C/85% Relative Humidity (RH) conditions, the results are speed-up by factor of six to seven. (E.g. Preasure testing can slash the standard 1,000 hour test time to less than a week - a significant benefit for production activity). Autoclave testing procedure is suitable for compating different device lots. By comparison to standard conditions, autoclave testing usually is done at 125°C at 85% RH or 110°C at 90% RH.

Purpose of Package Preconditioning and Autoclave Performance is to find out in what way the package construction and materials used might affect IC's and their functioning. Usually in this test the focus is to examine what is the effect of different molded compounds on the moisture resistance and corrosion resistance of the package, passivation crack and damages cause by the filler edges at molding process. The factors affecting performance are:

* Package dimensions
* Lead frame metal to plastic ratio
* Die size
* Material used for assembly: Mold compound, Die attach
* Passivation type
* Environmental conditions prior to board mounting
* Temperature extremes during board mounting

The flow chart of this test is depicted in Figure 3, and Figure 6 shows that the metal to plastic ratio is becoming a very important factor in the modern IC's design, e.g. 1 Mega Bit DRAM. In the Metal To Plastic Ratio Experiment the objective was to determine the effect of various large die sizes in the SOJ package with respect to temperature cycle, shock and vapour phase testing. Figure 7, shows the places of the usual defect positions on the 28 Lead SOJ package. The results of the Temperature Cycle Test as a function of different Metal to Package values showed that at higher values of Metal/Package ratio the cracks which appears were only the ckracks in the 'Pocket" area. The reason of this failure is that the pocket is located below the edge of the flag and consequently makes the package thicker. This result lead to the modification of the package. (The pocet areas has been removed). This modification improved package reliability in such a way that it pases the prescribed tests.

The improvement of the package design is strongly connected with the molding material improvements. In Figure 8, there is shown the comparison between the "OLD" and "NEW" molding materials. New molding materials show quite different particle structure when compared to the "OLD" one. Their particles are considerably smaller and of different shape. The prevailing shape of particle is a spherical shape. This enables this material to a more flexible and consequently less prone to the crack failures. During our tests with the "NEW" molding compounds we found considerable improvements of the pacakge performance during the autoclave test of the surface mount devices.

METHODS OF REDUCING PACKAGE FAILURES DURING SURFACE MOUNTING.

There are different ways for package failures reduction. Four of them are mentioned here:

1).Mold compond improvement: Low stress mold compounds have been proven to have less cracking than standard mold compounds. Parts which fail with old mold materials do not fail when using low stress materials.

2.)Decreasing stress initiation points: Cross section photos of cracked PLCC parts showed cracks initiating at leadframe burrs. Burrs which are created in the leadframe stamping process can be stress concentration points. Cracks propagate from the burr through the thinnest plastic region. Therefore reversing the stamping direction to cause burrs at the top of the leadframe should reduce the amount of cracking. MOTOROLA evaluation using the PLCC packages showed that the above theory is correct. MOTOROLA is specifying burr side up on new packages.

3.) Improved diebond process: Package cracks sometime extend from die bond voids at die edges. Therefore void-free die bonding process is crucial to preventing failures.

4.) Reduce moisture content in the package: Insuring that the parts are mounted dry can be done by backing the parts immediately before the board mounting process, or by shipping dried parts in a sealed moisture barrier bag with desiccant.

SOLDER JOINTS AND LEAD TYPES RELIABILITY EVALUATION

Solder, as a bonding material between metallic surfaces of an assembly, provides electrical, thermal and mechanical functions. It is thus apparent that the integrity of solder joints is vital to the overall function of the assembly. The basic failure processes in metals and alloys, such as creep, mechanical fatigue, thermal fatigue, corrosion-enhanced fatigue, intermetalic compound formation, detrimental microstructure development, and joint void formation has to be examined during the production process development and evaluation.

Solder joints reliability in surface mount technology is strongly dependent on the coplanarity of the package leads and the planarity of the board. MOTOROLA specifies the coplanarity of the leads within ±50 μm from a given plane. PCB manufacturers specify the planarity of the bonds apprx. 1% at room temperature. When the board is heated to soldering temperature the bowing of the board is usually greater than 1%. If in the position where the board is bowed, there will be placed a lead with more that 50 μm lead deviation of coplanarity, the resulting solder joint might be of a small thickness, and when cooling down a constant force will be builded up. This force will force the joint to creep which at the end will lead to a degradation of an electrical contact. Since it is not easy to preserve the ±50μm specification of coplanarity deviations during the package testing shipment, MOTOROLA INC. together with NATIONAL SEMICONDUCTOR Corp. developed and obtained the JEDEC specification for the "Molded Carrier Ring Concept" for high leadcount devices, see Figure 9.
The Molded Carrier Ring Concept attempts to meet the need

of reducing the incidence of bent leads and coplanarity rejects with additional benefits of easy standarization for customers and equipment manufacturers, which is very difficult to achieve with a large number of "Fine Pitch Plastic Devices" which are available today on the market.

The Ring holds the thin, fine pitch leads of the package in place to protect them from handling damage during testing, burn-in, and shipping. Then just prior to placing the part in position on the PC board, the leads are cut, formed and the unit is removed from the ring as final step. The opportunity for lead damage is reduced, and consequently the solder joint misformation is prevented.

When a Surface Mountable Package is soldered to a PWB and the assembly is thermal cycled, complex stresses and strains are generated in the assembly. The strains created in the solder can cause cracks and can eventually caused electrical failure during such cycling. The thermal cycling is done usually in 1 hour cycle; 20 min. low temperature; 20 min. transient: and 20 min. high temperature.

In ordeder to accelerate the thermal cycling testing, the IEEE specified "Flex Test Method" for surface mountable devices. The Flex Test has been developed to simulate the testing of solder fatigue failures experienced in thermal cycling, and for study of lead forms. The devices are placed on the Flex Test Board, see Figure 10. The board is forced to flex arround curved mandlers, taking an arc of 40.1 inches of radius. The leads flex during each cycle, exerting force against the solder joint. This induces creep in the solder to relieve the stress. The flex fixture then reverses the direction of flex, forcing the board to take a 40.1 inch radius in the opposite direction, thus both possitive and negative stresses are applied to the solder connection. The standard flex test cycle is six minutes, three minutes positive and three minutes negative. This provides 240 cycles per 24 hour period. Which is 10 times faster then using a real thermal cycling test. Testing is done in an oven at a constatnt 60 °C to increase the solder creep rate, which accelerates the failure rate.

Daisy chained packages are mounted on matching daisy chained boards. The leads of each package are divided into two resistive nets. For instance a 68 lead PLCC is devided into two nets each contains 34 solder joints (17 on the side parallel to the bend axis and 17 on the side perpendicular to the axis). Thus the package is divided into two equal parts with the same forces coming into play for each net. Each net is connected to a separate channel of the Anatech Event Detector, which constantly monitors resitance increases. A resistance of >1,000 ohms for a duration of >0.02 microseconds is recorded as an "event". Figure 11 presents a test of process variation comparison where different solders were evaluated using the 60 mils flex board and assembled with 68 PLCC, J leaded packages. The comparison of different packages type in the flex test performance is shown in Figure 12. The failure mode for the Butt and J lead 68 pin PLCC was the same for both the thick and thin flex boards, broken solder joints. The PQFP and CQFP packages, both having Gull-Wing shape leads, failure mode was found to be broken leads. Many solder joints were badly cracked, several completely broken, but there is evidence that the first failure to occur is a broken lead followed by broken solder joints if flexing continues for some time after leads break. The 100 pin PQFP is a low profile package with leads only 0.055" long, compared to 0.100" for the 132 pin PQFP. One would expect it to be less compliant, but the package is smaller (only 0.750" square compared to 0.950" for the 132 pin CQFP) and it is assumed that the smaller foot print is the dominant factor providing longer life. Type of lead material has influence on the solder joint reliability as well, Alloy 42 is better when comparing to Copper, but copper has twice as good thermal conductivity. The improvement can be probably found in using - aluminium fiber strenghen copper lead material which has the thermoconductivity of copper and the strengh similar to Alloy 42.

MOTOROLA is undertaking a temperature cycle test program which is hoped will provide correlation to the flex test data. The fail mode of the PQFP and CQFP packages being broken leads instead of broken solder joints, says that we now have leads that are more fragile than the solder joint. If we consider the 68 pin PLCC an acceptable industry standard, the 132 pin PQFP, which is the same size as the 68 pin PLCC, took two times longer to start failing and 2.6 times longer to reach 50% failure as compared to 68 pin PLCC. With a little effort the lead material can be modified to provide longer life.

FUTURE PACKAGING TRENDS FOR AUTOMOTIVE APPLICATIONS

Increased circuit complexity, space limitations, and very strict requirements on future automotive electronics will be a driving force for the packaging development. In Figure 13 .is depicted a high density package progression for the future years. The development is heading towards a "Multichip Modules" solutions.

Since the solder joint is very week point in the system, it should be avoided as much as possible. One possible way is to use ASIC as much as possible, while the silicon technology is considerably more reliable than the SMT.

The general trend in semiconductor technology is towards "Hermetic Chip". Here the protective role of the package will be transfered to the chip level. This will require improvements in chip passovation methods, e.g. double passivation, and having not exposed interconnect such as TAB. Progress in Plastic Packages depends on Mold Componds development and Thermal Dissipation enhancement technologies. The development in Mold Compond is directed toward low stress molding componds. The mechanical stress in large complex packages results mainly from the mismatch in thermal expansion rates between the silicon chip, leadframe, and the plastic encapsulation. The attempt is to match the coefficient of thermal expantion of the plastic with that of leadframe. Thermal enhancement technology in plastic packages is takled by introduction of heat spreaders into the package and metal slugs.

The high complexity and high reliability requirements on future vehicle electronics applications will force designers to use more ASIC circuits to a large extend. This will make the transfer from the PCB utilization to "Multichip Modules" more economical. The Multichip Modules will be used in Motor Electronic for better thermal system design, e.g. Ignition, Injection. In Drive Electronic for improved packaging density, modularity and reliability, e.g. ABS, Transmission Control. In fact they are already used today in cars al in Car Radios and Car Telephones, see Figure 14, and 15.

ACKNOWLEDGEMENTS

We would like to thank Glenn Dody, Ruth Reinhardt, Joan Hamilton, Jim Casto and the others from Advanced Packaging Development Pilot Line staff for their help with the information contained in this report.

REFERENCES

1. Prometheus, "Pro-Chip White Book", April, 1988
2. H.Danielsson:"Will ASIC Technology Demand a New Interconnection Technology instead of Soldering in Automotive Electronics?", Hybrid Circuits Number 19, May 1989, pp32-37

3. I.Fukuzawa,S.Ishiguro, and S.Naubu,"Moisture resistance degradation of plastic LSI's by reflow soldering," in Proc. 1985 IEEE IRPS, pp. 192-197
4. T.O.Steiner,D.Suhl:"Investigations of Large PLCC Package Cracking During Surface Mount Exposure" IEEE Trans. on CHMT-10 No.2, June 1987
5. R.Lin,E.Blakshear,G.May,G.Hamilton,and D.Kirby:"Control of package cracking in plastic surface mount devices during solder reflow process", IEEE/CHMT proceedings 1987.
6. G. Dody,P.Lin:"PLCC coplanarity measurement and improvements" IEPS Meeting, Nov.1986, San Diego
7. W.D.Smith:"The effect of lead coplanarity on PLCC solder joint strength", Surface mount technology, June 1986, pp. 13-17

ABREVIATIONS

Alloy42 Lead Frame Material composed of Ni/Fe, 42% of Ni
CLCC Ceramic Leaded Chip Carrier Package
CQFP Ceramic Quad Flat Pack Package
EIAJ Electronic Industries Association of Japan
I R Infra Red Radiation
JEDEC Joint Electronic Device Engineering Council
PDIP Plastic Dual in Line Package
SMT Surface Mount Technology
SOIC Small Outline Integrated Circuits Package
SOJ Small Outline J Lead Type Package

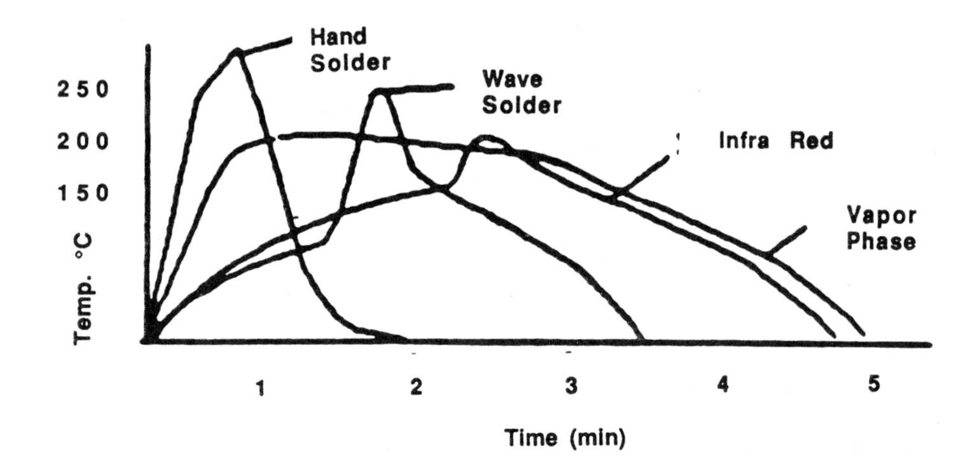

Fig 1 Different soldering methods and their time temperature characteristics

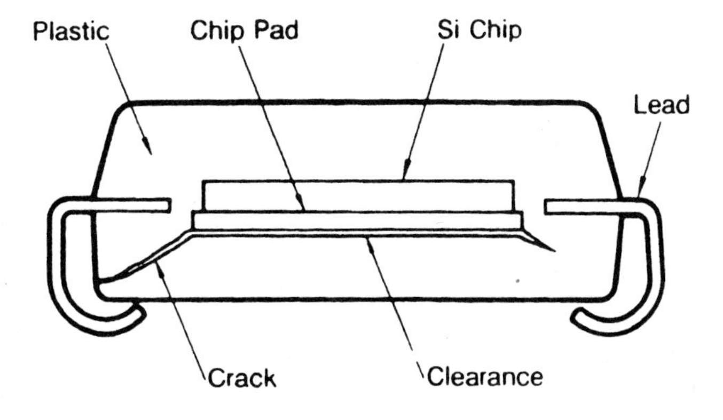

Fig 2 Cross-sectional view of cracked package

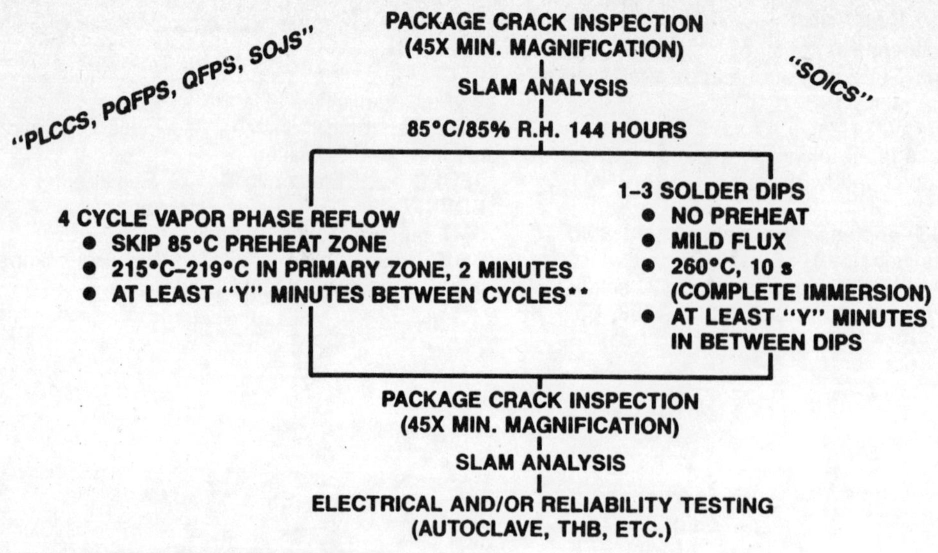

Fig 3 Flowchart – preconditioning/autoclave test for surface mount packages evaluation

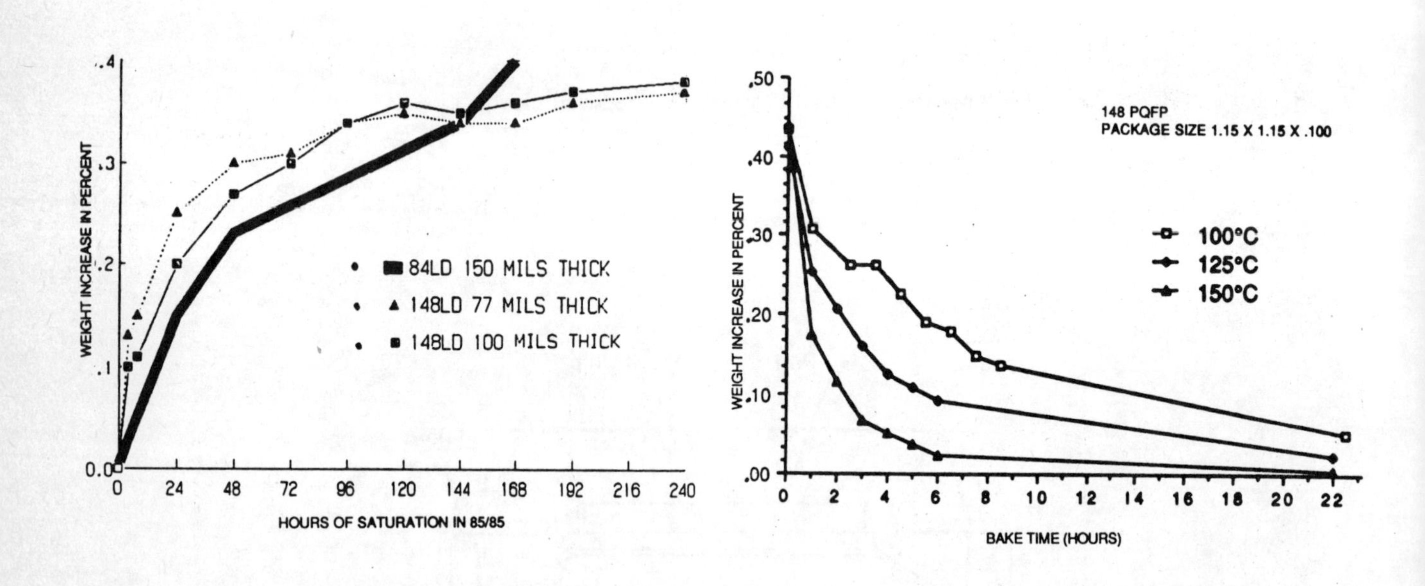

Fig 4 Moisture saturation test for 84 LD 50 mil PLCC and 148 LD PQFP 25 mil packages

Fig 5 Moisture loss during bake for the 148 lead FPPLCC

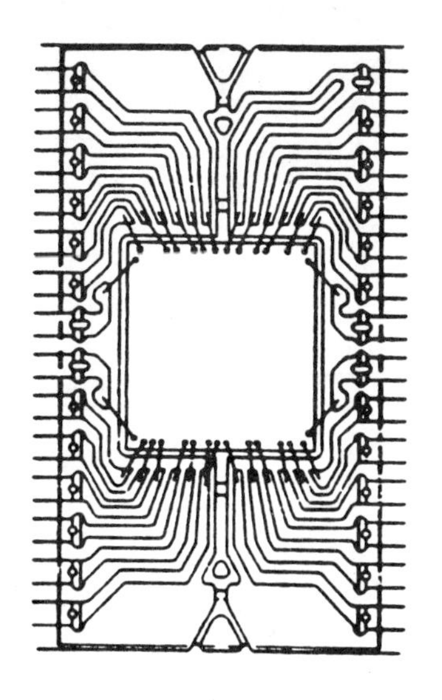

8K X 8 SRAM 28 - 400 SOJ
DIE SIZE .237 X 253
FLAG SIZE .255 X .270
PLASTIC BODY SIZE .400 X .725
METAL TO PLASTIC RATIO .484
DIE TO PLASTIC RATIO .206

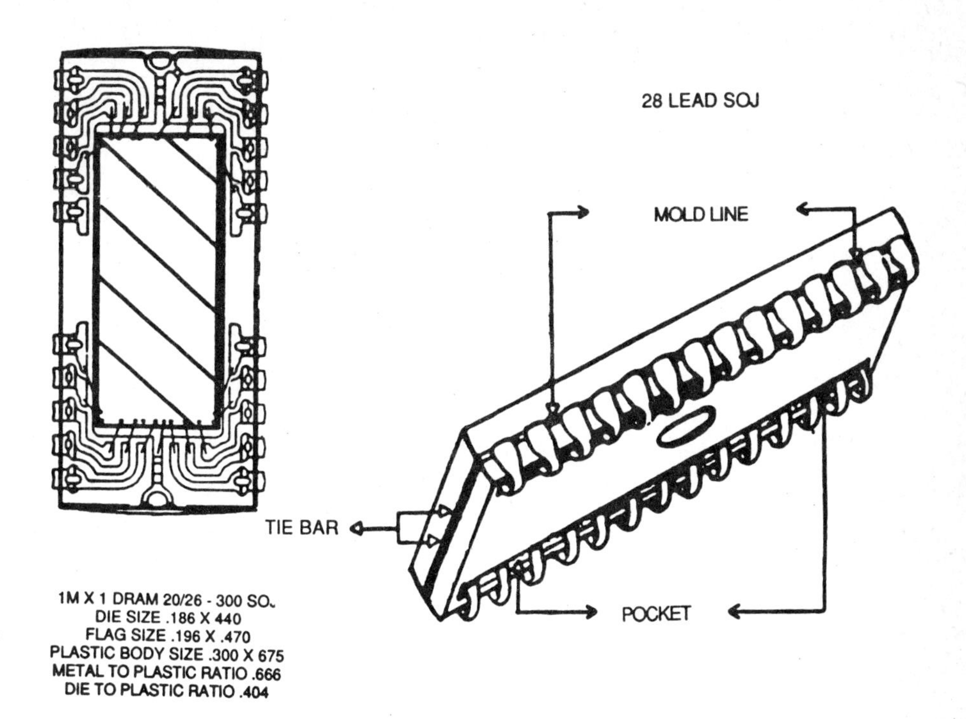

Fig 6 Package designs for 8 K X 8 SRAM, and 1M x 1 DRAM

Fig 7 Possible failure locations on the SOJ package

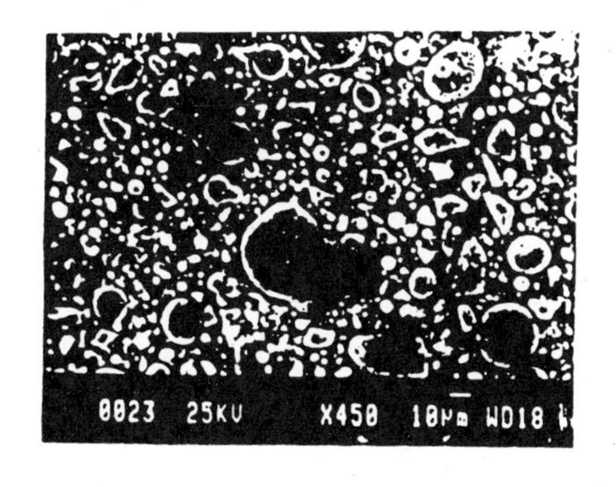

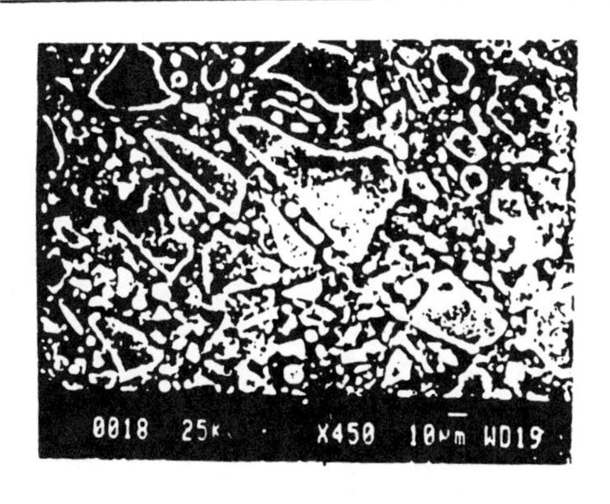

Fig 8 Comparison between the 'NEW' (upper part) and 'OLD' (lower part) molding compounds

0.050

0.025

Fig 9 Molded carrier ring design features bottom view

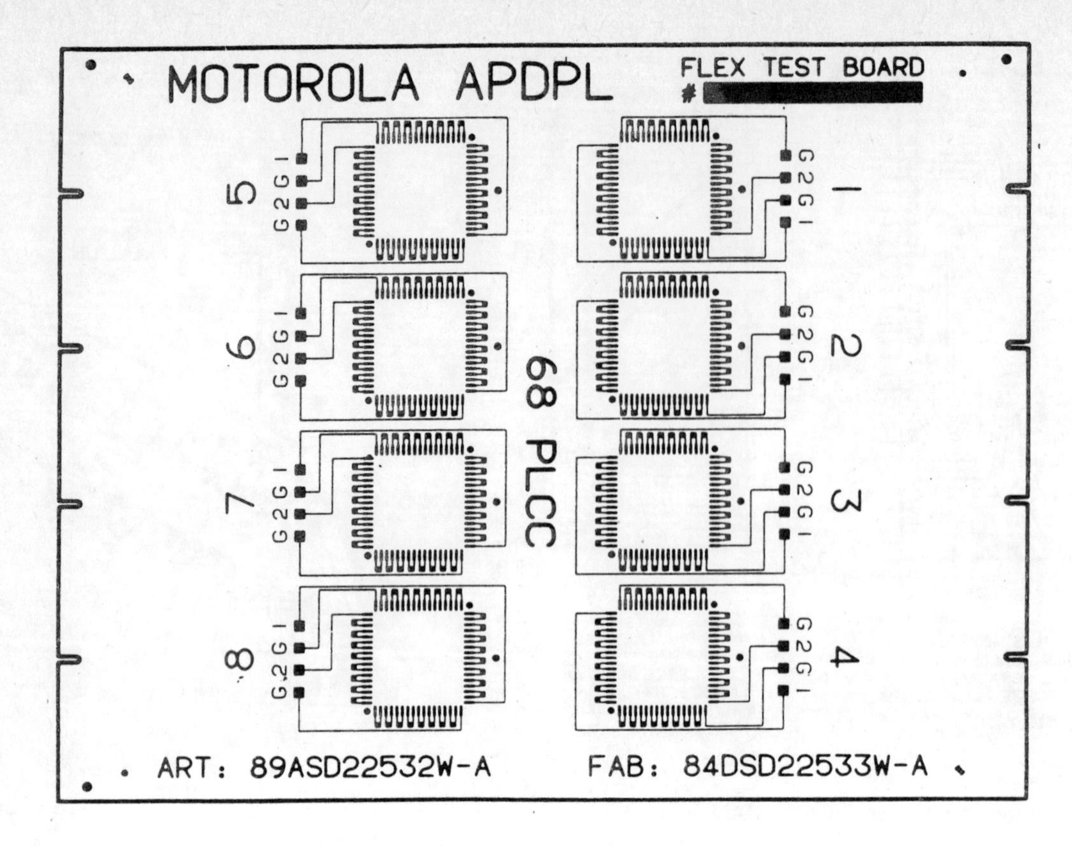

Fig 10 Flex test board layout

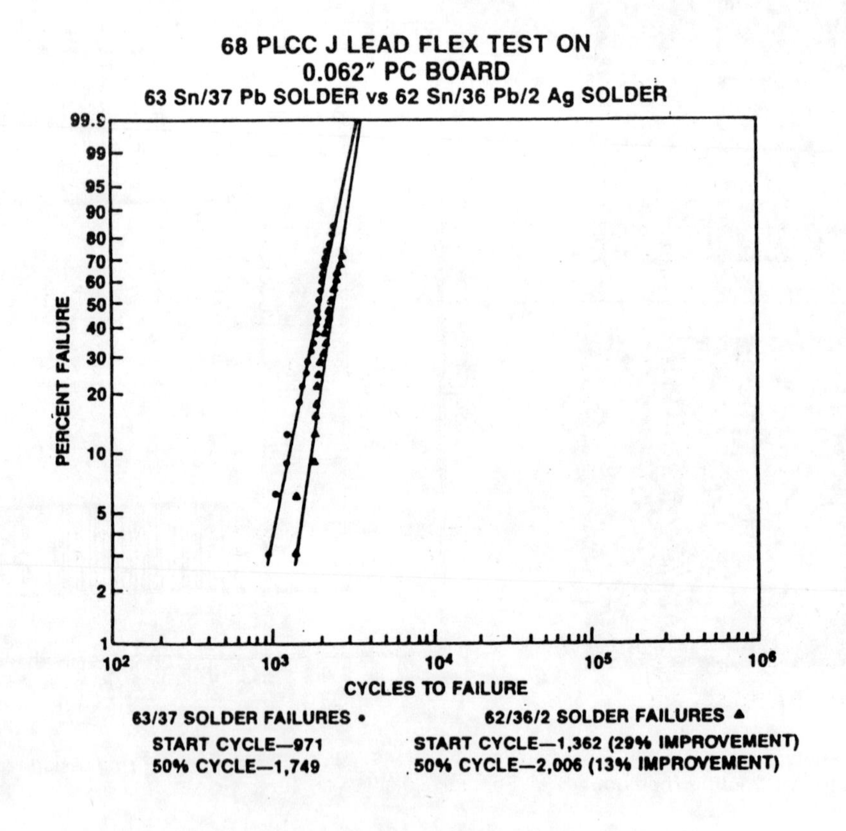

Fig 11 Comparison of different solder types

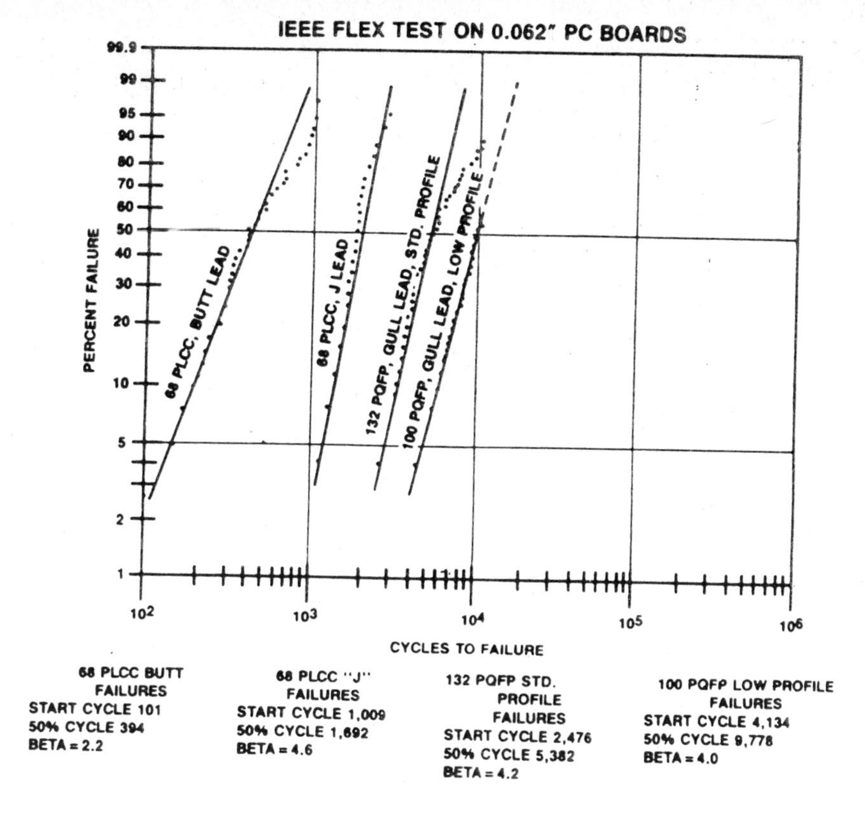

Fig 12 Weibull chart of 62 mil thick board flex test

MOTOROLA SEMICONDUCTOR PRODUCTS SECTOR
ADVANCED PACKAGING DEVELOPMENT PILOT LINE (APDPL)

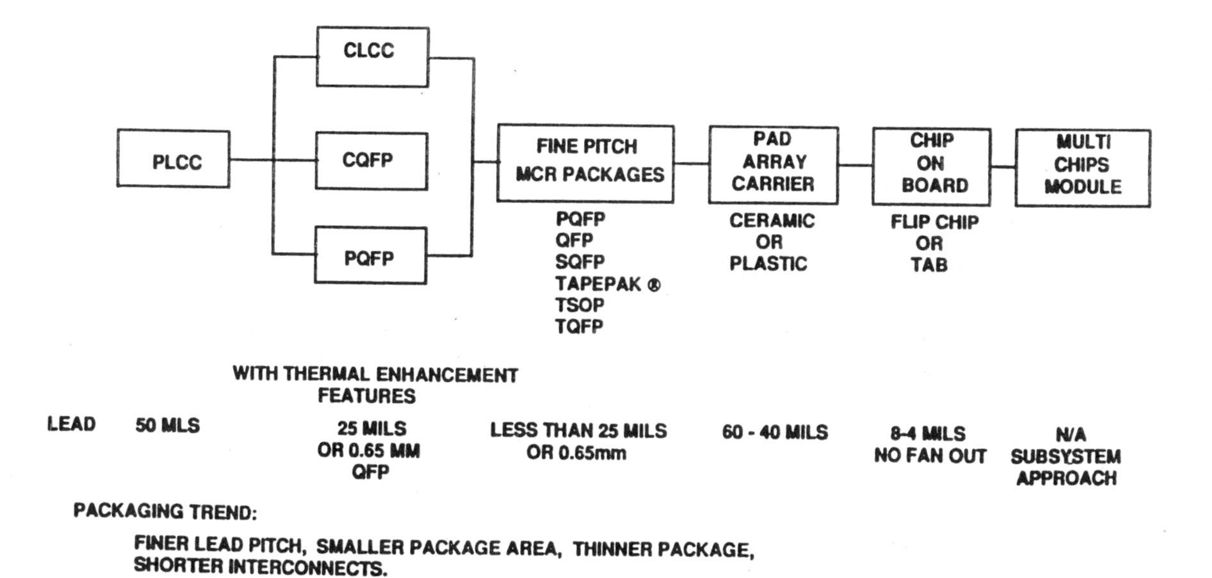

® REGISTERED TRADEMARK FOR NSC.

Fig 13 Surface mount package progression

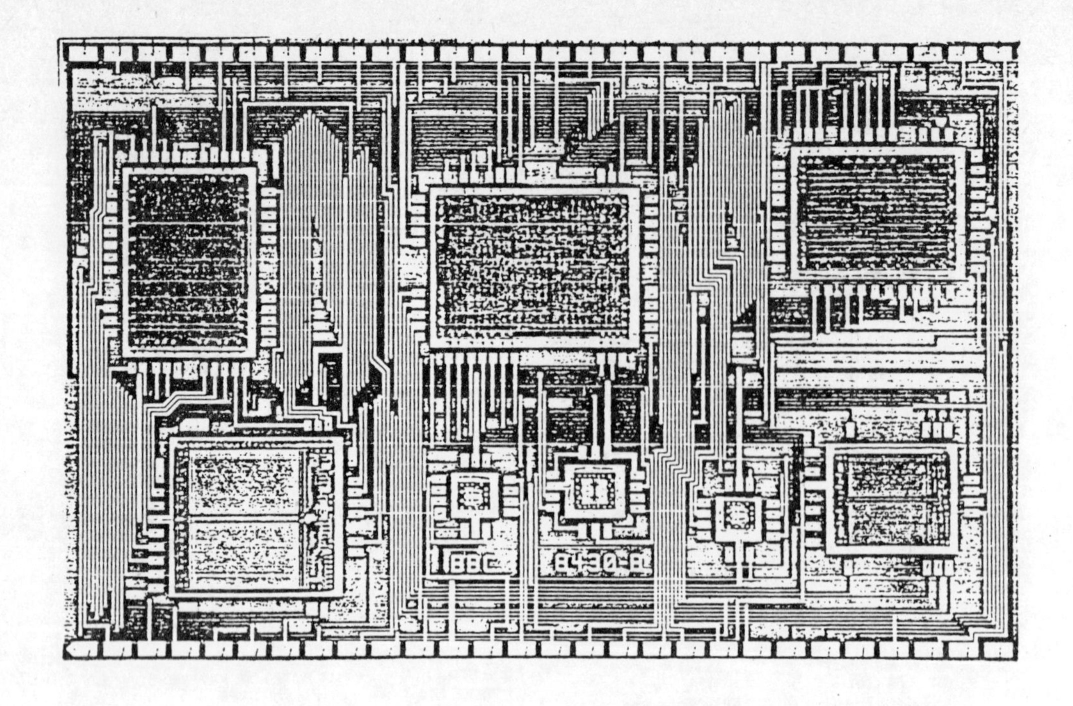

Fig 14 Multichip module example – 3 x ASIC; 2 x RAM; 3 x CMOS ICs; polyimide and Cu on ceramic substrate; 39 mm x 27 mm (substrate produced by Contraves AG, Zurich)

MULTICHIP CIRCUITS IN AUTOMOTIVE ELECTRONICS

Electronic for:	MOTOR	DRIVE	CAR BODY
	ignition	ABS	car radio
	injection	automatic dif. locking	car telephone
	fuel level sensor	transmission control	air bag
Why :	heat problems thermo cycling reliability	packaging density mudularity reliability	packaging density mudilarity reliability

Fig 15 Possible applications of multichips in automotive electronics

C391/017

Production testing of anti-lock-equipped vehicles

C F ROSS
Grau Girling Limited, Redditch, Worcestershire

SYNOPSIS From October, 1991 commercial vehicle manufacturers will be faced with the mandatory requirement to install anti lock brake systems onto certain classes of vehicle. For many manufacturers it will be their first encounter in installing a computer based control system on a production basis. How far must the vehicle manufacturer go to ensure the integrity of the anti lock system to the extent that when the vehicle leaves the production facility that the anti lock system will react correctly to imminent wheel lock conditions.

1 INTRODUCTION

While anti lock systems have been installed onto commercial vehicles in the UK since the late 1970's the installation has been carried out by specialist fitting centres or off line by the vehicle manufacturer. The introduction of Directive 88/194 EEC making anti lock braking a mandatory requirement from October 1991 will change this. The types of vehicle affected by the Directive as defined in Figure 1 will make it necessary for such systems to form part of the vehicle production specification.

Consideration was given as to why it was necessary to test the ABS system as the majority of components making up the system are pretested by the OEM supplier.

However the most important factor that needs to be addressed is the correct installation onto the vehicle. Errors on assembly such as incorrect relativity of wiring and pipework could result in vehicle instability or the shutting down of the anti lock system. To ensure that this possibility is minimised consideration must be given to three separate functional areas:-

1 Installation design.
2 On line testing during installation.
3 End of line testing.

This paper describes the installation and test procedure jointly developed by Grau Girling and Leyland DAF to ensure system integrity.

2 ANTI LOCK PERFORMANCE REQUIREMENTS

Before elaborating on the production test procedures it is necessary to understand the performance requirements of anti lock equipped motor vehicles defined by the mandatory anti lock Directive. From Figure 1 it can be seen that the level of anti lock performance is defined for motor vehicles as Category 1. To meet the required performance on split friction surfaces it is necessary to utilise an anti lock system with a control philosophy of individual wheel control. This philosophy is satisfactory for drive axle logic but unsatisfactory for steered axle logic when braking on split friction surfaces. The geometry of commercial vehicle steering and brakes is such that pressure and hence torque variances across the steered axle can result in excessive steering correction, vehicle yaw and instability. To ensure vehicle stability is maintained a philosophy of modified individual control (MIC) is applied to the steered axle.

Figure 2 illustrates a typical 4 channel Category 1 anti lock system. It can be seen that each wheel has its own wheel speed sensor and pressure modulator, all of which are connected to a central electronic control module (ECU) which computes the respective wheel speeds taking anti lock action where appropriate. Integrity of function is dependent on the correct orientation of all wiring and pipework.

3 DESIGNING FOR PRODUCTION FITMENT

From the above it is clear that the vehicle manufacturer must take precautions to ensure correct orientation of all connections. The Grau Girling DGX anti lock system has an on board diagnostic facility which can be invaluable when servicing but without additional testing cannot tell whether a sensor, solenoid or pneumatic pipe is correctly or incorrectly orientated. Only when over braking occurs can the ECU diagnose a fault as the wheels do not react correctly to the respective solenoid energisation.

Minimising possible production problems must be considered at the design stage of system installation. The axles will be supplied direct from the manufacturer with the sensors and exciters installed. Sensor output checks having been performed. A suitable interface would then allow the connection of the sensor to the main vehicle wiring loom which will transmit the wheel speed signals to the ECU. To minimise the number of connections in the anti lock loom Leyland DAF required the anti lock loom to be produced in one piece. The only connectors were then:-

1 35 way connector to the ECU.
2 4 sensor connectors.
3 2 valve solenoid connectors.
4 Power/diagnostic interface.

Cable lengths were such that the sensor cables were of a length that they would only fit one way. Positioning of the pressure modulators on the vehicle chassis was such that a front and rear interchange was not possible. Additional safeguards included labelling of the respective cables.

Valve position for the front axle ensured that only the left side of the valve could connect to the left wheel etc. The pneumatic pipe being too short to reach the opposite side of the valve.

Rear axle installations require a different approach as the normal pipe run is from the chassis mounted pressure modulator onto the differential casing and then across the axle to the respective brake chamber. As the flexible hoses run side by side the potential for crossing the pipes is high. This possibility was eliminated by having different connections at the valve and differential casing with the appropriate end fitting on the respective hoses.

4 'ON LINE' PRODUCTION TESTS

As components are assembled onto the vehicle chassis, it is always beneficial if the completed system can be checked prior to CAB assembly and final test as rectification is easier. With the axles, pressure modulators, brake system and anti lock wiring loom installed it is possible to carry out a full orientation and sensor output checks. As the anti lock wiring loom is in one piece a complete check is possible prior to final connection to the ECU when the cab is fitted.

A suitable control unit which is capable of energising the solenoids in the pressure modulator and measuring sensor output is connected to the anti lock wiring loom at the 35 way connector. First the brake system would be pressurised so that all the wheels are locked. The control unit is then operated which will energise an exhaust solenoid in the pressure modulator. The effect of this would be to release the brakes from one wheel. As the brake pressure has been exhausted that wheel should be able to rotate, if the wiring and piping are correct. When rotated an output from the sensor would be measured at the control unit. This procedure is repeated on all the wheels which are fitted with sensors and exciters. Having completed this check satisfactorily ensures that all the chassis wiring and piping is correct.

5 'END OF LINE' FINAL TEST

It is always possible that between the on line check of the anti lock system and the vehicle final test that a fault or damage has been caused that could affect the anti lock integrity. It is therefore essential that an end of line test be carried out. Leyland DAF's objectives from the outset of the project was to cycle the anti lock via the ECU without actually driving and braking the vehicle. The facilities available for test purposes through which every vehicle passes are:-

1 Engine power dynomometer.
2 roller brake tester.

The engine power dynomometer is able to drive all four vehicle wheels at speeds in excess of 10 kmh. (Maximum permitted sensor check out speed) and therefore that the anti lock warning lamp function is satisfactory and that the sensor signals are uncorrupted at higher speeds.

Following on from the engine dynomometer test each vehicle is brake tested on a roller brake tester to monitor performance and torque variances across an axle. The speed at which the roller rotates the vehicle wheels is approximately 3-4 kmh. At this speed the level of sensor output may be below a level which the ECU requires for operation in which case the diagnostic output from the ECU would not indicate the presence of a sensor signal. Therefore an alternative method of measuring the sensor output must be sought. This was achieved by 'piggy backing', the sensor wiring at the ECU interface which would terminate in a suitable connector which when coupled to an appropriate measuring device the actual voltage from the sensors could be measured. With the Grau Girling 60T exciter and sensor with a gap of 1mm the minimum voltage that would be recorded would be 0.75 v. Measuring the voltage relative to the rotating wheel combined with the earlier sensor test on the engine dynomometer has ensured total sensor integrity and relativity.

The sensors constitute only part of the vehicle anti lock system. The Leyland DAF objective was to completely test the whole anti lock system on the vehicle. It was unacceptable to remove the ECU and energise the solenoids in a set routine as the integrity of the ECU and its connection has not been checked. The only possibility of energising the pressure modulator solenoids via the ECU would be by the ECU reacting to changes in wheel speed. To achieve this the sensor 'piggy back' interface is connected to a computer facility capable of generating wheel speed signals and also measuring the braking force output from the roller brake tester. The wheel speed signals that the ECU could see would be the signals generated from the wheel on the roller brake tester and the signal generated within the external computer facility. The impedence of the signal from the external source was such that the signal from the actual vehicle sensors was completely swamped. Testing of the brakes is carried out on one axle at a time, but for the anti lock system to check out and function all four wheel speed sensors must generate a signal of approximately the same vehicle speed. Independent of which axle is being tested the ECU must see and monitor wheel speed signals from each wheel.

As explained in Section 2 the ECU has two levels of control logic.

Front axle - Modified Independent Control (MIC).
Rear Axle - Independent Control.

It is the modified logic which controls the torque variances across the steered axle under split friction braking. For the purpose of the production test the torque variances generated must be such so as to meet the following criteria:-

1 Variations in performance from 'Green' linings.

2 Limits the torque to ensure the vehicle remains on the roller brake tester.

To achieve the above the wheel speed signals needed to be tuned to obtain a given reaction by the ECU and hence brake force measurement. A signal was generated which represented a vehicle acceleration from rest up to 35 kmh. Having attained this speed one front wheel is decelerated at a rate which will obtain a reaction from from the ECU. To control the vehicle yaw during split friction braking the wheel on the high coefficient surface has the brake torque controlled by a sympathetic solenoid energisation based on the solenoid energisation of the low friction wheel. The level of modification applied to the front axle to maintain dynamic vehicle stability was defined during vehicle preparation prior to Type Approval and is integrated into the anti lock software algorithm. However for the purpose of the production test, wheel speed signals are tuned to a level which would be unrepresentative of real conditions but provide the brakeforce differences needed to check system function and integrity. The wheel speed analogue wave form generated has four variables. See Fig. 3.

1 Wheel deceleration.
2 Constant Slip (T1).
3 Wheel Acceleration.
4 Constant vehicle speed (T2).

Wheel decelerations and accelerations remain constant as does the level of wheel slip. Variances in brake force magnitude and side to side differences can be controlled by varying the times T1 and T2. Increasing T1 increases the difference in pressure and hence brake force across an axle while increasing T2 increases the skid pressure of the low friction wheel. During development of the process times T1 and T2 were optimised and the objectives achieved. Having generated the appropriate wheel speed signals the brakes of the vehicle are fully applied, because the anti lock solenoids are already reacting to the induced wheel speed signals the pressure and brake force difference increases over a period of approximately one second to reach the optimised level. Figure 4 illustrates the wheel speed wave form being induced onto the left hand front wheel with the associated pressure and brake force for the respective wheels. It can be seen that the brake force follows the pressure characteristics relative to the respective wheels. During production test the external computer will only be monitoring the brake force. The area under each brake force curve can then be compared. From Fig. 4 the characteristics are such that the external monitor can readily detect the brake force difference. The monitor is not purely checking that a brake force difference is seen but that the lower brake force is associated with the wheel which has has the wheel decelerations induced. Provided the two characteristics are related then the front axle has been checked for:-

1 Ability of the ECU to react to wheel decelerations.
2 Internal ECU circuitry is correct.
3 Wiring between ECU and solenoids is correct.
4 Piping between valve and brake is correct.

Should the relationship of wheel speed and brake force not follow each other then a fault would exist and the vehicle would fail production test. Further checks can be carried out by inducing the wheel slip condition into the front right hand wheel sensor wiring. The wheel deceleration/ brake force characterics would then be reversed. The rear axle of the vehicle can now be tested. Monitoring of the actual wheel sensor outputs is carried out by the same method as used on the front axle. However it was necessary to develop a different induced wheel speed signal that would be appropriate to the rear axle control logic of independent control. With this logic it was necessary for each wheel to have induced wheel decelerations. By modifying the two variables T1 and T2 from Fig. 3 a pattern of wheel speeds was developed appropriate to this logic. The right hand wheel having a cycle rate of 1.3 Hz while the left hand wheel required 3.5 sec to complete a cycle. Of this 3.5 sec, the wheel remained in slip for 2.5 secs. Figure 5 illustrates the induced wheel speed signals and associated pressure and brake force response. As there is a significant difference in the resulting brake force characteristics from each wheel it is easy to differentiate and relate a given wheel to a brake force.

Having completed the rear axle the complete anti lock system, vehicle wiring and piping have been totally validated.

When carrying out a vehicle Type Approval the anti lock system is optimised to the vehicle to give optimum performance. Grau Girling offer Category 1 anti lock systems for operation with 60T or 100T exciters, within each family there are several options. It is important that not only the correct family of ECU be fitted to a vehicle but also the option. To establish which ECU is fitted to the vehicle the external computer system is linked with the diagnostic output from the ECU. One of the serial codes transmitted down the diagnostic line defines the family and option level. Therefore when the external computer is connected to the vehicle and the vehicle type entered as part of the test information the computer will expect to see a serial code from the diagnostics which should be correct for that vehicle type. Regardless of part numbering, colour code or other external identification by monitoring the actual option which is programmed within the ECU a check against the correct option for the vehicle is made.

6 CONCLUSIONS

The development of this method of test has greatly reduced the possibility of a vehicle leaving a production facility, wrongly wired, piped or with the incorrect ECU. By utilising existing facilities any expenditure is limited to that of providing the external computer interface and relevant software. At the time of writing this paper only one vehicle type has been subjected to the test. However other vehicles are planned for the future. To accommodate other vehicles the monitoring software may require tuning to suit a given vehicle. It is considered, however, that any changes would

be minimal.
If this procedure was adopted by all vehicle manufacturers, where a sensor and diagnostic interface was standard on a vehicle this could provide the basis of a solution to the annual testing of commercial vehicles equipped with anti lock systems.

ACKNOWLEDGEMENTS

The author wishes to thank Leyland Daf Ltd and Mr. E. J. Slevin of Microface Computer Systems for their assistance in development of the system on which this paper is based.

VEHICLE CATEGORY	TYPE OF VEHICLE	VEHICLE MASS	ANTI-LOCK CATEGORY
M3	TOURING COACHES AND INTERCITY BUSES	>12T	CAT. 1
N3	ALL TOWING VEHICLES EQUIPPED TO TOW AN 04 TRAILER	>16T	CAT. 1
04	FULL, SEMI & CENTRE AXLE TRAILERS	>10T	ANNEX-X

Fig 1 Mandatory anti-lock

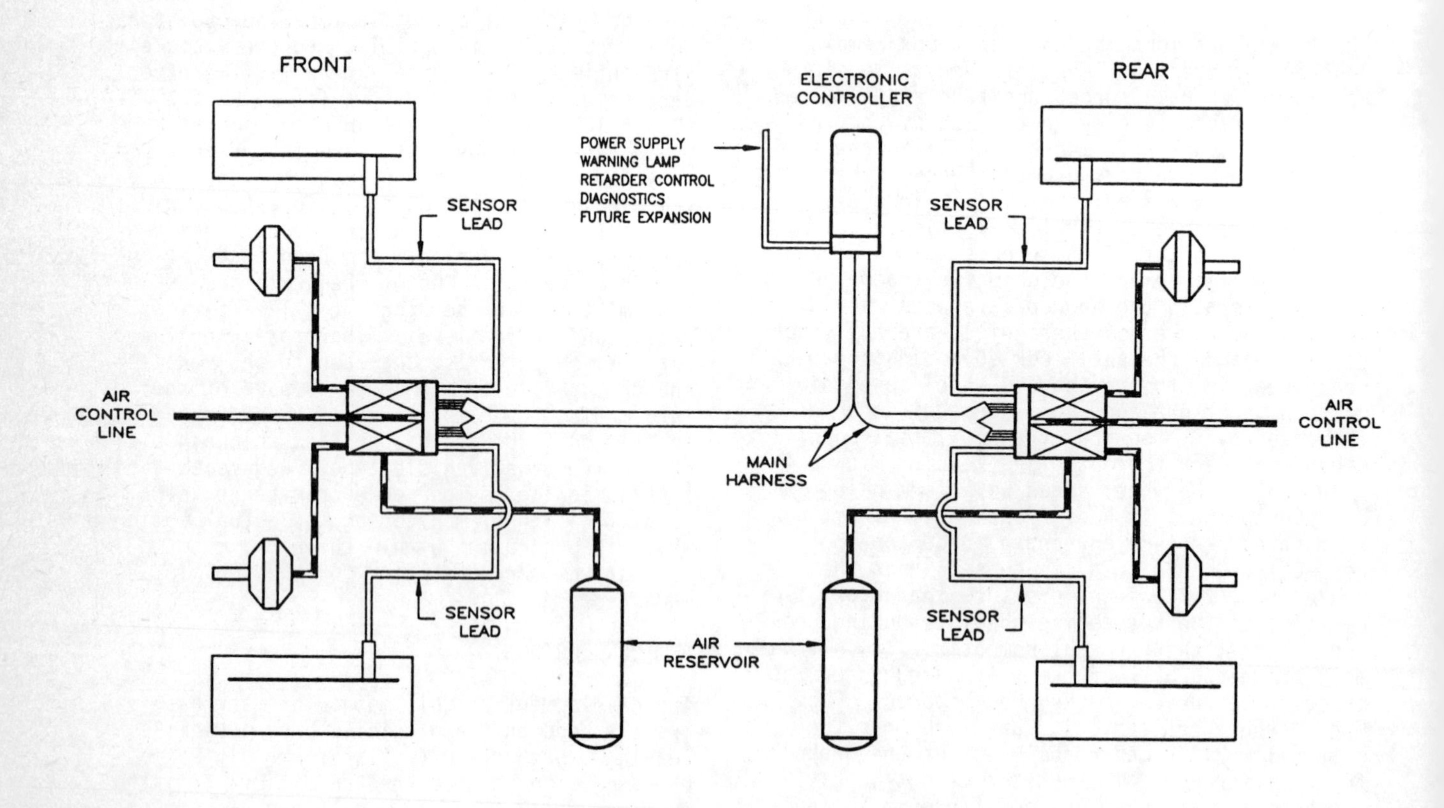

Fig 2 Category 1 anti-lock system

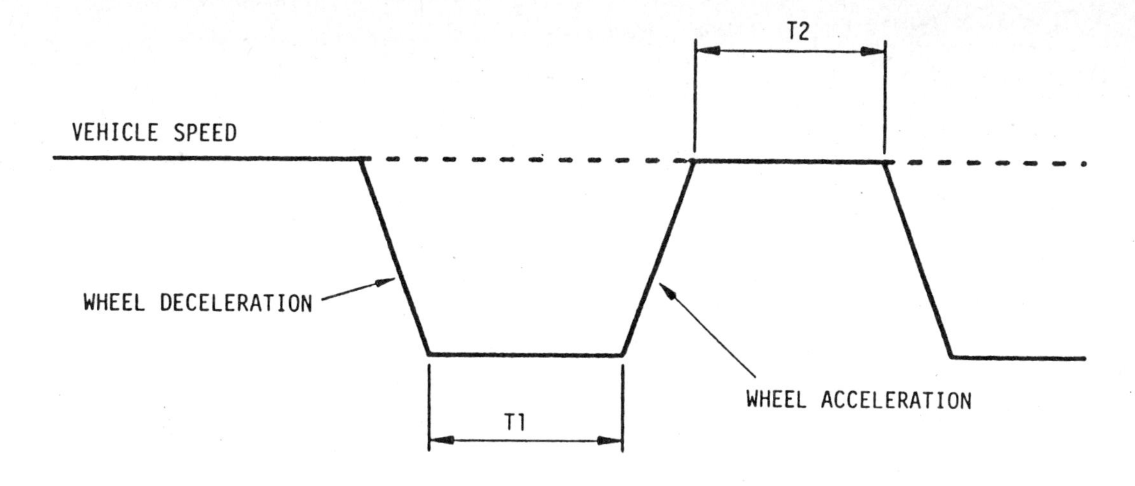

Fig 3 Induced wheel speed wave form

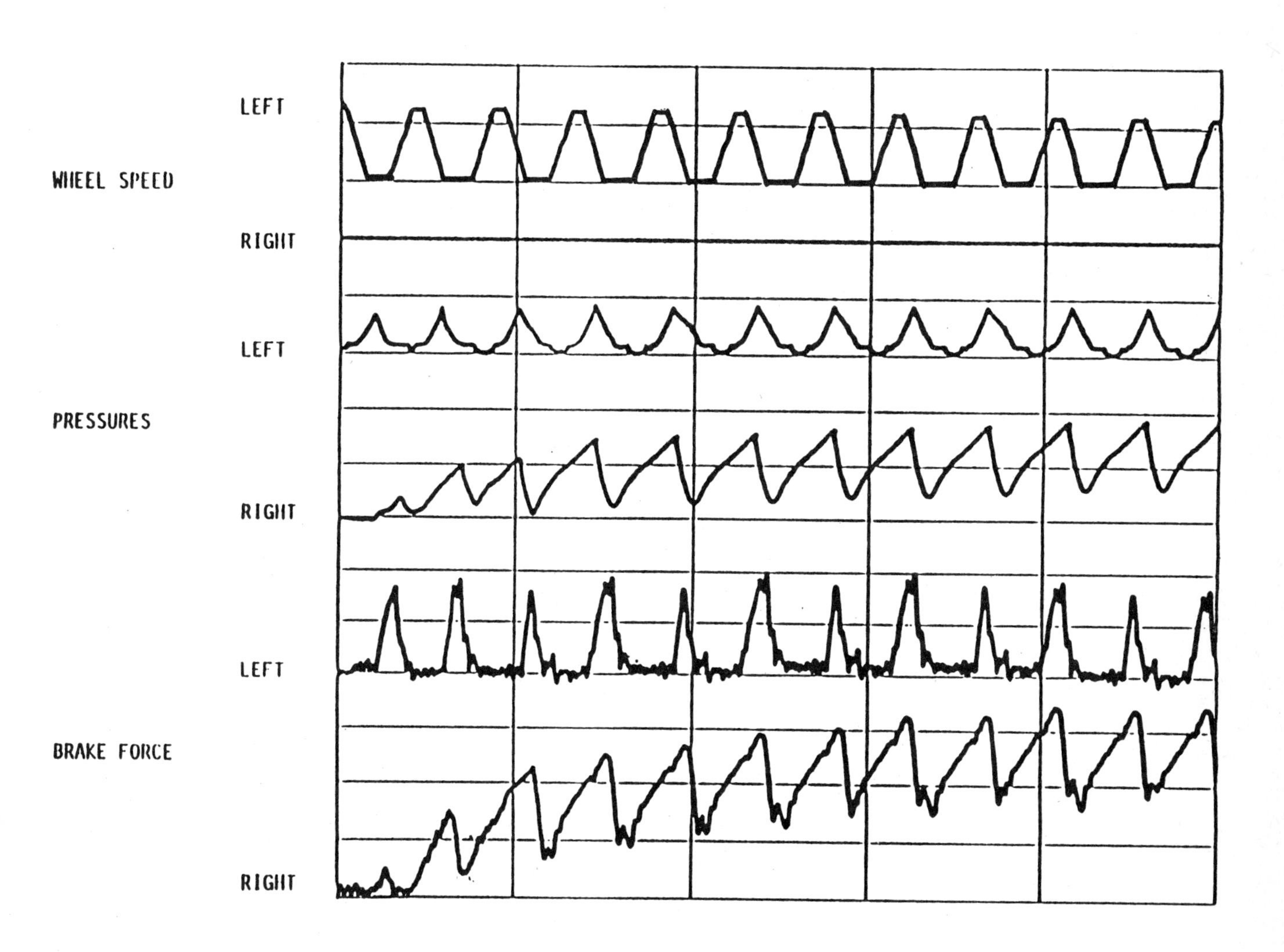

Fig 4 Front axle; modified individual wheel control verification

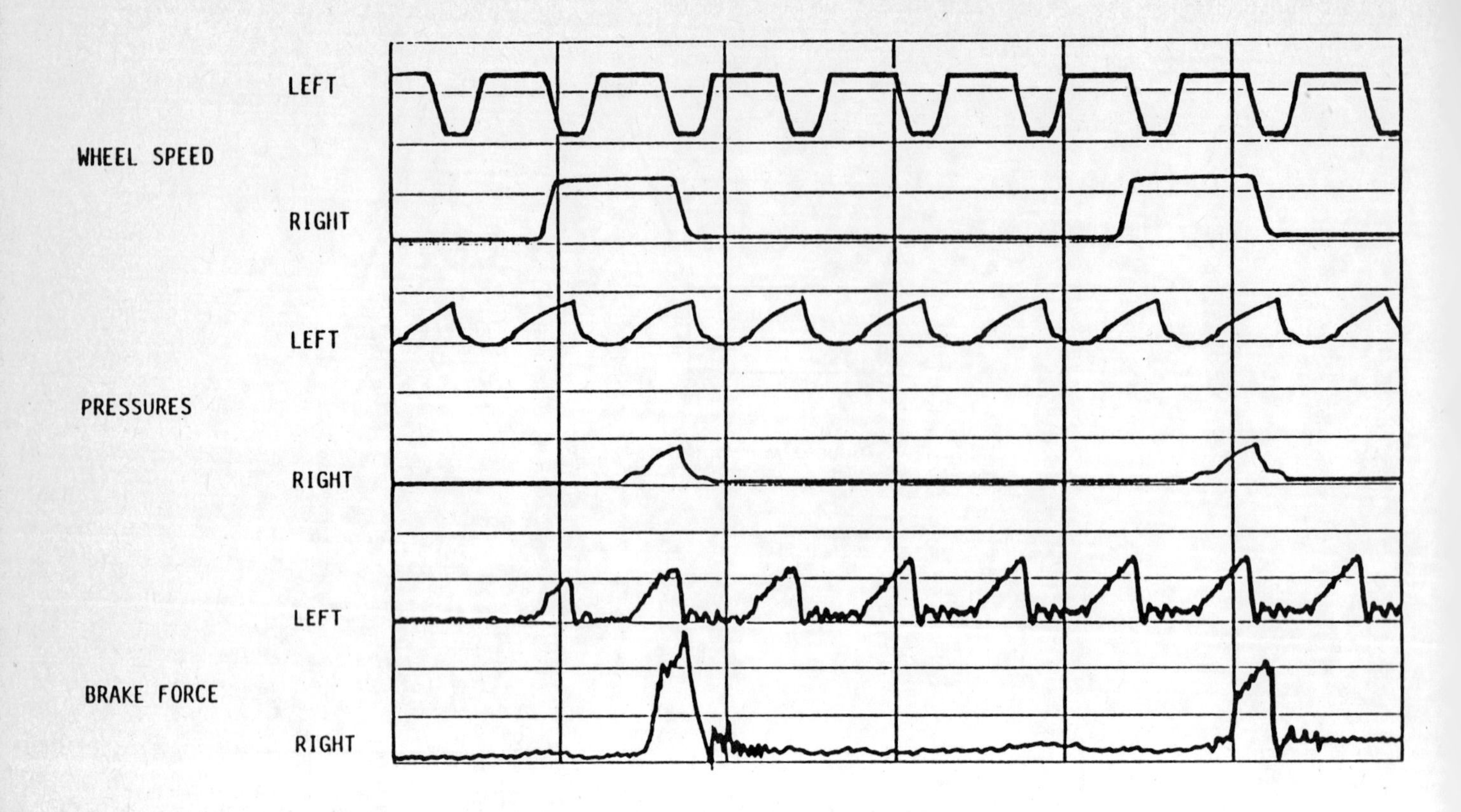

Fig 5 Rear axle independent control verification

C391/007

A comparison of component and whole vehicle radiated susceptibility testing

R S PORTER, MSc, CEng, MIEE
Grau Girling Limited, Redditch, Worcestershire

SYNOPSIS A specific anti lock controller, made to be especially susceptible to radiated electromagnetic fields, was characterised in a parallel plateline component test. Further test results were then obtained by determining the susceptibility of the same system when fitted to four different vehicles. A relationship between the test methods was established.

1 INTRODUCTION

It is a concern of any commercial organisation that new product development time scales are maintained or reduced as new technologies are introduced. This can only be so if component development is adequately done in advance of complete product assessment, where new technologies increase the system complexity.

Definition of adequate development testing for electromagnetic compatibility (e.m.c) has historically been difficult. The growth of automotive electronics has accelerated interest in e.m.c. matters and their solutions but as yet, no definitive relationship exists between the results of whole vehicle tests in anechoic chambers and individual component tests. This means that potentially, e.m.c. development should be on a complete vehicle in an anechoic chamber. Even if this philosophy is put into practice, there are still technical problems associated with producing real-life electromagnetic measurements of large vehicles. The commercial vehicle business embraces a multiplicity of vehicle configurations for any one particular model of basic form. Customer specified body type and wheelbase produce production line variants which will have e.m.c. performance differing in detail. Relating the importance of such differences to the e.m.c. specification of the vehicle components at the development stage is not a straightforward matter. Costs and logistics are a further consideration in any development programme. Therefore it is clear and well known that a component test which could be shown to represent a whole vehicle test would be an advantage to the development of automotive electronics. In 1983 there was no facility suitable for testing the e.m.c. performance of antilock controllers fitted to large vehicles, but the Department of Transport had to interpret a statutory requirement placed on electronic anti-lock systems. The framework within which is was necessary to make an interpretation was defined in a very firm style. Annex 13 of Economic Commission for Europe (ECE) Regulation 13 (1) specifies the requirements of electronic antilock brake systems exactly:- The operation of the device must not be affected adversely by electric and magnetic fields.

The task of producing a test to interpret this regulation was taken on by the Department of Transport, using a group of experts from the automotive industry, together with scientists within the Electrical Research Association. The resulting test method (2) uses a parallel plate line, with a plate separation of 800 mm. This provided a method of subjecting an electronic controller of moderate size (say up to 200 mm maximum dimension), together with a representative set of connecting cables, to electromagnetic fields. The configuration used during the test is as shown in Figure 1.

It was necessary to produce a test which it was believed would represent the electromagnetic environment found in practice as no facility then existed in Europe in which it was possible to make e.m.c. evaluations of antilock controllers when fitted to articulated vehicles. Therefore this component test had to be of a standard which would achieve satisfactory performance assessed against the requirements of safety - critical equipment. The test of time has been passed by this measure. The test, as originally defined, has been accepted by road transport authorities in other European countries as an adequate interpretation of the ECE regulation.

However the question still remained unanswered 'How does this component test relate to a whole vehicle test?'

Some experts would argue that a whole vehicle test is always more representative of the actual environment, and other experts would say that testing in some other way would produce the required result.

The purpose of this paper is not to propose the exact polynomial which relates component level testing to whole vehicle testing, and thus solve the question for all time, but to present the results of a programme of research into the topic of comparison between component and whole vehicle testing. The rate of change of the electromagnetic environment is such that the use of cellular telephones in commercial vehicles has become accepted since this work began. The frequencies used by such equipment is not included in the plateline component test specification at the present time.

2 THE RESEARCH OBJECTIVE

A joint work effort was commissioned by the Transport and Road Research Laboratory to determine the degree of agreement between susceptibilities as measured in the Regulation 13 plate-line test, and a whole vehicle test. The resources of Grau Girling, Lucas Automotive and the Motor Industry Research Association were brought together to execute the programme of work.

3 THE EQUIPMENT UNDER TEST

The Regulation 13 test was devised for testing antilock controllers, so it was a natural choice to use that type of electronic controller in the research. Several different types of controllers were used throughout the testing, the production versions of which are currently used in commercial vehicles. The production versions were, naturally, filtered to such an extent that e.m.c. failures during testing would be unlikely. Therefore the filtering was removed, to produce a convenient level of failure, thereby ensuring an adequate amount of data for correlation purposes.

4 THE PARALLEL PLATE LINE

The plateline test uses two parallel plates with a separation of 800 mm. Each end of the plates is terminated by resistors, and excitation is fed to the plates through a resistive matching network. The voltage between the plates is measured through an attenuator matched into a radio frequency (r.f.) voltmeter. The field strength between the plates is determined from the quotient of the voltage measured and the plate separation. The theoretical analysis of simple parallel plate lines indicates that resonances above 100 MHz would produce errors in the field strength sufficient to make the method unusable. In practice the field strength measured within empty lines of this design stays within $\pm$ 3dB of that predicted from the r.f. voltimeter reading, up to frequencies of the order of 400 MHz.

4.1 The arrangement of the equipment within the parallel plate line

Figure 1 shows how the equipment under test was arranged within the parallel plate line. The upper plate was connected to ground at the r.f amplifier, via the outer conductor of the coaxial drive cable. The lower plate was excited through the resistive matching network. Cabling to the unit under test was kept close to the upper, earthy plate and then went directly to the controller under test. The antilock warning lamp was remotely located so that it could be observed during the test, as were lamps which showed the state of energisation of the anti lock air control valve solenoids.

4.2 The measurement of the susceptibility profile

To produce an antilock operation cycle, simulated wheel speed sensor signals were electronically generated, and sent to the electronic controller via transformers. This wheel speed regime is shown in Figure 2, and is as specified in part 2 of the Regulation 13 test. The two sharp decelerations produce antilock valve solenoid energisations, which in the on-the-road case, would cause a release of braking before the point of wheels locking. During the first part of the test, a constant wheel speed of 40 km/h, as specified in the test, was used.

Two types of modulation are specified in the test. In the first part 50 per cent amplitude modulation, at a frequency corresponding to the output of a wheel speed sensor at a road speed of 40 km/h was used. In the second part of the test two types of modulation were used in succession. Firstly 50 per cent amplitude modulation at a frequency corresponding to a wheel speed sensor output at a road speed of 90km/h was used. Secondly 100 per cent pulse modulation at a frequency of 1 Hz was used.

The range of frequency of the applied radio frequency field specified by the test is 10 kHz to 500 MHz. This is to be continuously swept in the first part, and discrete frequency points are to be used in the second part of the test. In order to provide exact points for comparison, discrete frequency points only were used during both parts of the test in this research.

The criterion for recording a point of failure was taken as the illumination of the antilock warning lamp. This gave the simplest, repeatable, non-subjective measure of the point at which the antilock controller's fault checking circuits had decided to suspend operation of the controller.

4.3 Worst case results of the parallel plate line test

The worst case results (3) for one unfiltered controller are shown in Figure 3. Repeating the test with either the controller or its cabling in different positions with the parallel plate line produced a similar susceptibility profile.

5 WHOLE VEHICLE TESTS

The whole vehicle tests were conducted within the anechoic e.m.c. chamber at the Motor Industry Research Association. This chamber is 22 m by 10m by 7m in height and lined with 1.8m absorbers. A 6m diameter turntable and a 2 axle adjustable wheel base dynamometer within this facility were used to produce polar responses of the equipment under test, and cycling of the antilock system, respectively.

It is essential to define the calibration procedure used, when a susceptibility profile obtained from an anechoic chamber test is quoted. In these tests the anechoic chamber was calibrated by the substitution method. That is to say, the field existing in the chamber, for known antenna power, was measured without the vehicle in the chamber. The vehicle was then subjected to the field produced by the same amount of antenna power. The vehicle was in effect substituted for the field strength meter. During the testing the antenna power was varied to obtain the susceptibility profile, and linear interpolation produced the corresponding field strength values.

The separation between the antenna and the vehicle, and the position of the calibrating field strength meter form a critical part of the test configuration. The distances used for the the results quoted here were 1 metre between the antenna and the vehicle front skin, and 4 metres between the antenna and the field strength meter. The field strength meter was therefore positioned approximately at the position which became the centre of a large commercial vehicle, once the substitution had taken place.

5.1 Worst case results of the whole vehicle tests

Figure 4 shows the worst case results (3) for one commercial vehicle. This was achieved with vertically polarised radiation having 1 Hz pulse modulation applied.

6 COMPARISON OF THE TYPES OF TEST

It is not a straight forward process to make a comparison of the profiles of susceptibility as measured by the two different types of test. Peaks and troughs of susceptibility occur at different points in the frequency range, so a simple coefficient relating the two tests is not available.

However, over the frequency range 30 MHz to 200 MHz the whole vehicle profile has a mean level almost double that of the parallel plateline test. The minimum level of susceptibility during stripline testing was however, approximately one fifth of that recorded during the whole vehicle test.

7 CONCLUSION

The results indicate that a factor of five could be applied to the susceptibility results of a parallel plateline test to give the order of susceptibility to be expected in a whole vehicle test, when the calibration of the test volume has been done as described.

The operational immunity of equipment is governed by minimum levels of susceptibility, and therefore the ratio of minimum levels is presented as the most important factor obtained from the research results. The minimum level of susceptibility recorded during the whole vehicle test was approximately five times that of the plateline test.

The research produced a large volume of data obtained from testing two fire tenders, an articulated tractor/trailer combination, and a coach capable of seating in excess of 50 people, and is therefore seen as representative of the radiated susceptibility performance of electronic equipment in commercial vehicles in general.

8 ACKNOWLEDGEMENT

The author wishes to acknowledge the support of the Transport and Road Research Laboratory in carrying out the programme of work.

REFERENCES

(1) UNITED NATIONS ECONOMIC COMMISSION FOR EUROPE.

Uniform provisions concerning the approval of vehicles with regard to braking, 1988, ECE regulation no. 13/05.

(2) MADDOCKS, A. J. Draft specification of the measurement of immunity of road vehicle anti-lock braking systems to electromagnetic radiation, 1983, ERA project report no. 5043/4RG/4.

(3) GIBBONS, W. A correlation of EM1/EMC test procedures for road vehicles, 1988, MIRA Research Report, 1988.

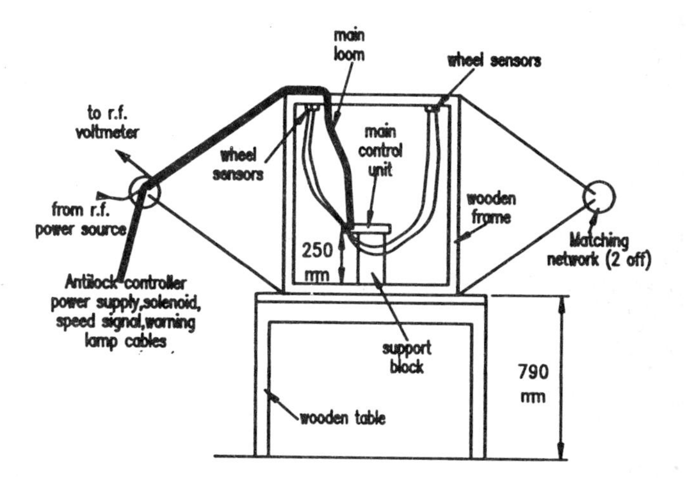

Fig 1 General arrangement of the plateline test

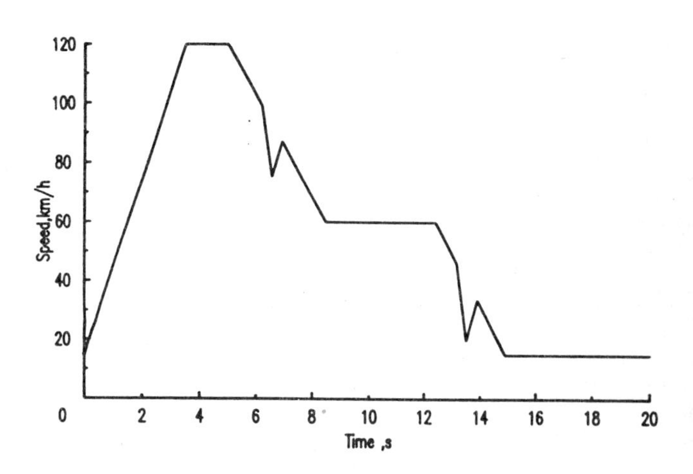

Fig 2 Simulated wheel speed profile

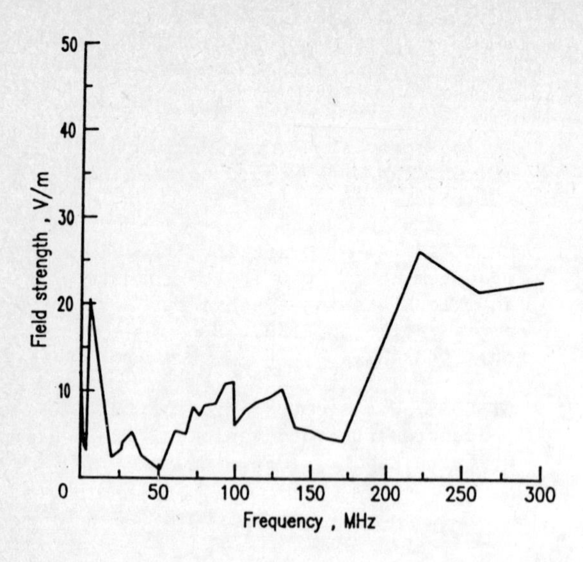

Fig 3 Susceptibility determined using the plateline, with 484 Hz modulation

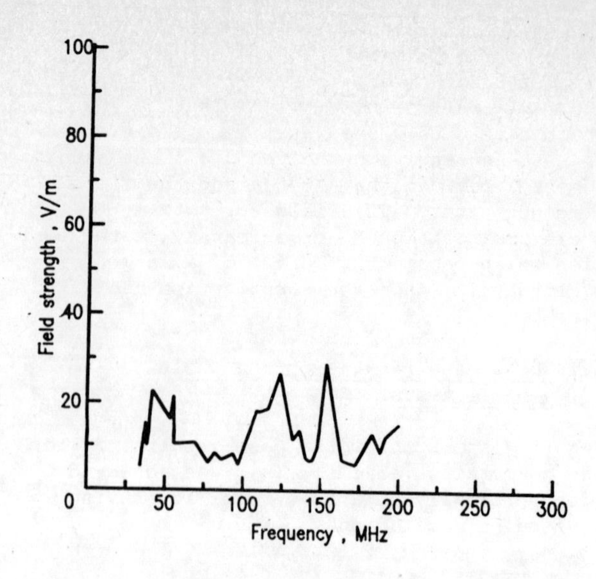

Fig 4 Susceptibility determined by the whole vehicle test, 1 Hz pulse modulation

Correlation of radio emissions from bench and car tests

M T CROWTHER, BSc, CEng, MIEE
Jaguar Cars Limited, Coventry

SYNOPSIS Prediction of on-car emissions from bench results is necessary when optimising the application of EMC principles at the beginning of an ECU's design cycle. A method is described which allows prediction, within certain limits, from empirically obtained data.

It is also important to apply the correct Bench Limit for a particular component to optimise development of the component.

1 INTRODUCTION

The tendency of digital electronics to emit radio frequency interference must be limited not only to eliminate inter and intra-system interference but also to ensure unimpaired radio reception. This tendency can be controlled by acting on EMC guidelines at the beginning of the design cycle.

Since representative cars are unlikely to be available at such an early stage, a bench test from which on-car emissions can be predicted is a necessity. This technique allows optimisation of EMC principles.

2 OBTAINING THE RESULTS

The Bench data was obtained using VDE 0879 part 3. The Car data was obtained directly via the car's own antenna. During the Car test only the ECU under test was powered and the engine was not running. For both tests the instrument used was a computer controlled receiver reading in dBµV through its CISPR detector.

All the comparative data has been generated by comparing emissions attributable to ECU clock harmonics at the same frequency from the same type of ECU in the FM Band (88 -108 MHz). That is to say an emission at 90 MHz will be compared only with emissions from other ECU's of the same type at 90 MHz.

The frequency range has been limited to the FM Band because it has been assumed, for our on-car test, that the car antenna is matched to 50 Ω over this range. The bench test was also restricted to this range.

3 DATA ANALYSIS

The question arises of whether to analyse the data in µV, as a ratio, or in dBµV, as a difference. It was resolved by plotting two graphs of frequency of occurrence of comparative result against comparative result. Each graph shows two types of ECU.

The first, Figure 1, shows the comparative result as a ratio (Bench µV/Car µV). The second, Figure 2, shows the comparative result as a difference (Bench dBµV - Car dBµV).

In order to resolve this problem quickly, only a few ECU's were analysed. This accounts for the irregular shape of the graphs.

Examination of Figures 1 and 2 shows that the distribution appears to have the form of a Log-Normal graph.

As a result of this, all work was carried out in dB. This also made further manipulation of the data simpler.

4 PREDICTING CAR EMISSIONS FROM BENCH TESTS

If, at each emission frequency, the difference (Bench dBµV - Car dBµV) is plotted against frequency, graphs such as Figure 3 are produced. The points comprising the graph have been joined for clarity.

Figure 3 clearly shows two distinct populations. Lines of least squares regression have been drawn to show the general trend of each population and the difference between these trends is about 10 dB.

Without a predictive technique, the worst case situation may have to be assumed and the optimum solution is unlikely to be found.

Given that the equation for the line of least squares regression is known and that the levels from the bench test and frequencies of emission are also known, it is possible to predict emissions on the car from emissions on the bench.

$$\text{Bench dB}\mu\text{V} - \text{Car dB}\mu\text{V} = m * \text{Frequency} + c$$

thus

$$\text{Car dB}\mu\text{V} = \text{Bench dB}\mu\text{V} - m * \text{Frequency} - c$$

Where c is the intersection of the line with the y axis and m is the gradient of the line.

Predictions of on-car emissions have been made from bench tests and the results compared both with the actual on-car results of the parts concerned and with on-car results obtained from similar parts. Correlation tests demand rejection at the 1 per cent level of the hypothesis that predicted and actual results originate from the same population. Further correlation tests demand acceptance at the 1 per cent level of the hypothesis that predicted results belong to the same population as predicted results for a number of similar ECU's.

The two cases shown in Figure 3 represent extremes of harness layout. The harness associated with ECU type 2 has little contact with any other harness and is at the opposite end of the vehicle from the radio aerial. The harness associated with the ECU type 1, however, is extensive both in its contact with other harnesses and in its layout around the car.

5 CHOOSING A SUITABLE LIMIT FOR THE BENCH TEST

When an acceptable level of on-car emissions has been chosen, it is possible to calculate the corresponding bench level for each value of (Bench dBμV - Car dBμV).

In the light of the preceding section, a Bench Limit should be chosen for each ECU as appropriate.

Bench Limits can be chosen taking into account harness layout as described above. If the harness associated with an ECU represents an intermediate case, an average Limit could be considered. In case of doubt, the worst case should be chosen.

If the acceptable limit for the car, 7 dBμV, is added to the (Bench dBμV - Car dBμV) data, a set of data showing what could have been emitted on the Bench is created. Figure 4 shows ogives for ECU's 1 and 2 from which the proportion of emissions exceeding the Car Limit for a chosen Bench Limit can be found.

It is also possible to choose a Bench Limit from these ogives by accepting a known risk that the Car Limit will be exceeded.

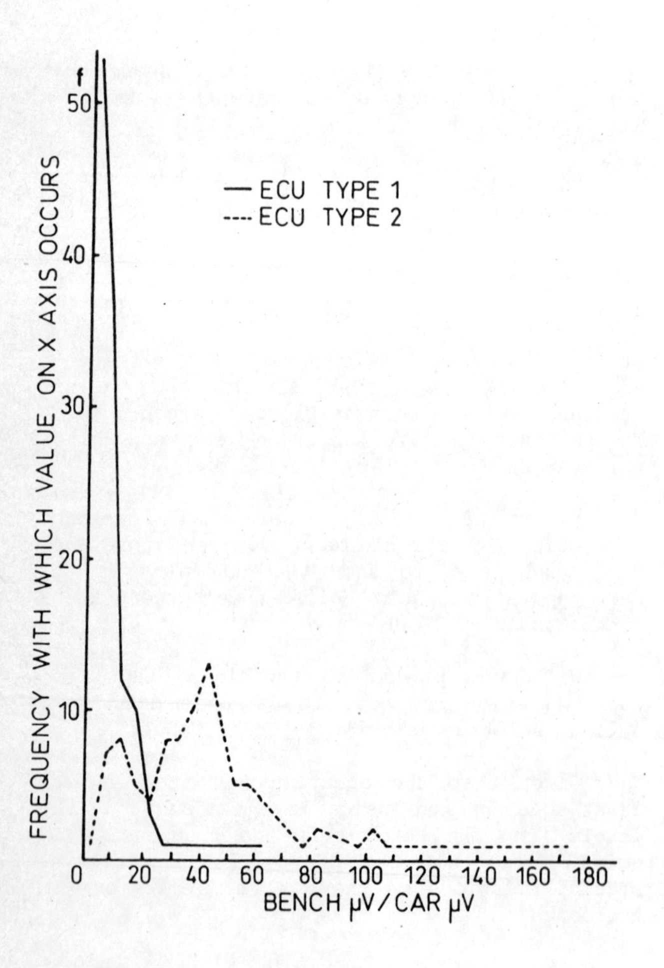

Fig 1

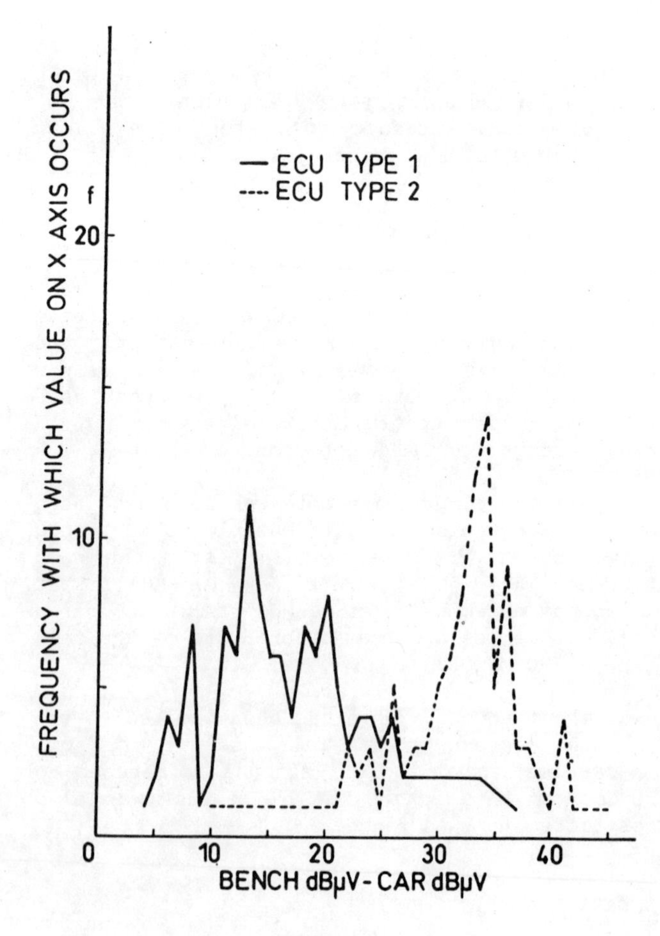

Fig 2

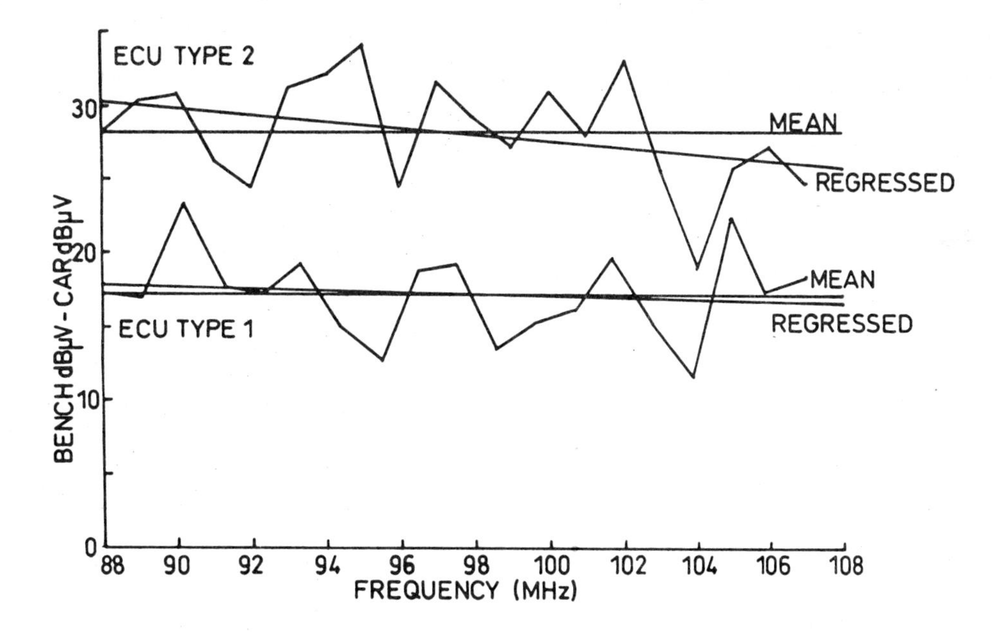
ECU TYPE 2
MEAN
REGRESSED
MEAN
REGRESSED
ECU TYPE 1
BENCH dBμV - CAR dBμV
FREQUENCY (MHz)
0
10
20
30
88
90
92
94
96
98
100
102
104
106
108

Fig 3

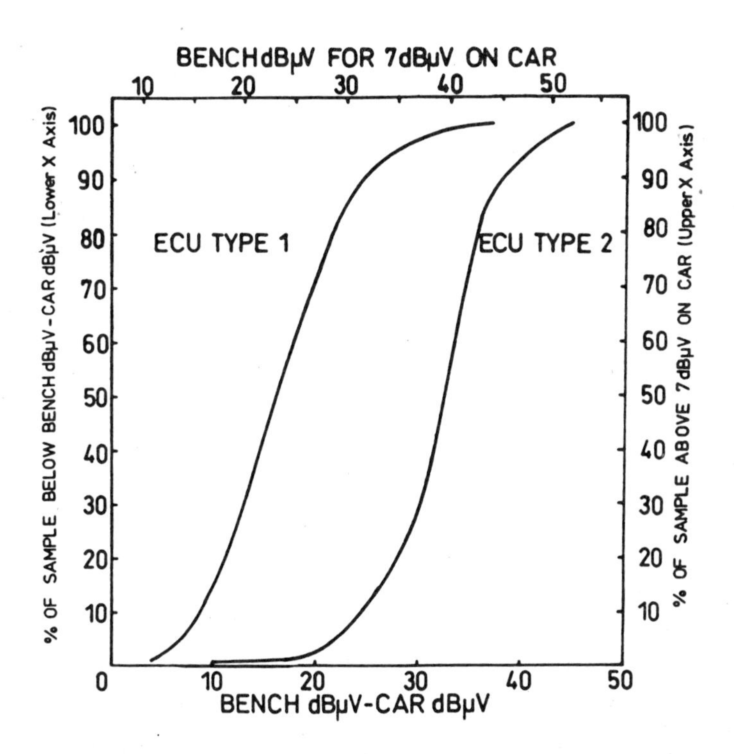
BENCH dBμV FOR 7dBμV ON CAR
10
20
30
40
50
% OF SAMPLE BELOW BENCH dBμV - CAR dBμV (Lower X Axis)
% OF SAMPLE ABOVE 7dBμV ON CAR (Upper X Axis)
ECU TYPE 1
ECU TYPE 2
0
10
20
30
40
50
60
70
80
90
100
BENCH dBμV-CAR dBμV

Fig 4

C391/033

Electromagnetic compatibility (EMC) comparison of data bus media

N J CARTER, BSc, AMIEE and **C R BOYCE**, BSc, AMIEE
Lucas Automotive Limited, Solihull
J A PHILPOT, BSc
Rists Limited, Newcastle, Staffordshire

SYNOPSIS A prototype vehicle network developed within Lucas Automotive uses the Controller Area Network (CAN) protocol and the Intel silicon implementation. A four station system has been installed on a vehicle together with two physical (interconnection) layers, one using fibre optic cables and the other using a copper wire twisted pair.

This paper covers a comparison of the two bus media with respect to their electromagnetic compatibility (EMC).

THE REQUIREMENT FOR EMC TESTS

The electromagnetic environment within a vehicle is known to be a severe test of electrical and electronic systems. Its severity is increasing as system voltages rise and electrical equipment proliferates. In terms of vehicle safety, EMC is likely to become prominent within a few years. As more vehicle functions are electronically, rather than electrically or mechanically controlled, EMC takes on a new significance. Unlike the effects of temperature or vibration, which are generally predictable and gradual, electromagnetic interference (EMI) may strike without warning and have instant effect.

The wiring harness in a vehicle acts as an aerial, that is, electric currents may be induced in it by electromagnetic radiation. As the harness is made up of many wires of different lengths it is susceptible to radiation at many frequencies. The harness in itself is not affected by this interference; the problem occurs because it is connected to electronic modules which are susceptible. Screening of the modules cannot totally protect them from interference via the harness, although this interference can be filtered to some extent. The radiation environment in and around a vehicle is extremely complicated and almost impossible to predict accurately. Parts of the vehicle act as waveguides and raise the field strength locally; other parts are relatively quiet. The best available way to find out the effects of interference is to get experimental data from real vehicles. Even this is not particularly repeatable, but it does give a reasonable indication of the system's performance.

Optical fibres are not affected by EMI, since they are not electrically conductive. Unfortunately the signals on these fibres have to be converted first from, and then back to, electrical signals. The transceivers themselves are susceptible to EMI. This applies especially to the receivers as they have to provide high signal amplification and are therefore sensitive. Again, the positioning of these receivers within the vehicle is a significant factor in their performance. There must, in addition, be some form of power distribution, again using a metallic harness.

In addition to the susceptibility problem, a wiring harness acts as a radiating aerial for any electrical signals it carries. This is largely caused by power switching, especially if the electrical loads (motors, lamps, etc.) are not optimally suppressed. As digital techniques, such as pulse-width modulation, become widespread more electrical noise is generated. This adds to the external EMI environment caused by radio transmitters, lightning, other vehicles, etc.

Automotive specifications in the UK generally require equipment to tolerate radiated fields of up to 75 V/m over a frequency range from 10 kHz to 400 MHz for safety critical functions (2). In the US, where more electrical equipment is standard, up to 200 V/m has been proposed (2).

It should be noted that this test programme was not designed to validate a system to a certain field level, but to measure the envelope of tolerance to radiated EMI.

The two transmission media - polymer optical fibre and copper twisted pair - were selected on the basis of potential customer requirements, reasonable cost and data transmission rate.

SYSTEM DESCRIPTION

Vehicle

The vehicle used for these tests is a Rover 820Si, equipped with fuelling, ignition and dashboard electronic control units (ECUs) as standard. This vehicle has been fitted with four CAN nodes. These interface with engine sensors, dashboard displays, and fuelling and ignition controllers. The tests were carried out with two different bus media: a fibre optic active star network and a copper wire twisted pair bus. Two twisted pair harness routes were used independently; this was to assess the effect of routing on their EMC properties. During the tests on the optical system, the twisted pair harnesses were removed to ensure that they did not affect the EMC properties of the vehicle. Each twisted pair harness was tested individually, with the other twisted pair removed. The optical interface modules were removed during the tests on the electrical system.
A schematic of the system is given in Figure 1.

For a discussion of the optical data bus, (1).

Physical Layers

The ECUs are connected to the CAN bus via interface modules. This involves very little modification to the existing ECUs as the interface unit handles the transfer of system variables between the data bus and the ECU. This approach was seen as the fastest way of evaluating the CAN protocol on a vehicle system.

At the sensor node, the analogue sensor signals are digitized and put onto the data bus. At a receiving node, these signals are converted back to analogue form and fed into the ECU inputs.

To enable a quick turn-round in the EMC test chamber, the twisted pair transceivers are designed to be physically compatible with the fibre optic ones. Changing between physical layers involves replacing the transceiver units and connecting the relevant data bus medium.

Fibre Optic

This system was first implemented as a separate project (1).

For the optical physical layer, the nodes are star coupled to a repeater mixer, which boosts the optical signal levels on the data bus. Power to the stations is provided by a wire running parallel to the optical fibres. This 12 V supply is fused and switched at the optical mixer.

The fibre optic system uses polymer fibres of 1 mm diameter, and light-emitting diodes (LEDs) which emit in the visible red (660 nm) region. The transceivers were designed in-house, as commercially available types were not suitable for automotive use.

The active mixer converts input optical signals from any node to electrical signals. These are amplified and converted back to optical signals for re-transmission to all nodes. As the mixer forms a vital part of the system it has been made dual redundant.

Twisted Pair Bus

Two twisted pair buses were used independently to find out what effect, if any, the physical layout would have on EMC properties.

The twisted pair data bus uses unscreened 120 ohm cable, together with in-house designed transceivers operating from a 5 V supply. Active bias and line termination are incorporated into the two nodes at the ends of the bus. The other two nodes are purely transceiver nodes.

Test Tools

During susceptibility testing in the chamber, the bus traffic was monitored using two tools.

The first is a commercially available CAN bus analyser tool. This enables bus traffic to be logged, and statistics produced - such as bus loading, number of errors occurring, and identifiers present on the bus.

The second tool is an in-house designed unit card which interfaces a Personal Computer to a CAN data bus. The PC may then interact with traffic on the CAN bus - transmitting and receiving messages. This tool has been of use in the development phase of data bus systems.

These tools allow monitoring of the effects of electromagnetic interference on the data bus system, rather than monitoring the end results (e.g. engine stalls). Measures were taken to ensure that the test tools were isolated from the test chamber.

TEST PROGRAMME

Design

The potentially susceptible components were carefully designed with the aid of powerful analogue design tools. Particular attention was paid to the grounding and screening aspects of the printed circuit boards. The boards were validated in prototype form before final production.

The vehicle was initially designed to contain an optical data bus and was driven, prior to this programme of tests, for several thousand Km in both urban and open road conditions. During this period it was not possible to attribute any faults directly to electromagnetic interference. The twisted pair data bus was originally developed to evaluate the CAN protocol using a copper wire physical layer. The two systems use common interface units (hardware and software) but differ in the transmission media and transceivers. The power to the modules was supplied via the same cables for both systems.

Overview

The tests were all performed using the same monitoring equipment (described above), using optical fibres as the diagnostic transmission media. The monitor transmitters fitted to the vehicle were isolated from the power by using separate rechargeable batteries. One external test facility was used throughout the susceptibility testing programme with the same monitoring equipment used for each system. The emissions tests were also performed on the two systems using the same internal facility and equipment throughout. The interfaces to the monitoring equipment were, necessarily, different for the two types of transmission media.

Susceptibility tests were made on two twisted pair systems with the interface modules in the same location but with the harness following two routes around the vehicle, and with the nodes at different points on the harness. This was to discern whether cable layout and positioning of modules on the harness were responsible for differences in the susceptibility of the system.

The tests were made to compare the three systems (two twisted pair and one optical) built to achieve the same function. It was intended to perform exactly the same tests on each system, rather than to obtain a complete picture of any one system. It was decided, therefore, to concentrate on frequencies up to about 200 MHz during both the emissions tests and the susceptibility tests, with brief excursions above this frequency.

Radiated Emissions Test Programme

Initially the emissions work programme concentrated on a bench system. The work was then extended to include some open air tests on the whole vehicle. This was to ensure that the low emissions of the data bus and power lines would be measured in the presence of the varying ambient environment. This was achieved by gaining prior knowledge of those frequencies of interest during the bench test.

The emissions tests were performed on the optical system and on one of the twisted pair systems. The emissions of the twisted pair system were measured before the susceptibility tests, and those of the fibre optic system afterwards.

Conducted interference was first ascertained using a spectrum analyser. Uncalibrated measurements were made on the power harness in both cases and the twisted pair harness. This was to determine those frequencies which would be interesting to investigate. Calibrated readings using a measuring receiver were then taken to confirm these results.

Radiated emissions from a bench system were then measured with the modules arranged on a ground plane in a screened chamber. The available bench space made it impossible to use a full harness layout, so a shorter representative harness was used. The harnesses were arranged in a similar way for each of the systems, having marked their exact position on the bench. The investigation was concentrated on frequencies below 220 MHz, using an active antenna to cover the lower frequencies and a passive biconic antenna to cover the higher frequencies.

The vehicle was then taken to an open area test site, where whole vehicle 10 metre measurements could be made. Emissions tests were performed at the spot frequencies conforming to BS 833 (3). Tests were also made at those frequencies determined by the conducted interference method mentioned above. These open air measurements were made with the system powered (ignition on) both with and without the engine running. The ambient electromagnetic environment was recorded several times during the tests to ensure that errors caused by external influences could be eliminated from the results.

Susceptibility Test Programme

The susceptibility tests were performed on the vehicle in a large screened anechoic chamber (3), specifically designed for vehicle susceptibility testing shown in Fig. 2. The vehicle was positioned on a rolling road throughout the tests. Again the twisted pair vehicle was tested first.

The vehicle was monitored using the test tools described in a previous section. The twisted pair bus was monitored via a fibre optic link using a small internally powered electro-optical converter. The fibre optic bus was monitored via an output of the optical mixer. This optical fibre link extended into the personnel chamber.

The susceptibility tests concentrated on the frequency range up to about 200 MHz, with only brief excursions above this level. This was to ensure that as complete a comparison as possible could be made between the systems.

CONCLUSIONS

The results and conclusions will be given during the presentation.

REFERENCES

1 BOYCE, C.R. "A Four Station Controller Area Network"
IEE Symposium: "Vehicle Networks for Multiplexing and Data Communication"
1988 Digest No. 1988/138

2 ROVER GROUP "EMC - Susceptibility"
BLS.62.21.626
EMC - Susceptibility to External Electromagnetic Field.
BL Engineering Standard

3 British Standards Institution "Specification for Radio Interference Limits and Measurements for the Electrical Ignition Systems of Internal Combustion Engines"
BS 833 : 1985

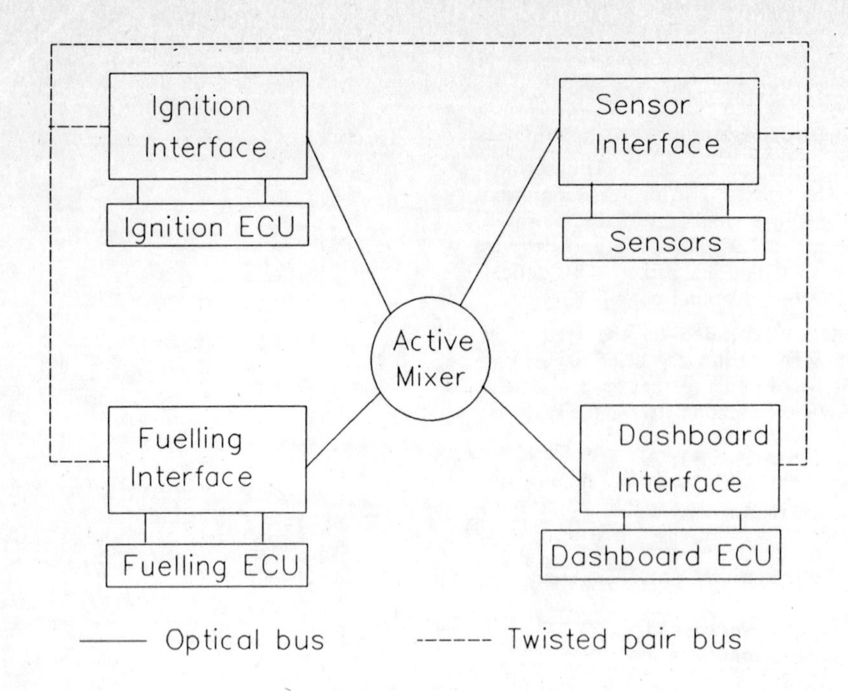

Fig 1 Schematic diagram of system

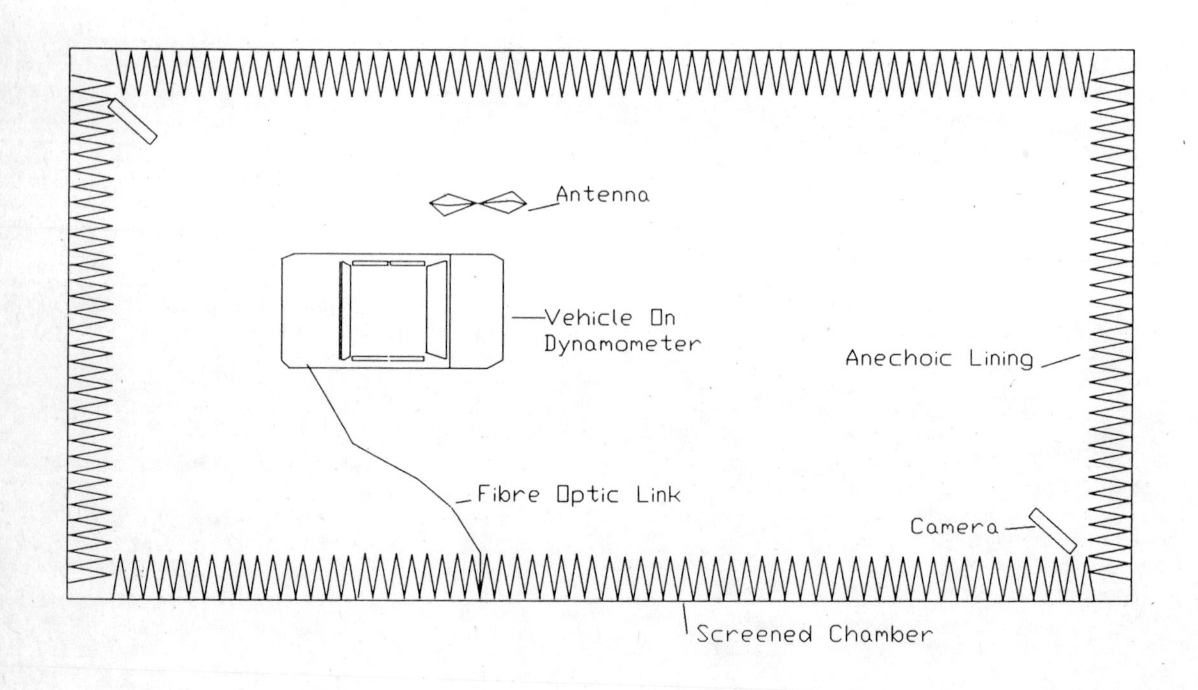

Fig 2 Vehicle arrangement during susceptibility tests

C391/030

An overview of a range of novel automotive sensors

E N GOODYER, MSc, CEng, MIEE
Sira Limited, Chislehurst, Kent

SYNOPSIS This paper is a relatively non-technical summary of sensors developed at Sira Ltd which may be put to automotive use. The following sensors as developed for various companies will be presented:

A range of low cost steering wheel position sensors which offer resolutions better than 0.5 degrees.

An optical fuel flow sensor that detects light scattered by particles moving along with the fuel.

A highly accurate electromechanical road tanker liquid level sensor and its' derivatives including a low cost sensor for use in car petrol tanks.

A "Dim Dip" which optically detects an oncoming car or the tail lights of a preceding car and automatically dips the car's main beam.

An optical linear displacement sensor, which uses a handful of widely available optical components. Possible applications include suspension arm extension, and wheel to road measurements.

1 STEERING WHEEL POSITION SENSORS

Sira Ltd has developed a range of steering wheel position sensors for First Inertia Switch Ltd. These sensors have been specifically designed for a number of European car manufacturers to meet their differing specifications and requirements.

High and low angular resolutions from 2 degrees down to 0.3 degrees can be achieved. Absolute zero position signals can be included for one or more turns of the steering column. A variety of signal outputs can be provided depending upon the input specification for the unit that accepts data from the sensor. The basic unit provides a pair of digital pulsed outputs that represent left and right turns of the steering column, but existing systems may prefer to receive quadrature signals and to derive rotation, which can also be supplied. Analogue outputs can also be added if required, and basic signal processing (eg rate of change of angle) can be incorporated with the addition of a microcomputer.

1.1 Confined Space Option

An early requirement was to design a steering position sensor that would fit into the confined space between a 23mm diameter steering column with an outer sleeve of 35mm diameter, with a resolution of better than two degrees. This allows only 4mm total space between the column and the outer sleeve for the sensor optics. The technique employed in currently available sensors is to obtain quadrature signals from two adjacent detectors viewing an LED through a slotted wheel (see fig 1.a). The wheel would require 45 slots 1.2mm wide at the edge, down to 0.8mm wide next to the steering column inner shaft. The detectors would also need to have a similar geometry and be carefully positioned. Such an approach is unlikely to provide a low cost solution.

Sira overcame the confined space problem by using a wedge of light guides as shown in figure 1.b and a chopper wheel with only 24 2mm wide slots. The light guides overcome the problem of mounting the detectors accurately in the confined space, as the wedge can be assembled separately and 'slotted' into place. The light is then taken externally and can be detected by any reasonable detector regardless of it's geometry. Using 4 light guides effectively quadruples the resolution available from the chopper wheel, which can be more easily manufactured because of the wider slot width.

Simple thresholding electronics provide a set of pulse trains that is decoded by a PLD to provide a pair of pulsed outputs that represent left and right rotations of the steering column, so that the car does not have to decode quadrature signals. An error signal is also available should a fault be detected.

The sensor resolution is 1.875 degrees, with an accuracy of better than 0.2% of full scale, and can operate up to 1500 degrees per second.

1.2 Use of Photodetector Array

Later models of the sensor use a photodetector array (figure 1.c). instead of the light wedge. Use of the array reduces the mechanical complexity of the sensor, and improves the electronic performance. The active elements are laid down to match the the geometry of the light guides, thus retaining the advantage of increasing the slot width resolution. In addition the detectors are inherently matched which simplifies the electronic design and setting up procedure, and light losses are minimised, improving detector response.

The sensor package size has been further reduced by incorporating some of the signal processing electronics in the array package.

1.3 High Resolution Solution

Higher resolutions are usually obtained by either reducing the slot width or increasing the diameter of the chopper wheel. Ultimately however this approach is limited by the actual geometry of the steering column assembly. Smaller slot widths will also eventually stop the sensor functioning reliably, if at all, usually because there is insufficient light available to obtain a response from the detectors. As the geometries get smaller so too the manufacturing difficulties will increase.

Sira improved the resolution of the basic sensor concept by an optical technique that has the added advantage that the collecting area of the detector can actually be larger than the slot width. Figure 1.d shows the optical arrangement of the high resolution sensor. The concept is the same in that a chopper wheel is placed between a light source and a detector, but in addition a second identical chopper wheel is added and a lens is used to image the first wheel onto the second wheel. Because the lens inverts the image of the first wheel as the wheels move in one direction the image moves in the opposite direction, effectively doubling the resolution of the slot pitch. The signals seen by the detectors are a pair of zero order Moire fringe patterns as used by many optical linear displacement sensors as opposed to a series of ON & OFF pulses.

The detectors can now collect light from the whole of the fringe pattern. This means that large sensor areas can be used, which has the dual advantages of low price and greater sensitivity.

2 FUEL LEVEL SENSOR

The fuel level sensor was originally developed for Drum Instrumentation Ltd for use in petrol delivery road tankers and is now in full scale production. The success of this development has resulted in the design of a number of spin-off products, most notably a robust and accurate density probe for use with a range of fluids, and tank contents gauges for use with different fuels such as LPG.

The road tanker sensor is approved by the United Kingdom Weights and Measures Department and is now in every day use. It is accurate to 0.15% and is available in a range of sizes capable of measuring tank volumes up to 1500 gallons.

Recently the Ford Motor Company sponsored Sira to investigate whether or not the design could also be used as the basis of a low cost level sensor for use in car fuel tanks. This study was successful, and a new low cost sensor is now under development.

2.1 Operating Principle

The sensor's operating principle was derived from an existing concept developed by Marconi for measuring the level of electrolyte in aircraft batteries. The sensor is a tuned electromechanical resonator, which is achieved by attaching pairs of piezo electric crystals to the orthogonal axes of an aluminium rod. One crystal vibrates the rod whilst the other acts as a pickup. The loop is closed electronically by a phased lock loop circuit. When the rod is immersed in a fluid the resonant frequency changes. As the amplitude of the flexural vibrations created is extremely small (less than 1 micron), the change in resonance is primarily due to the mass of fluid that adheres to the rod, and not to sheer (or viscous) forces in the fluid. The sensor therefore measures true mass. Tank contents is calculated by an integral microcomputer using this data and information obtained from a number of other sensors that monitor fluid density and temperature. A calibration table of the tank shape is stored in a look up table, and is used to convert the level reading into a volume. Figure 2.a shows a schematic of the sensor, figure 2.b is a picture of a real sensor.

The basic design gives rise to an extremely accurate sensor; this is needed in order to meet the strict Weights & Measures requirements applying to road tanker operation. However variations of this design are available at a lower cost and can also boast excellent performance specifications. The density probe can be obtained as a separate item and has found a wide range of applications, most notably in oil exploration drilling operations.

3 DIM DIP SENSOR

First Inertia Ltd engaged Sira to develop a Dim Dip sensor that would be capable of automatically dipping a car's headlamps at night whenever an oncoming vehicle is seen on the opposite side of the road, or when the car comes close to a preceding car. In this respect the sensor is in advance of its competitors in that it is sensitive to both white headlamps and red tail lights.

3.1 Principle of Operation

Detecting an oncoming bright light source in itself is not a problem. The difficulty with this application is that the sensor has to 'look' at the correct place, and respond to both white headlamps and red tail lights in a different direction. The bulk of energy radiated from a tungsten lamp is in the infra red, and as the sensor is also required to respond to red tail lights a silicon detector (which has its sensitivity peak in the near infra red) is used. An infra red detector is also less sensitive to ambient visible light. The correct field of view is obtained by a low cost lens and a barrel shaped mask that takes account of the lens aberrations. Simple electronics threshold the detected signal and provide a switch output. Figure 3.a is a schematic of the sensor operation, and figure 3.b is a picture of an actual production sensor.

4 FUEL FLOW SENSOR

Sira developed the Fuel Flow sensor for the Ford Motor Company. The first prototype was constructed in 1986 , and was handed over to Sira's independent Instrument Evaluation Division for evaluation. The results of these tests showed that the sensor's optical principle is viable, but highlighted problems with the performance of the electronic signal processing unit, which is the subject of further research. The fundamental measuring principle is however sound, and can be used in other similar applications.

4.1 Operating Principle

Small particles of contamination are always present in fuel, and are carried along with it through the fuel lines. It is the presence of these particles that is the key to the sensor's operating principle. A transparent tube is placed in the fuel line and is illuminated by off axis light sources (see figure 4.a). such that no light will normally appear at the collecting lens. Particles carried along in the fuel will scatter the light so that some will fall onto the lens. The lens focuses the image of the illuminated particles onto a Fresnel beam splitter, which sends light alternately to the two collecting channels. The outputs from these channels are sinusoidal signals produced by particle images travelling along the length of the beam splitter. These signals are in antiphase.

The difference between the outputs of the two collecting channels is determined by the electronic signal processing unit. This will enhance the information obtained from the antiphase flow signals and remove common mode noise. The resulting sine wave output is squared up and used to generate a pulse train, the frequency of which is directly related to the fuel flow rate.

4.2 Sensor Assembly

Figure 4.b is a schematic drawing of the prototype sensor construction. This schematic shows fibre optic light guides as the means to bring the illumination into the sensor body and to take the collecting channel return signals away. This permits the electronics to be housed remotely from the sensor, away from the harsh engine environment. A lower cost solution was also developed which replaced the fibre assemblies with direct illumination and sensing using automotive light bulbs and low cost silicon detectors. Either approach is viable.

Figure 4.c shows the finished prototype sensor.

All optical components other than the tube are of plastic construction. The tube is glass, but could be substituted with plastic in a final model. The sensor does not use costly components or materials, and is a good example of how optical sensing can provide a low cost solution that is inherently immune to the electrical noise problems that are always present in the engine compartment.

4.3 Sensor Evaluation

A series of tests were carried out on the prototype sensor primarily to establish whether or not the optical sensing principle was viable in practice. Of particular interest in this application would be the repeatability and transfer function of the sensor.

Figure 4.d shows the results of the repeatability tests. It can be seen that the sensor is nonlinear, which is probably due to the fact that the flow profile front varies with velocity when under laminar flow conditions. The sensor is repeatable, but the margin of error is about 5%.

These results gave sufficient confidence in the optical technique to justify further development of the electronics unit, which was considered to be the main source of the 5% uncertainty on the measurement.

5 LINEAR POSITION SENSOR

Linear position detection has a wide range of applications in the automotive field, such as suspension arm position and wheel to road position. Sira has recently developed a prototype low cost optical sensor that could solve some of these difficult measurement problems. As yet no manufacturer has come forward to exploit this principle.

5.1 Operating Principle

Referring to figure 5, the sensor consists of a pair of detectors one of which is located at the focal point of a lens, the other is located in the same plane but shifted vertically. A light source is placed at the other end of the system, and is free to move with respect to the pair of detectors and lens. The detector at the focal point is chosen such that it is overfilled by the light that is focused onto it by the lens. The other detector is placed such that no light falls on it when the source is furthest away, as the light source moves closer the amount of out of focus light falling on this detector increases. It can be shown that the signal from the detector placed at the focal point is independent of light source distance from the lens (see appendix). This signal can be used as a normalising reference. The output from the other detector is related to the inverse square of the separation between light source and lens, and therefore is used as the basis of the output signal.

This principle is simple and low cost, but has the disadvantage that a light source and the detectors have to be fitted to different ends of the moving part. This can be overcome by using a mirror or a retroreflective system, but such an arrangement must be kept clean and free of contamination.

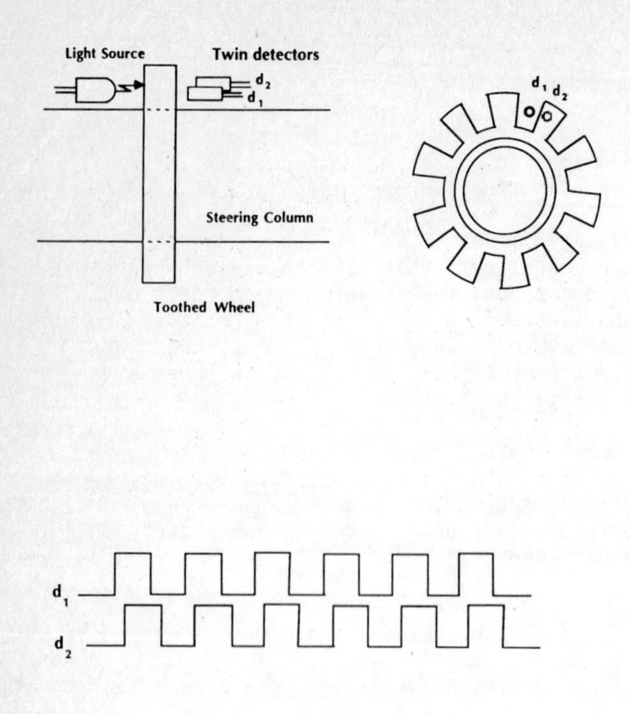

Usual Quadrature Type Angular Position Sensor

Fig 1a

Fig 1c

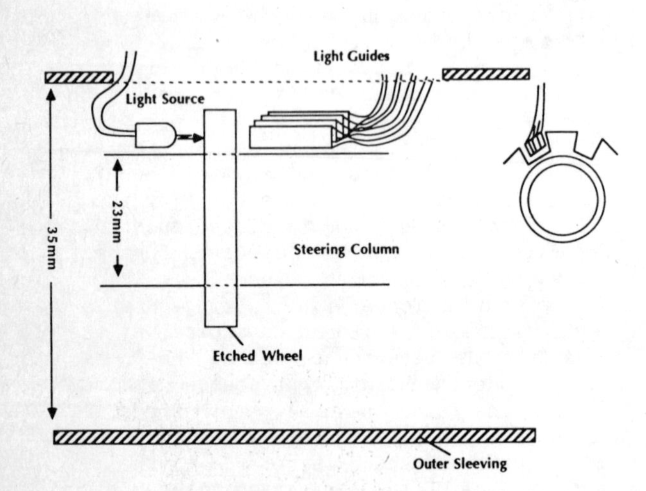

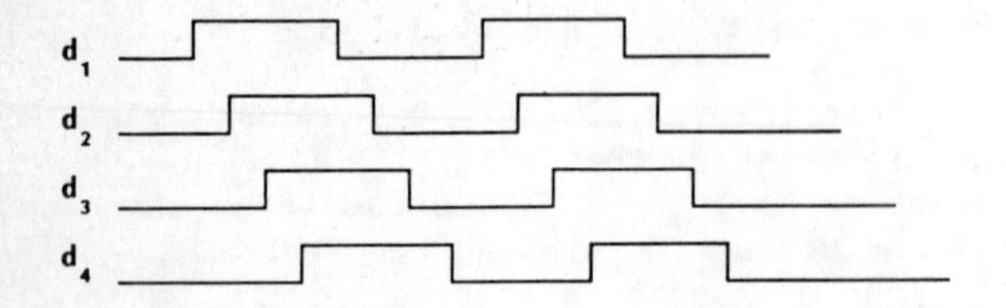

Light Guide Method

Fig 1b

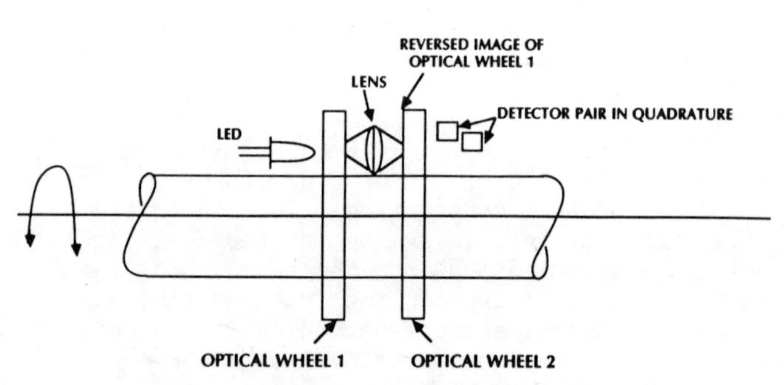

HIGH RESOLUTION POSTION SENSOR

Fig 1d

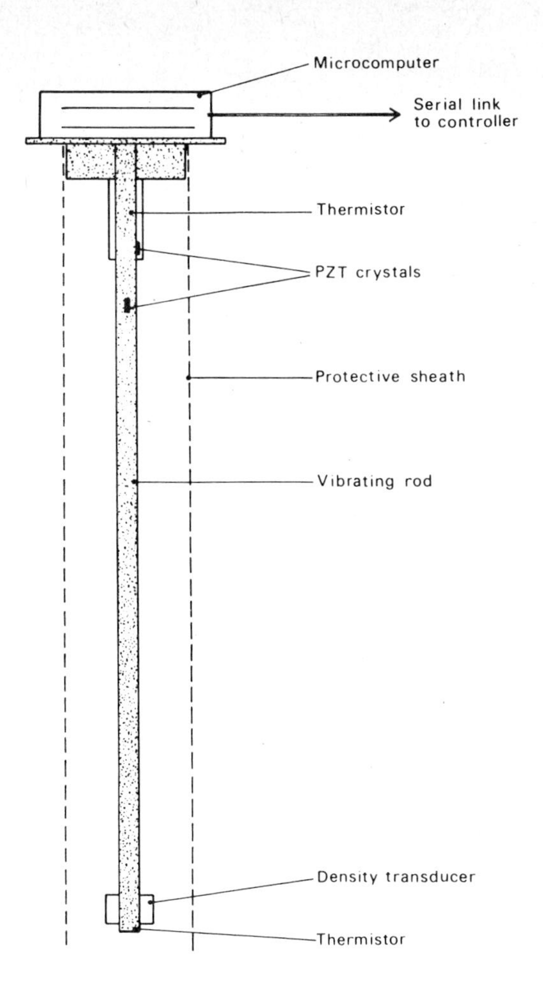

Fig 2a

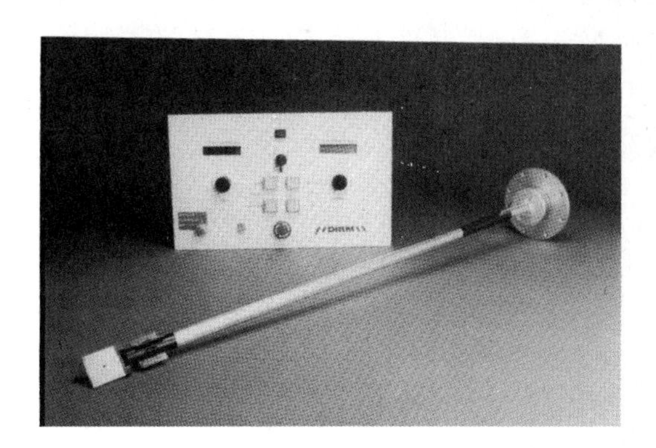

Fig 2b

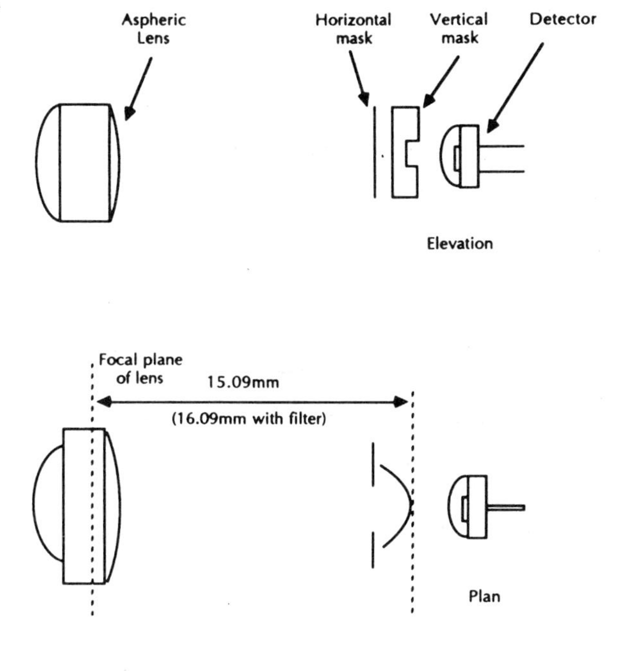

Optical layout of the Sira/FIS Autodipper

Fig 3a

Fig 3b

light sources
fuel flow
scattered light
collecting optics
collecting channels
fresnel beamsplitter

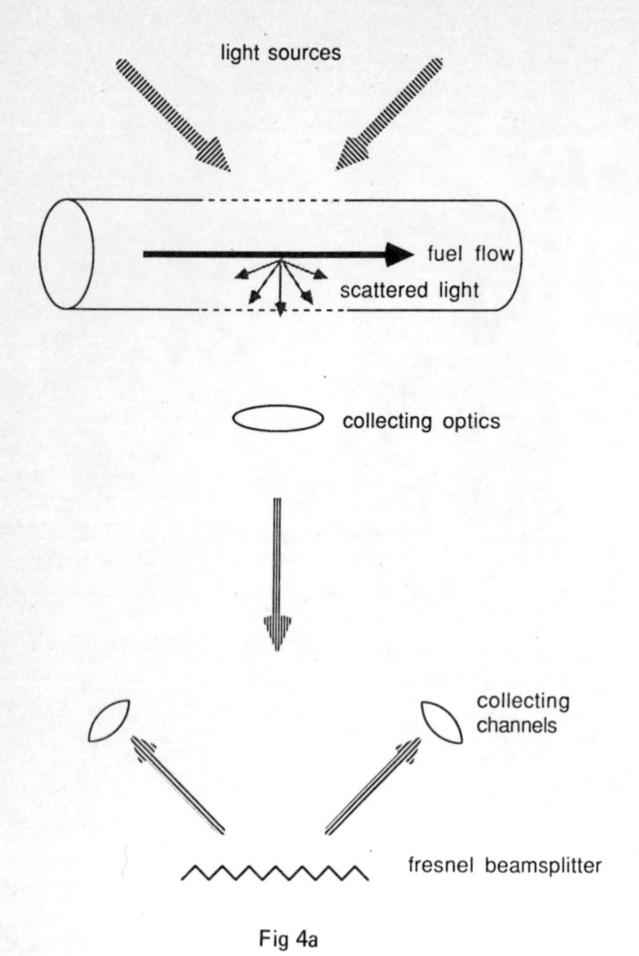

Fig 4a

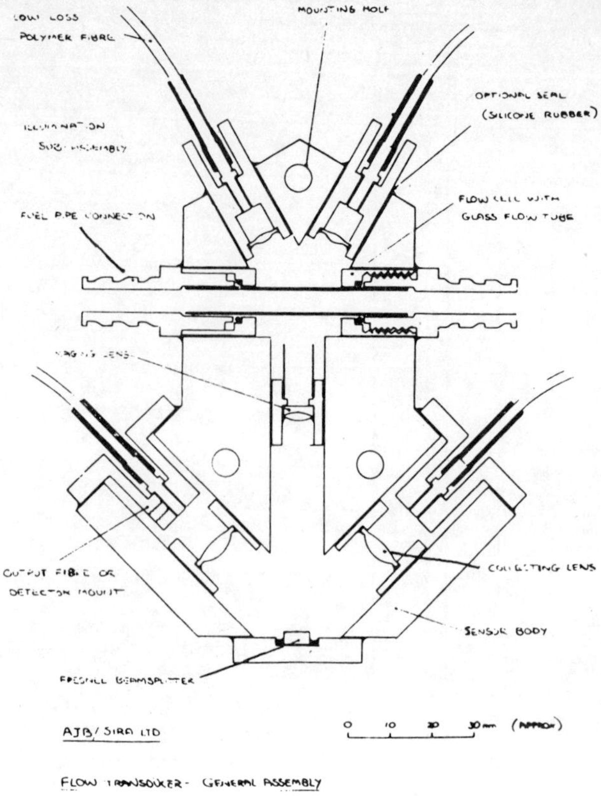

LOW LOSS POLYMER FIBRE
MOUNTING HOLE
OPTIONAL SEAL (SILICONE RUBBER)
FUEL PIPE CONNECTION
FLOW CELL WITH GLASS FLOW TUBE
COLLECTING LENS
SENSOR BODY
FRESNEL BEAMSPLITTER
AJB/SIRA LTD
0 10 20 30mm (APPROX)
FLOW TRANSDUCER - GENERAL ASSEMBLY

Fig 4b

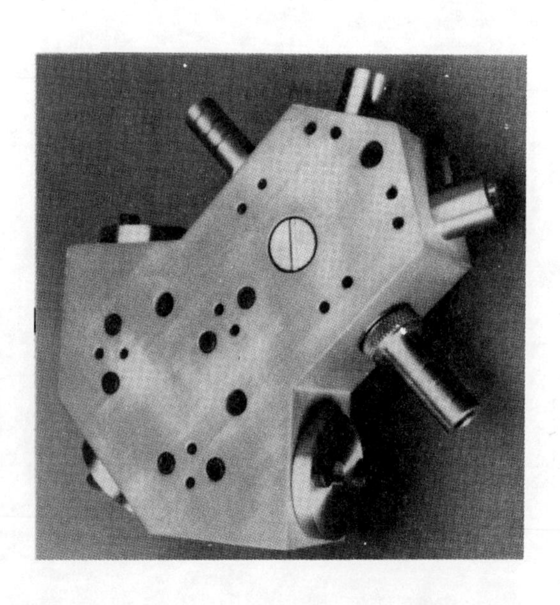

Fig 4c

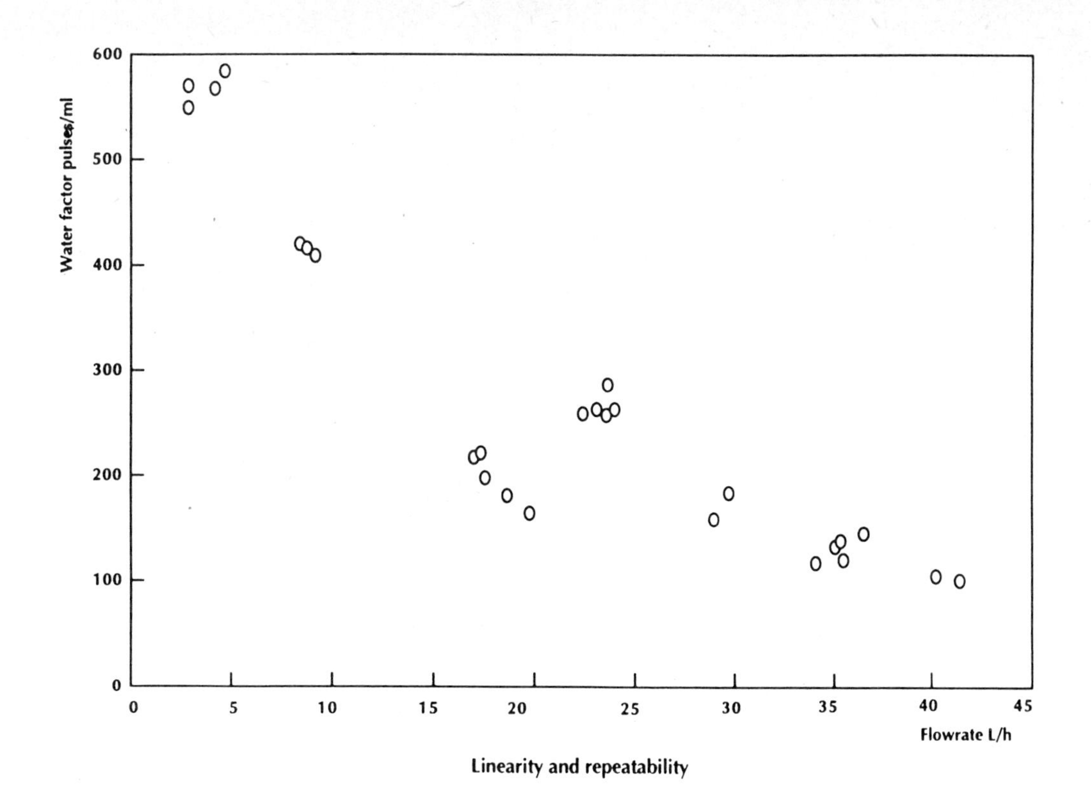

Linearity and repeatability

Fig 4d

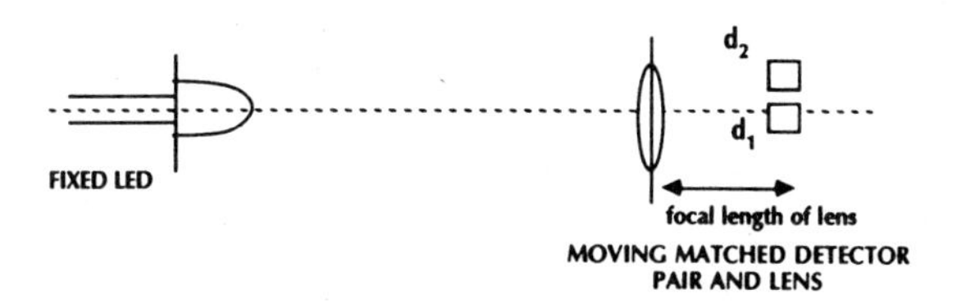

LOW COST OPTICAL POSITION SENSOR

Fig 5

APPENDIX

Requirement: detected power independent of source/detector distance

simple approximation

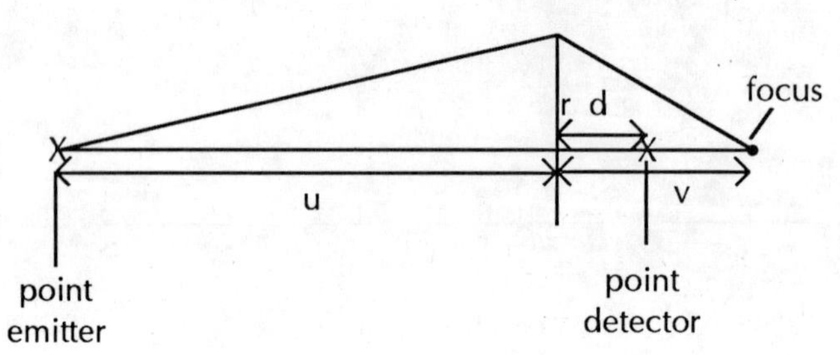

equivalent radiated power of emitter: P (ie power if radiated uniformly in all directions)

power intercepted by lens: $Q = \frac{P\pi r^2}{4\pi u^2}$ $(u >> r)$

radius of illuminated area in detector plane: $r_1 = \frac{(v-d)\, r}{v}$

power per unit area at detector: $i = Q/\pi\, r_1^2$

$$\Rightarrow i = \frac{P\pi r^2}{4\pi u^2}\left(\frac{v}{(v-d)r}\right)^2 \frac{1}{\pi} = \frac{P}{4\pi u^2}\,\frac{1}{(1-\frac{d}{v})^2}$$

$$\frac{1}{f} = \frac{1}{u} + \frac{1}{v} \quad \Rightarrow \quad i = \frac{P}{4\pi u^2}\,\frac{1}{(1-d(\frac{1}{f}-\frac{1}{u}))^2}$$

$$\Rightarrow i = \frac{P}{4\pi}\,\frac{1}{(u(1-\frac{d}{f})+d)^2}$$

if set $d = f$

then $i = \frac{P}{4\pi f^2}$

(independent of u)

Hence place detector at focal distance from lens

note: with real emitters and detectors, emitter must overfill lens which in turn must overfill detector. Also $u >> r$.

C391/028

Magnetoresistive sensors for navigatio

A PETERSEN
Philips Components Application Laboratory, Hamburg, West Germany

SYNOPSIS For car navigation systems two sensors are necessary. A compass sensor measures the orientation of the car relative to the earth. Wheel sensors deliver information on movement and direction. Sensors based on magnetoresistive effect can be used with advantage for both tasks.

1 INTRODUCTION

Car navigation systems are of increasing interest to many companies (1), (2), (3), (4). In contrast to naval or aircraft navigation, automotive navigation need not be so highly sophisticated because of the possibility to compare maps with visible position and road signs. Especially in connection with high road and traffic density, however, or in foreign countries, this common kind of visual navigation consumes much time and fuel.

In the past no economical medium for the mass storage of topographical information was available. The situation has now changed. With the introduction of CD-players into cars, it will be possible to read-out road data stored in CD-ROMs with very large storage capacities (3), (4).

2 POSITION DETERMINATION

The most elegant system is satellite navigation, but during the next few years a system such as the one described here (Fig. 1) will be optimal. It is based on drivers' initial input of start and destination points. Later on, current positions are the result of sensor-based calculations compared with stored road information.
A compass system measures the direction of the earth's magnetic field and so determines the car's direction of movement. The magnetic compass can be based on flux gates or magnetoresistive sensors; the latter will be described here. At the moment, only two-dimensional systems are under consideration.

A wheel sensor measures the car's progress and should be able to distinguish forward from backward driving. The compass may sometimes be disturbed by irregular magnetic conditions near iron structures like bridges. For this reason a second sensor at the other wheel of the same axis is desirable. The difference between the two signals would indicate direction variations, and allow short-distance position calculation.

Two magnetic principles may be used. The first one operates with permanent magnets or better with a multipole strip of plastic magnetic material bonded to the wheel rim (3). Two magnetic field sensors aligned with each other can determine the direction and revolutions of the wheel.

A second method uses a moving iron part with periodic structure, for example a so-called pulse ring or a teeth wheel. A combination of sensor and permanent magnet detects field variations caused by the movement of the periodic structure. This method is very common in antiskid systems. For direction detection, two sensors would be necessary. Sensors may be based on inductive pickups (variable reluctance), Hall-elements, field-plates, or magnetoresistive elements. The last will be discussed in more detail.

Continuous comparison of calculated and stored road data reduces integration errors and additionally compensates for the disturbing and unfortunately erratic magnetic field of the car itself.

3 MAGNETORESISTIVE SENSORS

The magnetoresistive effect describes a relationship between the magnetic fields influencing a body and its resistance. The sensor under consideration makes use of this effect in permalloy, a ferromagnetic material consisting of iron and nickel (5), (6). Here resistance depends on the angle between the direction of magnetization and that of the measuring current. If the sensor is designed with a large number of parallel strips to increase absolute re-e to practicable levels, and are no external magnetic fields, these strips will be magnetized tly in the length direction (Mo, g. 2). This result is due to a manufactured anisotropic structure and geometric configuration.

When a magnetic field Hy is applied perpendicular to the length, a plain element demonstrates nonlinear behavior as shown in the upper part of Fig. 2. According to the strength of the field Hy, the magnetization M moves in the Y-direction until an angle of 90 degrees is attained. After that, no further movement is possible. The two possible directions of original magnetization (+Mo, -Mo) and of the external field (+Hy, -Hy) do not influence the resistance.

A very different behavior with a linear response can be obtained from an additional layer of much better conducting, relatively thick aluminum strips deposited on the permalloy elements. In this case, the current I does not flow parallel to the length. Instead, it takes a path as shown in the lower part of Fig. 2. This linearizing technique results in an asymmetrical behavior as well. The relation relation between resistance and field Hy depends on the Mo-direction, which is very useful for the compass application.

To get a good sensor, four groups of parallel strips with two different geometries are arranged to form the four arms of a Wheatstone bridge. This is enclosed in a small plastic envelope.

For most applications, it is necessary to establish an auxiliary magnetic field Hx along the length of the strips (Fig. 2). Otherwise well-defined behavior and response would not be certain. Typically, the output signal is about 1 - 2% of the bridge supply voltage. Different geometries of the sensing elements also make different sensor sensitivities possible. Thus magnetoresistive sensors can be used for a highly sensitive compass as well as for wheel measurements.

4 COMPASS

The most sensitive bridge-type magnetoresistive sensor has a measuring range of about + 500 A/m and a sensitivity of about 0.1 mV/A/m with a supply voltage of Vb = 5 V. From the earth's horizontal field maximum of 15 A/m, a signal of about + 1.5 mV is thus available. With a typical offset drift of + 3 uV/VK for the sensor itself and about + 7 uV/K for a standard operational amplifier, an uncertainty of about + 0.88 mV is possible in a temperature range of -20/+60 degrees C with Vb = 5 V. Relative to the sensor range this is a reasonably small value, but it is insufficiently accurate without extensive compensation for temperature drift. To improve this result, the use of magnetic field concentrators (antennas) has been suggested (7).

A more elegant way to overcome the problem is to use the two possible sensor responses described in Fig 2. When the "no signal" magnetization Mo is switched between the two directions, offset voltages can be separated from the magnetic field signals. The switching can be accomplished with periodic positive and negative current pulses through a coil around the sensor (Fig 3). While the offset voltages remain constant, magnetically caused signals alternate between the two responses. This yields rectangular ac-voltages, and separation from the drifting dc-voltages is electronically possible.

Normally the sensor dies are designed so that the sensitive Y -direction is parallel to the pin direction as shown in Fig 6. For compass applications, another design with the X-direction parallel to the pin direction is preferable. This makes arrangement of the coils much easier (Figs. 3, 4).

A workable sensor needs a field strength of about 3 kA/m for switching. This can easily be achieved with a coil arranged around the sensors as in Fig. 4. Another possibility, which is optimal for very short pulses and/or reduced stray fields, is a winding on the sensor capsule itself. This allows pulse times of about 1 us, in contrast to about 100 us with larger coils (Fig. 4). The main advantage of short pulses relative to the pulse frequency is that their interference with the square-wave output of the sensor can be neglected. Thus electronic suppression or reduced precision need not be considered.

Fig. 4 shows a block circuit and mechanics for two-dimensional field measurement. A square-wave generator with a frequency of say 100 Hz delivers short alternating current pulses to a coil connected with a coupling capacitor C. Capacitive coupling ensures that no small but disruptive dc current appears between pulses.

Another very important component in this circuit is a damping resistor R, which prevents oscillations. Only aperiodic pulses switch optimally.

The resulting sensor signals have to be amplified and separated with a capacitive coupling in an amplifier. Afterwards a synchronous detector controlled by the square-wave generator rectifies the ac-voltages into dc-voltages V(H1) and V(H2) at the output.

When both sensing lines are calibrated to the same sensitivity, by perhaps an adjustment to one amplifier, the magnetic field direction can be derived from the relation between the two signals.

A compensation for temperature drift is unnecessary in case of similar values from both sensors.

Normally the earth's magnetic field is superimposed with other constant fields, especially from the car's bodywork but also from magnetic components like iron wires in the sensor circuit. In that case it is necessary to use electronic compensation. The field signal to be neutralized can be found by driving the car around a traffic circle, or better, because always possible, by comparing measured and stored road data. This method makes temperature compensation superfluous as well, except when high temperature stability of sensor signals is required. A change from constant voltage supply to constant current supply reduces the corresponding coefficients of sensitivity drift from -0.4%/K to about -0.1%/K. With temperature dependent signal-amplification, full compensation is possible.

This method yields a ten to hundred times higher accuracy than ordinary measurements, with chances for further improvement. Limiting factors are the tolerances in temperature behavior and the two magnetization directions Mo. Due to less than ideal material properties, some slight deviations appear. These are visible in fluctuating signal levels, a kind of noise with an equivalent field strength uncertainty of about 0.1 A/m. Another limitation comes from the generation of magnetic fields in the sensor itself. The fields of the measuring currents are in principle compensated for, but tolerances may increase by several tenths of an A/m. In connection with the general compensation for all car or circuit internal fields, this influence will be suppressed.

If the fields to be measured are larger than about 250 A/m, this measuring principle will not work satisfactorily. The usable range is therefore about 0.25 to 250 A/m, sufficient to measure the earth's maximum horizontal magnetic field of 15 A/m with an error of a few percent. This is equivalent to a direction accuracy of a few degrees, sufficient for road navigation based on low-priced equipment.

5 WHEEL SENSORS

Wheel sensors are common in modern cars. They are used in antiskid systems. Until now they have been based on inductive signal generation. Coupled with the wheel is a ferrous pulse ring often similar to a teeth wheel. It moves in the stray field of the inductive sensor and generates signal voltages. This is a very rugged technology with reasonable costs, but as Fig. 5 shows, it has a big disadvantage. At low velocities (a few km/h), the signal diminishes. This behavior and a high sensitivity to vibration are the reasons for major efforts to develop static sensors for antiskid systems. Static sensors detect field variations caused by the pulse ring's position and not by its velocity like inductive sensors.

Fig. 6 shows how to use magnetoresistive sensors for this purpose. The sensor is attached to a magnet and connected with a comparator. Away from the pulse ring the magnetic field is distributed symmetrically, and in the sensor plane there are two small Hy field components in two different directions, which are compensated for in the Wheatstone bridge. Iron parts like the pulse ring passing in front of the sensor bend the field lines. This causes the sensor to deliver a signal which may switch the comparator.

A necessity for this kind of wheel sensing is a high accuracy in the magnetic, geometric, and electronic design to guarantee that the measured signals at the comparator input are larger than all possible offset signals. For direction determination two sensors at one wheel separated by a quarter of the teeth period are necessary. Thus four sensors form an optimal wheel sensing system.

Another possibility for wheel speed measurement is to attach permanent magnets or plastic moulded permanent-magnet strips with alternating polarity to the wheel rims. Then a direct measurement of the field is possible. Arrangement and circuits are similar to those described above. The advantage is that this system can be installed after production of the car.

REFERENCES

(1) Haeussermann, P.
On-Board Computer System for Navigation, Orientation and Route Optimization
SAE Paper 840485

(2) Heinrich, B.F., Vrabel, R.J.
Vehicle Compass and Outdoor Thermometer System
SAE Paper 840316

(3) Thoone, M.L.G.
CARIN, a car information and navigation system
Philips Techn. Rev., 1987, Vol. 43, No. 11/12, 317 - 329

(4) Fernhout, H.C.
THE CARIN CAR INFORMATION AND NAVIGATION SYSTEM AND THE EXTENSION TO CARMINAT
Sixth Int. Conf. on AUTOMOTIVE ELECTRONICS, London 1987, 139-143

(5) Dibbern, U.
Magnetic field sensors using the magnetoresistive effect
Sensors and Actuators, 1986, Vol. 10, Nos. 1 & 2, 127 - 140

(6) Petersen, A.J.
1983, Magnetoresistive sensor for automotive applications
Fourth International Conference on Automotive Electronics, London 1983, 8 - 13

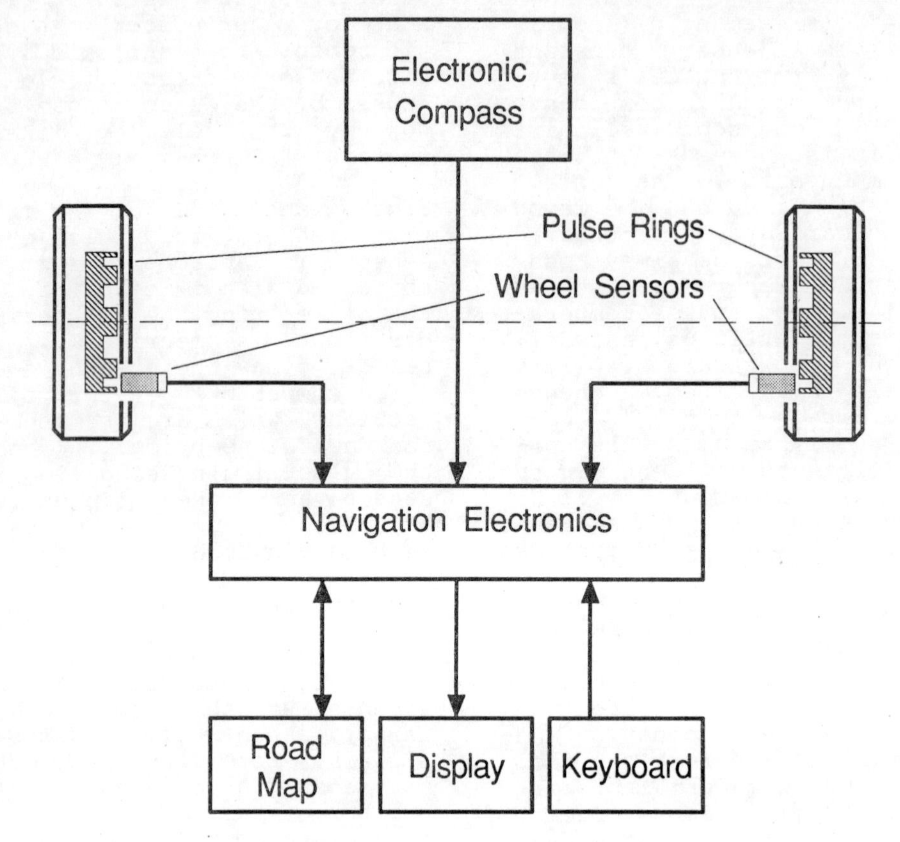

Fig 1 Block diagram of a navigation system

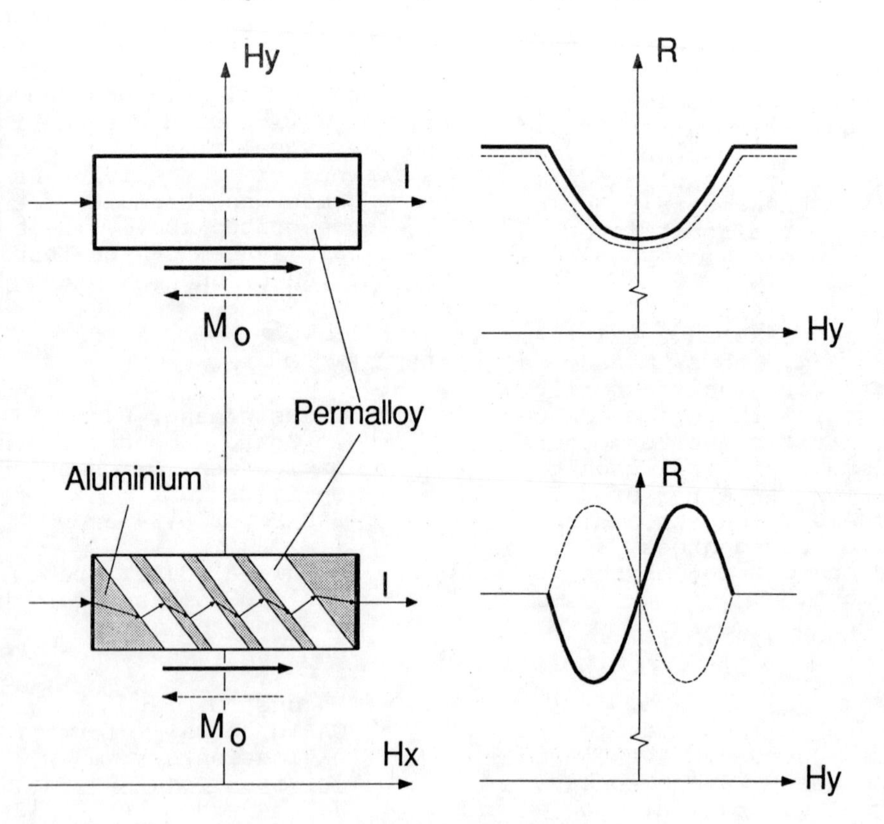

Fig 2 Magnetoresistive sensor elements

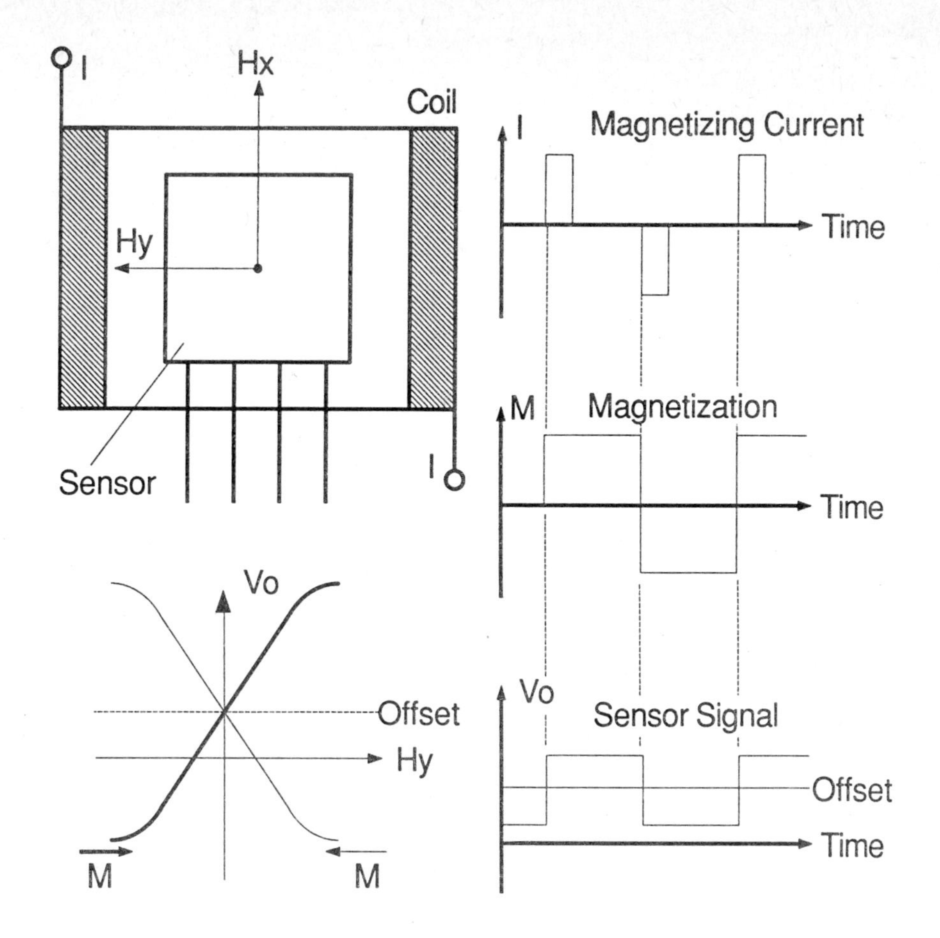

Fig 3 Switched magnetoresistive sensor

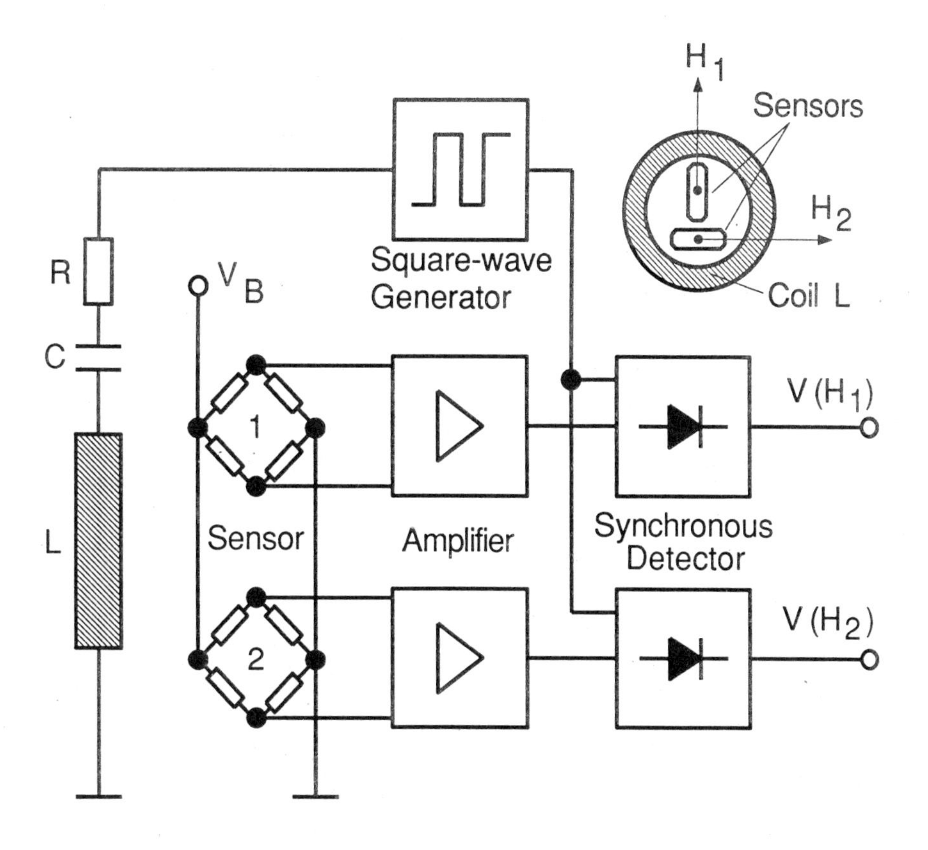

Fig 4 Sensor arrangement and block diagram of a two-dimensional compass system

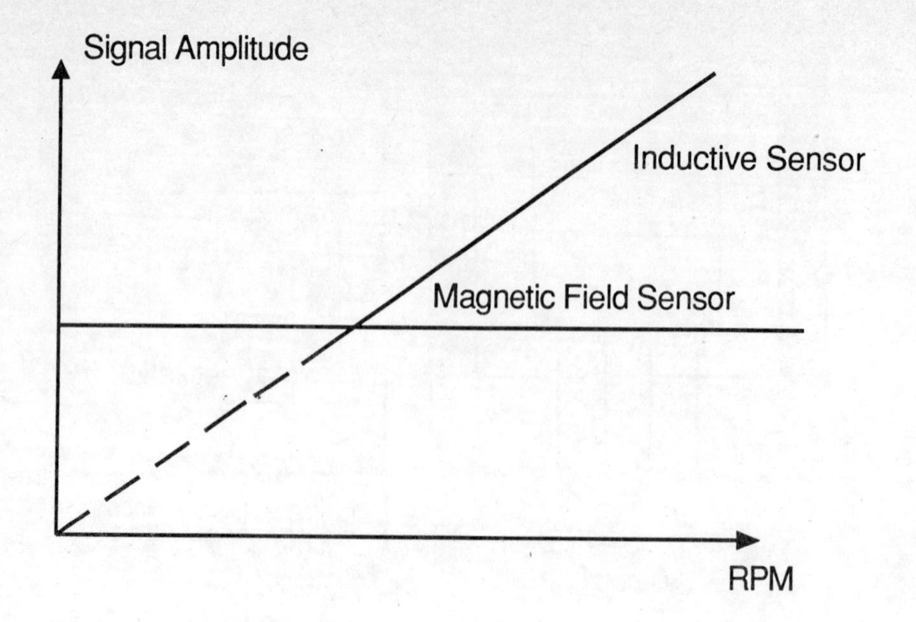

Fig 5 Principial behaviour of inductive sensors and magnetic field sensors for wheel sensing

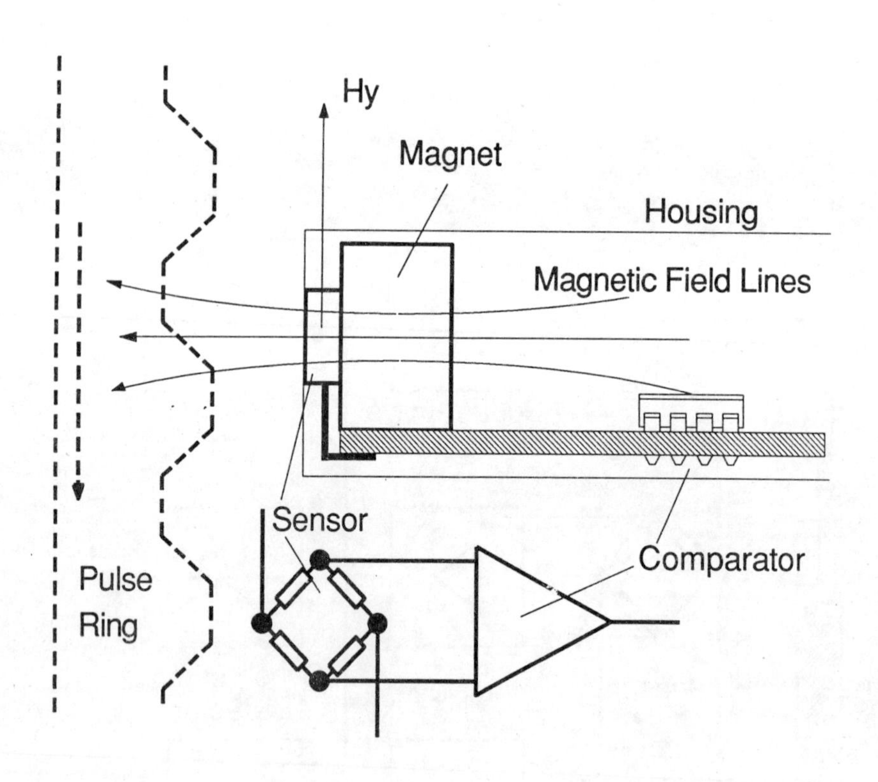

Fig 6 Wheel sensor with magnetoresistive element

C391/059

A differential Hall IC for geartooth sensing

R PODESWA, PhD
Siemens AG, Nuremburg, West Germany
U LACHMANN, Dipl-Ing(FH)
Siemens AG, Munich, West Germany

SYNOPSIS Over the years, various rotation sensors have been developed and applied in automotive systems. These can be classified into two groups according to the sensing technology: passive sensors and active or solid-state sensors. Most of these sensors in common use are of the electromagnetic induction type, otherwise known as variable reluctance. Such sensors exhibit some well known disadvantages which can be avoided by the use of solid-state sensors with either differential hall-effect or magnetoresistive cores as the sensing element.

This paper discusses the progress in the development of solid-state sensing elements and their application. In particular, the newly developed differential hall sensor IC (TLE 4920) offers very economical solutions for various sensing tasks. The advantages of this sensor are high sensitivity, high surge operating temperatures, static or dynamic operating mode and complete input and output protection circuitry. This allows new concepts in rotation sensing to be considered, for example the integration of the sensor within the sealed part of a bearing.

1. INTRODUCTION

The increased use of electronic automotive systems requires a change in the type of sensor employed. Automobile manufacturers have used non-contacting rotational speed sensors in applications such as anti-lock brakes, traction control, engine and transmission management, power steering, electronic speedometers and on-board instrumentation for navigation and fuel consumption calculation. Sensors for these fields of application must meet very stringent requirements such as high operating temperature, long-term stability, operating ability with large air gaps, and immunity against voltage peaks and electromagnetic influence. In addition, zero speed sensing capability is often required.

Over the years, various rotation sensors have been developed and applied, each exhibiting its own merits and disadvantages. Table 1 shows the well-known principles for currently-used sensors with performance comparisons indicating the general advantages and disadvantages of each.

The basic sensor technologies listed in this table are subdivided into three main categories:

- Cat. I: Wheel speed sensing technologies most commonly used,
- Cat. II: Special devices used for niche solutions,
- Cat. III: Magnetoresistors and Hall-Effekt ICs increasingly used in automobile manufacture.

Since category I type sensors have already been discussed in numerous publications, this paper will consider only category II and III type sensors.

2. CATEGORY II SENSORS

2.1 WIEGAND EFFECT

The Wiegand Effect occurs in magnetically bistable conductors. The construction of such a conductor is very similar to that of a coaxial wire. The core is made of a soft ferromagnetic material shielded with a harder ferromagnetic material. Under the effect of a strong magnetic field, both parts of the wire are magnetized to their saturation levels. It is then possible to switch the polarity of the soft inner core by using a reversed magnetic field strength which is not strong enough to change the polarisation of the outer material. The resulting magnetic pulse can be used to induce a voltage up to five volts in a pick-up-coil wound around the Wiegand Element (or "Impulsdraht, VAC"). A specially designed target wheel / sensor configuration can be used to obtain a variation in magnetic field strength according to the rotational speed which is sufficient to switch the element. The resulting voltage peaks are independent of speed and of near-constant amplitude. Wiegand Effect sensors are the only ones that provide zero-speed information without the need for an external voltage supply. Their main drawback is, that the voltage peak duration of approximately 20 to 50 microseconds leads to rise times which are filtered by typical automotive ECU's.

2.2 EDDY CURRENT

Sensors based on this technology are the only ones which are able to detect rotational speed without the need to use a ferromagnetic target wheel. In a typical configuration the flat end of the sensor probe contains a ferrite E-core (or bell-core) which holds an outer excitation coil and an inner detector coil. By using the outer coil the sensor produces a high-frequency magnetic field which induces eddy currents in any conducting material placed in front of the sensor. The resulting phase shift between the coils can be used to detect three different situations: a gap in the sensing wheel ($\Delta\phi = 0$ degrees), a ferromagnetic tooth on the wheel ($\Delta\phi = 90$ degrees) or a ferrite tooth ($\Delta\phi = 180$ degrees). Another concept of eddy current devices uses the modulation or damping of radio frequencies in the coil (or coils) caused by eddy current generation in a target wheel. They all allow zero speed detection but their principle disadvantage is the use of RF and the need for complex signal conditioning circuitry.

Table 1

Comparison of the various types of Speed Sensor

	Basic Technology \ Features	Zero Speed	Maximum Frequency	Resistance to Electrical noise	Resistance to Mechanical noise	Target Wheel Requirements	Square Wave * Output	Anticipated * mature costs
I	Variable Reluctance	No	< 100 kHz	Low	Low	High	No	Good
I	Reed Switch	Yes	< 1 kHz	High	Low	Low	Yes	V Good
II	Wiegand Effect	Yes	~ 20 kHz	Low	Low	High	No	Moderate
II	Eddy Current	Yes	< 500 kHz	Moderate	High	Low	Yes	Moderate
II	Optical	Yes	≫ 1.000 kHz	V High	V High	Moderate	Yes	Moderate
III	Hall Effect IC's	Yes	~ 1.000 kHz	High	High **	Low	Yes	Good
III	Magnetoresistors	Yes	~ 50 kHz	Moderate	High **	Moderate	No	Moderate

(*) For typical automotive sensors
(**) For differential type devices

2.3 OPTICAL DEVICES

These devices are rarely used in automobiles due to the inherently dirty environment. They are only chosen when there is a requirement for extreme high resolution.

All the above mentioned rotational speed sensors offer several special advantages but are not commonly used in automobiles.

3. GALVANOMAGNETIC SENSORS

The most frequently used solid-state magnetic sensors are semiconductors based on the Hall principle and magnetoresistive elements. Both types detect changes in the strength of a magnetic field by measuring the voltage caused by the deflection of moving current carriers due to the magnetic field. Semiconductors based on an alloy of indium antimony show the largest usable physical effect for solid-state magnetic sensor elements, due to their very high electron-mobility of 80.000 [cm^2/Vs]. This technology also offers high-temperature capabilities. For example, Siemens differental magneto resistors in tape automatic bonding packages are specified for temperatures as high as 200°C. The main disadvantage of these elements is the strong temperature dependence (0.4 percent/°C). At high temperatures their output signal becomes very small and therefore has to be amplified. Since it is not possible to combine both indium antimony and silicon technologies, the required signal conditioning cannot be integrated on the indium antimony wafer. Therefore additional circuitry within the sensor package is necessary. This reduces the maximum operating temperature. Most of these disadvantages can be avoided by using integrated hall sensors.

3.1 SILICON HALL SENSORS

Hall-generators can be manufactured on a normal bipolar IC-process without any additional processing steps. Usually the epitaxial layer is used for the construction of the hall-elements. This weakly doped n-type silicon produces an acceptable hall-effect due to its carrier mobility of about 1.200 cm^2/Vs. Recently developed hall-generators show sensitivities as high as 400 VA/T. This is sufficient for many applications in the automotive field. The best-known integrated silicon hall-sensor is the vane switch which has been widely used for ignition timing since about 1982.
The sensing of ferromagnetic gearwheels requires a more sophisticated chip design. Compared to the vane switch, the magnetic signal is much smaller, more care must be taken in amplifying and Schmitt-triggering of the analog signal.
Figure 1 shows the principle of operation: A soft magnetic geartooth wheel causes a magnetic imbalance between sensor 1 and sensor 2, which is transformed into a digital output signal by means of the integrated signal processing circuit.

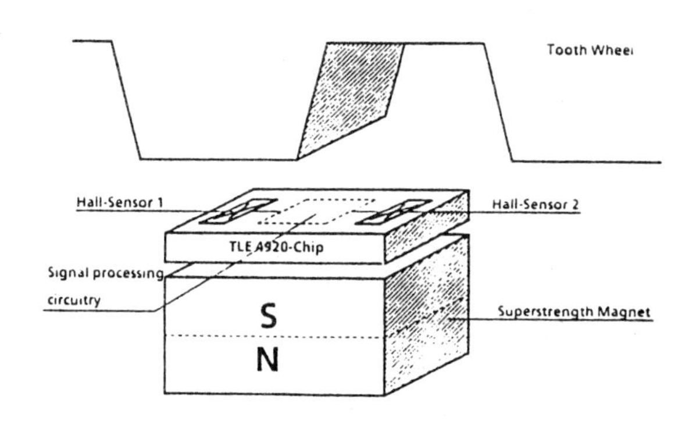

Fig 1 TLE 4920 operated by a gearwheel

The biasing magnetic field is not essential for the function of the sensing element in this case. Therefore the TLE 4920 can also be operated by a simple alternating field as shown in figure 2.

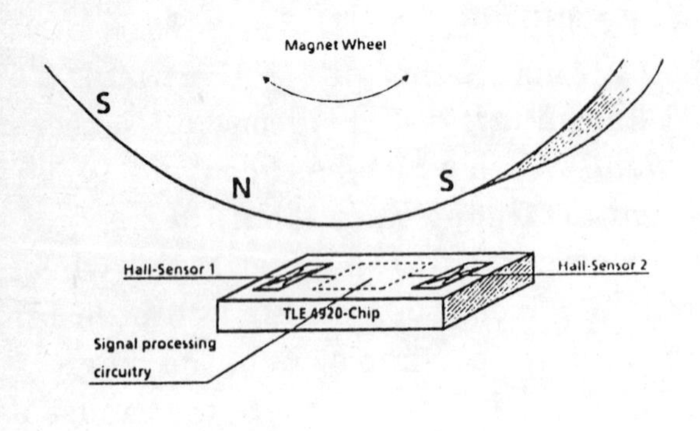

Fig 2 TLE 4920 operated by a magnet wheel

3.2 TLE 4920 INTEGRATED CIRCUIT

The TLE 4920 is manufactured on a standard bipolar technology line. Figure 3 shows the functional block diagram.
Intended for the use in automotive environment, the IC is protected against EMI and RFI transients. For this purpose special clamping structures at the supply pin as well as at the output pin are implemented. The only external component required is a resistor in the supply line. The maximum voltage is limited to about 38 Volts.

The clamping structure consits of an array of zener diodes which drive a relatively large npn transistor. The latter amplifies the zener-current in the event of overvoltage.
The sensing of the magnetic field is achieved by two arrays of eight hall sensors located opposite to each other with a separating distance of 2.5 mm.

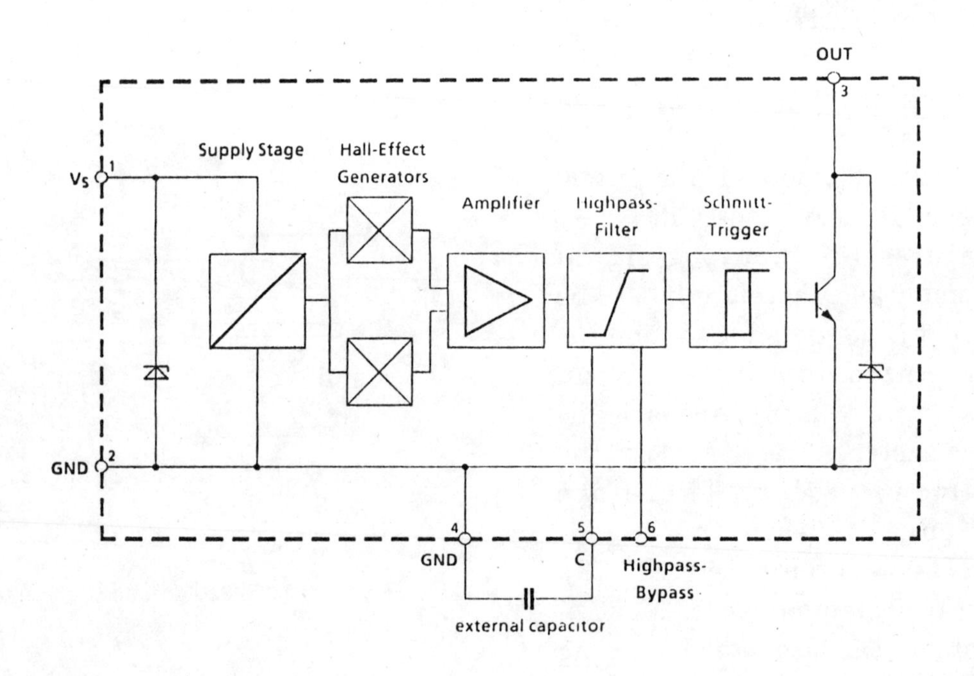

Fig 3 Block diagram

An optimum differential signal is therefore achieved using a geartooth wheel with a pitch of 5mm or larger. Decreasing the pitch will diminish the differential hall signal and the maximum possible airgap. Only gearwheels with a pitch of about 2.5 mm cannot be used as they will only produce a common mode signal.
The bias current for the hall-elements is supplied by a conventional band gap reference regulator. The voltage across the hall cells is kept constant over the whole temperature range. Therefore the offset drift is kept as small as possible. The hall voltage itself, however, decreases markedly with increasing temperature. This effect is caused by the decrease of the carrier mobility in the weakly doped bulk silicon. To compensate for this effect, the amplifier's temperature coefficient is made positive.
Silicon hall sensors generally suffer from offset problems caused mainly by the so-called piezo-resistive effect. Using an economic plastic package, mechanical stress cannot be avoided completely. In a single hall cell large DC-errors may occur, which may vary over the lifetime of the sensor. To overcome this problem, each of the 16 hall cells is connected to an individual pair of NPN transistors in such a way that this effect is compensated for. Additionally the statistical spread of each offset is averaged by the amplifiers, and the resulting offset is very small.

3.3 TRIMMING CIRCUITRY

With a given biasing magnet, the maximum tolerable airgap between sensor and geartooth wheel depends mainly on the ratio between the hall signal amplitude and the remaining offset of the hall cells.To improve this ratio, the offset is reduced on the wafer by a trimmable resistor network which consists of four stages providing 15 trimming steps.

3.4 DYNAMIC OPERATION

For applications requiring very large air gaps the remaining offset may still be too large. A well-proven method to avoid DC-offset problems is to insert a highpass-filter into the signal path. Thereby signals below a certain cutoff frequency are suppressed.
This method can also be utilized with the geartooth sensor. However, for the highpass filter an external capacitor becomes necessary. The construction of a highpass filter is rather difficult because the cutoff frequency should be as low as possible, whilst the external capacitor should not be too large. Only ceramic capacitors provide reasonable performance at high temperatures. Therefore the charging current has to be very small which requires special care.
The highpass filter of the TLE 4920 can be bypassed by external wiring. This causes the sensor to operate in a static mode.

3.5 APPLICATION CIRCUITS

The output is an NPN open collector which can sink up to 40 mA. The sensor accepts both three wire and two wire connections which means that the output can either be a voltage or a current signal. Three schmitt-trigger loops provide a proper digital signal even if the supply voltage is affected by transients.

3.6 PACKAGING

One main advantage of integrated hall sensors is their small size which allows new rotation sensing concepts such as integrating the sensor within the sealed part of an automotive bearing.
For this as well as for many other applications, the sensor provides an operational temperature range up to about 180 C. Practical measurements done by

several automobile firms have shown, that these high temperatures only occur a few times and for short periods during a normal automobile lifetime. So a normal, cost-effective plastic package is sufficient. Being designed for various applications, this single basic package allows a highly increased production volume and decreased unit cost.

CONCLUSION

This new integrated sensor is therefore destined for a wide range of applications in car sensor systems.

ACKNOWLEDGEMENT

We would like to thank our collegues at the technological development department for their cooperation.

REFERENCES

(1) Joseph M. Giachino
Smart Sensors for Automative Application

(2) R. S. Müller, D.L. Polla, R. M. White
"Integrating Sensors and Electronics-New Challenge for Silicon" 1984 IEEE Electrotechnology Review 1985

(3) K. S. Peterson "Silicon Sensor Technologies" 1985 IEEE International Electronic Devices Meeting Technical Digest p.2-7

(4) L.Halbo, and J.Haraldsen; The Magnetic Field Sensitive Transistor-A New Sensor for Crankshaft Angle Position, SAE Paper 800122 (1980)

(5) T.Usuki, S.Sugiyama, M.Takeuchi, T.Takeuchi and I. Igarashi; Integrated Magnetic Sensor, Proc. 2nd Sensor Symposium P.215 (1982)

(6) Ulrich Lachmann:
Analog Hall- Effect Circuit and its Application

(7) Ulrich Lachmann:
Function and Application of the Hall-Effect Vane Switch HKZ 101

C391/046

Fast switching PWM-solenoid for automatic transmissions

W BREHM and **K NEUFFER**
Robert Bosch GmbH, Stuttgart, West Germany

Abstract

Electronic control systems for automatic transmissions have been gaining more and more ground over the past few years. An important component of such systems are electrohydraulic actuators which either function constantly to modulate pressure or are used as inlet/outlet valves for shift-point control. This paper describes a newly developed solenoid valve with seat which can be used as a continually functioning actuator by means of timed electrical activation. Design was primarily geared towards achieving maximum operating safety and precision as well as economical manufacture. Characteristics of the construction and of the electrical control are explained and possible applications are listed.

1. Introduction

Proportional valves are used in electronically controlled automatic transmissions to set the system pressure. In such cases, the pressure is adjusted according to the characteristic curve (pressure versus current). These actuators function either as pressure relief valves or as pressure regulators. The force resulting from the controlled pressure, the magnetic force and the spring force are in equilibrium. Adjustment is effected by way of changes in current induced by the solenoid of the proportional valve.

This principle enables the pressure to be set with a high degree of accuracy and without the need for additional pressure sensors. These elements must however be manufactured with extremely high precision if they are to be able to satisfy the requirements for high accuracy, low hysteresis and maximum service life. Additional filters may be required in heavily contaminated oils. The expenditure for the electronic controller is relatively high because of current regulation requirements. The electronic control technology is characterized by the use of digital computers. It is therefore beneficial to be able to actuate the electrohydraulic converters directly by digital means. In view of the fact that switching valves of the poppet type can only actuate two switching states, they are particularly well suited to direct actuation by microcomputers (1). The mechanical configuration is simpler than that of proportional valves. But that means that it can be manufactured at reduced costs.

2. 3/2-way poppet valve

This type of valve can assume two switch positions and has three controlled ports (P, A, T). A distiction is made between the "open when deenergized" and "closed when deenergized" versions in line with the actuation logic resulting from the d.c. magnetic system. Consideration is only given in the following to the "closed when deenergized" type.

3. Pulse modulation

If such a solenoid-operated valve is switched on and off at a constant supply pressure P with a current or voltage pulse, an average pressure results in the power port A in line with the opening/closing time ratio. The pulse duration is made up of the opening process, the dwell time in the open position and the closing process. The actuation pulse is established by shape, duration and amplitude (Fig. 1).

Pulse-duration modulation and pulse-frequency modulation can be considered as actuation methods. With pulse-duration modulation, the pulse duration T_1 is varied whilst maintaining a constant pulse repetition period. Pulse-frequency modulation refers to actuation with a constant pulse duration, but varying pulse repetition period. As a general rule, microcomputers operate on the basis of discrete-time, equidistant setpoint selection. This means that direct actuation of the switching valves with a constant pulse frequency is the most favourable solution. Fig. 2 illustrates the pressure profile occurring as a function of the on/off ratio given a constant pulse frequency.

4. Design

Rapid actuation of the valve element is the most important task of the electromechanical converter. Further criteria are repeatability and low power input. Consideration is also to be given to aspects which facilitate economical manufacture. Environmental conditions such as temperature, pressure, service life and contamination of the hydraulic fluid should likewise not be forgotten. Fig. 3 illustrates the design of a pulsed solenoid-operated valve of the "closed when denergized" type for use in automatic transmissions.

The 3/2-way solenoid valve is hermetically sealed for this application both when current is flowing and when there is no current flow. The spherical armature is also a closing element. Due to this fact there is no need for additional switching elements.

Short switching times can be realized since the moving mass is relatively small. The conical opposite pole and the spherical armature result in a shallow magnetic-force characteristic over the stroke (Fig. 4). This results in large energization forces and a low retention force when energized. It is therefore possible to achieve favourable flow values in both directions even with short strokes. The enlarged sealing diameter and the associated reduction in compressive load per unit area minimize poppet wear in conjunction with the short armature stroke. An additional advantage of the actuator ball is that the rolling friction is relatively small. Moreover, the tendency to tilt caused by transverse magnetic forces is not as pronounced as with cylindrical components. A further aspect is that the ball - as a standard component with very low dimensional tolerance and excellent surface quality - can be produced extremely economically.

The magnetic circuit was optimized using a computer program (2), so as to be able to establish a favourable geometry for the winding window as well as for the shape and position of the working air gap. Care was also taken to ensure that in the magnetic circuit no cross section had an impermissible high field strength.

The materials used for the magnetic circuit feature advantageous magnetic properties thanks to corresponding selection and heat treatment. The choice of the production method and the materials used provide further benefits in terms of function and economy. The base plate and housing are simple drawn stampings. This type of production gives rise to a further advantage. The core and base plate is pressed from the plug side through a punch against a die inserted from the flange side, thus simultaneously forming the shoulder A. This guarantees, that the seat at the zero drain and the conical opposite pole are always in the same position with respect to the shoulder A. The high degree of accuracy of the ball is such that only the manufacturing tolerance h from the armature to the contact shoulder A forms part of the stroke and the working air grap. Additional adjustment, for example in the form of an adjusting screw, is not required.

The residual air gap is formed by the seat at the zero drain and is always in the same position with respect to the conical opposite pole (toolinduced). There is thus no need for a residual-air-gap disk. The scatter in the deenergization current and deenergization time is therefore very slight. Noise generation can also be minimized on account of the small armature mass, the short stroke and the plastic valve seats which feature a high level of intrinsic damping.

5. Electrical actuation

The considerations to date apply to valves with an ideal time behaviour. In a real system, only the electrical setpoint selection can be viewed as being ideal. The volumetric flow rate and mean pressure will lag behind the setpoint on account of the magnetic, mechanical and hydraulic time delay. The simplest method of actuation is to apply voltage to the valve. Given actuation with an identical on/off ratio, this will however result in considerable differences in terms of volumetric flow rate and mean pressure on account of the temperature dependence of the ohmic resistance of the coil and fluctuations in supply voltage.

In view of the fact that permanent connection may also be required, a relatively high winding resistance would have to be selected to prevent thermal overload as a consequence of the electrical power loss. This can be achieved with a larger number of turns per unit length, but this increases the inductance of the magnetic circuit and reduces the maximum pulse frequency at which a mechanical valve reaction still takes place. Insufficient actuation frequencies can however result in undesirable pressure pulsations. To avoid this effect, use is made of a special signalling measure, namely the "peak and hold" method. The current required to "hold" the armature in the end position is less than that at the start of the armature travel on account of the profile of the magnetic-force characteristic (Fig. 4). After reaching the end position of the armature, the current value can thus be reduced (Fig. 5).

This in turn reduces the mean power loss in the winding. It is therefore possible to realize smaller coil resistances and lower inductances. At the same time, there is a significant reduction in the scatter of the deenergization time, since deactivation is always effected from the same current level. It is also possible in this way to ensure that the coil current is completely disconnected at the end of the mechanical movement whatever the on/off ratio and that no magnetic hysteresis occurs. At the same time, there is a complete reversal of the direction of movement of the armature at every operating point on the characteristic curve with the result that the mechanical hysteresis is also negligible.

The scatter of the energization time caused by fluctuation in the winding temperature and supply voltage can be corrected within certain limits by measuring these parameters and appropriately altering the setpoint selection, so as to ensure good characteristic-curve repeatability. In contrast to proportional valves, there is however no correction of disturbance variables caused for example by system leakage, air inclusions or fluctuations in supply pressure. Such actuators are thus suited as a reasonably priced alternative in particular to applications where the control pressure does not directly represent the controlled variable, but rather is used as a manipulated variable.

One application in automatic transmissions is, for example, gear ratio control (3) (Fig. 6) or the reduction of torque-converter slip losses by influencing the contact pressure of the lock-up clutch.

Closed-loop pressure control using a frequency valve has considerable advantages. The dynamic behaviour of the proportional valves is essentially determined by their design and by the properties of the hydraulic medium (in particular its viscosity). The hydraulic damping decreases at higher operating temperatures due to the lower viscosity.
At the same time, the rate of flow of the hydraulic oil increases as a result of this effect. This means that such valves may be subject to sustained oscillation which clearly impairs the control quality. At low oil temperatures there is accordingly a pronounced increase in response time.

Dynamic behaviour is also dependent on the upstream and downstream transmission elements. The demands for a short switching time at low temperatures and stability at high temperatures can only be realized with considerable outlay. Self-induced oscillation cannot occur with a pulsed solenoid-operated valve. In view of the fact that the controlled variable is detected by the sensors (e.g. speed sensor, pressure sensor) and processed in the control unit, the dynamic behaviour of the closed loop can be optimally adjusted as a function of operating temperature by adapting the setpoint selection. A further advantage is that compensation can also be provided for other changes in parameter in the controlled system, such as scatter in the coupling friction values.

6. Conclusions

Electrohydraulic converters in automatic transmissions represent the transmission elements from the electronic control unit to the hydraulics. In the majority of applications use is made of proportional valves for pressure modulation. An alternative to the above can be provided by pulsed, poppet-type solenoid-operated valves which are characterized by a simple design and a high degree of insensitivity to contamination. The use of such actuators can bring benefits in terms of reliability, costs and function particularly in closed-loop applications.

References

(1) Lühmann, B.
Digital gesteuerte Hydraulikventile und ihre Anwendung
Dissertation Technische Universität Braunschweig 1983

(2) Müller, W. u.a.
Numerical solution of 2- or 3-dimensional nonlinear field problems by means of the computer program PROFI
Archiv für Elektrotechnik 6 (1982) s. 299-307

(3) Hiramatsu, T., Naruse, T.
Shift quality control of an electronically controlled four speed automatic transmission. SAE-Paper 86 5149

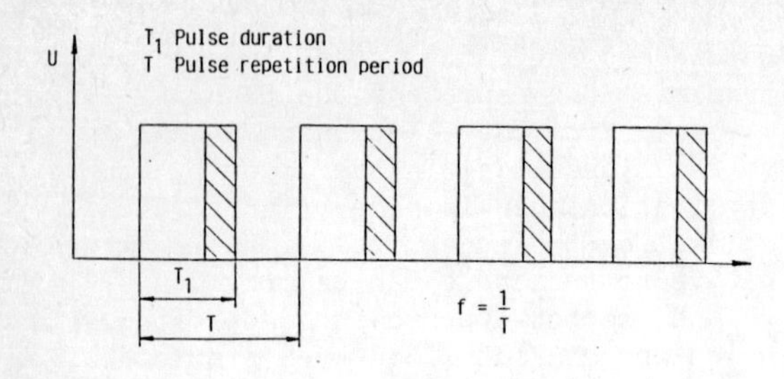

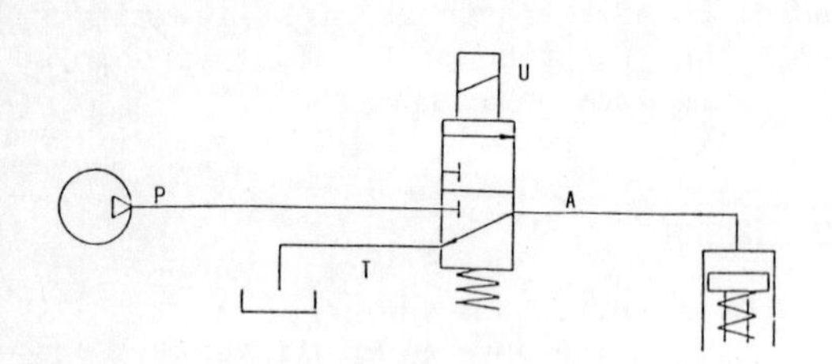

Fig 1 Pulse-duration modulation

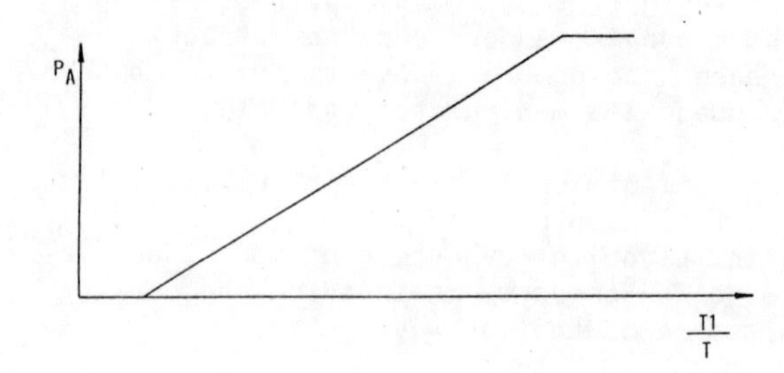

Fig 2 Characteristic curve of pulsed solenoid-operated valve

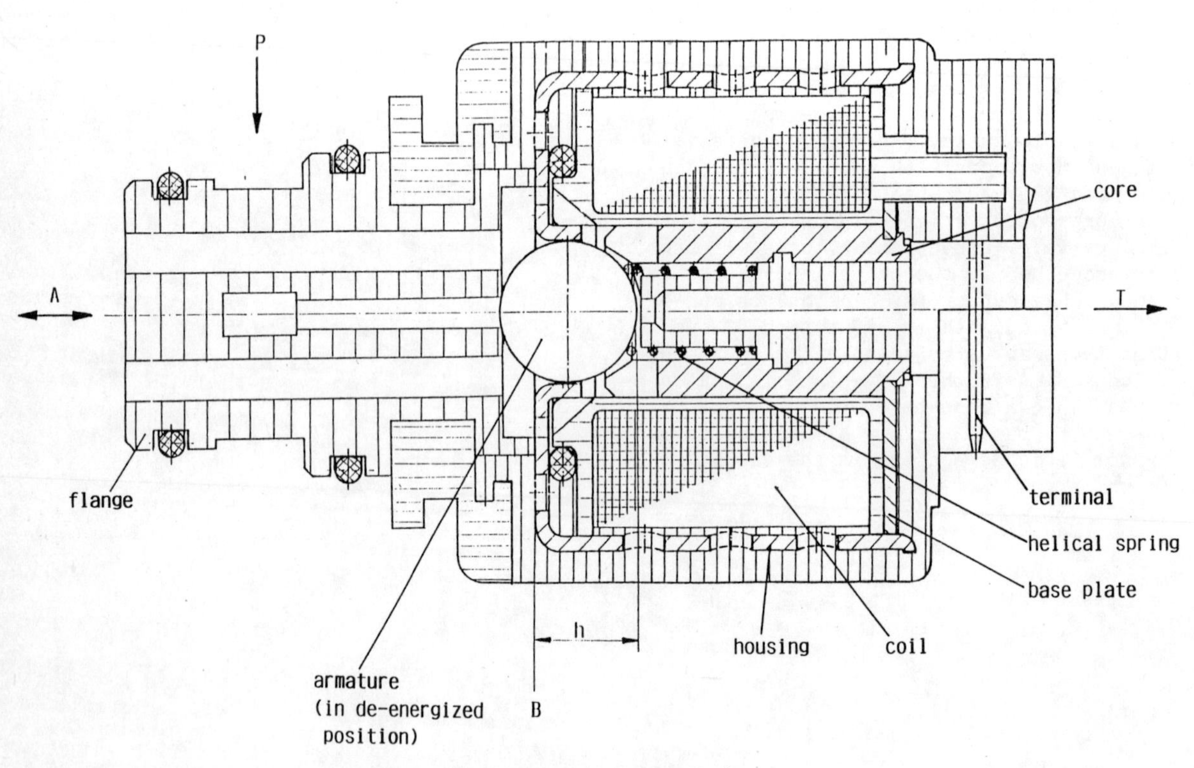

Fig 3 Pulsed solenoid-operated valve

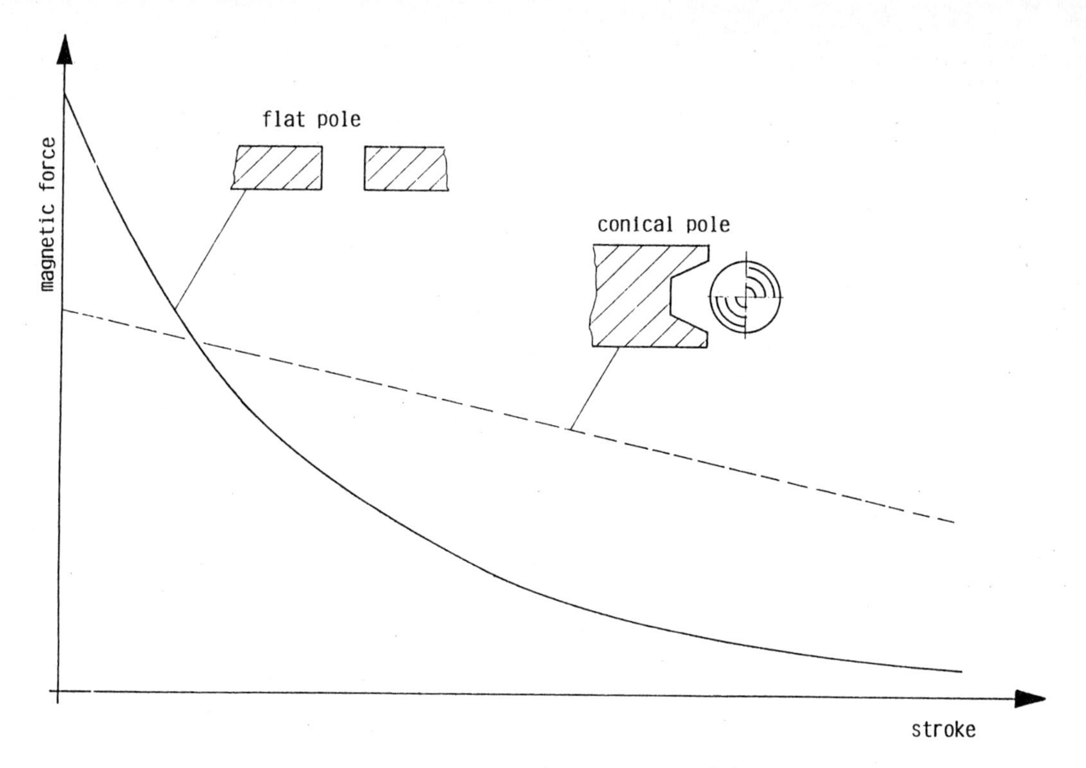

Fig 4 Magnetic force/stroke characteristic curve

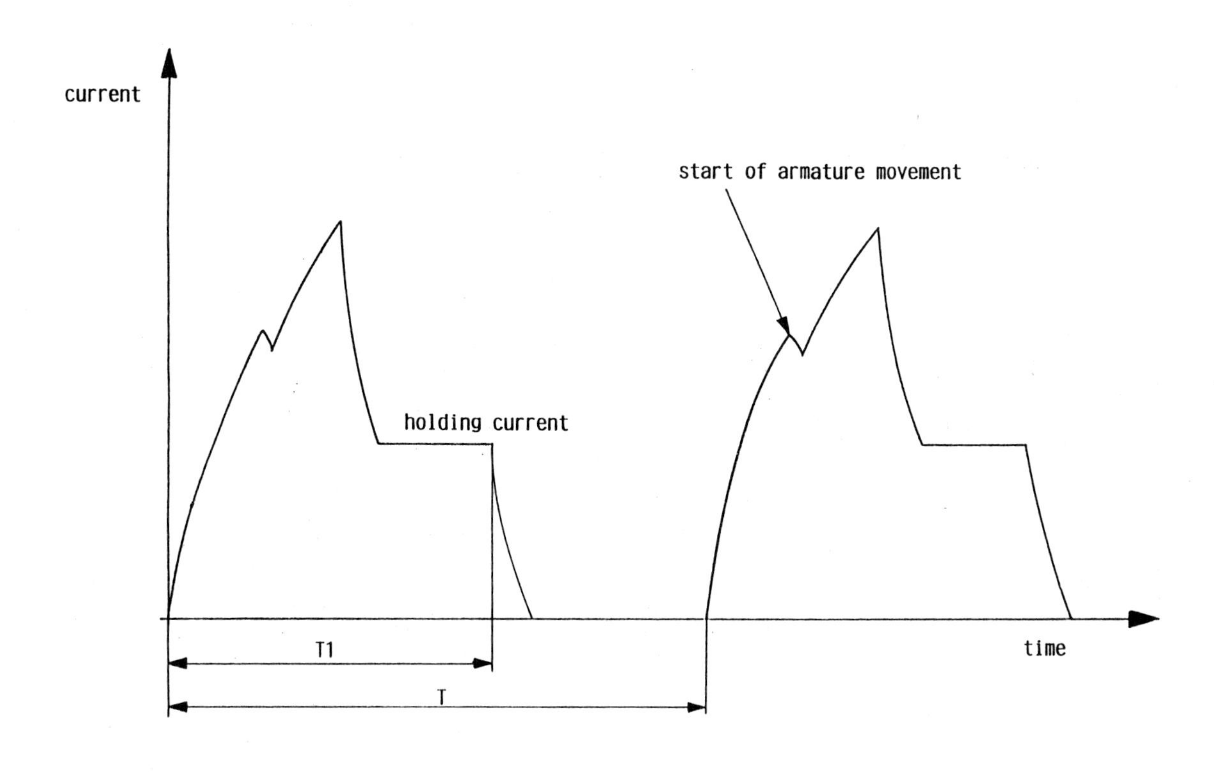

Fig 5 'Peak and hold' actuation

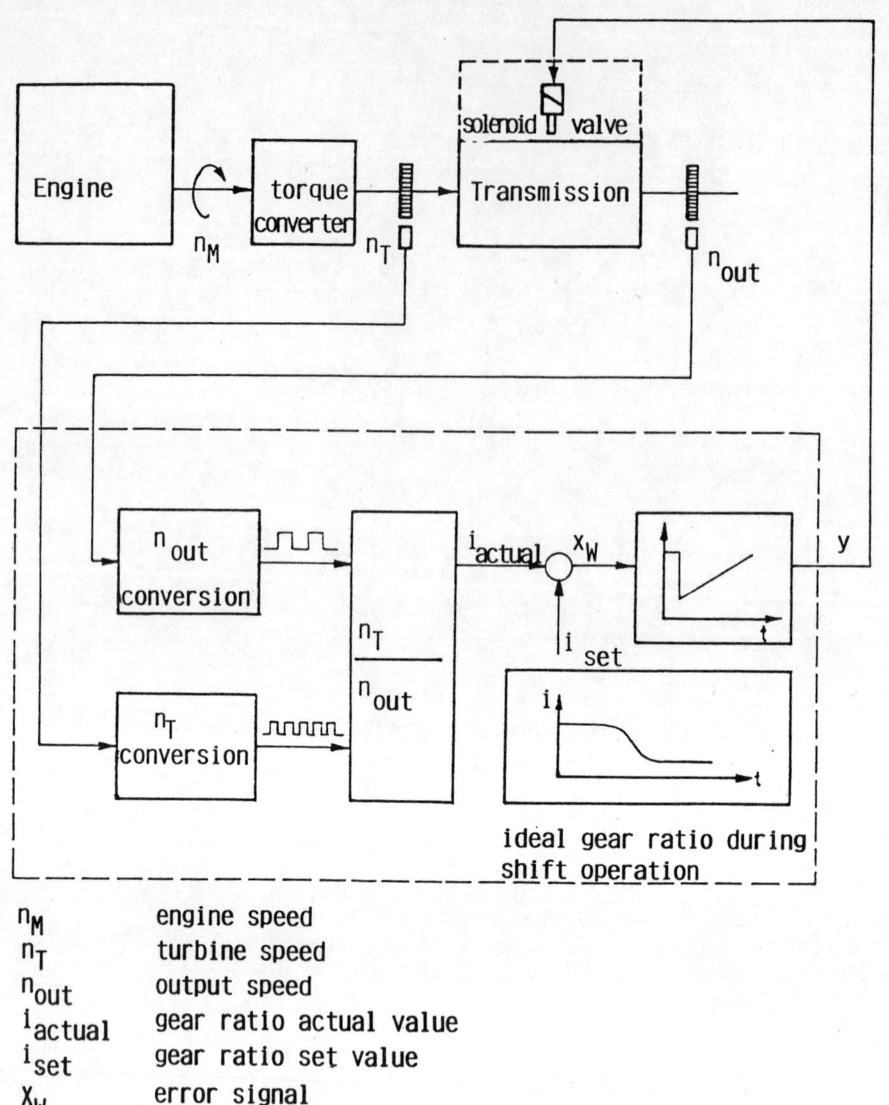

Fig 6 Gear ratio control

C391/036

A Hall effect accelerometer

R E BICKING, BEE, MSEE, ISA, SAE, IEEE
Micro Switch Division of Honeywell Inc, Freeport, Illinois, USA

SYNOPSIS Combining a cantilever beam with a magnet acting as the pendulous mass and a linear Hall effect sensor results in a simple accelerometer capable of meeting the requirements for electronically controlled suspension systems, anti-skid braking systems and passive restraint deployment systems. Silicone oil damping eliminates any tendency of the beam to ring at its natural frequency. The fact that d.c. response is unnecessary permits a.c. coupling of the electrical output with the resultant freedom from null stability concerns.

1 INTRODUCTION

Recent developments in automotive systems have created a need for low cost accelerometers. Applications exist in electronically controlled suspension systems for accelerometers to measure body motion and in anti-skid braking systems and passive restraint systems to measure deceleration.

The first two applications involve low gravitational unit (g) range measurements and both sets of requirements could be satisfied by the same device. The passive restraint application requires a much higher measurement range and somewhat higher frequency response. Requirements for these applications are summarized in Table 1.

The common characteristic of all these applications that facilitates a low cost, open loop measurement technique is that only dynamic acceleration has to be measured. This means that a high pass network can be used to a.c. couple the accelerometer output. Therefore, long-term null stability of the accelerometer and its null stability over temperature are unimportant, obviating the need for complex, closed loop accelerometer mechanizations.

2 TECHNOLOGY CHOICE

A number of technology choices exist, including piezoelectricity, piezoresistivity, the Hall effect, etc. Preliminary work done on piezoelectric ceramic indicated that it was difficult to achieve scale factor stability over the entire automotive temperature range and that packaging to minimize mounting stresses was difficult. Also, the electronics interface is made difficult by the need to achieve a response down to 0.1 Hertz. For a typical piezoelement capacitance of 1000 pico Farads, the load resistance must be on the order of 10^9 ohms or greater. Choosing a piezoelement with greater capacitance helps solve this problem but leads to a larger accelerometer and greater packaging challenges. Piezoresistivity has some attractive advantages in simplicity, ease of integration and cost, especially for high g devices. Substantial development effort would be required to realize these advantages, however.

Availability of a high performance linear magnetic sensor using the Hall effect, combined with a cantilever beam which has a magnet located at its unsupported end, resulted in a simple accelerometer capable of satisfying the requirements of automotive applications.

3 DESCRIPTION OF LINEAR HALL EFFECT IC

The linear Hall effect sensor used incorporates both thin film on-wafer trim and thick film (application specific) trim to achieve high sensitivity, low null and sensitivity, temperature coefficients and unit to unit interchangeability.[1] A specification summary is given in Table 2. For automotive applications, a version of this device capable of operation from a 5VDC supply is completing development and will be available by the time this paper is presented.

4 DESIGN OF HALL EFFECT ACCELEROMETER

Accelerometer design always involves a tradeoff between sensitivity and baudwidth because of the inverse square law relationship between movement of a spring-mass system and its natural frequency, since the displacement is given by:

$$x = -\frac{m}{k}\ddot{x}$$

whereas the natural frequency is given by

$$fn = \frac{\sqrt{k/m}}{2\pi}$$

where fn = natural frequency
k = spring constant
m = mass

and sensitivity) is ing the mass or lowering ant, both which lower the y. It is desired to operate eter well below its natural o achieve a flat response. In gn, the response of the Hall sensor magnet follows the inverse square law, own in Figure 1. Therefore, the magnet d beam) movement must be kept small to approximate a linear response. The accelerometer is shown in cross-section in Figure 2.

The ceramic substrate supports the Hall IC, as well as thick film trim resistors and additional circuitry used to amplify and a.c. couple the output. The beam, formed of a high modulus of elasticity material such as Elgiloy or stainless steel, is clamped to the ceramic. The magnet is located at the end of the beam and acts as the proof mass as well as providing magnetic excitation of the Hall sensor.

One of the potential problems with a spring-mass accelerometer is ringing at the natural frequency due to shock excitation. It is, of course, aggravated by the choice of beam material, which needs a high modulus of elasticity for long-term stability and linearity. Silicon oil damping was incorporated to eliminate any tendency to ring. Both the viscosity and volume of the oil changes with temperature. Closed cell plastic foam is used to account for volume changes as a function of temperature as shown in Figure 2. Provided that the device is operated well below the natural frequency, the damping has a negligible effect. The output as a function of damping fluid viscosity is esentially constant as shown in Figure 3. This range of viscosity corresponds to operation from -40°C to 120°C. Since viscosity change is the major temperature effect in this design, the output should be essentially constant and test data confirms this. Figure 4 illustrates the linearity as a function of acceleration. Figure 5 shows a prototype device. Scaling this design for various acceleration ranges may be done by changing beam parameters. Table 3 illustrates calculated beam dimensions of length (l) and thickness (t) and calculated natural frequencies for a constant output of 100mV and a constant beam width of 0.375 inch. One package can easily accommodate all these ranges with minor changes in the beam's dimensioning and sensor IC locations.

Table 1 Automotive accelerometer requirements

Application	Range (±g)	Frequency Response (Hz)	Temperature Range (°C)	Accuracy (%)	Transverse Sensitivity (%)
Anti-skid	1	0.1 - 15	-40 - 85	5	5
Suspension	2	0.1 - 50	-40 - 120	5	5
Passive Restraint	100	10 - 100	-40 - 105	10	5

Table 2 Linear Hall IC specifications

Supply Vdc)	6.6 to 12.6
Response Time (μS)	3 typical
Span (Gauss	-500 to 500
Sensitivity (μV/Gauss)	5.00 ± 0.10
Null output (Vdc)	4.00 ± 0.04
Temperature Coefficient of null & span (%/°C)	±0.02 max

Table 3 Calculated beam design parameters

Range	l (in)	t (in)	fn (Hz)
2g	0.615	0.0045	281
10g	0.485	0.006	617
50g	0.405	0.009	1413
100g	0.400	0.012	2021

REFERENCES

(1) MICRO SWITCH Catalog 20, Specifiers Guide for Solid State Sensors, p. 25, 1988.

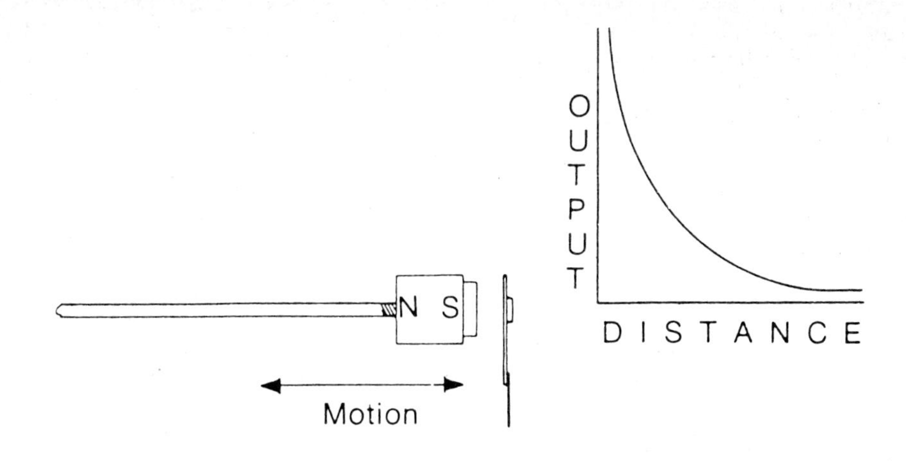

Fig 1 Hall sensor response to magnet movement

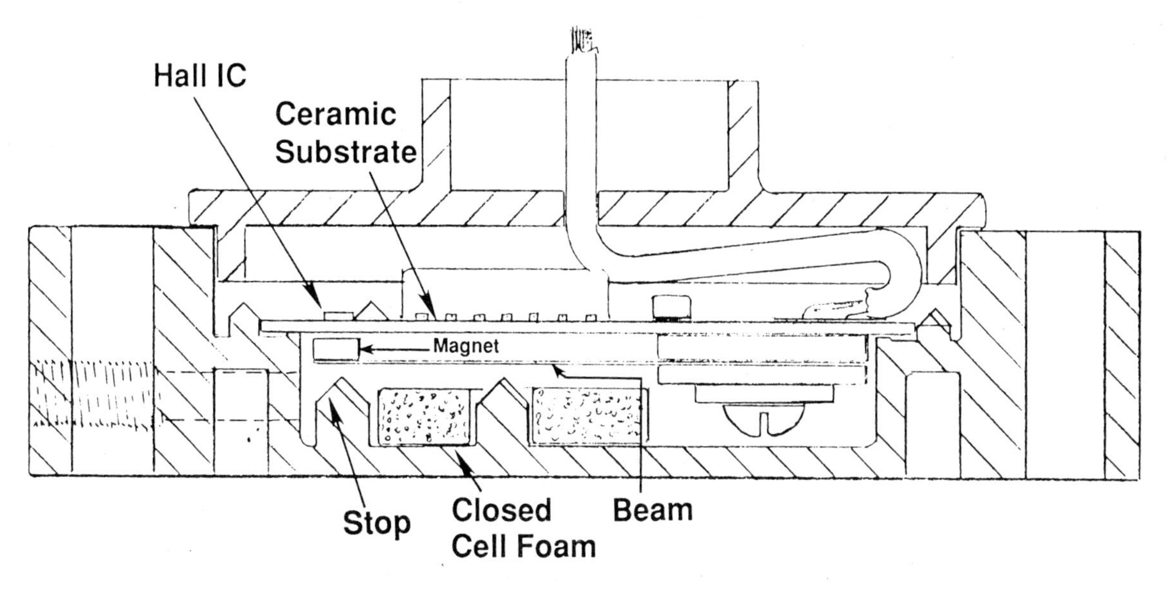

Fig 2 Cross-section view of accelerometer

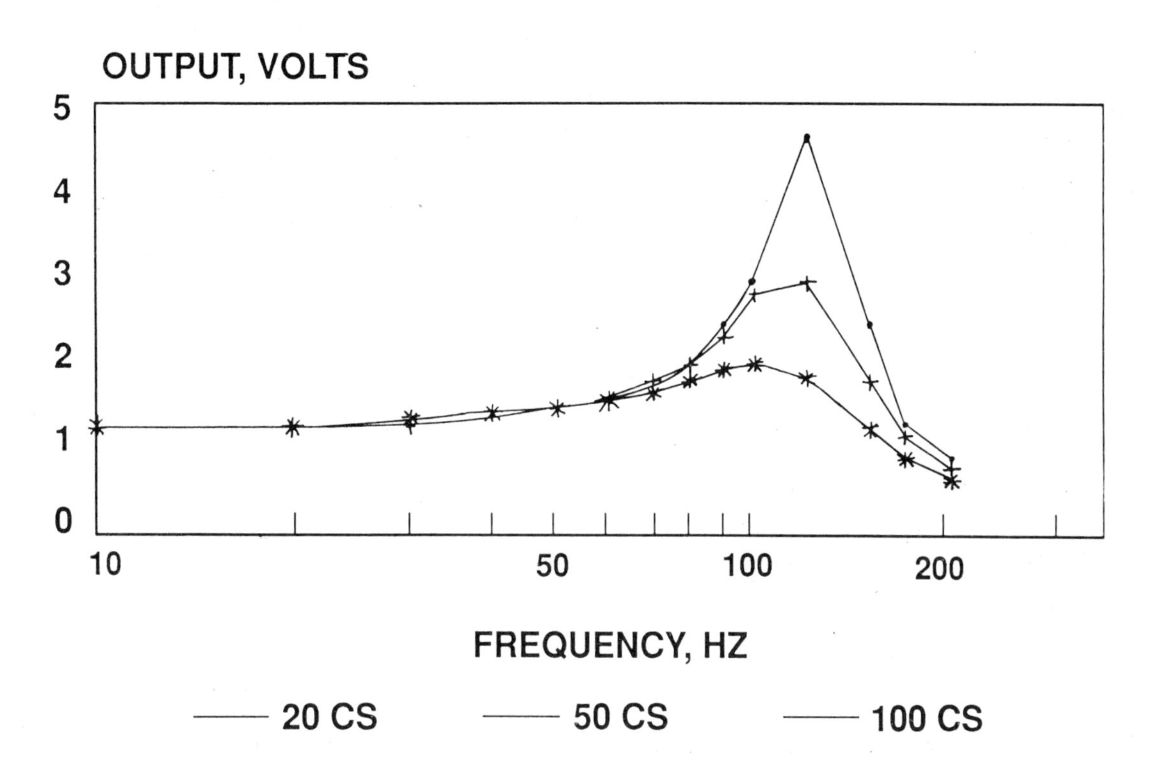

Fig 3 Frequency response of accelerometer

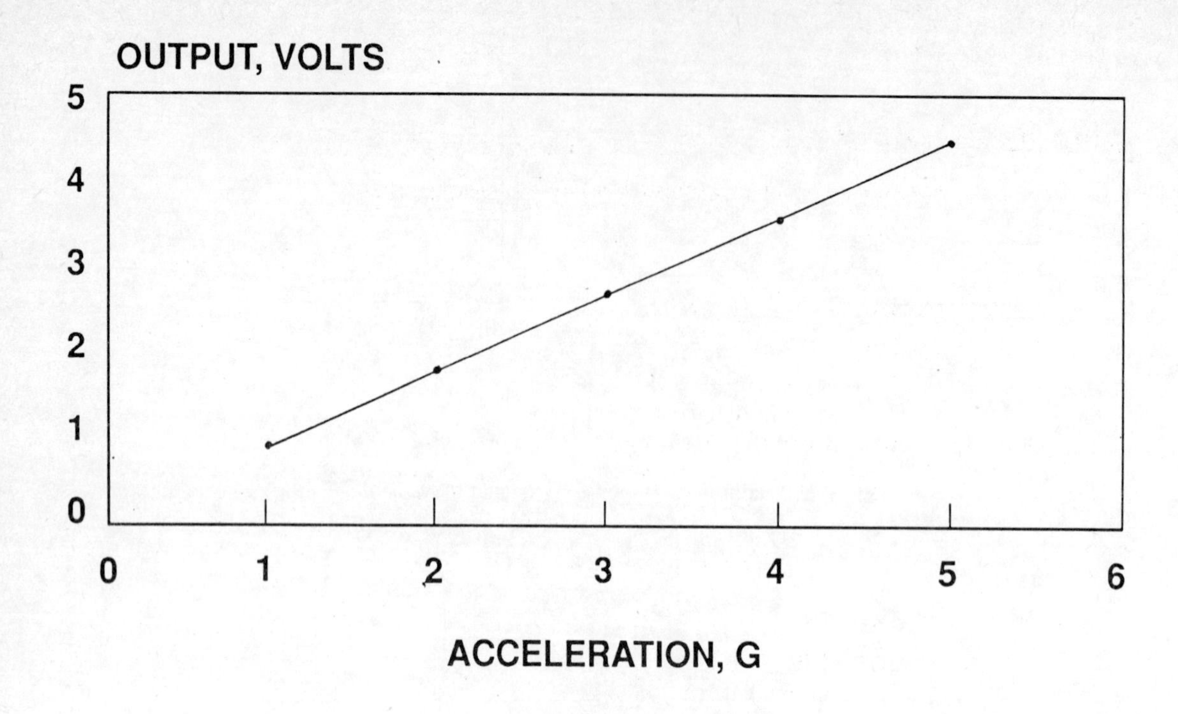

Fig 4 Linearity of accelerometer

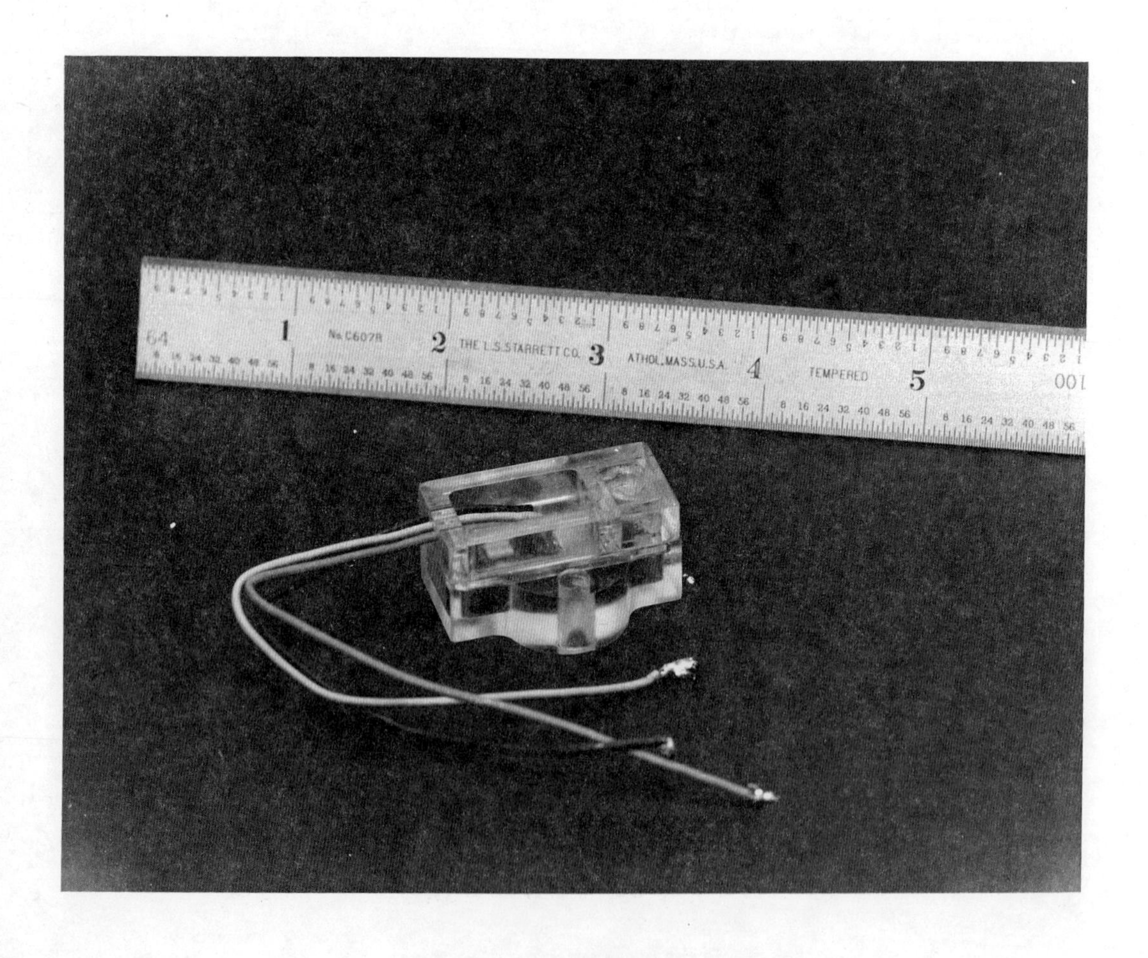

Fig 5 Prototype accelerometer

C391/072

The properties and applications of rare earth – iron magnetostrictive alloys

R D GREENOUGH, PhD, **A J WILKINSON**, PhD, **A JENNER**, PhD, **A PARVINMEHR**, MSc and **M SCHULZE**, MSc
Department of Applied Physics, University of Hull

SYNOPSIS Rare Earth alloys are now available which exhibit extraordinary magnetoelastic and magnetomechanical properties. The application of magnetic fields ~ 40 kA/m (~ 500 Oe) can generate strains ~ 1500 ppm thereby converting magnetic energy to mechanical energy with great efficiency. This transducer action is sustained and even enhanced by the application of uniaxial stress and large loads of up to 200Kg (440 lbs) can be moved by a single rod of material with a diameter of 6mm (0.25 in).

These properties are described together with non-destructive techniques for evaluating the magnetomechanical coupling in different samples of material. Its behaviour in AC fields will be considered. Applications currently being developed, such as dynamic vibration control, and future R & D opportunities are reviewed.

1 INTRODUCTION

Piezoelectric materials have been frequently employed in transducers during the last few decades but their magnetic equivalent, magnetostrictive materials such as nickel based alloys, have been used less. However magnetic alloys and compounds containing Rare Earth (RE) elements have now been developed as transducer materials with properties which can not only equal but surpass those of piezoelectrics.

In the following paper we briefly describe the origin of RE magnetostrictive materials, their outstanding properties together with techniques to assess their performance and quality. The following discussion then indicates how the materials may be put into practise and the types of applications which are currently being researched and developed.

2 MAGNETOELASTIC EFFECTS IN TERFENOL

Large magnetostrictive strains ~ 2200 ppm, have been observed in the RE element Terbium (Tb) but only at temperatures, T $\lesssim$ 220K [1]. Although room temperature ferromagnetism and magnetostriction could be produced by alloying Tb with iron [2], there remained one major obstacle which had to be overcome before the material could be considered useful for devices. The magnetic anisotropy associated with the crystal structure was too large and demanded magnetic fields ~ 2 × 10^6 Am^{-1} to generate the strains. With the controlled addition of Dysprosium (Dy) these fields were reduced to ~ 40 kAm^{-1} while preserving the magnetostriction from the Tb ions. Hence the emergence of Terfenol-D*, for example $Dy_{0.73}Tb_{0.27}Fe_{1.95}$ [3], as a potentially useful material. The magnetomechanical response of Terfenol has now been enhanced by the fabrication of rods with oriented crystal grains [4] so that the application of static stress $\lesssim$ 100 MPa (1400 psi) causes the magnetisation to lie perpendicular to the rod length. Subsequent application of a magnetic field ~ 40 kA m^{-1} parallel to the rod axis causes the majority of magnetic moments to rotate abruptly through ~ 90 degrees, a change in magnetisation direction which is principally responsible for the realisation of magnetic strain ~ 1500 ppm. An abrupt change in length, referred to as the 'burst effect' [5], is seen in materials such as $Dy_{0.7}Tb_{0.3}Fe_{1.95}$ (Figure 1) and this transducer action, in which magnetic energy is converted to elastic energy (or vice versa), is highly efficient. Just as this conversion is gauged by an electromechanical coupling coefficient for piezoelectric materials, so too is the magnetomechanical coupling represented by the coefficient k_{33}. The fraction of input magnetic energy that is converted to elastic energy is given by k_{33}^2 and with $k_{33} \gtrsim 0.7$ (compared with ~ 0.6 for piezoelectric materials) we now have a material which develops sizeable strains under large loads with relatively high efficiency. A comparison with piezoelectric material is given in Table 1.

* a name compounded from Ter (terbium) fe (iron) nol (Naval Ordinance Laboratories), the laboratories where the material was developed.

Table 1 Properties for Terfenol compared with PZT

Magnetomechanical Coupling	$\gtrsim$ 0.7	0.65
Magnetostriction/ Electrostriction (ppm)	$\gtrsim$ 1500	100-300
Relative permeability (at max. strain)	~ 5	-
Sound velocity (ms^{-1}) (constant field)	~ 1729	~ 3100
Density (Kgm^{-3})	9.3 × 10^3	7.5 × 10^3
Compressive strength	700 MPa	-
Thermal expansion coeff. (/°C)	12 × 10^{-6}	13.3 × 10^{-6}

3 CHARACTERISATION AND PERFORMANCE ASSESSMENT

For an intercomparison of samples with nominally the same composition and to assist in gauging how useful such material might be in devices, certain parameters are studied. These are

magnetostrictive strain (λ), usually measured by bonding a strain gauge directly to the Terfenol rod,

magnetostrictive strain coefficient d_{33} = $d\lambda/dH$), the rate at which the magnetostrictive strains are developed with applied magnetic field, and

magnetomechanical coupling coefficient (k_{33}), determined using one of two commonly used techniques. The results of separate experiments to measure the incremental magnetic permeability at constant stress (μ^{σ}), the elastic compliance at constant field (S_H) and strain coefficient, d_{33}, are combined [6] through the equation

$$k_{33}^2 = d_{33}^2/\mu^{\sigma}S_H \qquad (1)$$

Alternatively, the frequency of a small oscillatory field is swept until the rod of material goes into resonance (~ kHz for a rod ~ 6" in length) and from the resonance and antiresonance frequencies, f_R and f_A respectively, the coupling coefficient can be estimated [7] from

$$k_{33}^2 = \frac{\pi^2}{8} \left(1 - \left(\frac{f_R}{f_A}\right)^2\right) \qquad (2)$$

The natures of these two determinations of the coupling are physically different and apparent differences in the results need to be reconciled. Although the first method, conducted at low frequencies, is more directly related to the material properties, the latter measurements are more easily made. The authors are currently improving on the corrections that have been proposed by Meeks and Timme [8] to compensate the apparently low coupling values for the effects of eddy currents (skin effect) and rod geometry.

With the ultimate use of the materials in actuators in mind, a sample holder has been designed and constructed to enable measurements of λ and k_{33} to be made under different stresses, provided by springs, and magnetic bias fields. The results (Figures 2 & 3) illustrate the field and pressure dependences of λ and k_{33} in $Dy_{0.7}Tb_{0.3}Fe_{1.95}$, the positions of the maxima (optimum coupling) being related mainly to the field dependences of d_{33} through equation (1). The strain coefficient for $Dy_{0.73}Tb_{0.27}Fe_{1.95}$ is smaller in magnitude but the strains develop more linearly with applied field and the hysteresis is less marked. In the present work, a maximum value of $k_{33} \simeq 0.72$ has been observed in this latter composition.

4 DISCUSSION

With reference to the schematic diagram shown in Figure 4, the material is typically to be incorporated in a device which provides a steady magnetic field, either from a DC solenoid or permanent magnets, so defining a mean operating point, H_b. Dynamic variations in applied field supplied by a solenoid then create +ve or -ve strains. Excursions in applied field which drive the field in the opposite direction ($H < 0$) cause the material to expand again and at the other extreme, large fields merely drive the rod to magnetic saturation with no additional strain and a loss of coupling. Operation with an oscillatory drive field and no bias field causes frequency doubling.

The choice of magnitude for H_b is governed by the need for either optimum coupling efficiency or maximum strain levels. The maximm coupling efficiency can only be maintained if the dynamic variations in applied field are small, otherwise the field enters regions in which k_{33} is reduced. If the operation of a rod is designed to utilize the full extent of available strains and obtain maximum displacement, non-linearities will be evident. Variations in dynamic loading need consideration in case the applied bias stress is annulled. For instance the frequency of vibration for a given mass under gavity can, through negative g, oppose any applied stress and alter the material behaviour during part of a cycle.

The most obvious difficulty in employing Terfenol in devices is the presence of the hysteresis in the strain vs applied field, shown for example in Figures 1 and 2 and with currently available materials, this presents a problem in practise. For applications in which positioning is of critical importance, for instance valve actuators or antivibration systems, the control system has to satisfy the usual criteria of speed and power control but also contend with the problem of hysteresis.

Quite often potential users of the material are looking for maximum displacement and although the strains that are developed are by no means too small to be used directly, it is the combination of displacement and load bearing capability which makes Terfenol attractive. The movements provided by a 5" long rod are typically 0.007 ins (~ 0.18mm) but this movement can be amplified by a lever system because the load bearing capability is high.

So far in this discussion the guidelines and obstacles to the incorporation of Terfenol in a device have been summarised and to this extent may give the impression to potential users that they might be faced with insurmountable difficulties. However research and development at Hull University has produced a successful prototype active antivibration system using a unique combination of a Terfenol device and an intelligent control system, providing a reduction of vibration amplitude ~ 25dB. Also a device has been constructed with movement amplification provided by a lever system which results in movements ~ 0.015 in, suitable for valve actuation.

The future of Terfenol lies with improved material and in the majority of applications, intelligent control. Links between the manufacturers of material and the non destructive testing and evaluation facilities at Hull, together with the development of prototype applications are helping to harness the full potential offered by Rare Earth-Iron alloys as actuator materials.

ACKNOWLEDGEMENTS

The Departments of Applied Physics and Electronic Engineering at Hull University would like to take this opportunity to thank the Royal Aircraft Establishment Farnborough and Johnson Matthey plc for their support. Also the authors wish to thank the technical staff at the University of Hull for their valuable assistance.

REFERENCES

[1] LEGVOLD, S., ALSTAD, J. and RHYNE J. Giant magnetostriction in Dysprosium and Holmium single cystals. Phys. Rev. Lett., 1963, 10, 509.

[2] CLARK, A.E. Magnetic and magnetoelastic properties of highly magnetostrictive rare earth-iron laves phase compounds. Am. Inst. Phys. Conf. Proc., 1974, 18, 1015.

[3] ABBUNDI, R. and CLARK, A.E. Anomalous thermal expansion and magnetostriction of single crystal $Tb_{0.27}Dy_{0.73}Fe_2$. IEEE Trans. Mag., 1977, MAG-13, 1519.

[4] VERHOVEN, D.J., GIBSON, E.D., McMASTERS, O.D. and BAKER, H.H. The growth of single crystal Terfenol-D crystals. Met. Trans. A, 1987, 18A, 223.

[5] CLARK, A.E., TETER, J.P. and McMASTERS, O.D. Magnetostriction 'jumps' in twinned $Tb_{0.3}Dy_{0.7}Fe_{1.9}$. J. Appl. Phys., 1988, 63, 3910.

[6] DAVIS, C.M. The properties of conventional magnetostrictive materials for use in underwater transducers. US Navy Journal of Underwater Acoustics, 1977, 27(1), 39.

[7] SAVAGE, H.T., CLARK , A.E. and PAVERS, J. Magnetomechanical coupling and ΔE effect in highly magnetostrictive Rare Earth-Fe_2 compounds. IEEE Trans. Mag., 1975, MAG-11, 1355.

[8] MEEKS, S.W. and TIMME, R.W. Acoustic resonance of a Rare Earth-Iron magnetostrictive rod in the presence of eddy currents. IEEE Trans. on Sonics and Ultrasonics, 1980, SU-27 No. 2, 60.

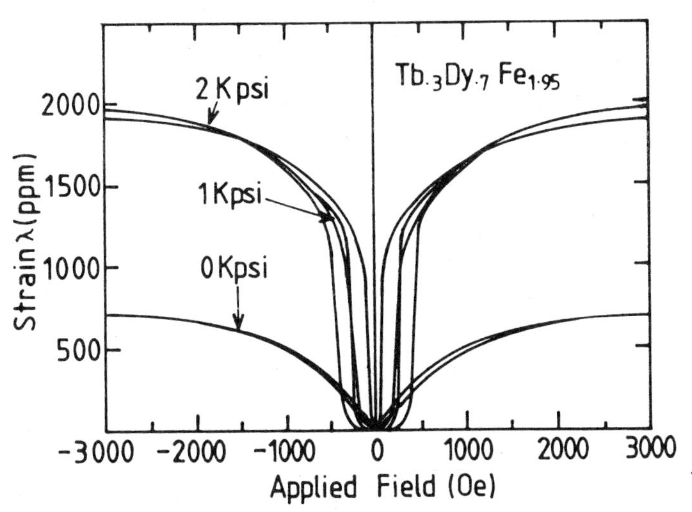

Fig 1 Manufacturers data for the strain generated in $Tb_{0.3}Dy_{0.7}Fe_{1.93}$ as a function of applied magnetic field for different applied stress

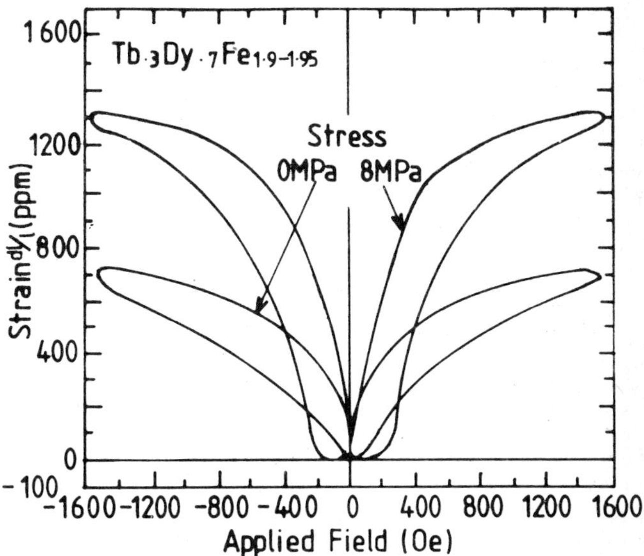

Fig 2 Present experimental data for the field dependence of the magnetostriction in $Tb_{0.3}Dy_{0.7}Fe_{1.93}$ with applied stress

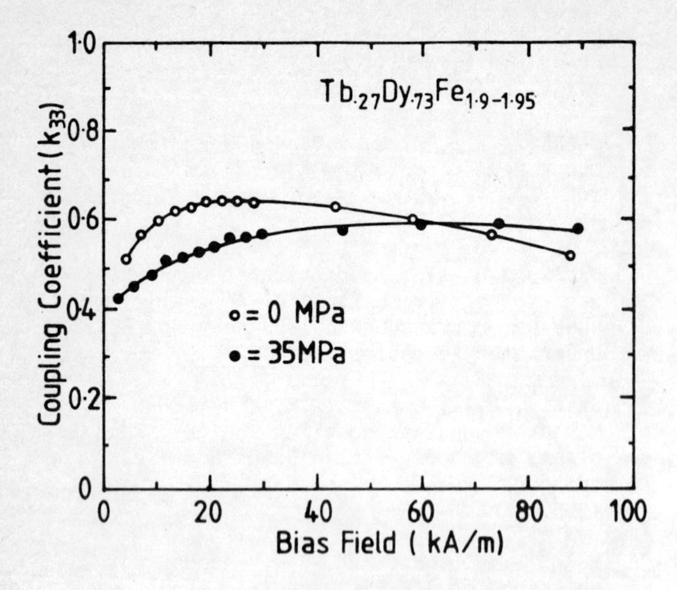

Fig 3 Data for the field dependence of the coupling coefficient for 0 and 35 MPa applied stress

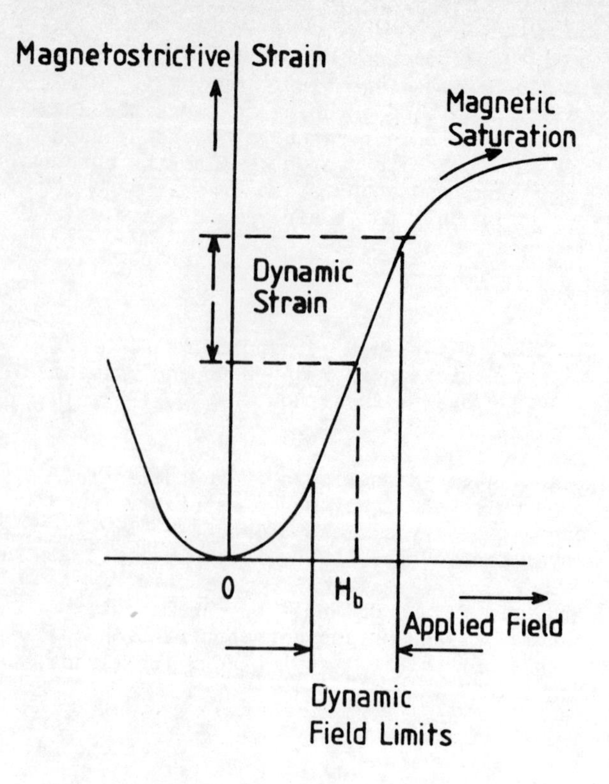

Fig 4 Schematic diagram illustrating the use of an applied d.c. magnetic field to provide a 'bias' working point H_b and strains resulting from the subsequent application of a dynamic field

C391/054

An electrically driven automotive coolant pump

J W McBRIDE, BSc(Eng), PhD and **M J REED**, BEng
Departments of Mechanical and Electrical Engineering, University of Southampton

ABSTRACT

This paper reviews the state of development of a project aimed at the more efficient cooling of the spark ignition engine using a variable speed, electrically driven coolant pump. The more efficient cooling of the engine would result in improved combustion, with the associated benefits.

The use of brushless dc and conventional dc motors are compared as suitable drives for the system. The system described uses a brushless dc motor, with temperature sensing at a number of points in the engine. An important measurement is shown to be the valve bridge temperature. This temperature is sensed by thermocouples, and the voltage produced used as an input to a microprocessor based control system. The system is then used to control the speed of the motor, and thus the degree of cooling. Control strategies are discussed, and some initial results presented for the laboratory based environment.

1. INTRODUCTION

In the conventional automotive coolant system the coolant pump speed is a direct function of the engine speed, but the heat flux from an engine is a function of both the engine load and speed, hence the belt driven system does not give the optimum coolant flowrates, for all conditions, and indeed for most running conditions a typical engine is overcooled.

In this paper a system based on a continuously variable speed motor is used to control the pump speed to give the optimum conditions.

A typical cooling system is designed as a compromise between maintaining a stable cylinder wall temperature, and cooling the engine to protect the materials and lubricants. If the cylinder wall temperature of an engine is allowed to increase there will be two effects. Firstly the viscosity of the oil will decrease, and secondly there will be an increase in the air inlet temperature which will act to reduce the engine's volumetric efficiency, [1]. Hence a prime factor of the cooling system is to keep the oil at an optimum temperature to prevent wear and to minimise losses; and at the same time balance out the temperature related effects of Knock, pre-ignition, volumetric efficiency and emissions.

The range of local heat fluxes within an engine is very large. Small areas of very high heat flux exist, and these areas can lead to the onset of film boiling. This condition is shown in figure 1, three regions are identified, free convection (Sub-cooled boiling), nucleate boiling and film boiling above the critical heat flux. When the heat flux reaches the critical value vapour bubble formation creates a situation where the liquid is no longer in contact with the wall. The value of the critical heat flux is dependent on the flow velocity and the properties of the liquid. The value of this temperature at the wall is referred to as the "coolant side metal temperature" (C.S.M), and for a standard pressurised coolant system it has been shown that 140°C is a typical maximum value before the onset of film boiling, [2].

For a small automotive engine with a pressurized EG/water mixture, the highest heat fluxes occurs in the area between the exhaust and the inlet valves, known as the valve bridge. In order to investigate the performance of the standard system consider the data produced in figure 2. The graph shows the theoretical response of the coolant side metal temperature against engine speed and load for a small automotive engine. The areas corresponding to driving at 70 mph on a level road, and the EEC standard ECE 15 cycle for urban driving have been added to the figure. The figure shows that under urban driving conditions the C.S.M is 25°C lower than the ideal of 140°C. For motorway driving it is 12°C too low. Tholen [1] states that an increase in the coolant temperature of 60° C will lead to a 20% decrease in fuel consumption at no load. Hence if a temperature of 140°C were maintained an improvement in fuel consumption would be expected; this will also be aided by the lower oil viscosity and reduced power loss in the pump drive because of the lower speed.

A system base on an electrically driven pump has a number of advantages over the existing system. It can allow the pump to run at lower speeds thus reducing the effect of overcooling, and saving power. In addition the location of the coolant pump is more flexible and indeed it might be advantageous to mount both the drive an the pump away from the engine, if coolant volume is also optimised. The over heating of an engine after the engine has stopped due to the thermal inertia of the system could be reduced by maintaining the speed of the pump.

The main disadvantages of cost and reliability could be accounted for. The former is always likely to be more than the standard system but the advantages of the more efficient engine could override this if the electric motor based system were not costly. The reliability of a motor drive could be obtained using a brushless dc motor but this type of motor is expensive at present, although with mass production this could be reduced.

2. MOTOR SELECTION

The basic motor selection choice with the 12 volt dc supply, is either a conventional dc or brushless dc motor. Although there are prospects of using switched reluctance motors, these are at an early stage of development, [3].

Table 1. shows a comparison of the basic operational principles of the two types of dc motor. The brushless dc motor is a clear favourite for this type of application, because of the inherent reliability, although this has yet to be tested in the severe underbonnet conditions, and a full implementation could be limited by the cost.

The brushless dc motor uses power electronics to control the commutation of the machine. The position of the rotor is detected with Hall sensors, which produce the control signals for the drive circuits. The motors use a permanent magnetic rotor to produce the field, with the phased stator windings producing a rotating magnetic field which is related to the rotor position by the position sensors. The inherent reliability of such motors comes from the fact that no brushes are used, [4]. In the system used here the power and control electronics are mounted in the motor casing, leading to a compact drive. The main concern of using such a motor in the "noisy" environment of a typical engine, was tested at an early stage and showed that with the motor mounted in the position of the alternator, and with the power supply coming from the 12 Volt battery no interference with the power electronic commutation was monitored.

To control the speed of the motor the 12 volt supply is pulse-width modulated (PWM) to give a continuously variable voltage to the motor.

3. SYSTEM DESIGN

There are a number of options available for the choice of measurand used to control the motor/pump speed:

(i) from the engine speed and throttle position measurements. This would lead to an open loop control system which would fit conveniently within current engine control strategies, but safety margins would be required to prevent engine damage and these could negate the beneficial effects.

(ii) from the coolant side metal temperature. This measurement would give the most direct measurement, but requires the careful mounting of the sensors.

(iii) from the bulk coolant temperature rise through the engine. This system would be easily implemented but would suffer from the inherent time lags due to the thermal inertia of the coolant

(iv) Hi-lo control of the motor using a thermostatic switch. This would be the simplest system to implement, but would suffer the same disadvantages as (iii), with the additional problems of the accuracy and reliability of the switch over long time periods.

The strategy investigated in this paper is based on option (ii). Six thermocouples were mounted on the test engine a laboratory based Morris 1100 engine. Four were mounted between each of the four inlet and exhaust ports. With the additional two mounted in the inlet and outlet coolant ports. The overall control system used is shown in figures 3 and 4, with the power electronic circuit shown in figure 5. The four temperature signals taken from the thermocouples measuring the CSM temperature, are summed and amplified before entering the differential amplifier. This is used to set the control reference temperature. The PID unit is used to improve the response of the system, after which the output is storedin the sample and hold circuit before entering the ADC. The digital output is then directly related to the error signal. The microprocessor uses the error signal to control the PWM signal which is then output to the control electronics used to control the motor speed.

4. EXPERIMENTAL ANALYSIS AND RESULTS

The system was implemented on a Morris 1100 Laboratory based engine to prove the principles of the design before full testing on a modern engine.

The initial results evaluate the pumping requirements for the engine used.

Figure 6 shows the power requirement of the water pump, based on the evaluation of the electrical input power, for a full range of pump speeds. This data has been collected by driving the pump via a belt by a conventional dc motor. The input volts and current were measured to give the power. The characteristic of the pump is shown by the loaded line. The maximum power at 5000 rpm being 800 watts. The power used is also dissipated in the motor windings hence the motor was tested disconnected from the belt, and this characteristic is shown by the No-load line. To account for the full losses in the case of the loaded curve the copper winding losses should also be accounted for because of the increased current in the loaded case, and these have been evaluated for the known resistance. The result is the additional power loss, which added to the no-load case gives the total losses characteristic.

Having evaluated the total losses, the full power and torque requirements of a motor to drive the pump can be established, and the results are shown in figures 7 and 8. The Pump requirements in terms of power are the difference in the two characteristics in fig 6. Also shown in the figures are the power and torque curves of a brushless motor supplied for the driving of a recently designed pump. This new pump was designed for lower power and low flow rates. Although both pumps use a mechanical face seal consisting of a sealing rubber held in place by a retaining spring, the Morris pump uses the principle of hydrodynamic lubrication to reduce the friction in a seal with high static friction. This relies on water being forced underneath the seal to lubricate the join between the stationary seal and the rotating metal water pump shaft. At low speeds the system pressure is not sufficient to force water into the join and thus the friction losses are high. This creates a relatively high starting torque as shown in figure 8. The new pump uses a much lower friction seal and does not depend on the hydrodynamic principle.

Clearly the brushless dc motor selected for the new pump cannot match the requirements of the laboratory engine, hence for the purpose of the experimental analysis a conventional dc motor has been used, and operated to give the desired control characteristics. Although the electronic system has been used, the complete control system has not been implemented in this work. The electronic system and brushless dc motor have been shown to produce a variable speed drive based on the measurand, however these results have not been confirmed in this paper, on the complete system, and this should constitute the final stage of development.

The results presented are therefore preliminary and demonstrate the principle rather than the full implementation.

It can be deduced from fig's 6,7 and 8 that at a speed of 4000 rpm the efficiency of the dc motor is 30%, compared with that of a typical brushless motor at this speed with a bipolar drive of typically 75%.

Figure 9 shows the map of the conventional cooling system taken for the laboratory engine, and based on experimental data. The average temperature is the mean of the four thermocouple measurements, again this method of cooling leads to over cooling during most of the driving conditions. Hence by controlling the temperature the increased benefits of thermal efficiency and fuel consumption will be determined. The control temperature for this engine was set at 95°C, since this engine is unpressurised and neucleate boiling has be shown to occur at around 100°C, the value used therefore has an inbuilt safety factor.

Figure 10 shows the variation in the water pump speed for various engine loads, with the CSM temperature maintained at 95° C. For comparison the conventional cooling system speed is also identified. This clearly shows the reduction in the pump speed for all values of the engine speed. The reduction in power associated with this speed difference is shown in figure 11. This shown the increase in the engine power resulting from the speed control. As can be seen the greatest percentage gains in the power output occur with the lowest engine loads.

6. DISCUSSION

The experimental results show that a power saving is possible if the pump is controlled by an electric motor, however these results were taken for a system where the motor was controlled independently of the engine; hence for a full evaluation the overall system has to be considered. This has been undertaken with a theoretical evaluation, and the result is shown in figure 12. The calculation takes into account the brushless dc drive with 75% efficiency, the altenator with 50% efficiency and the belt drive with an estimated power transmission efficiency of 90%. The figure shows that the proposed system leads to an increase in power output over the vast majority of driving conditions. During the ECE15 cycle the power increases range from 1.1% to 3.5%. Only at high engine speeds and full load is the overall power output decreased. These power increases have been translated into percentage changes in fuel consumption shown in figure 13. As can be seen the new system will lead to improvements in fuel consumption over the ECE 15 test cycle of around 2%. Although initially this may not appear to be a considerable margin, when it is considered that average fuel consumption for passenger cars has increased only marginally over the last 10 years then the use of the electrically driven coolant pump would give a significant advance in vehicle economy. The values identified in this paper are open to re-evaluation for a more advanced engine and cooling system, however all of the indications show that a more significant improvement could be achieved.

As discussed in the introduction the raising of the CSM temperature would raise the combustion temperature improving the combustion, and therefore reducing emissions.

Additional benefits will come from the improved warm up times, and the possible reduction in the coolant volume

7. CONCLUSIONS

The results presented in this paper show that;

(i) By controlling the average coolant side metal temperature to 95° C, fuel economy improvements of up to 4% have been achieved on a laboratory based Morris engine.

(ii) The brushless motor driven coolant pump, has been shown to be compatible with environment of the engine used.

(iii) Further work should be undertaken to investigate the benefits to combustion, and fuel economy with the full implementation of the system on an new engine with improved coolant pump.

ACKNOWLEDGEMENTS

The author would like to acknowledge M J Reed, R.S Poll, and I.P Part, [5], [6], [7], for their useful work over the last three years. I would like to thank the Ford Motor company for their financial support

and finally to Southampton University for the use of facilities.

REFERENCES

[1] P Tholen "Coolant and Lubricant temperature effects fuel consumption" Automotive Engineering, the Society of Automotive Engineers, Vol 91 No 12, Dec. 1983, p 53.

[2] Hai Wu, R A Knapp, "Thermal conditions in an internal combustion engine." ASME Conference, San Francisco, Dec 1978, Pub ASME New York, 1978, pp 79-87.

[3] P.J Lawrenson etal, "Controlled-Speed Switched Reluctance Motors", IEE Conference on Drives, 1985 pp 23-26.

[4] Kenjo "Permanent magnet and brushless dc motors", Oxford University Press.

[5] M.J Reed. "An Electrically Driven

Automotive water pump", Southampton University internal report, May 1989.
[6] I.P Part "Controlling an electrically driven automotive coolant pump", Southampton University internal report, May 1988.
[7] R.S Poll "An Electrically Driven water pump", Southampton University internal report, May 1987.

	Conventional dc	Brushless dc
Mechanical Structure	Field Magnets on the Stator	Field Magnets on the Rotor
Distinctive Features	Quick response Controllability	Reliability
Commutation	Brushes on the Commutator	Power Electronics
Rotor position Detection	Brushes	Position Sensors
Reversing	Terminal Voltage	Logic Sequence

Table 1. A comparison between conventional and brushless dc motors.

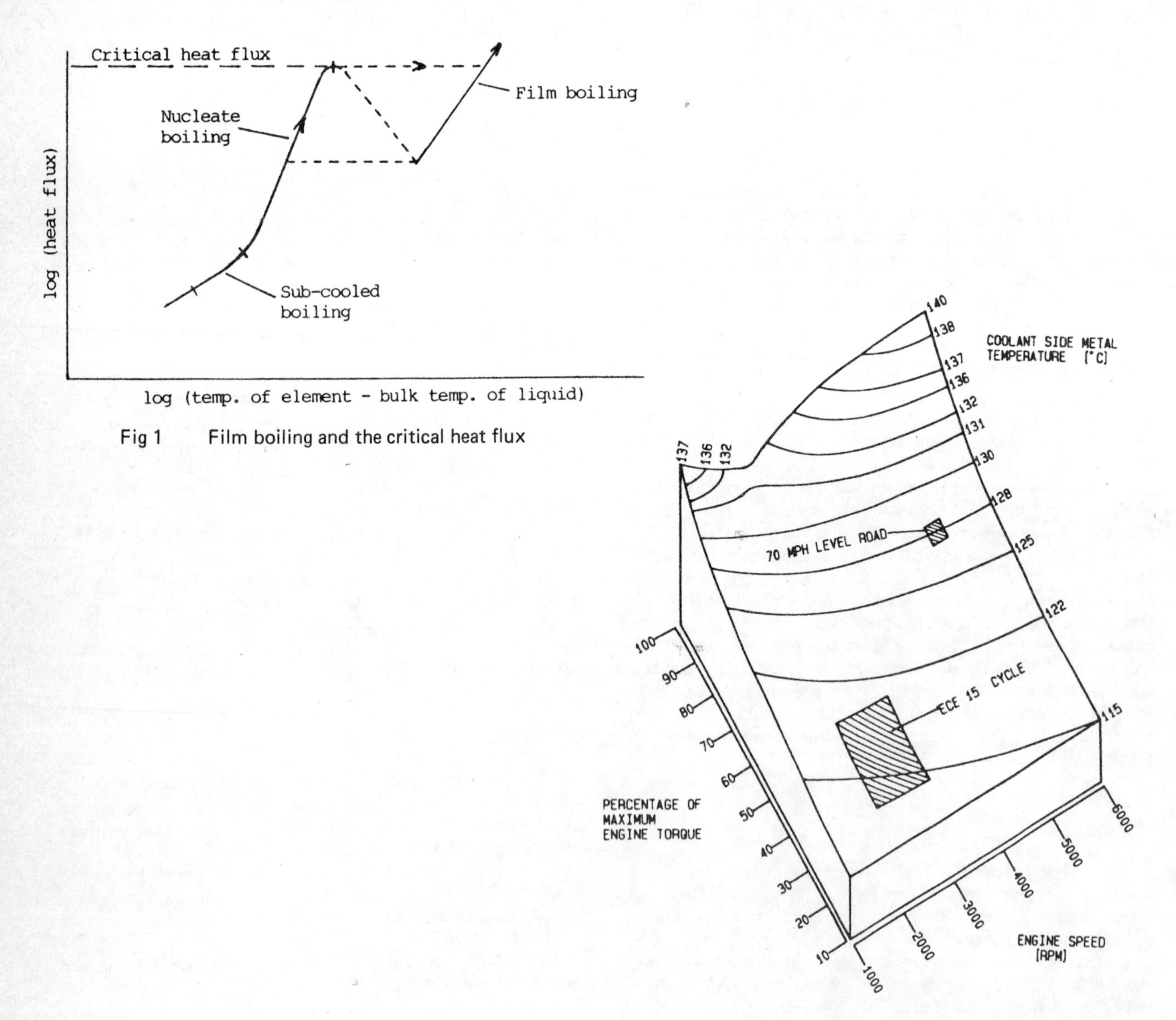

Fig 1 Film boiling and the critical heat flux

Fig 2 (right) Coolant side metal temperature at the valve bridge of a conventionally cooled engine

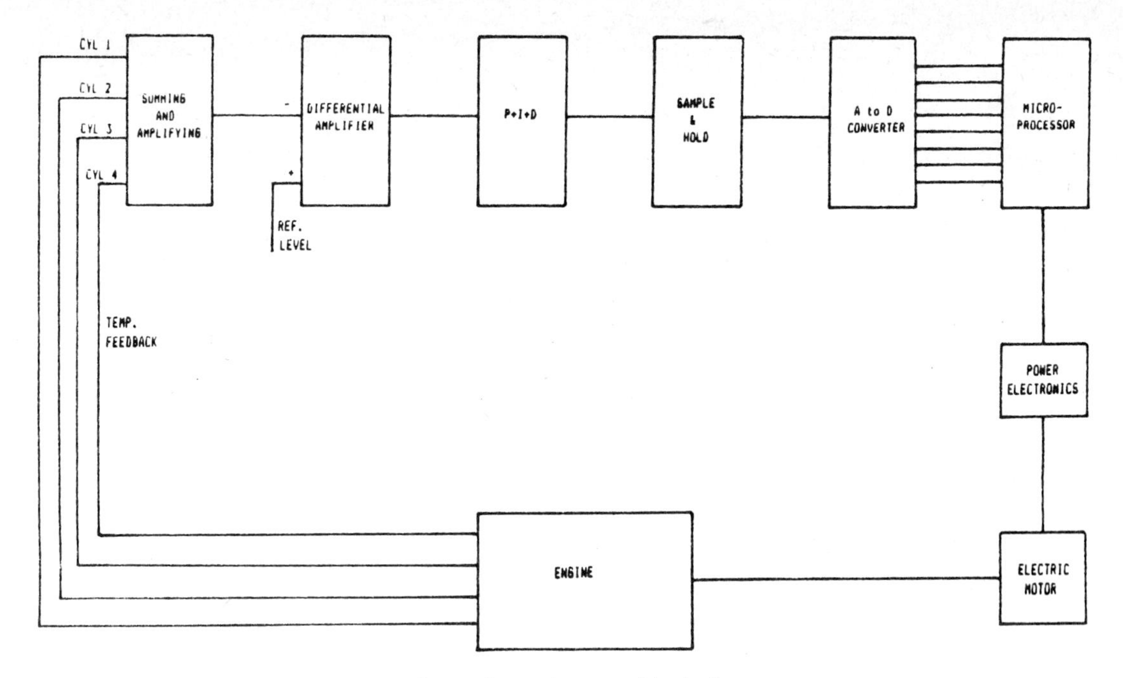

Fig 3 Control system block diagram

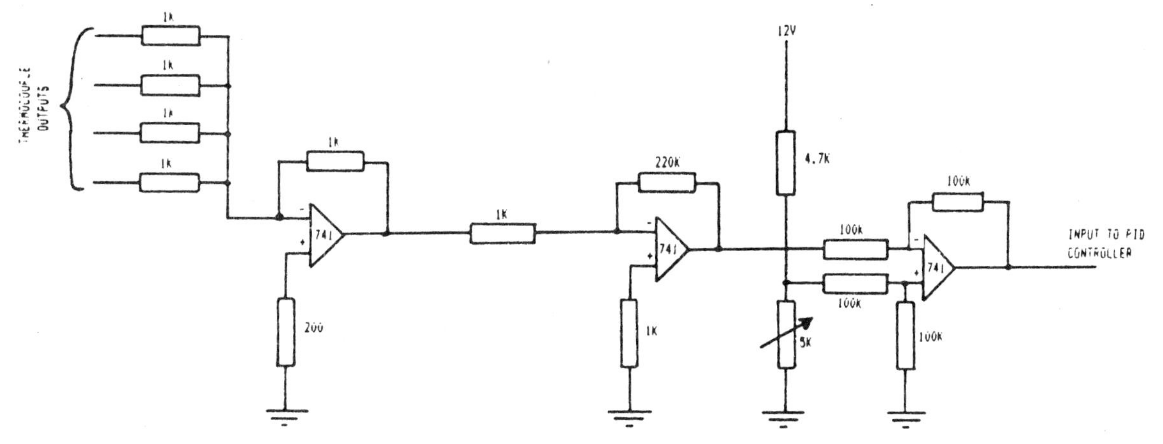

Fig 4 Control system electronics

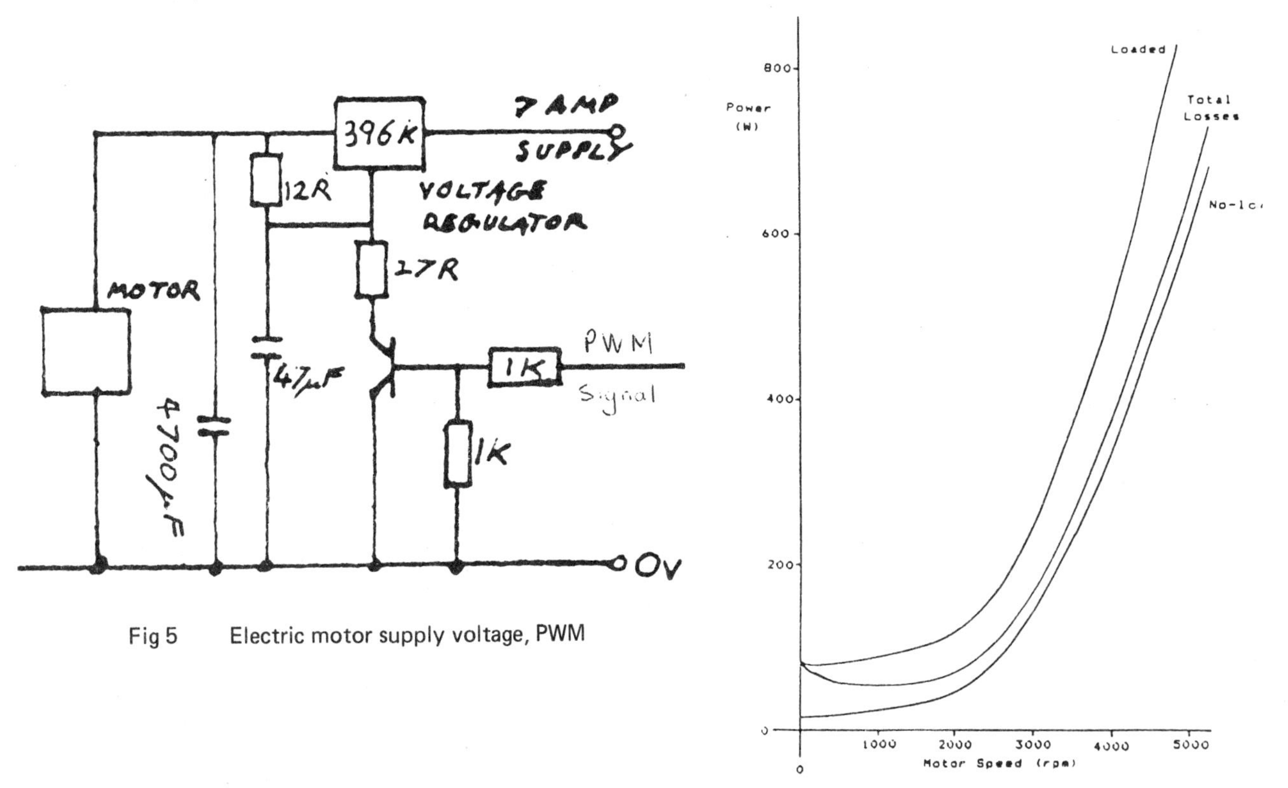

Fig 5 Electric motor supply voltage, PWM

Fig 6 Power requirement of test engine coolant pump

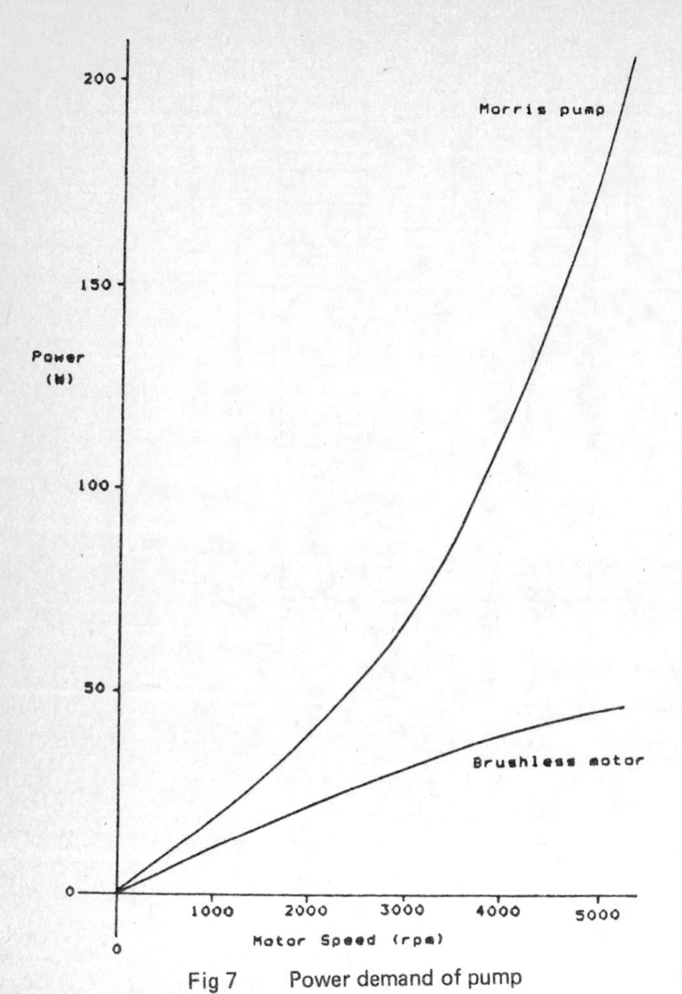

Fig 7 Power demand of pump

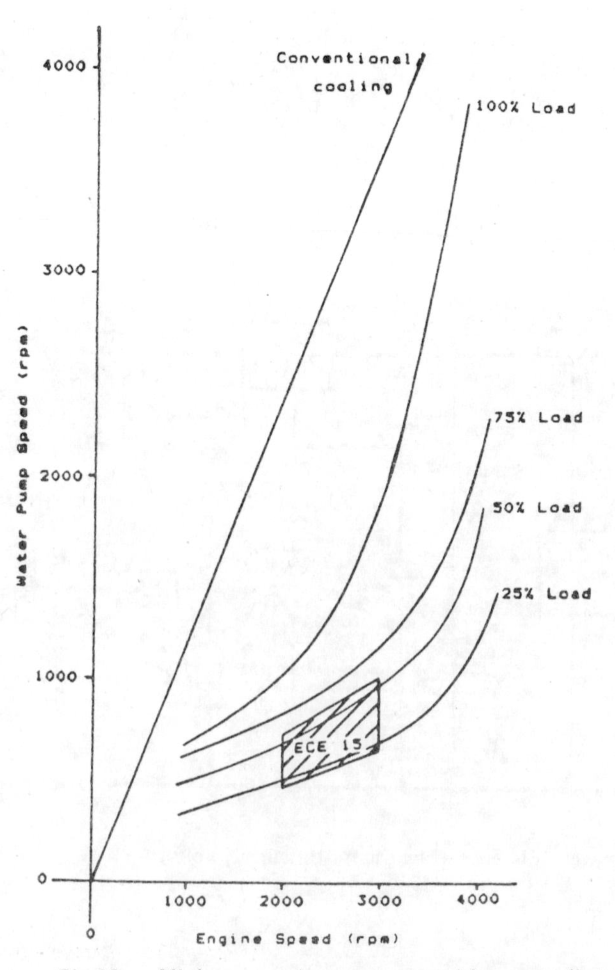

Fig 9 Test engine temperature map, with conventional cooling

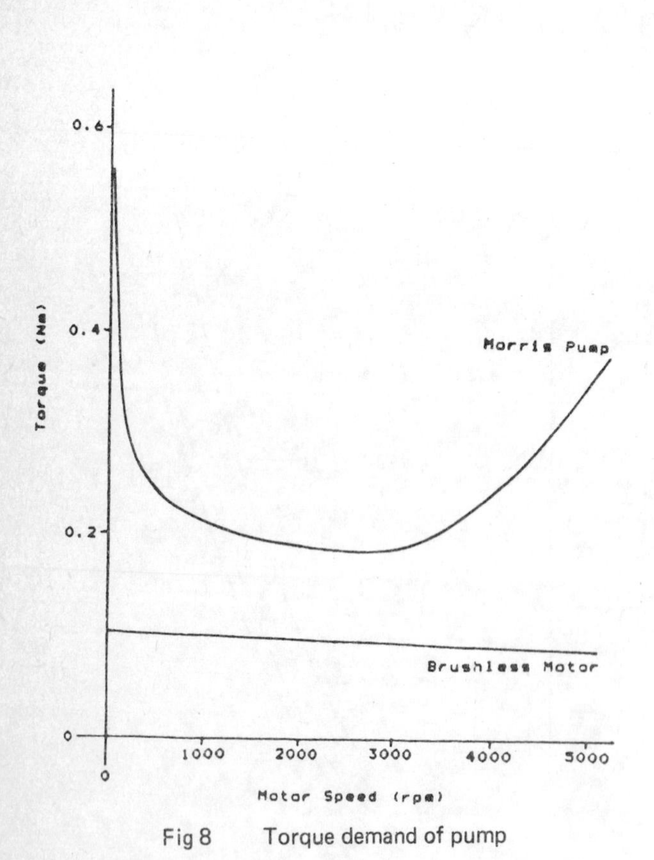

Fig 8 Torque demand of pump

Fig 10 Minimum coolant pump speed for controlled cooling versus engine speed

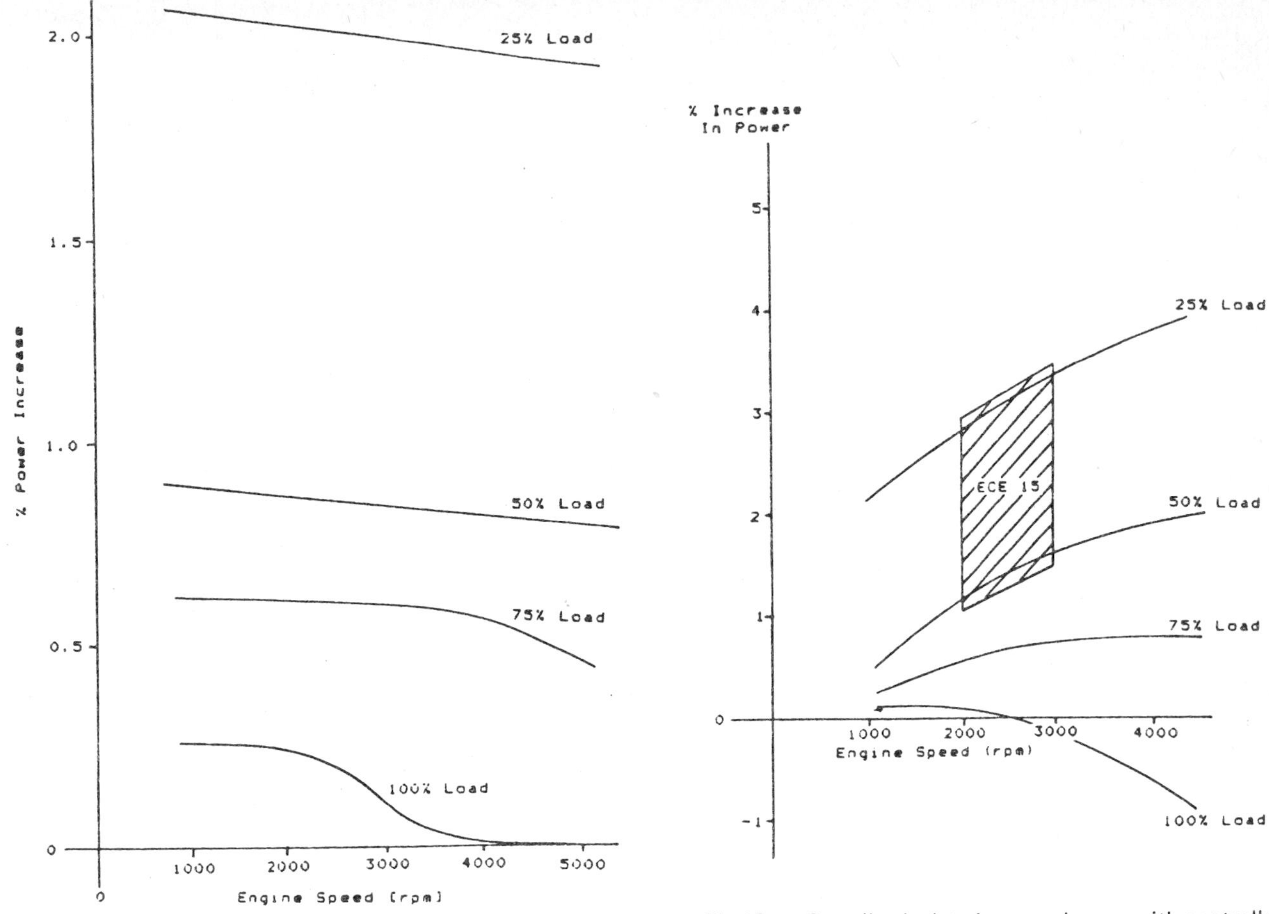

Fig 11 Changes in engine power output with controlled cooling versus engine speed

Fig 12 Overall calculated power changes with controlled cooling versus engine speed

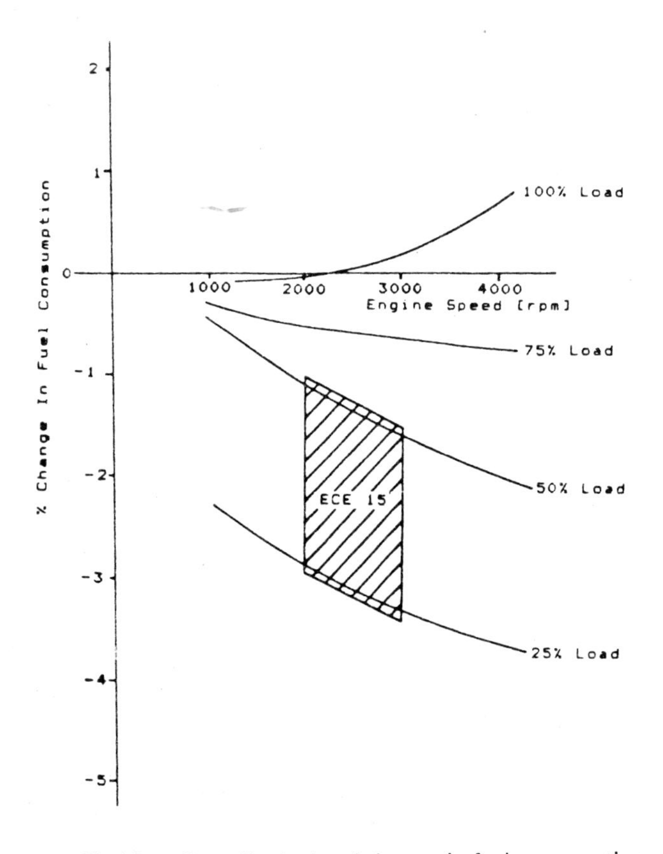

Fig 13 Overall calculated changes in fuel consumption with controlled cooling versus engine speed

C391/064

Power IC technologies

C CINI and **R FERRARI**
SGS-Thomson Microelectronics, Gentilly, Italy

ABSTRACT

This paper explains how and why developments in power IC technology affect all automotive engineers, not just specialists in electronic system design.

Reviewing the state of the art, it describes three power IC technologies suitable for automotive applications and explains how these technologies cover various system configurations. The relative merits of these technologies are compared and some criteria for selection are described.

Several existing integrated circuits are presented and proposals for a complete family of multiplex switches are examined. This family shows how different technologies are used for similar functions. The importance of packaging for power ICs is underlined and all of the important package types reviewed.

INTRODUCTION

There is no single power IC technology that is superior to all others in every aspect. For each application the choice of technology must be a tradeoff depending on the type of load and on the circuit complexity required. Consequently, the availability of many power technologies is an advantage, allowing the best choice in terms of both technical characteristics and cost.

The technologies used to manufacture ICs that can be defined smart power for the automotive sector are divided into two main families: bipolar and mixed. These are divided in turn into horizontal and vertical structures.

The difference between horizontal and vertical structures concerns how the collector or drain current is collected and brought out of the IC. The following definitions are valid for all process types, bipolar and mixed.

Bipolar technologies are used when the main characteristics of the IC to be realized are characterized by exclusively analog functions, while mixed technologies are characterized by both analog functions and digital functions that are rather complex. Other great advantages of mixed technologies are the absence of second breakdown, fast switching speed and very high conversion efficiency.

Mixed technologies are derived from bipolar IC processes by adding components like CMOS and DMOS. The increase in the components available on the same chip in mixed processes compared to traditional bipolar types brings an increase in the complexity of integration but also an increase in cost. For this reason it is important to choose the most suitable technology for each application.

A HOSTILE ENVIRONMENT

Apart from the process choice, for an integrated circuit the electrical and physical environment of a vehicle is extremely hostile. Any electronic component designed for automotive applications must operate over a very wide temperature range, it must operate on a wide range of supply voltages, it must be able to survive accidental battery reversal and it must be robust enough to withstand the high energy transients which sometimes occur on the battery rail. When the battery is accidentally disconnected from the alternator during a high current charging phase, for example, a transient is generated which can exceed 100V. Transients of up to -100V can also be generated if an inductive load is de-energized suddenly.

HORIZONTAL & VERTICAL STRUCTURES

In a horizontal structure the current enters and leaves the chip through the upper surface of the chip. In this case, too, an isolated structure can be added that contains the control circuitry. This structure is derived from that adopted in standard power integrated circuits where the substrate is simply a mechanical support and a heat conductor. In the horizontal structure several variations can be distinguished on the basis of the current and voltage required.

Using DMOS as a power element, operating at low voltages (~<20V), the lateral structure offers the best performance.

In mixed horizontal/vertical structures the current flow is vertical in the power device structure but the current is brought to the surface through buried layers and sinkers. This approach allows the integration of any number of power devices connected in any way; in addition it offers medium currents and high voltages.

In a vertical structure the current flows through the device from the top to the bottom, across the substrate and the die attach area to the package itself. This scheme, with the substrate above ground potential, is like that of discrete power devices with the addition of an isolated structure containing the control circuitry. The vertical structure allows very high current densities but there is the limitation that in one chip it is only possible to integrate one power transistor, or several with the collectors or drains commoned.

SMART POWER PROCESSES

Smart Power processes using bipolar technology have been known for many years; more recently Multipower-BCD and VIPower processes have been industrialized. Multipower-BCD is a family of smart power processes with a horizontal structure, covering the range from 20V to 100V and beyond, particularly suited for the automotive sector.

The name "Multipower" indicates the possibility of making one or more power devices, isolated from each other and connected, by the metal, in any configuration. "BCD" means that Bipolar, CMOS and DMOS elements can be integrated on the same chip. Figure 1 shows a cross section of the Multipower-BCD process.

VIPower processes are, on the other side, characterized by the power devices of vertical type; VIPower stands for Vertical Intelligent Power. While the power DMOS transistors have a well defined structure, the bipolar and CMOS signal elements have characteristics and performance similar to that of Multipower-BCD processes. The main difference lies in the P type in the Multipower-BCD process and N type in the VIPower vertical process.

Figure 2 shows a cross section of the VIPower process; figure 3 shows a cross section of a bipolar process with a horizontal structure.

PROTECTIONS

Overvoltage protection

The overvoltage protection circuits permit devices to withstand positive and negative overvoltages.

The choice of the best solution (limit of the transient by zener diodes, and switch off of the device by adequate sensing circuits among the the other) is strictly related to the energy associated to the transient. External components such as varistors and clamp zeners may sometimes be necessary.

Current limiting

To protect the device and the load a current limiting circuit is provided. This circuit reduces the base current of the power output device and limits the output current. The current limitation circuit must be designed to prevent oscillations during short circuits.

Thermal Shutdown

To prevent damage from overtemperature conditions a thermal shutdown circuit turns off the power device when the chip temperature exceeds 160 degrees. Overtemperature is generally caused by short circuits at the output, but may also depend on mechanical troubles, such as a poor thermal contact between the IC case and its heatsink.

SMART POWER ICs

In vehicles the loads that must be driven can be divided into four basic types: motors, solenoids, resistors and lamps.

The evolution of power technologies, mainly high current, makes it possible to drive the different loads choosing the most appropriate process, following the price-performance concept.

The integrated circuits commonly known as smart power of the first generation (Bipolar process) make it possible to drive loads very simply where, in addition to a power stage, there are protection circuits such as thermal shutdown and current limiting in addition to level shifting circuits to make the input TTL compatible.

Smart power devices of the latest generation (BCD or VIPower processes) are much more complex and in addition to driving two or three low-current loads (less than 1A), have current protection, thermal shutdown and overvoltage protection and are able to report the load status and to process a signal from a microcontroller.

MOTOR DRIVERS

DC motors in vehicles range from the 300mA used for rear view mirrors to 20-30A for window lifts and windshield wipers. For the rear view mirror the current is low but there are two or three motors to be driven and the best solution is to use 6 to 8 power transistors in a full bridge configuration, allowing the motors to rotate in either direction but one at a time; this allows eight power transistors to drive three motors. The most appropriate process is the horizontal type, both bipolar and mixed, but able to integrate on the same chip all of the power transistors and control circuits.

Figure 4 shows the L6202, a motor driver realized in Multipower-BCD technology [4]. This device has a chip area of 20mm2, is assembled in a 20-pin DIP package and delivers 1.2A. The same die assembled in the Multiwatt 15-lead package delivers up to 3A at 48V.

The L236 is an excellent example of a motor driver where the current delivered to the load exceeds 20A. The solution adopted is a half bridge in a bipolar process [5] because the product is simple, therefore the bipolar process is more appropriate than more complex processes. And since the total drop must be less than 2.5V at 20A the chip size is very large. In addition, the power dissipated is high therefore it is preferable to have two half bridges in two separate packages which aid dissipation. (Fig 5).

SOLENOID DRIVERS

Integrated circuits for solenoid driving are divided into two types: low side drivers and high side drivers. For the driving of solenoids in low side configurations the use of a process with NPN power transistors, is recommended in case of bipolar process; N-channel power MOS if mixed technology is used.

For high side driving it is necessary to have either a power PNP or PMOS transistors (or PMOS equivalents). It is necessary to know what type of load will be used and then choose the process.

Figure 6 shows the block diagram of a low side driver realized with a VIPower process able to stand up to 470V. The circuit, called VB020, is a driver for electronic ignition. It drives directly the load and is controlled by a microcontroller without other interfaces. The key element in this process is the NPN output transistor which has application conditions for high voltage and high current (450V/7A) in addition to integrating the driving and control components.

Another IC for solenoid driving is L9350 shown in figure 7. This IC is realized with horizontal bipolar technology and is a high side driver which has a 1A isolated vertical PNP transistor as the output stage. The process guarantees that VCEO>=60V but it is able to withstand positive and negative voltages for several milliseconds of 120V (load dump and field decay). Another important characteristic of the L9350 is that the standby current is less than 100uA and a fast turnoff time is guaranteed by the recirculation voltage of -30V. Protection against overvoltages, thermal shutdown and current limiting are built into the chip. No external components are needed in the application circuit [6].

LAMP & RESISTOR DRIVERS

In vehicles there are many lamps with powers ranging from 3W to 60W and a smaller number of resistors that we can consider a subclass (simpler to drive) of lamps.

Integrated circuits for driving these loads are generally high side drivers because in modern vehicles the common side of the supply is the ground, while the process can be divided as a function of the load.

The process used is of the mixed type in both the VIPower and Multipower-BCD versions, the choice is in favor of VIPower for high powers because it has a lower ON resistance. When the power or current supplied to the load is low or there are functions with more power transistors or a package with the case connected to ground rather than Vcc is required, a horizontal structure can be chosen because ON resistance is not the primary criterion.

The block diagram of a lamp driver is shown in figure 8; a chip photo is shown in figure 9. This device, type VM200, has an ON resistance of 40 milliohms and is realized with the VIPower process [7].

Figure 10 shows a chip photo of the L9801, which is realized with the Multipower-BCD process and has an ON resistance of 100 milliohms. The difference between the two devices, apart from the ON resistance, is the pin configuration because in VIPower technology the case is at the supply voltage, while in Multipower-BCD technology the case is grounded [8].

POWER PACKAGES

Together with the general improvement in performances made possible by the extended use of Smart Power processes, it is more and more necessary to have power packages, strong, reliable and capable of dissipating in the best way the thermal energy.

Power DIP Packages

A typical example of the power DIP package is illustrated in Figure 11. Externally this package is identical to the standard DIP but the leadframe is made of 0.4mm thick copper and designed so that the four center pins conduct heat from the die island to the PCB. Various leadframe configurations are manufactured. For these packages in many cases the only heatsink required is an area of PCB copper connected to the heatsink pins. However, an additional heatsink can be added if required.

Power Tab Packages

TO-220 style packages are available in 5, 7, 11 and 15 lead versions (figure 12). The five and and seven lead types are 1 cm wide, like the original 3-lead TO-220; the 11 and 15 lead types are 2 cm wide.

These packages have a maximum junction-case thermal resistance of 3'C/W and are suitable for the most powerful ICs. An external heatsink is normally bolted, clamped or rivetted to the tab.

Surface Mounting Packages

Surface mounting packages for power ICs are still rare. One type available today is a medium power package based on the plastic chip carrier (figure 13). Externally the package is identical to a 44 lead PLCC, but the lead frame is designed so that 11 of the leads are used to conduct heat from the die flag to the substrate.

For higher power ICs new surface mounting packages are being developed, like the example shown in figure 14. These packages are designed to exploit special substrates with a low thermal resistance, such as copper backed PCBs, to eliminate the need for add-on heatsink. That exploits in the best way the resources of the Smart Power processees.

TO-3 Packages

Widely used for Power transistor, TO3 packages (Fig. 15) are metallic cases, hermetically sealed; with their thermal resistance (junction case) of less than 2'C, they are particularly suited for very high current (50 Amp and more), high power and high junction temperature (>150'C).

CONCLUSIONS

The actuators used in vehicles are very different from each other in terms of functions, current levels and complexity. This differentiation leads to the use of processes to make power ICs that are very different from one another such as bipolar or the new mixed processes both horizontal and vertical. The latest generation of processes, mixed, allows a greater integration of functions on the same chip -- analog or digital -- in addition to the capacity to deliver very high output currents (20-30A); however it is not always true that high current and high complexity are incompatible, but often there are solutions -- usually economical -- where high complexity has a lower current capacity and vice-versa.

REFERENCES

1) C. CONTIERO, A. ANDREINI, P. GALBIATI & S. STORTI.

IEDM Conf. Digest, pp 766-769, Dec 1987.

2) R. ZAMBRANO, G. FERLA, S. MUSUMECI & P. PAPARO

17th, ESSDERC Conf DIGEST, pp653-656, Sept 1987.

3) S. RACITI & M. PAPARO

MOTORCON 1987 Conf. Digest, pp 142-151, April 1987

4) C. CINI, C. DIAZZI & D. ROSSI

A High Efficiency Mixed Technology Motor Driver ic, PCI 85, pp 267-277,Oct 1985.

5) S. STORTI & F. CONSIGLIERI

A 30A 30V DMOS Motor Controller and Driver, ISCC 88, pp 192-193, Feb 1988.

6) F. MARCHIO', P. MENNITI & F. FERRARI

High Current Source Driver for Automotive Applications IEE Automotive Electronics, Birmingham G.B. 29 Oct - 1 Nov 1985

7) M. ZIZA & S. STORTI

High Side Monolithic Switch in VIPower Technology ATA Events 1988

8) C. CINI, C. DIAZZI, D. ROSSI & S. STORTI

High Side Monolithic Switch in Multipower BCD Technology International Microelectr. Conference,Munchen,Nov 86

9) C. CINI, C. CONTIERO & B. MURARI

Intelligent Power Technology A Reality in the 1990s Proceedings of 1988 Intern. Symposium on Power Semiconductor Devices, Tokyo, pp 88-95

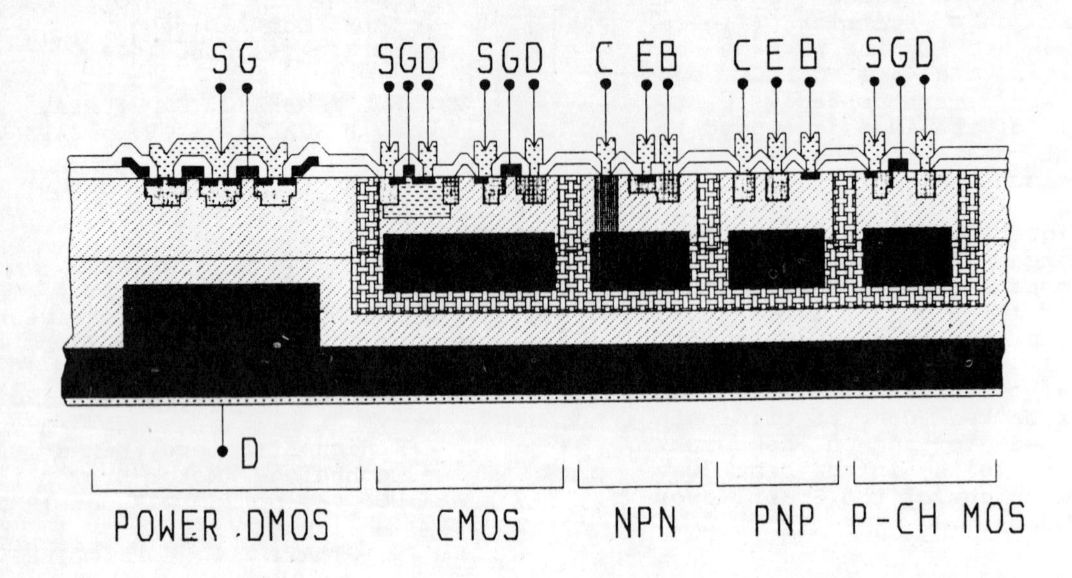

Fig 1 Cross-section of the multipower BCD technology

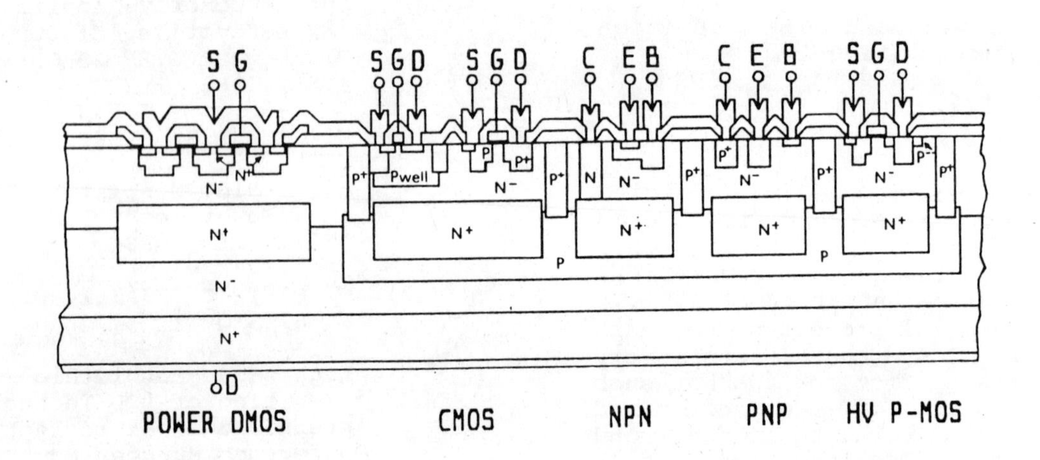

Fig 2 Cross-section of the VIPower process

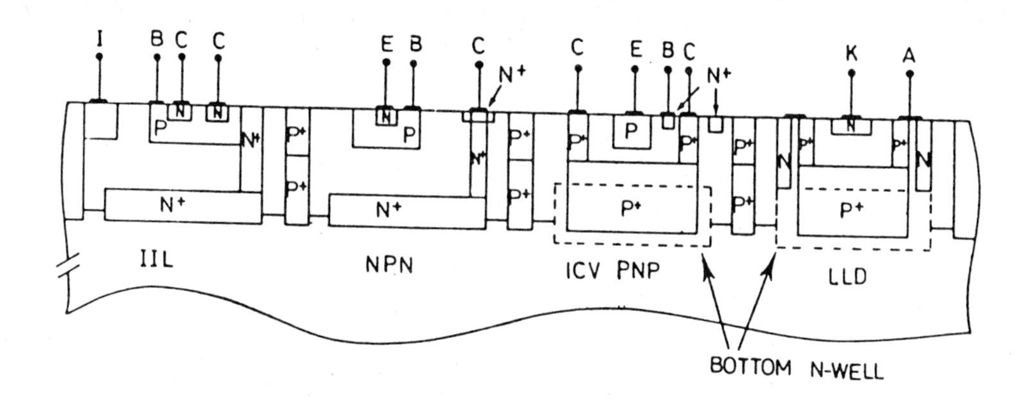

Fig 3 Cross-section of the horizontal bipolar process

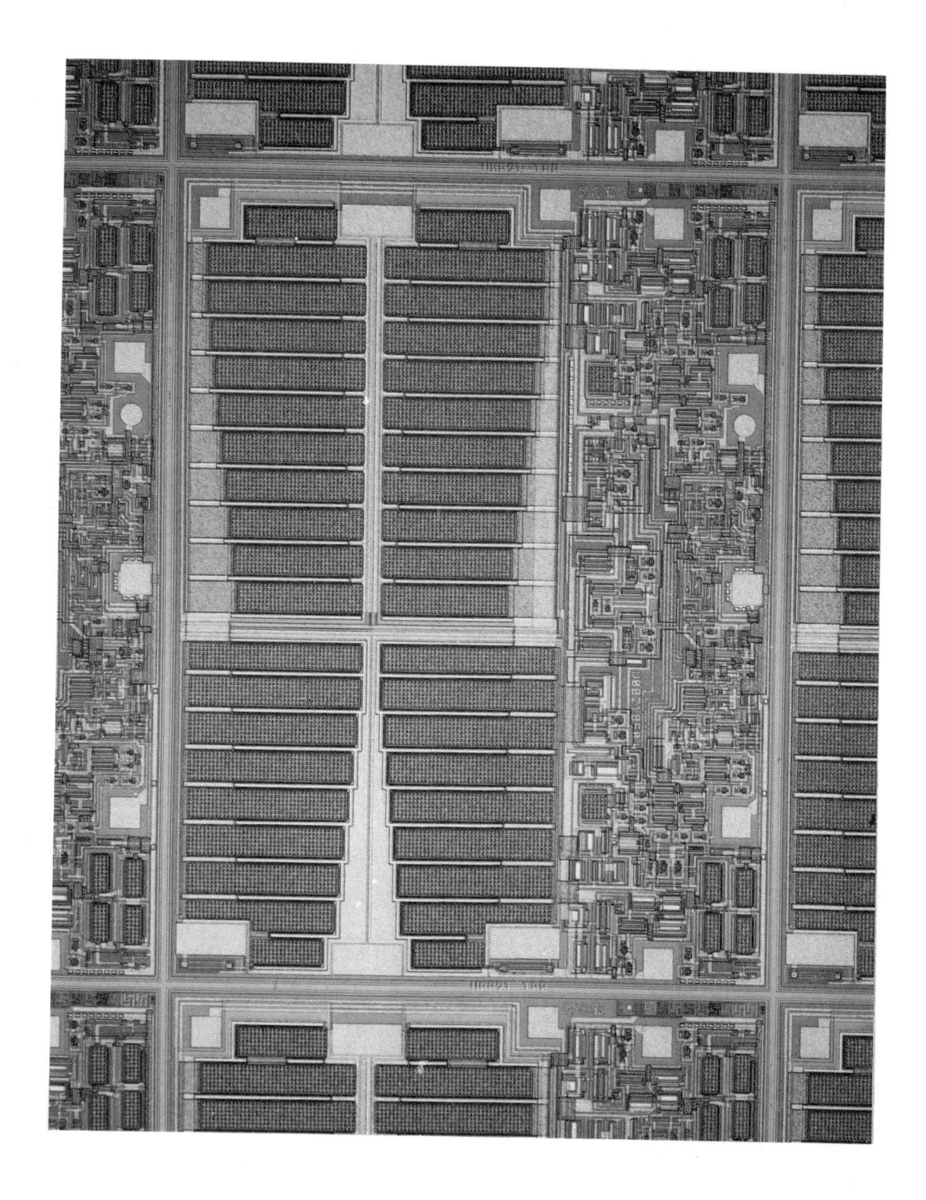

Fig 4 L6202

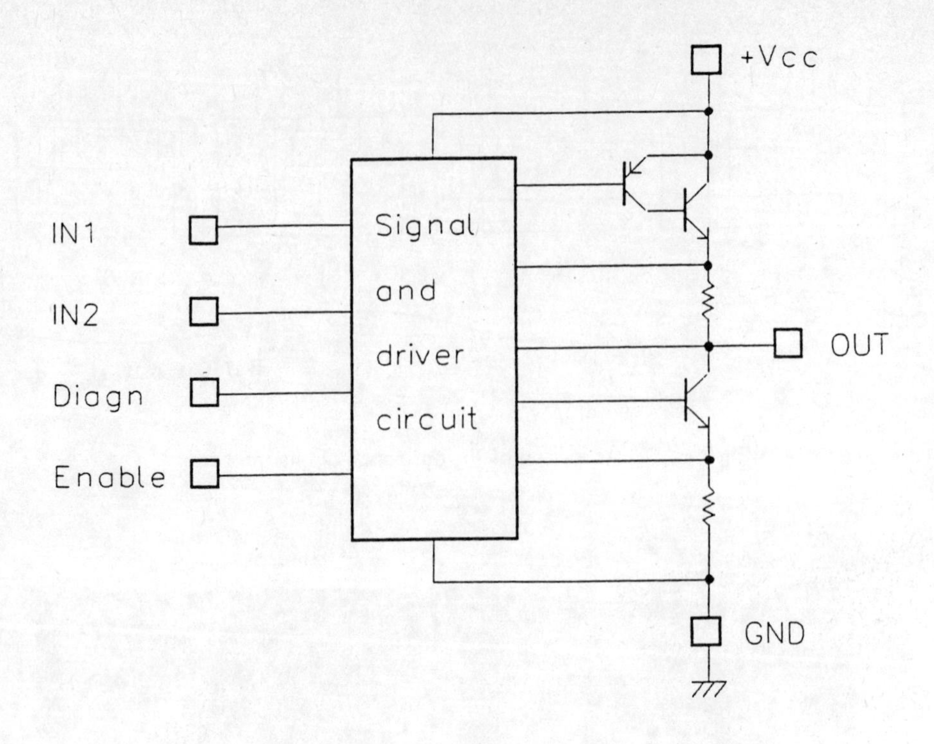

Fig 5 Block diagram of L236, half-bridge motor driver

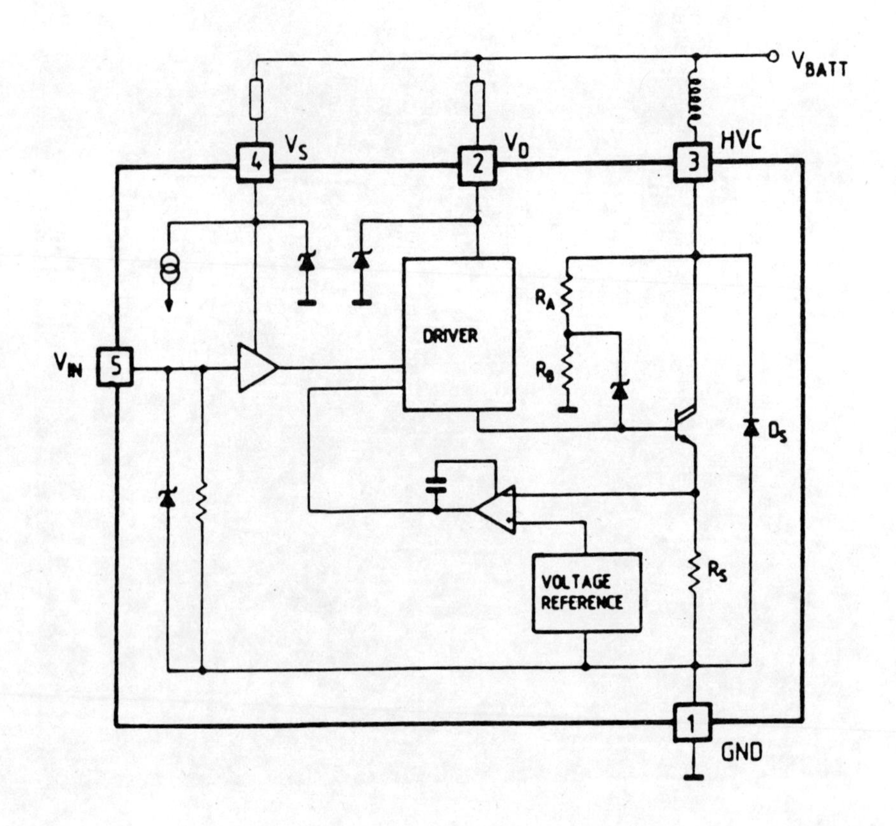

Fig 6 Block diagram of VB020

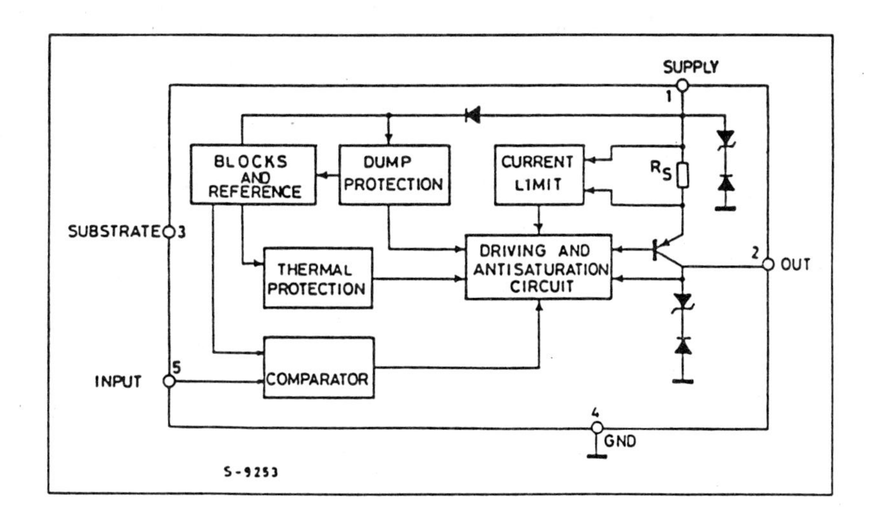

Fig 7 Block diagram of L9350

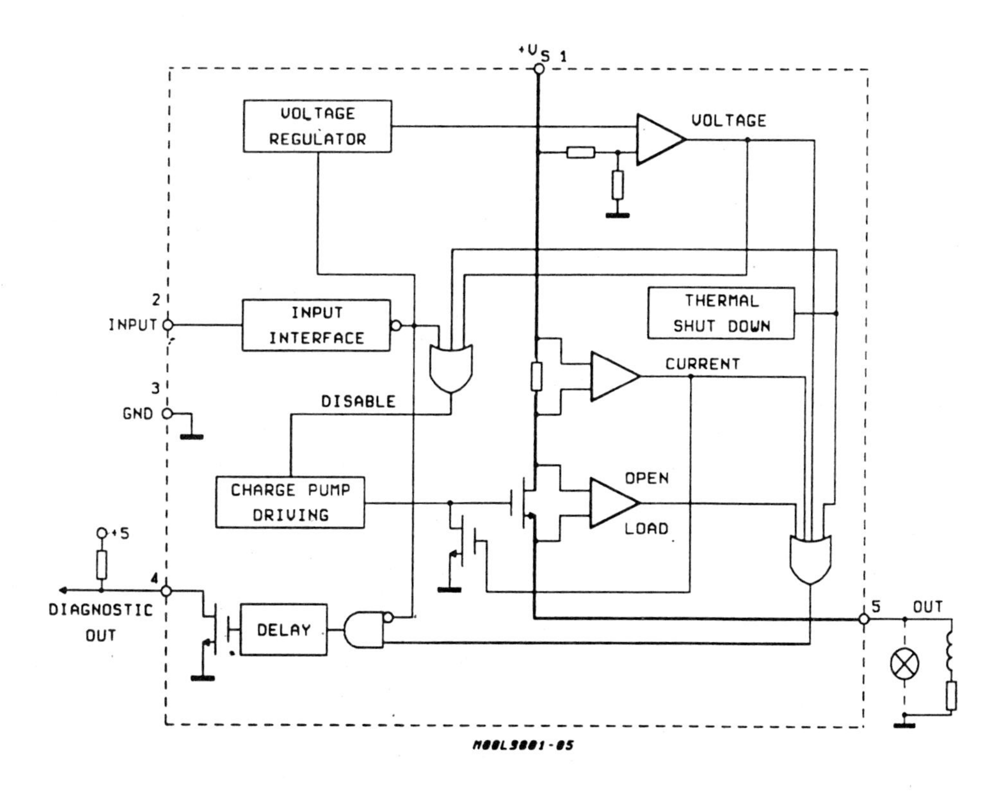

Fig 8 Block diagram of L9801 lamp driver

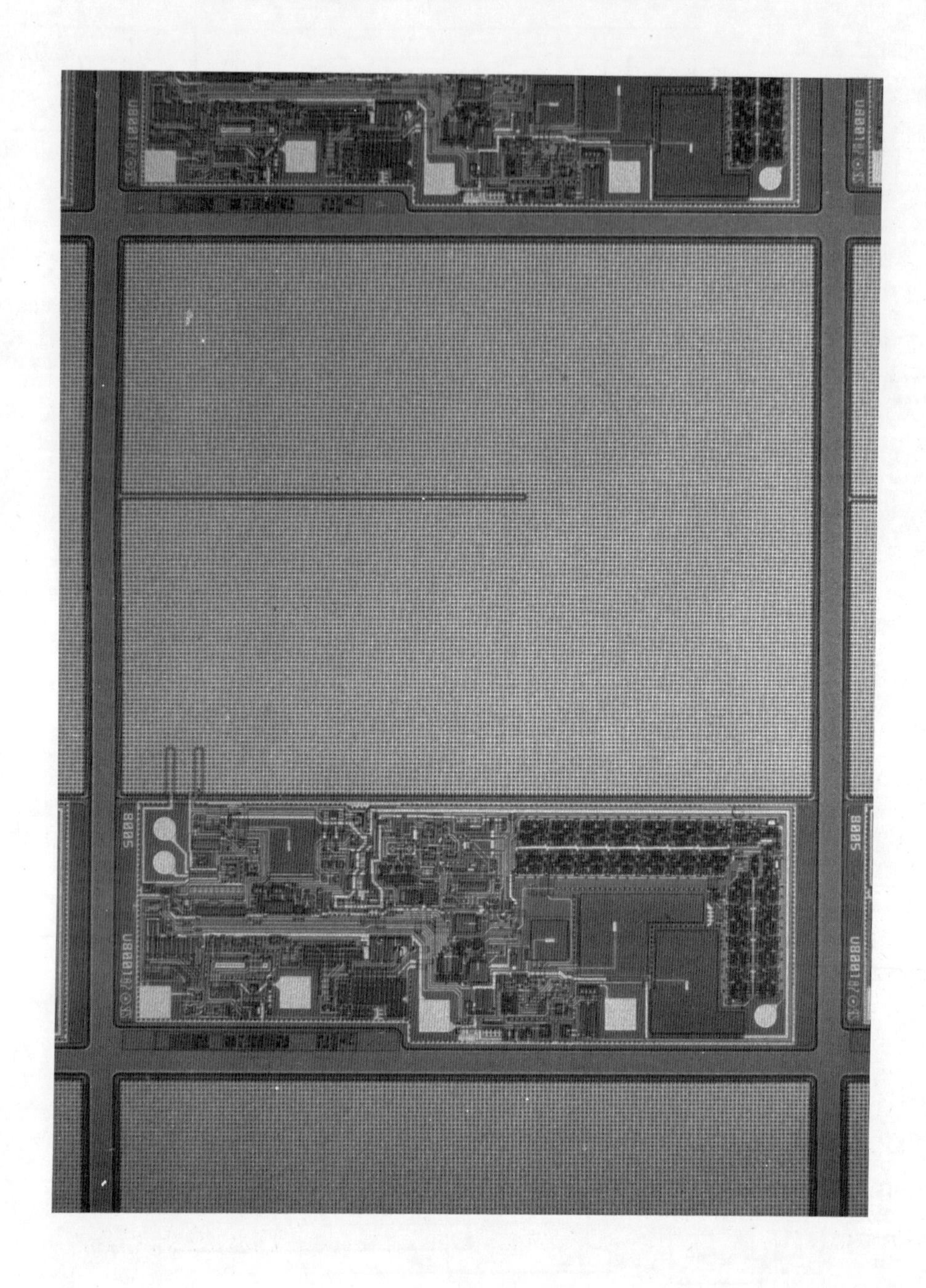

Fig 9 VM200

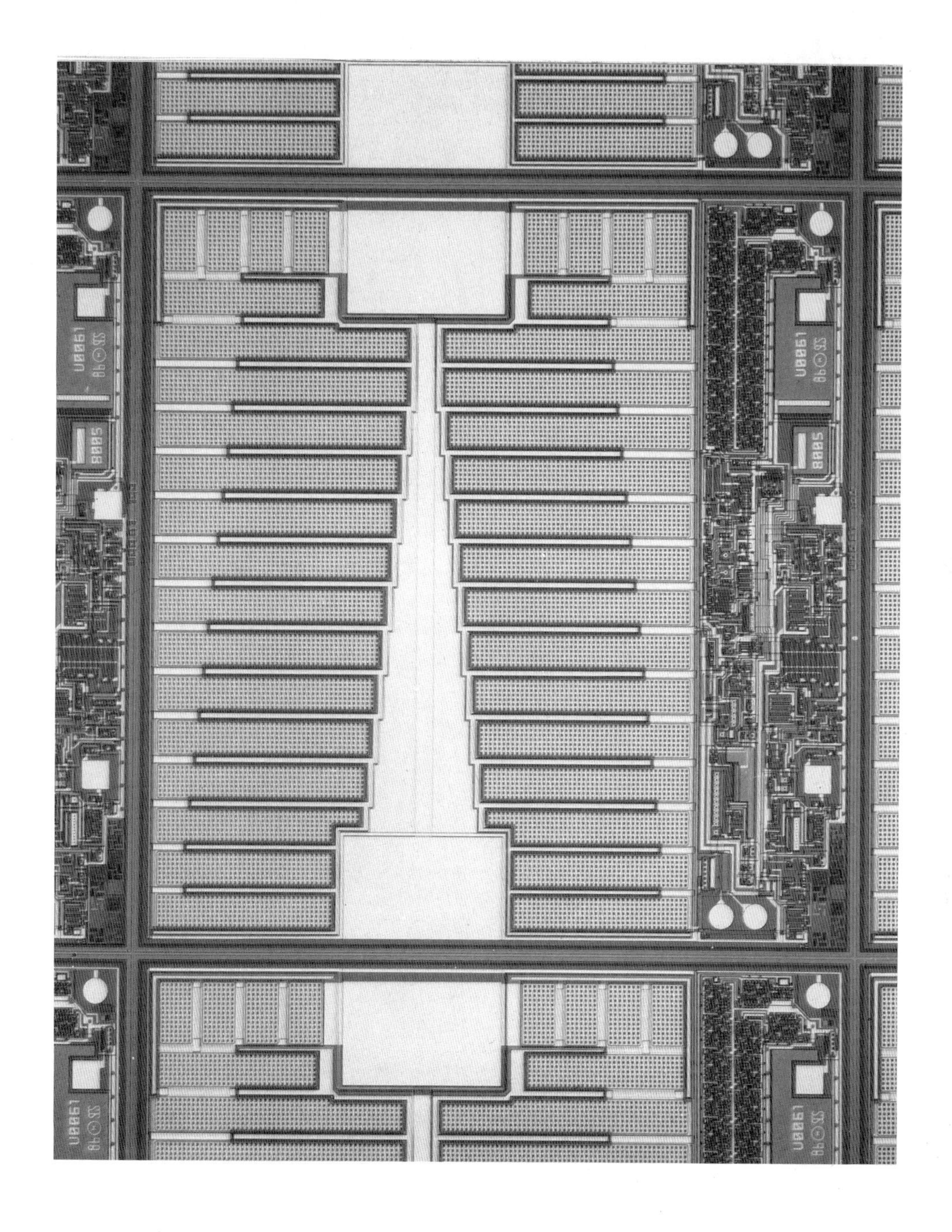

Fig 10 L9801

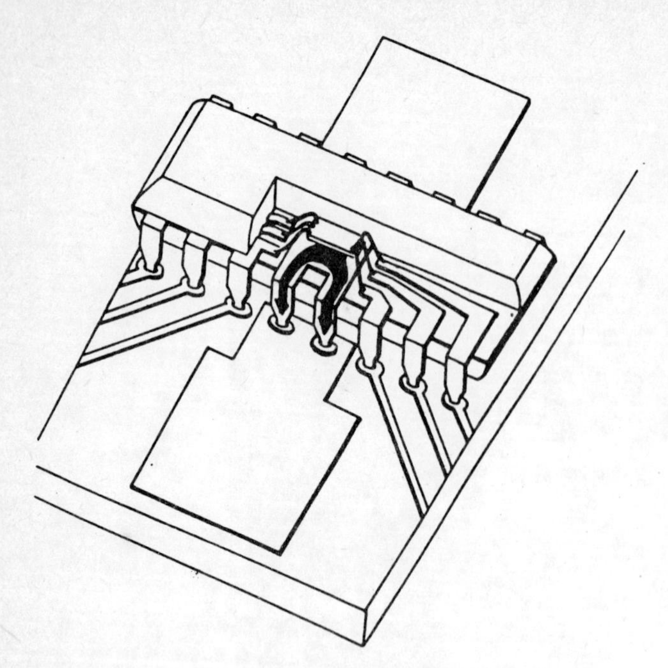

Fig 11 Cutaway drawing of a power DIP package showing how heat is conducted from the die to the printed-circuit board

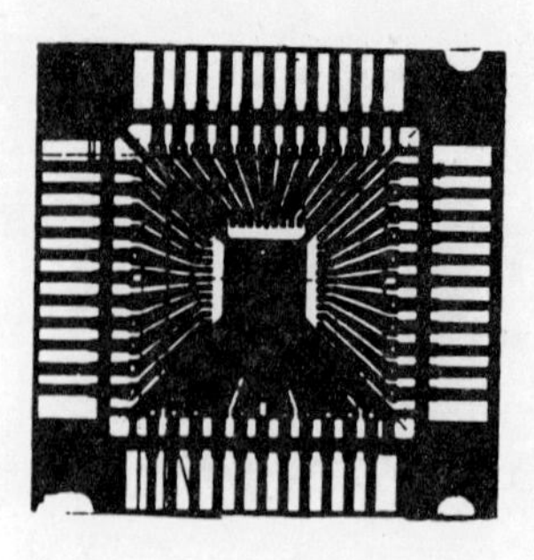

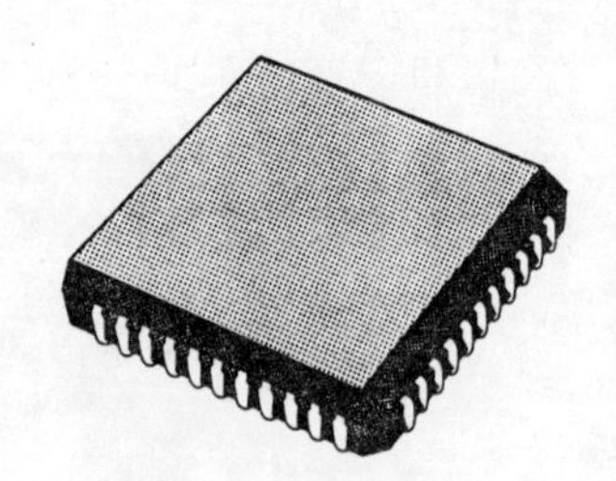

Fig 13 Lead frame and external view of the PLCC 33+11 medium power surface mounting package. The external dimensions are identical to those of a standard 44-lead PLCC but power dissipation capability is 2W, compared to 1W for the standard type

PENTAWATT

1975

- ❑ THERMAL RESISTANCE JUNCTION – CASE MAX 3°C/W
- ❑ THERMAL RESISTANCE JUNCTION – AMBIENT MAX 50°C/W
- ❑ 5 PINS

HEPTAWATT

1984

- ❑ THERMAL RESISTANCE JUNCTION – CASE MAX 3°C/W
- ❑ THERMAL RESISTANCE JUNCTION – AMBIENT MAX 50°C/W
- ❑ 7 PINS

MULTIWATT 11

1979

- ❑ THERMAL RESISTANCE JUNCTION – CASE MAX 3°C/W
- ❑ THERMAL RESISTANCE JUNCTION – AMBIENT MAX 35°C/W
- ❑ 11 PINS

MULTIWATT 15

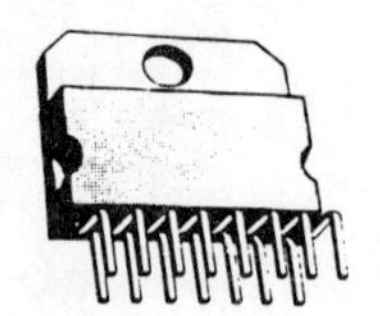

1980

- ❑ THERMAL RESISTANCE JUNCTION – CASE MAX 3°C/W
- ❑ THERMAL RESISTANCE JUNCTION – AMBIENT MAX 35°C/W
- ❑ 15 PINS

Fig 12 The family of T0220-based plastic power packages

C391/056

Semiconductor technologies for automotive power switching

S A WRIGHT, BSc, CEng, MIEE
Siemens plc, Sunbury-on-Thames, Middlesex

SYNOPSIS Over the past few years, several device technologies have been presented as candidates for solid-state switching in automotive systems. More recently, some of these technologies have established themselves into vehicle development programmes and, in some cases, production vehicles.

This paper reviews the progress of these technologies, and examines how and where they are being applied. The relative benefits of both bipolar and MOS technologies are examined by application. Where appropriate, the same arguments are considered for monolithic and hybrid solutions.

Areas of application include both high-side and low-side switching, and also reversible drivers for motor control. Various degrees of intelligence are considered, ranging from standard semiconductor switches to fully integrated self-protecting, self-diagnosing, logic compatible devices.

1 INTRODUCTION

Semiconductors are becoming increasingly specified as power switches for automotive applications. This does not, however, imply the imminent demise of electro-mechanical means of switching - there are numerous applications for which the relay remains better suited - but many application areas can also benefit from recent advances made in semiconductor switching technology.

Semiconductor power switches cover an extensive range of technologies, from the most basic single switching transistor to the fully intelligent, self-protecting and self-diagnosing, logic compatible device. This latter case provides a single component interface between an on-board computer and the load to be controlled.

Devices at both ends of the intelligence spectrum can find applications in automotive systems. Large concentrations of switches could use simple, economical devices, with any intelligence and/or protection located centrally. Other application areas will benefit from the integration of all necessary intelligence and protection within the device. However, it should be noted that all semiconductor switches require some degree of protection against damage from electrical transients generated within the automotive environment. Whether that protection is provided externally or internally within the device is dependent on application and cost.

In those situations where intelligence and/or protection are required within the device, both monolithic and multi-chip solutions are available. The correct choice is again dependent on application. Similarly, the relative advantages and disadvantages of both bipolar and MOS technologies must also be considered on an application basis.

2 SEMICONDUCTOR V. ELECTROMECHANICAL SWITCHES

When considering the choice between electro-mechanical and semiconductor switching for a given application, a number of parameters should be considered.

In general, semiconductors can offer advantages in the following areas:

- wide supply voltage range
- low drive power
- high switching frequency
- unlimited switching cycles
- switch bounce
- short-circuit protection
- RFI emissions
- status feedback
- reliability

Relays, however, can offer the following advantages:

- voltage drop
- galvanic isolation
- high currents
- normally closed switches
- reverse polarity protection
- reversible drives
- RFI susceptibility
- cost (dependent on specification)

Hence the choice for any one application depends on the priorities of the above.

3 SEMICONDUCTOR TECHNOLOGIES

3.1 Bipolar v. MOS

3.1.1 Voltage rating

In considering the relative merits of bipolar and MOS technologies for a given application, the choice of integrated or discrete solution should also be examined. For example, the high voltage rating required of an ignition coil switch limits the use of power integrated circuits. This particular application is largely the domain of the discrete bipolar transistor, which offers improved conductivity compared to a MOSFET of equivalent voltage rating. However, the emergence of the IGBT (Insulated Gate Bipolar Transistor) which combines the advantages of a bipolar output stage with MOS input characteristics is clearly of interest.

For lower voltage ratings MOSFETs are more applicable, but have always required protection from overvoltage transients. The introduction of more rugged avalanche rated MOSFETs means that the devices can survive overvoltage transients provided that the specified avalanche energy is not exceeded.

3.1.2 Current rating and drive requirements

For the majority of automotive switching applications much lower voltage ratings apply, and the choice of technology is perhaps more dependent on current. For low currents bipolar output stages are cost effective as they require a smaller silicon area than MOSFETs for an equivalent saturation voltage. Additionally, a low output current will require a correspondingly low drive current, thereby imposing minimal demand on the drive circuitry.

For higher output currents, the situation is somewhat reversed. Power MOSFETs are voltage driven rather than current driven, and so only low power drive circuitry is required. The advent of logic-level and intelligent MOSFETs permits drive directly from logic stage outputs. The voltage drop across the MOSFET is a function of the forward current, since the device exhibits a certain on-resistance. The value of this on-resistance when the device is fully enhanced is a function of silicon area, and compares favourably to the saturation voltage of equivalent bipolar transistors, particularly where the voltage breakdown is in the 50-60 volt region.

The MOSFET is also a more rugged device as it has no second breakdown mechanism and consequently exhibits an improved safe operating area compared to that of the bipolar transistor.

The temperature dependence of both bipolar saturation voltage and MOSFET drain-source voltage is shown in figure 1.

3.1.3 Switching speed

In this respect the MOSFET shows clear superiority over bipolar transistors. The switching time of a MOSFET for high frequency chopping operations is determined by the time taken to charge and discharge the internal gate and Miller capacitances, and is normally measured in nanoseconds. In most automotive applications, however, it is more important to optimise switching speed to avoid generating RFI. The MOSFET permits direct control of the switching speed by dimensioning of the gate resistor.

3.1.4 Reverse polarity

The ability to withstand a reverse polarity connection is an important requirement for most automotive systems, and there is a significant difference in this respect between MOSFETs and bipolar transistors. As part of its construction, a MOSFET has an integral inverse diode connected between drain and source. This diode is frequently usefully employed as a re-circulating diode when switching inductive loads.

In the event of a reverse polarity connection, however, current will flow through the diode, thereby allowing reverse current to flow through the load. Normally this situation could only occur under service conditions, but this must be borne in mind when controlling potentially hazardous loads such as a fan motor ! In this instance some form of reverse polarity protection must be employed.

The issue of reverse polarity is also of concern in half-bridge and full-bridge configurations using MOSFETs, since the two inverse diodes in series form a short-circuit path across the battery. Again, some form of reverse polarity protection is necessary.

The technology of bipolar power ICs with pnp transistors permits the design of internal reverse polarity protection. This can also be included in MOS power ICs, but will obviously only apply to the logic stage. The reverse diode of the MOSFET power stage will still conduct as described above.

3.2 Bipolar Devices

3.2.1 Bipolar power integrated circuits

For low current power switching applications a bipolar technology called DOPL has been developed and is suitable for loads up to about 4 Amps. The main feature of this technology is that the insulation wall is doped from both sides during fabrication. This offers the advantages of a narrower insulation wall, thereby giving greater packing density, reduced leakage current to ground, and also allows the inclusion of other components on the chip such as fast recirculating diodes.

A number of logic and protection features can be included in these integrated circuits. For example, the inputs can be driven directly by a CMOS/TTL compatible signal. Internal logic will then direct current from the main supply line to the base of the power transistor. Hence no drive power is required of the control circuit. If desired, the input stage could also be configured for an analogue threshold voltage.

The voltage, current and junction temperature at the power stage are all monitored and a computation of safe operating area is made. If this exceeds a certain threshold, then the circuit will shut down in order to protect itself. Status feedback logic is also available to indicate that a shutdown has occurred, due for example to a short-circuit or over-temperature fault. This status logic can also detect if a load is open circuit. The possibility of inherent reverse polarity protection is another feature of this technology.

A wide range of power integrated circuits are already available using this technology. They include power operational amplifiers, reversible motor drivers, stepper motor drivers, dual power switches (high-side and low-side) and voltage regulators (1) (2) (3).

3.2.2 Discrete bipolar devices

Bipolar power transistors have become well established in automotive electronics, particularly in the area of high voltage ignition coil drivers, where power darlingtons are frequently used. MOSFETs, whilst easier to drive and control, exhibit high on-resistance at high voltage rating, and this on-resistance is itself highly temperature dependent.

A recent technology to emerge is the IGBT (Insulated Gate Bipolar Transistor), which combines the advantages of MOS input characteristics with bipolar output characteristics. An IGBT can be considered as an extension of a vertically structured MOSFET by the addition of a p-emitter on the rear side of the chip. This allows positive charge carriers to be injected in a controlled way into the n-region responsible for conduction behaviour. In this way, the conductivity of a high current, high blocking voltage MOSFET can be increased manyfold. The structure is shown in figure 2 (4).

3.3 MOS Devices

3.3.1 Discrete MOSFETs

The MOSFET has now established itself as the main technology in medium power semiconductor switching for automotive applications, due to the advantages expressed above. N-channel devices are mainly used as P-channel devices are significantly less silicon efficient, although some P-channel devices do find applications as high-side switches in situations where moderately high on-resistances can be tolerated.

The MOSFET requires a certain voltage from gate to source in order to turn on, and by increasing that voltage until the device is fully enhanced a minimum on-resistance will be achieved. In a conventional MOSFET the required voltage from gate to source is about 10 V. In "logic-level" MOSFETs the gate channel is narrowed such that 5 V from gate to source will fully enhance the device. If true logic-level compatibility is required, i.e. sub 3 V, then an integrated power IC should be used.

When driving the MOSFET in a low-side switch configuration, the battery line can be used to derive the necessary gate-source voltage, since the source is held at ground. In a high-side switch, however, since the source must rise almost to battery potential, a higher voltage supply is required to drive the gate. This is most commonly achieved using a pump circuit.

Another recent development with MOSFETs is the introduction of an avalanche rating. This means that the device can withstand overvoltage transients provided that the specified avalanche current and energy are not exceeded.

Any power semiconductor must not be allowed to run beyond a certain junction temperature. In most cases, for optimum reliability, this temperature is about 150°C. If it is desired to protect the device against damage from overtemperature then some means of junction temperature monitoring is necessary. In power ICs such a feature is quite common, but in more basic discrete devices temperature protection can be a problem.

The TEMPFET (5) (6) (7) is a hybrid concept where a temperature triggered sensor is mounted directly onto the active area of a MOSFET chip. The sensor forms a normally open switch across the gate and source, shorting them together in the event of the junction temperature rising above the permitted value. Hence with the gate-source voltage forced below the minimum turn-on threshold of approximately 1.5 V, the device turns off and so avoids damage. This principle can be applied at time of manufacture to almost any MOSFET chip. If desired, the sensor contacts can be brought out externally using a five pin package.

Because the TEMPFET is driven in the same way as a conventional MOSFET, it can be used as a direct replacement for unprotected devices in existing designs. It is also possible to introduce short-circuit protection by using an external resistor and zener diode to limit the forward current for a short period until the junction trigger point is reached.

3.3.2 Intelligent power MOSFETs

Intelligent power MOSFETs can be either monolithic or hybrid, and basically form a self-contained interface between logic and load of a system, and perform comprehensive protection and diagnostic duties in addition to switching the load.

One example of an intelligent power MOSFET family is the PROFET (5) (6). This is configured as an automotive high-side switch, using an N-channel MOSFET as the power stage. The required gate drive is derived from an on-chip charge pump circuit, which is controlled from a logic input stage. Hence true CMOS/TTL logic compatibility is achieved.

The structure of the PROFET is a combination of lateral CMOS logic, high voltage CMOS and a vertical power MOSFET (figure 3). This combination can be either monolithic or hybrid. The power stage requires 4 production masks and the logic stage a further 10 masks.

Basic features of the PROFET family are protection against short-circuit, over-temperature and over-voltage transients, reverse polarity protection of the logic element, status feedback of any shutdown mechanism along with open load detection, protection against inductive voltage spikes at the output, and also a very low standby current.

The first member of the PROFET family, the BTS 412A, entered volume production in 1987, intended for automotive applications typically around 1 Amp. It has now been complemented by other devices of lower on-resistances down to 20 milliohms. This means that automotive applications for the PROFET in the range 15 - 20 Amps are viable.

3.3.3 Monolithic v. multi-chip

The production technology used in the manufacture of the TEMPFET can also be used to mount a more complex control chip (including temperature sensor) on to a MOSFET chip to produce a two-chip PROFET. It can be seen from figure 4 that the monolithic approach remains cost effective down to about 100 milliohms, but for very low on-resistance the multi-chip approach becomes cheaper.

A further advantage of the multi-chip approach is the improved flexibility it offers, particularly where mixed technologies are concerned. Experience gained with the TEMPFET proves that there are no quality and reliability concerns in adopting a multi-chip solution.

4 APPLICATION AREAS

Power semiconductors can find application almost anywhere within the automotive environment, although in some areas the relay may still provide the more cost-effective solution (this is particularly so for very high currents where parallel semiconductors would be required).

Due to the relatively high quiescent current exhibited by bipolar power integrated circuits, their use is precluded from applications requiring a permanent battery feed to the power line, e.g. door lock motors, sidelights. They are, however, well suited for ignition line based loads requiring currents up to a few amperes. Typical examples would be mirror motors and headlamp levelling motors (reversible driver ICs), solenoid and relay control (dual high-side and low-side drivers) and fuel control (stepper motor driver).

For higher currents, the MOSFET is more applicable. A vast majority of electrical loads in a vehicle are under 10 Amps, and the very low leakage current of a MOSFET means they can be safely used in multiple installations from the battery line without affecting the vehicle standby condition. This is particularly significant in remote switching and multiplex wiring systems. The TEMPFET can provide a very cost effective and reliable solution here if the diagnostic requirement is minimal. For more sophisticated systems the PROFET offers some advantage.

Vehicle lamps exhibit significant inrush currents due to the cold resistance of the lamp filament. A MOSFET is capable of withstanding this short-term overload, and the PROFET has been specifically designed to allow for inrush currents before detecting any possible short-circuit fault. An on-resistance of 40 milliohms is sufficient for driving a 60W lamp with acceptably low voltage drop across the device.

Safety critical applications such as engine management and ABS systems require fast and reliable switching of inductive loads. The PROFET is also well suited, being able to diagnose and report any load failures which may occur, whilst protecting itself from the harsh environment in which it must perform.

Reversible motors requiring reasonably high currents, e.g. seat motors, window lift motors, can also be controlled by MOSFETs. An overload protected, logic compatible H-bridge using two PROFETs and two TEMPFETs is shown in figure 5. This offers advantages over a relay based design in terms of short-circuit protection, high reliability with repeated switching and quiet operation for applications within the passenger compartment. However, when compared to the cost of two changeover relays, the MOSFET solution remains more expensive, and further cost must be added to achieve reverse polarity protection.

5 SUMMARY

The technologies for semiconductor power switching are continuing to develop to meet the needs of the automotive system designer. No single technology can fulfil the varying requirements of automotive electronics and the designer must select that which is most suited to a given application. The advent of the multi-chip power switch and its potential for mixed technologies means that even closer tailoring of a switch to its load is now possible.

REFERENCES

1. Electronic components for the automobile
   - Hauenstein, Heuwieser; Siemens Components magazine 6/88

2. Cross-current free bridge IC for electric motor control
   - Lenz; Siemens Components magazine 2/89

3. TCA 1560/1561 control of bipolar stepper motors
   - Schwager; Siemens Components magazine 5/86

4. Switching performance of a new fast IGBT
   - Lorenz, Schulze; PCI June 1988

5. Smart SIPMOS : intelligent power semiconductors
   - Brauschke, Sommer; Siemens Components magazine 5/87

6. Smart power - the million $ automotive question
   - Record; 6th Int. Conference on Automotive Electronics, October 1987

7. Built-in protection makes TEMPFET resistant to catastrophic failures
   - Glogolja; PCIM March 1989

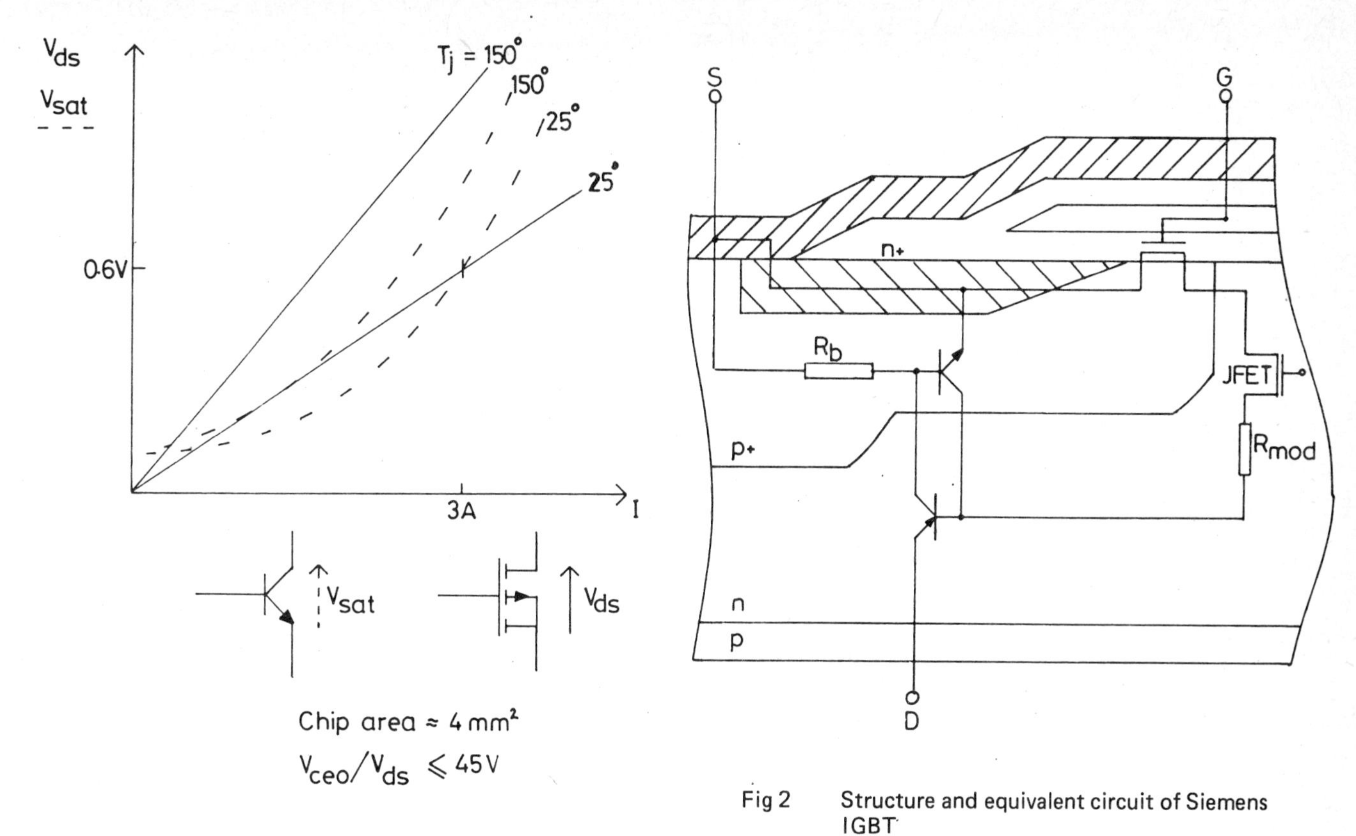

Fig 1 Temperature variation of bipolar saturation voltage and MOSFET drain-source voltage

Fig 2 Structure and equivalent circuit of Siemens IGBT

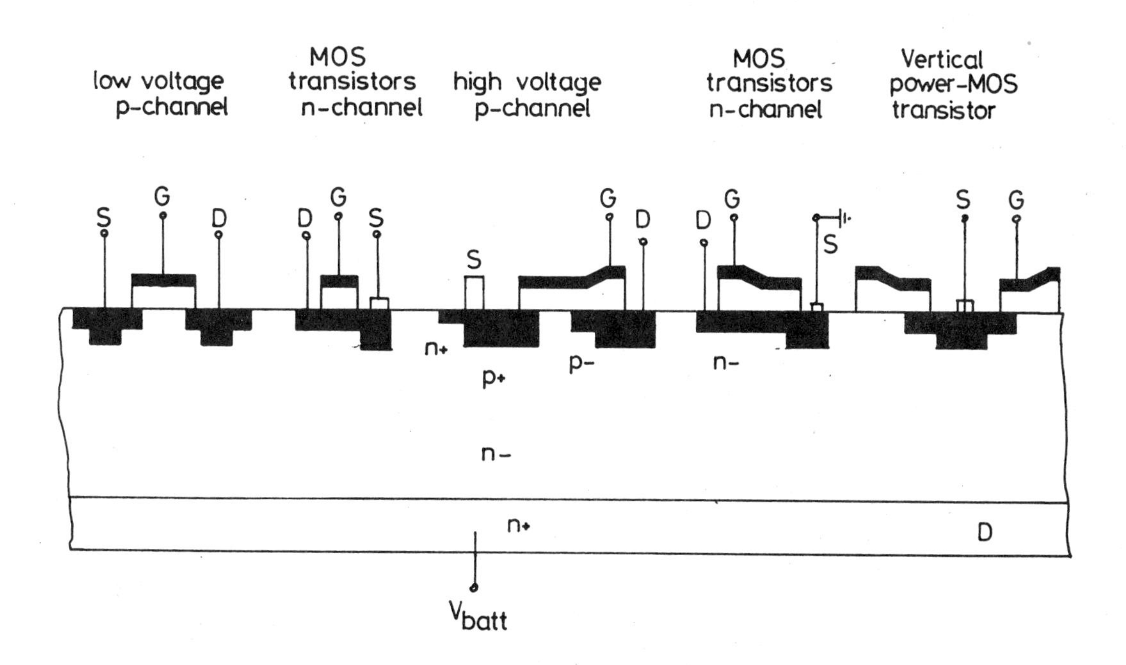

Fig 3 Structure of monolithic PROFET

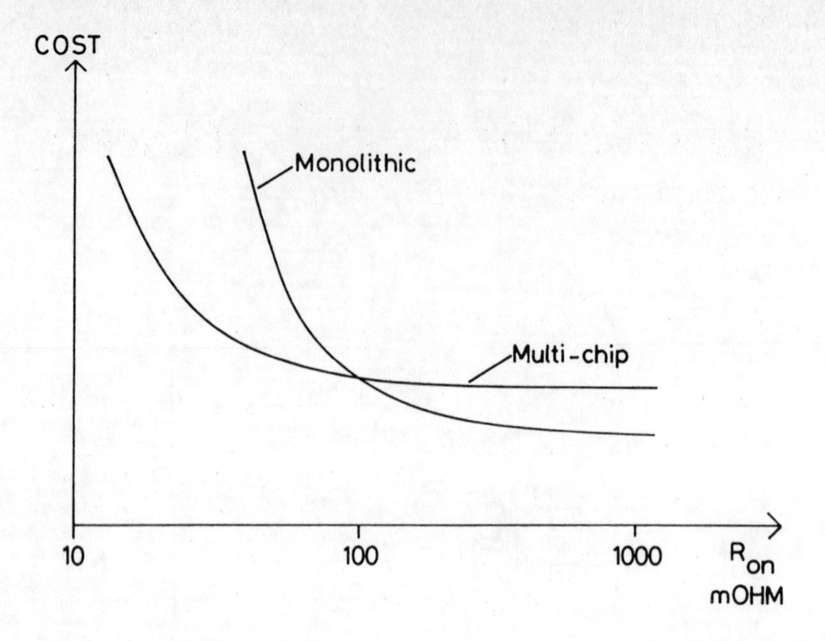

Fig 4 Cost comparison of monolithic and multi-chip PROFET (V_{DS} = 50 V)

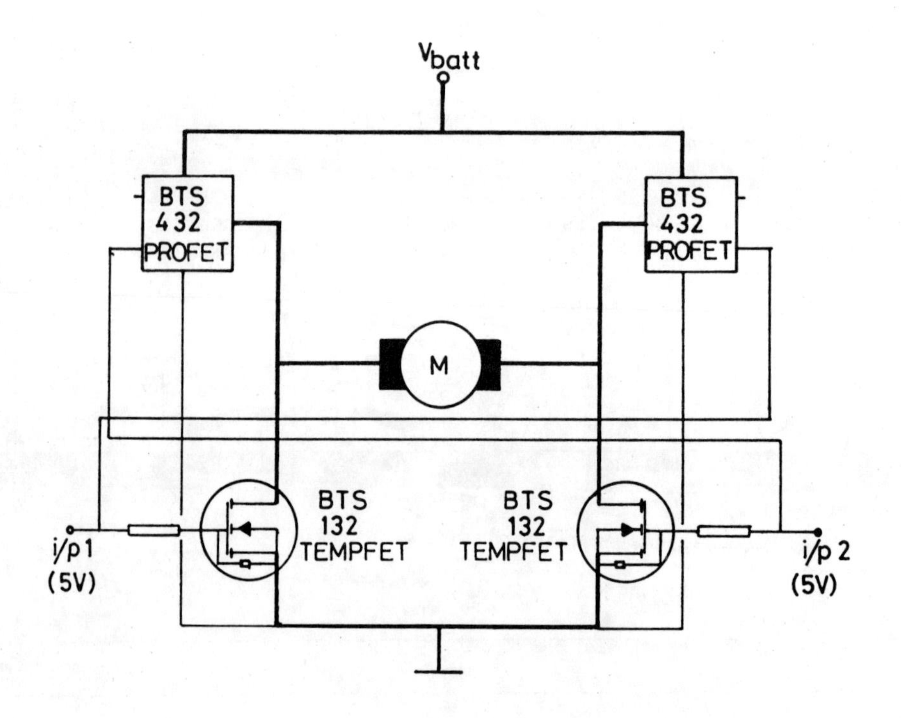

Fig 5 Motor bridge with logic-level drive and short-circuit protection

C391/018

A multidie approach to intelligent power switches for automotives

D COLMAN, PhD, CEng, MIEE, **E LEONARD**, BE and **A MARSHALL**, PhD, MIEE, MInstP
Texas Instruments Limited, Bedford

SYNOPSIS Multidie is a variation on the concept of hybrid packaging, but utilises the low cost and high volume production techniques of the semiconductor packaging industry. The result is packages that contain more than one die, but which are externally similar to conventional single die packages. The trends of multidie packaging are assessed for their impact on the rapidly developing automotive electronics market. The multidie approach offers flexibility not available from monolithic solutions. This approach also gives cost-effective power switching for higher current single-ended switches and most medium to high current bridge configurations.

1 INTRODUCTION

The traditional approach to load switching in cars is to use mechanical dashboard switches, through which the load current (eg: to the headlamp) passes. This requires wiring from the battery to the switch on the dashboard, then on to the load, and involves a heavy and complex wiring loom for all the vehicle's loads. Such a solution requires high current capability switches, and there is considerable voltage drop in the cable. As the complexity of electrical systems have increased this approach has resulted in congestion behind the dashboard and in other areas of the car, particularly in the door panels, leading to a need to explore alternative solutions which reduce the size of the wiring harness. Relays situated close to the load partially address this problem, but they are not ideal as they suffer contacts wear and a substantial wiring loom is still required. Furthermore, relays do not offer a low cost solution to the need for diagnostic and fault warning features. However, with this approach, the wiring loom can in part be constructed of low current cable. The trend in the automotive industry is now towards using semiconductor switching, with the aim of developing a semiconductor based multiplex wiring system (Fig 1) which simplifies the wiring loom and incorporates fault warning indicators. As a result there is an increasing requirement for high-sided, high current semiconductor actuators in automotive applications.

2 MULTIDIE VS HYBRID

Hybrid modules have for many years been used to create compact enviromentally protected circuits, including intelligent functions. To distinguish the approach discussed here from traditional hybrid techniques it will be referred to as 'Multidie'. Traditional hybrids have been costly when compared with single die packaging as practiced by the semiconductor manufacturers. The usual substrate for hybrids has been alumina, often with screen-printed metallization and resistors. The new "multidie" packaging technique has been developed as a low cost approach for the semiconductor industry to provide the intelligent functions now required by the automotive market and large volume consumer product manufacturers. Based on a new type of substrate material which is compatible with low-cost mass production techniques, it also has low thermal resistance and provides the benefit of an electrically isolated heatsink.

3 MULTIDIE VS MONOLITHICS

Monolithic intelligent actuators have begun to penetrate the automotive market. Starting from

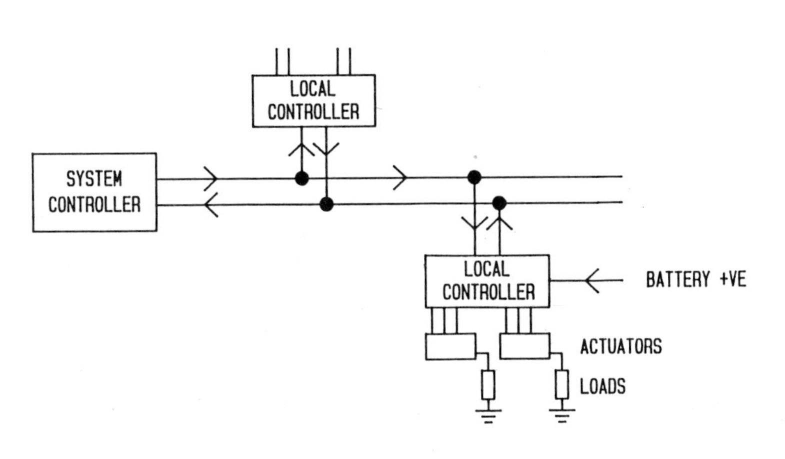

Fig 1 Typical multiplex wiring configuration, where the system controller is normally a microprocessor and the local decoder an ASIC array

below an amp, the capability of such devices has been steadily rising to the present maximum operating current capability of about 5 amperes. Multidie actuators are now being considered for some of the higher current applications (Fig 2).

Multidie packaging is more costly than packaging a single monolithic device due to the requirement for electrical isolation. However, balancing this, the costs of the individual dice are lower because simpler processes can be used with lower processing costs and higher yields. As a result, at low currents, monolithic structures are likely to have lower overall costs but at higher currents the monolithic die costs are greater due to the higher complexity of the processing and inevitably lower yields.

Therefore, just a brief description of the functionality of the power device and the analogue control IC will be given .

The power device is a pnp transistor structure with open collector output. It has been designed specifically for automotive high sided switching up to 6A with loads including lamps, solenoids, relays and motors. It features high current gain and low leakage characteristics, with a collector emitter saturation voltage at 6A of 0.25V maximum. It also features a high emitter base breakdown voltage to allow the device to withstand a reversed battery voltage. A temperature sense diode is built onto the chip.

The analogue control IC is designed to drive the output power device from a 5V logic input

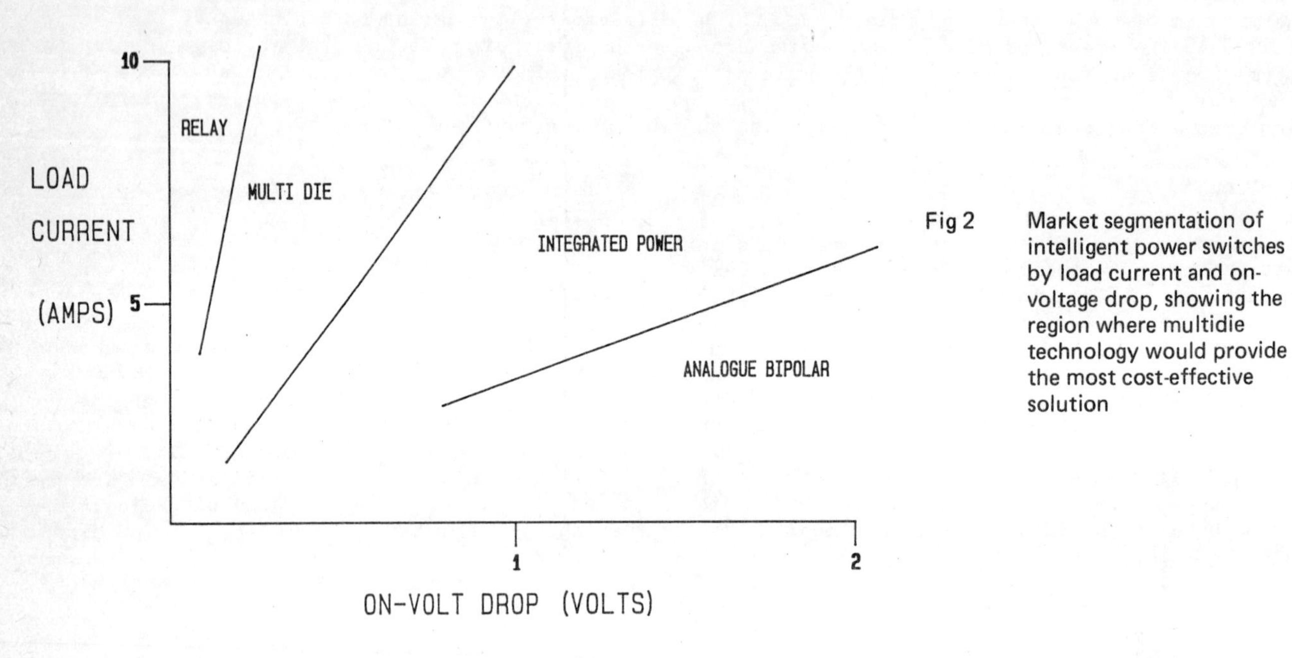

Fig 2 Market segmentation of intelligent power switches by load current and on-voltage drop, showing the region where multidie technology would provide the most cost-effective solution

4 MULTIDIE

Until recently only monolithic switches with a relatively low current handling capability have been available with the desired integrated fault detection. These have usually been in the form of intelligent MOS devices. Despite the advance of these devices to higher currents, there will always be a requirement for multidie devices to cover situations where monolithic actuators do not compete economically. This occurs where monolithic die sizes become excessive or due to package related problems such as thermal expansion and reliability and where yields become unacceptably low.

A two-die hybrid for headlamp actuation, the TLP410, (Fig 3) with a 6A current drive capability and a saturation voltage of 0.25V at 125 degrees C, is used here as a design example. The use of a multidie approach has involved co-ordination of several major aspects of the development programme. Such a design can conveniently be considered in three sections:

(a) Power device (output)
(b) Analogue IC (controller)
(c) Package

Details of the power device and system for this design are described elsewhere (Ref 1 & 2),

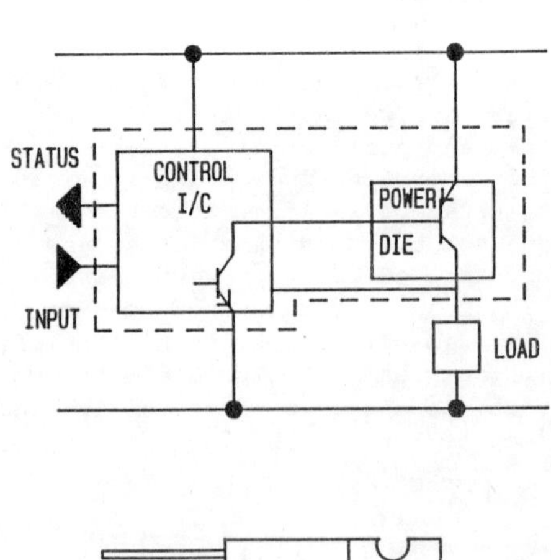

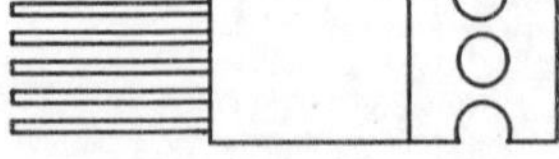

Fig 3 Schematic of the TLP410 6 A, 0.25 V two-die hybrid, designed for headlamp and motor actuation and showing its conventional SIP outline

signal. An active low status output, again compatible with 5V CMOS logic, provides diagnostic feedback of load and device conditions. The input and status output can both be interfaced directly to a microcontroller. The IC has protection circuitry which detects when the output power device goes over temperature or the supply voltage goes either under or over voltage. Under these conditions, the power device is switched off and remains off until either the supply return to a normal operating value, or the temperature of the device reduces such that it is again safe to drive the load. The status output provides a fault monitor for on-state short-circuit and open-circuit load conditions, over-temperature, under and over-voltage on the supply and also off-state open-load conditions.

Fig 4 shows the package layout of the TLP410. The package is a 5 pin SIP with standard 0.1" spacing. The smaller analogue IC is shown in Fig 4 positioned on the left with the power die on the right. The header is of aluminium, with a top layer of copper foil insulated from the aluminium by a thin layer of epoxy. Both epoxy and copper foil are coated to typically 50um. This 'header sandwich' is called "TISTRATE", and is supplied by Texas Instruments Metals and Controls division. The copper is etched using printed circuit techniques, to create a metal pattern which allows isolation of the two chips, as well as optimum bonding and thermal properties. Both power and analogue devices are soldered down to separate pads of the copper, the metal pad of the power die doubling function as the collector contact. The epoxy layer gives isolation allowing the package to be bolted directly to the grounded vehicle chassis if desired. Bondwires are required to connect between analogue IC and leadframe, and between the power transistor and leadframe. Further bonds are used to connect the two chips electrically. Low current bondwires are used for all connections except the high current emitter.

A similar concept is used to create a package which will house three high-side drivers. This is a fifteen pin package, developed on the same principles as the five pin package which houses a single hybrid as discussed above. This houses six chips, and might for example be used to supply a single dipped and double main beam that represents the headlamp count for one side of a typical car.

5 FUTURE OF MULTIDIE

The same aluminium-epoxy-copper technology could be developed for use with an intelligent H-Bridge, using four individual power devices plus a control die, for driving anti-skid braking systems etc.

The multidie package will permit a trend towards higher power capability within a single package and a higher level of integration. It also permits an application to rapidly take advantage of new technologies as they are introduced without the need to compromise any conflicting requirements between logic and power. In principle an outstation IC of a multiplex wiring system may be incorporated into the multidie package with control and power dice, reducing the package count still further, and minimising system and wiring costs.

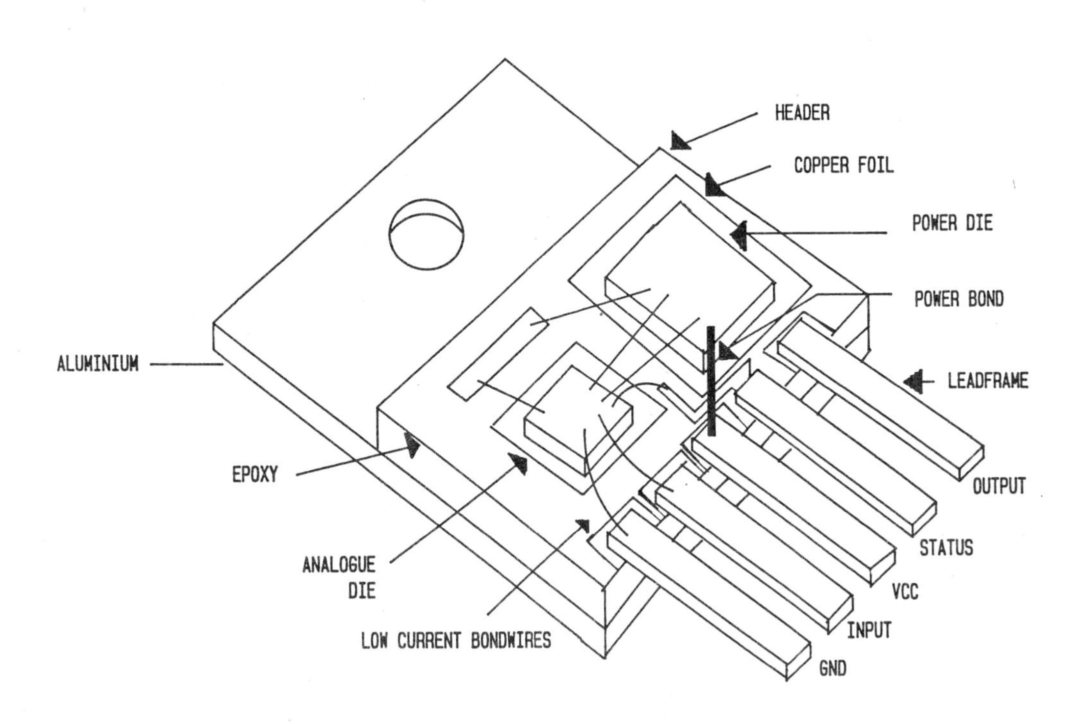

Fig 4 Package layout of the TLP410; the package is a 5-pin SIP on a TISTRATE substrate, with isolated power and analogue dice soldered to the etched copper foil

There is also a continued interest in the development of specialist packages designed for the requirements of the automotive industry. Examples are the end caps of alternators which could contain the alternator regulator chips. The aluminium-epoxy-copper sandwich is particularly useful in this type of application, as it can be shaped and formed easily, in contrast with traditional hybrid materials such as alumina, which are brittle and also have poor thermal characteristics.

6 CONCLUSIONS

We have described some current multidie developments and available products, together with future trends of these devices within the automotive environment. The multidie approach to packaging and the range of features available from this method of encapsulation have been compared to conventional hybrid techniques, and to monolithic intelligent actuators. The undoubted scope for multidie packaging and products make this area of automotive electronics development key to the future direction of multiplex wiring systems and intelligent switching in the car

ACKNOWLEDGMENTS

The authors wish to express their gratitude to Phil Cavanagh of Texas Instruments Ltd, Bedford, without who's assistance this work would not have been possible.

REFERENCES

(1) A. Marshall, A low cost intelligent power actuator for automotive multiplex wiring systems. ISATA, Florence, 1988 paper 88133

(2) A. Marshall, D. Colman and S. Collins, Power device design for a hybrid actuator. International Conference on Power Electronics & Variable Speed Drives, London, 1988, p 58.

C391/010

Predicting load balance and electrical system performance

M J HOLT, BSc, MSc
Lucas Automotive Limited, Shirley, Solihull

SYNOPSIS Arguments are put forward for treating the electrical power aspects of vehicles as a single system, rather than independent components. The decision making processes of system and component design are discussed, and a simulation tool presented to assist in these processes. Examples of its use are given, along with possible future developments in this field.

1 INTRODUCTION

The electrical power system on modern cars is not just a collection of isolated components. The loads consume power, the alternator provides it, and the battery buffers and stores it - almost obviously there will be a high degree of component interaction during operation, and therefore there should also be a high degree of interaction during component design. That is the basis of advocating an integrated systems approach to electrical power on vehicles.

That being so, who is to have the system design authority? An obvious answer is the vehicle manufacturer (VM). However, a supplier of key electrical components to the VMs would not be content with such a state of affairs and Lucas is encouraging interaction with the VMs at the system level, to assist in the decision making process. Throughout the system design stage decisions are being made - by marketing, by senior management, and by practising engineers. Three levels of design decisions can be identified:

- o system configuration, and system specification
- o component specification
- o component design

The VM has system configuration decisions to make at the top level about any new model. For example, will the car have heated front screen, heated seats, air conditioning, electric windows, electric locking, additional driving lights, any multiplexed wiring, and so on. The VM must also decide the system performance specification for the vehicle - can a user expect to use it for city commuting Monday to Friday, put it in the garage over the weekend, and carry on next week, or will the battery need additional charging. If he gets caught in a motorway traffic jam, how long can he continue with engine idling before the battery is too flat to restart the engine?

Having made these decisions, the component specification needs to be formulated - how big an alternator, how big a battery, any extra protection like timer switches for high power heating loads? Following this, the component suppliers will be required to concentrate on component design and manufacture to meet the required specification.

Of course there is not normally a clear cut distinction between these three areas, nor between who holds the responsibility. For instance, component suppliers will assist the VMs in selecting what electric devices are possible, and also in determining meaningful system performance specifications. The relevant suppliers should also be able to assist in the component specification task, even if they do not themselves supply all the components. At the component design stage, the responsibility rests more clearly with the supplier.

In order to assist at the middle stage above, and also to a certain extent at the top stage, Lucas Automotive has developed a computer simulation tool to predict performance of the almost universal lead-acid battery and alternator system found on today's passenger cars. Known as AEPSS for Automotive Electrical Power System Simulation, its development and use will be described briefly.

2 SYSTEM MODEL

Two ways of looking at any system are used here:

- o what is its purpose, and how does it interact with the "outside world"?
- o what is the system comprised of?

The purpose of the electrical power system is to provide for operation of electrical devices under conditions of:-

a) engine not running

b) vehicle stationary and engine idling

c) vehicle moving

Item a) includes the crucial starting requirement. Normal use, of course, involves mixed sequences of all three conditions. The interaction with the "outside world" includes as input effects what kind of traffic and road conditions the vehicle is being operated in; what the weather conditions are like - dark, light, cold, hot, wet, dry, etc. Output effects include voltages at various components - eg battery, alternator, electronic controller for brakes; and indirectly, state of charge (SOC) of the battery.

Consideration of this, and of the composition of the system led to a modular structure for the simulation tool as shown in Figure 1. The influence of traffic and weather is handled by the environment module - both vehicle and electrical system environment. The drivetrain links the vehicle usage with alternator speed, and the electrical system itself is considered to comprise loads, charging system, storage system and distribution system. A system controller is also allowed for in the modular structure. The modules have been coded in ACSL(1), and are used on PC compatible computers, and on Sun workstations.

The mathematical model for each module has been based on combinations of basic physical laws (eg Ohm's law, Kirchhoff's law), design data (eg alternator and regulator specifications), and observed characteristics (eg temperature effect on alternator output, transient response of battery voltage).

Some comparisons between measured system behaviour and model predictions are given in Figures 2 and 3. Figure 2 represents a vehicle stationary in motorway traffic on a wet winter's night. Lights and heaters are on, from time to time the radiator fan will be switched in to control engine coolant temperature. This can be seen in the top traces of battery current - where the battery is in discharge for almost the whole of the two hour experiment. The dotted line shows the simulated current, with very close agreement. The bottom traces show the battery voltage for the same situation. Apart from some transient errors, agreement is within 0.08 volts. The comparative figures for battery SOC at the end of the run are 65-70% measured, 68.6% predicted.

Figure 3 represents driving in city traffic, when the battery switches between charge and discharge as the engine speed moves between idling and running. Although some transient errors are apparent, there is again good agreement for both current and voltage.

3 MODEL USE

One trend in recent years is improved control of engine idling speed, resulting in lower speeds for a fully warmed up engine. Because alternator output is extremely sensitive to speeds at this low end, see Figure 4, it can have a major effect on the performance of the electrical power system. Figure 5 shows the simulation results for a family saloon in winter city commuting conditions. The battery state of charge during a half hour run with the original vehicle specification is shown by curve A. It can be seen that the battery is still in a fairly healthy state at the end of the run, ready for the next journey. A second run was made with a reduction in engine idle speed from 735 to 650 rpm. The battery is now depleted by about 3% at the end of the journey - curve B. After ten such journeys in a working week the battery may fail to restart the engine - Monday morning blues! The electrical system engineers need to respond to this change - two possible solutions would be to increase the alternator drive ratio, or to install a larger alternator.

The effect of changing drive ratio from 2.37 to 2.60 is shown by curve C, which gives an almost identical performance to the original specification. This may, however, not be a feasible solution due to the increased alternator speeds under motorway cruising conditions. The system performance with the original drive ratio, but with alternator rating increased from 60 amps to 65 amps is shown in curve D. This leaves the battery slightly more charged at the end than at the beginning - definitely a healthy state of affairs. For further interest, curve E shows the system with a 70 amp alternator.

The potential use of the model very early on in the design life of a new vehicle is thus demonstrated. Instead of relying on past experience, admittedly with a high degree of engineering intuition, AEPSS allows a thorough, scientific analysis of system performance well before complete systems are available for test and analysis. Similarly, proposed solutions to changes in system specification, both during the design stage, and during the production life of the vehicle, can be very quickly, easily and cheaply assessed on the model, before confirming with the recommended hardware implementation.

4 SYSTEM TRENDS

Another trend apparent in the automotive electric power field is the increase in installed generator capacity. Figure 6, taken from (2), shows the rated generator output on typical passenger cars in Europe and in USA through this century. The introduction of alternators in place of dynamos offered a step change that was very quickly made use of by increased fitment of electric devices. Even conservative predictions point to installed power of 3 kw in Europe by the end of the century.

Various options to supplying this increased power demand include:-

- fitting bigger and bigger 12 volt alternators
- fitting variable speed drives to alternators, to allow higher output at low engine speeds
- adding intelligent power management to the system
- changing some, or all, of the system to work at higher voltages

To make the decision between these alternatives, it is necessary to predict the performance of the system under each option. Tools like AEPSS will play a vital part in this process. A fictitious, but representative modern, well equipped car will serve as a case study. It has been a frosty night, with the car parked in the open. Start the engine, switch on seat heaters, rear screen heater, and blower fan. Assume five minutes warm up whilst scraping the screen, etc. Now switch lights and wipers on and set off into the city rush hour traffic for half an hours drive to work. Assume that the seats switch off after 12 minutes, and the rear screen after 15 minutes. The predicted effect on battery state of charge is shown by the lower trace of Figure 7. Over 10% of the total battery charge capacity has been depleted, so we cannot expect to make many such journeys in succession without taking precautions!

A number of proposed improvements will be examined in turn, not primarily as solutions, but to demonstrate the role of AEPSS in such a process. A simple power management function could be to switch headlights and fog lights off when the vehicle is stationary - about 30% of journey time. The effects are shown by the second curve - an improvement, but not ideal. Try instead the obvious solution of fitting a larger alternator - from 60 to 70 amp rating. The third curve shows the result - battery depletion now under 3%.

Yet another proposal is to fit a two-stage alternator drive mechanism, with a high ratio of 5:1 for engine speeds under 1200 rpm, normal ratio of 2.5 otherwise. As discussed above, this will have a big effect on alternator output at engine idle and hence on battery SOC - see the top curve in Figure 7. The battery is now recharged by the end of the run. As a comparison, the next curve down is with idle speed increased to give the same effect - it had to be raised to 1500 rpm to give nearly as good performance - hardly an acceptable solution.

AEPSS was originally developed to represent 12 volt systems, but is easily adapted for 24 volt systems, and is being developed to represent multi-voltage systems, as proposed by several parties[2,3].

Of course, these results are only part of the picture - economic, commercial and other technical considerations will have to be weighed up before even making a short list of possible solutions. Lucas are also active at this level, both in collating such information, and also in developing further simulation tools to assist with these decisions as well as the purely technical ones.

5 CONCLUSIONS

This paper has demonstrated the need for a system view of electrical power on vehicles. Three areas of decision making have been identified, and a simulation tool described to assist in two of these areas, with examples of possible solutions to the rising demand for electrical power on modern vehicles.

6 ACKNOWLEDGEMENTS

The author wishes to thank the directors of Lucas Automotive Ltd for permission to publish this paper; Dr K J B McEwan and Mr J A Bant of Lucas Yuasa Batteries, Mr J H Evans of Rists for their support, guidance and expert information; and Mr P W Birkett for his detailed modelling work.

For further information, please contact the author, Dr McEwan or Mr Evans.

7 REFERENCES

1 "Advanced Continuous Simulation Language (ACSL)", Reference Manual, 4'th edition 1986, Mitchell and Gauthier Associates, Concord, Mass. 01742, USA.

2 J G W West, "Powering up - a higher system voltage for cars", IEE Review, Jan. 89, pp29-32

3 M Frister, G Henneberger, "New Concepts for Vehicle Electrical Systems", IEE colloquium "Vehicle Electrical Power Management and Smart Alternators", London, April 1988.

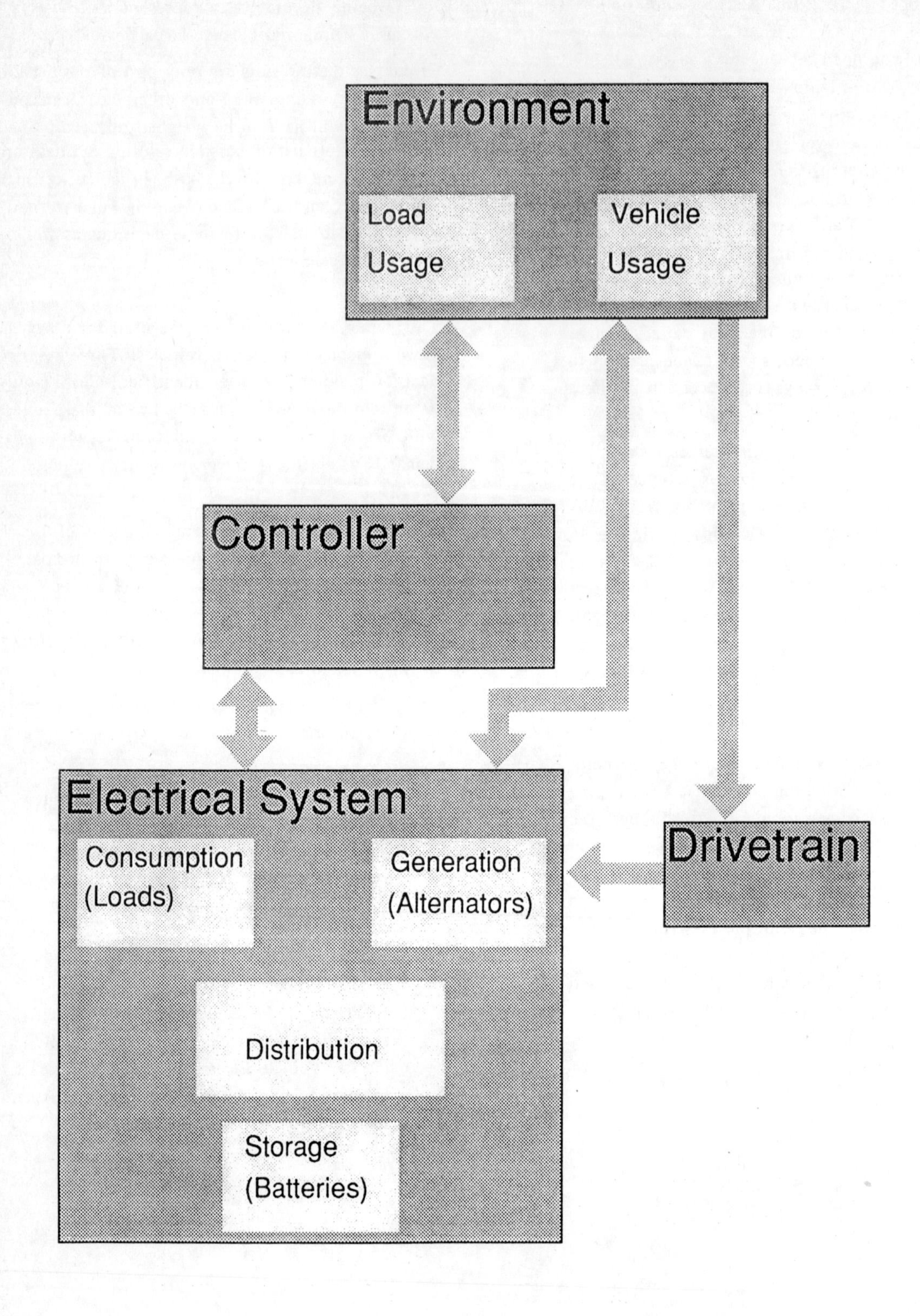

Fig 1 Modular structure of AEPSS

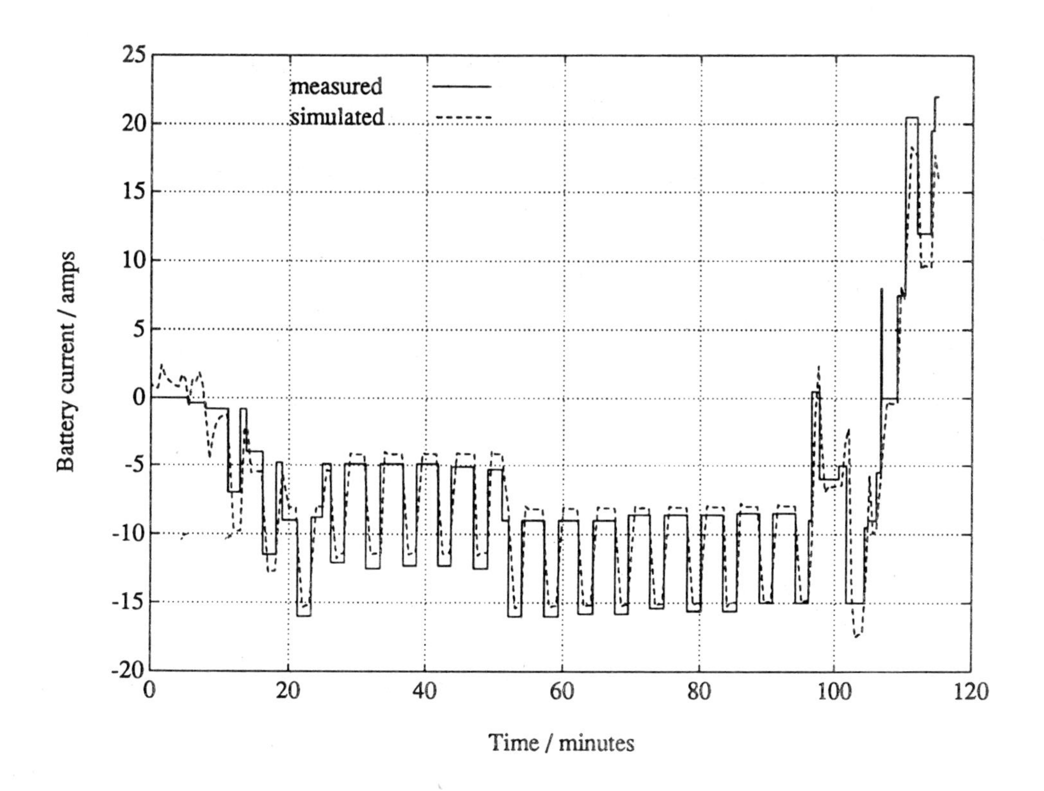

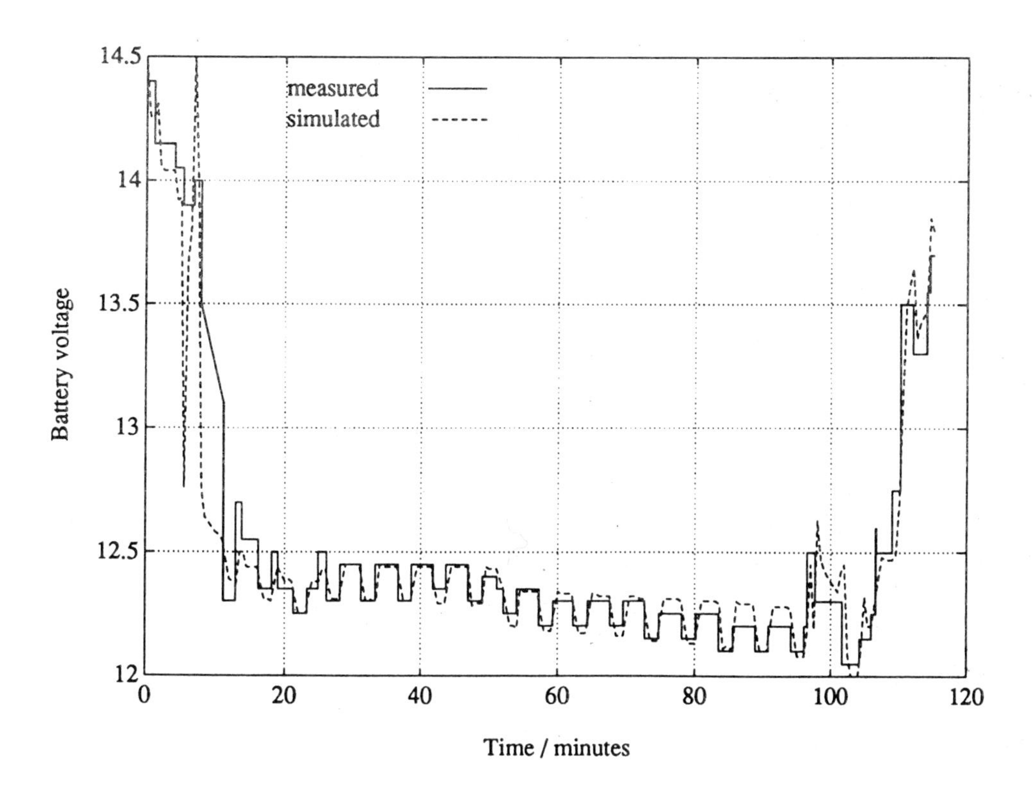

Fig 2 Validation results for motorway idle

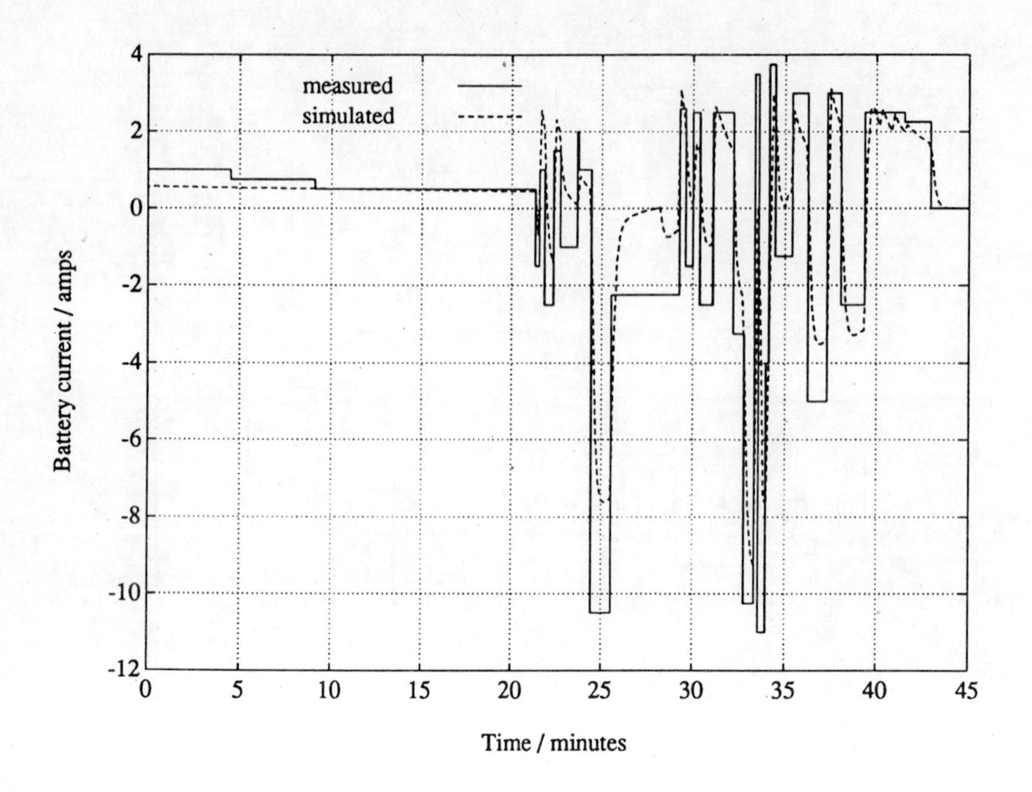

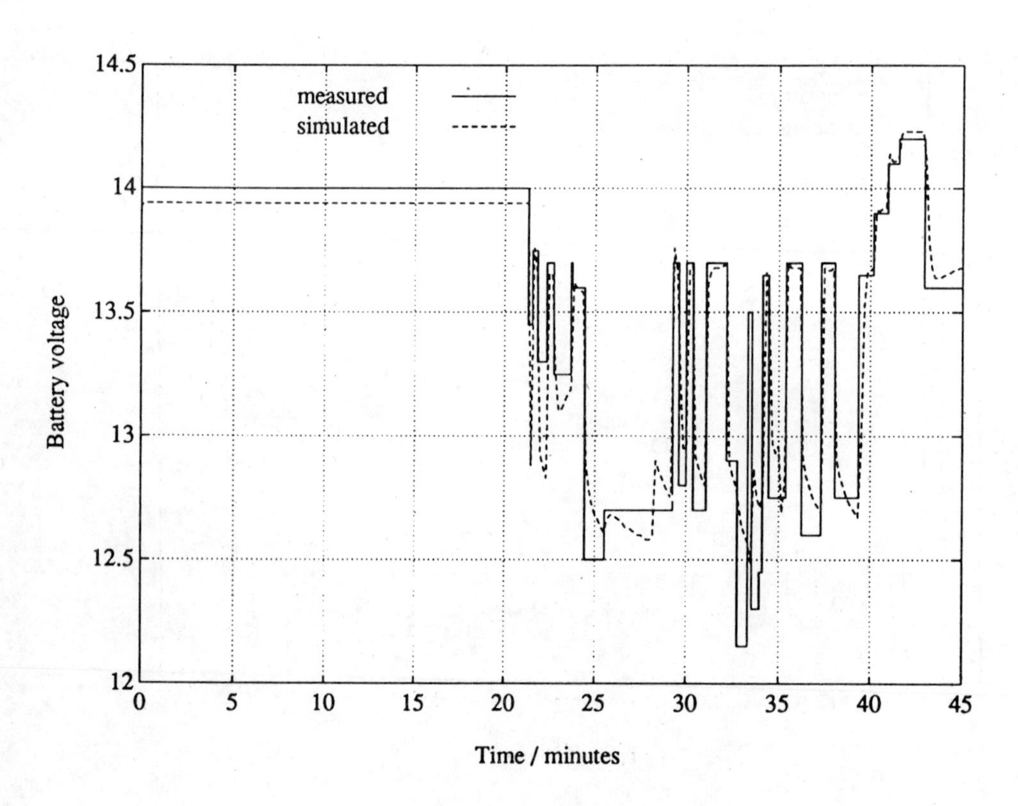

Fig 3 Validation results for city driving

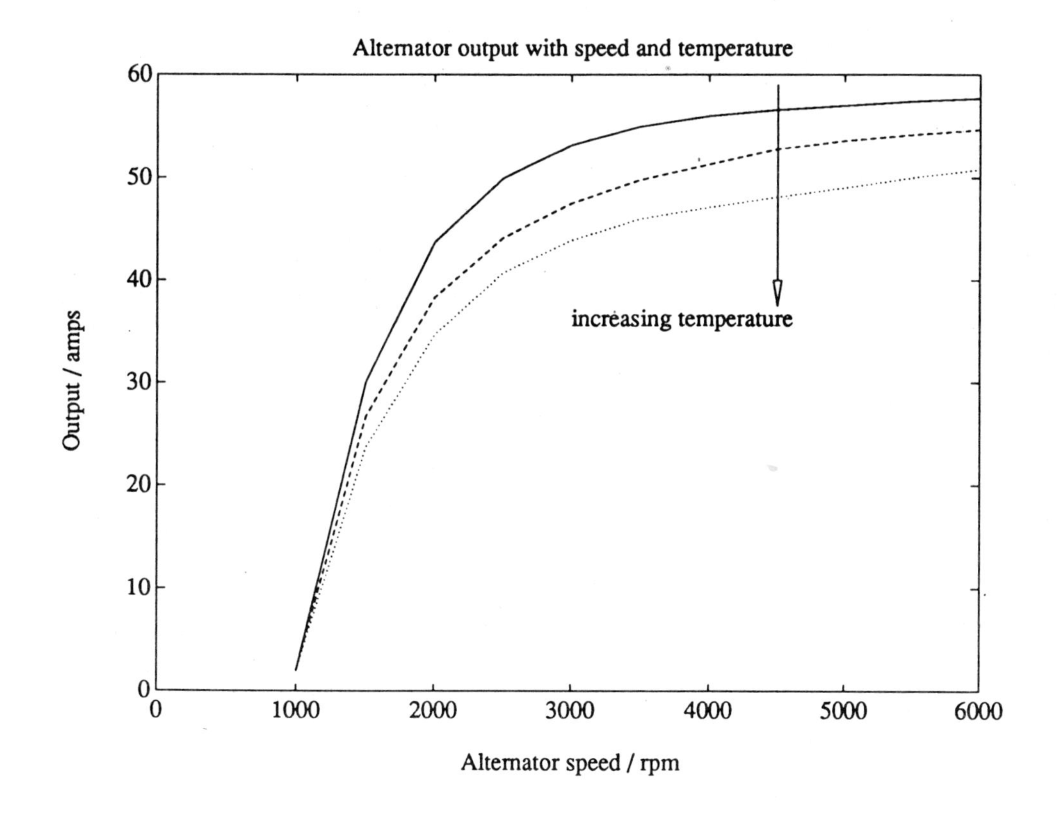

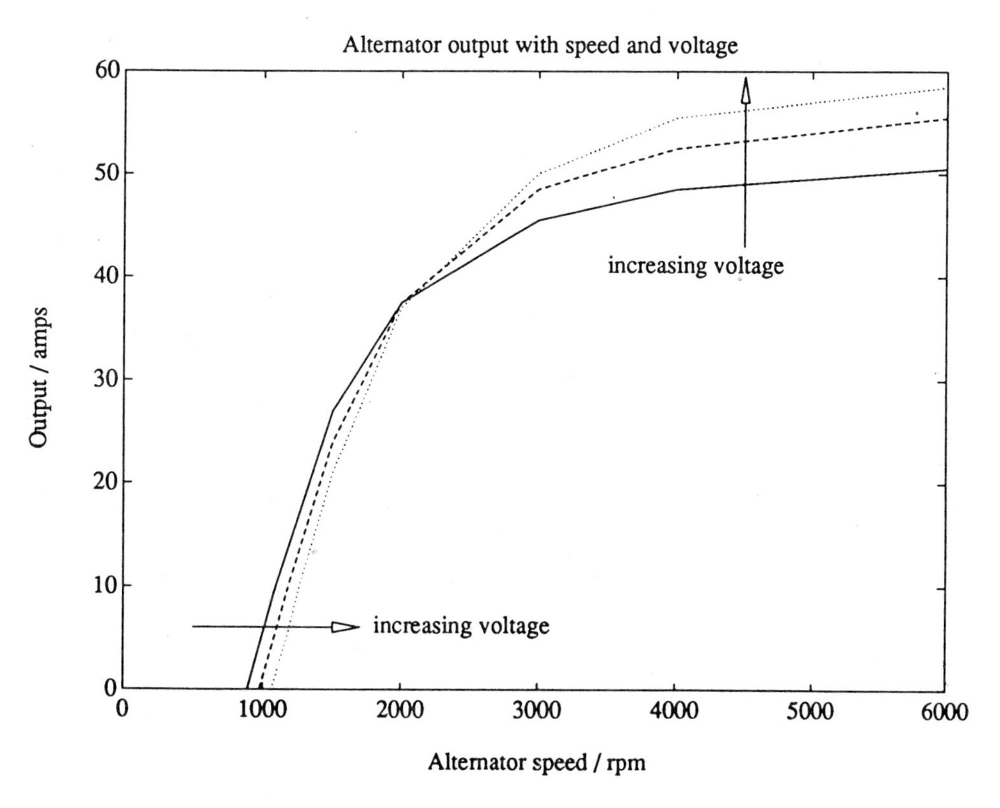

Fig 4 Typical alternator output characteristics

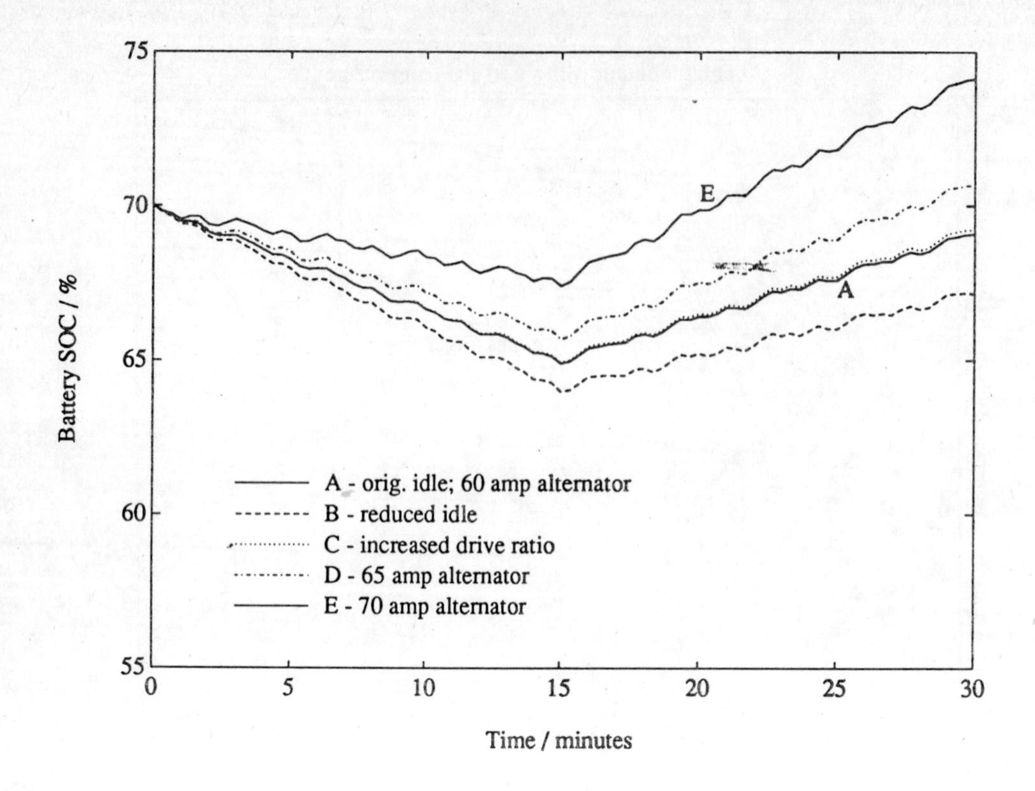

Fig 5 City driving, effect of changing idle speed

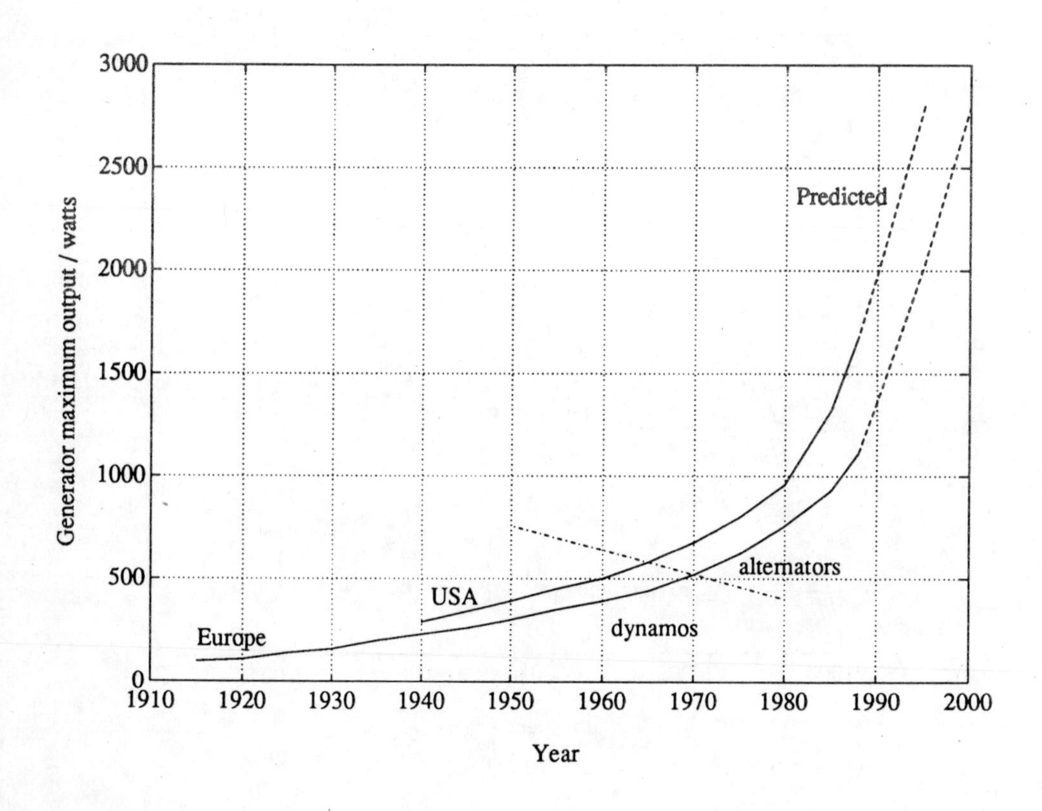

Fig 6 Trends in 12 V car generator output power

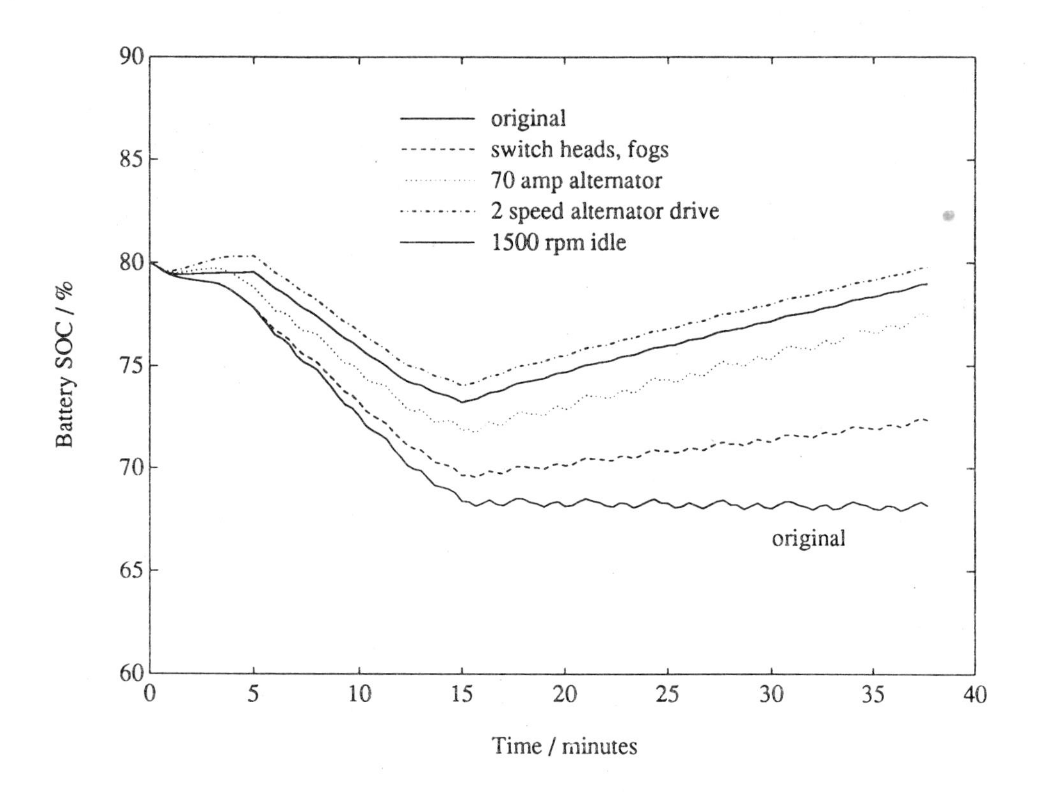

Fig 7 Winter city commuting – comparison between possible solutions

C391/082

Modular ASIC approach to vehicle timers

D MYATT, BSc
Lucas Body Systems, Witney, Oxfordshire
J BODLEY-SCOTT, BSc
Lucas Body Systems, Birmingham

BACKGROUND

Traditionally vehicle timers have been produced in 2 ways. The discrete approach, where conventional components are used to create single timing functions, or the integrated solution that relies upon a microprocessor performing multiple tasks.

These 2 methods have proved successful at either end of the complexity range, but have left a gap for the mid range multi function timer.

Discrete timers are normally only economical for 1 or 2 functions. Fitting more functions into a unit pushes the component count up to such an extent that the unit size becomes too large. The unit also becomes increasingly difficult to test and design leading to reliability problems. Long time constants such as Heated rear window timers (up to 15 mins) are difficult to measure quickly for testing purposes. Conventional testing methods often rely on setting the timer going and waiting for it to time out. This leads to long and unacceptable test times. These long time constants often rely on large value resistor and capacitors. This in itself leads to further reliability and tolerance problems. The tolerance on standard electrolytic capacitors is - 10 + 50 %. This will lead to large timing variations from one unit to the next. The large value resistors that also are necessary to generate the long time periods mean that the board is very prone to humidity related failures and must therefore be protected, normally by conformal coating.

Single chip microprocessor timers also suffer from a number of drawbacks. The microprocessor requires a reliable 5 V power supply with a low quiescent current. This necessitates the use of a good and therefore expensive regulator. The microprocessor inputs and outputs need protection from the harsh automotive environment; this often means that more integrated circuits are required to protect the microprocessor. Microprocessors are prone to software bugs that may only occur occasionally and are very difficult to trace and design out. All microprocessors are liable to data corruption, following for example, voltage transients or RFI and this can cause the processor to lock up into an endless loop; an external watchdog is therefore essential. This additional circuitry adds complexity, component count and cost to the unit without enhancing its operational performance. From past experience it was decided that units with less than about 8 functions could not be economically designed using a microprocessor.

WHY APPLY A NEW TECHNOLOGY

With the ever decreasing cost of complex electronic devices this has allowed a new technology to become competitive with the traditional approach. This new approach integrates all the logic and timing functions in to a single Application Specific Integrated Circuit or ASIC. With the ASIC containing all the complexity then the associated hardware design becomes relatively straightforward. This allows an elegant and simple solution to a complex timer unit. The simplicity of the hardware design allows standard switch input interfaces, output drive circuitry and power supply circuits to be designed and then used on every timer product. The input interface only consists of a pull up/down resistor and a series protection resistor in the order of 47K. This will protect the unit from all automotive spec. transients. The supply to the ASIC is derived from a high impedance zener power supply with a small smoothing capacitor. This offers a cheap and low quiescent power supply without the need for a voltage regulator. The ASIC also has excellent low voltage performance, in fact it will operate satisfactorily down to well below one volt. The problem of the ASIC locking up is also not possible so no watchdog circuitry is required. The use of standard circuitry for all the I/O means that only a few different values of component are required. This reduces the chance of incorrect components being inserted and reduces the number of components held in stores. This standard circuitry also has the advantage of having been tested on past designs, therefore giving improved customer confidence in new products. The timing functions are controlled from a single master oscillator in much the same way as for a microprocessor. This oscillator operates at a few KHz and can meet +/- 5% tolerance specifications. This means that all the internal timings are directly related to the master oscillator and can be toleranced accordingly.

Testing is also remarkably simple. The logic designed into the ASIC is tested by the ASIC vendor at every stage of production. This allows production units to be fully tested by checking the power supply, the I/O paths and the master oscillator frequency. Long time out periods needn't be tested for accuracy ; from determining the oscillator frequency then all timings can be accurately predicted. The simple and standard form of the hardware design means there is a reduction in component count and therefore solder joints. This enables easy testing of correct component placement and value through the units connectors.

Designs based round an ASIC are far more secure than discrete solutions. As all the 'intelligence' is hidden within the ASIC thus 'borrowing' of your design for use by competitors is not possible.

WHAT IS THIS NEW TECHNOLOGY

ASIC's fall into 2 main categories, Gate Array and Standard Cell.

Gate arrays consist of strips of logic gates formed on prefabricated wafers. The designer then specifies how the gates are connected up. Each gate array has a predetermined number of gates available, dependant on the silicon size and technology used. Regardless of how many gates are used the cost is always that of all of the gates. There is also always a degree of redundancy with this type of design.

Standard cells consist of specific components such as flip flops, and counters that can be integrated together to complete a total design. Only the components that are required are laid down in the silicon so unlike gate arrays there is no redundancy.

It was decided to develop a Modular ASIC design concept that could be implemented in either gate array or standard cell. This allows a structured approach in order to minimise leadtime and risk and maximise reliability and performance. Each timing function was designed as a self contained module. These library module can then be adapted to the customers specification as necessary.

LIBRARY MODULE CONCEPT

Using the modular design approach a set of standard modules have been designed to perform a wide variety of timer functions. These include ; variable front wipe, heated rear window timer, rear wash/wipe, burglar alarm, courtesy light delay and dead-locking.

PWM outputs have been designed for instrument illumination dimming and headlight dim-dip.

Each of these modules has been designed on Sun workstations using a Cadat simulation package and schematic capture software. IBM compatible pc's are ethernet linked into the workstations to create a multi-user environment. Designs are also produced on a VAX 11/750 using a different design package. This method uses netlist entry rather than the schematic capture package.

These individual library modules each have a functional description and test requirements written by the design engineer. New modules are designed when required. The new modules are built up from the basic gates and macros offered by the ASIC vendor. A wide number of macros are available such as 4 bit ripple counters, 4 bit compares etc. This makes designing of complex functions relatively quick and easy.

The master oscillator is used at a very low frequency compared to the maximum allowable frequency of about 10MHz. This means that the library modules are therefore independent to variations in gate delays associated with different vendors silicon implementation. This gives greater freedom of choice of vendors.

As each module is designed it is then tested by passing through test patterns that stimulate the inputs, the outputs are then monitored for correct operation. Once the module has been fully tested and documented it is then added to the existing library of working modules. The modules can then be 'plugged' together to give a multifunction timer with only a limited amount of additional work. As the library of modules is built up it becomes quicker to turn around each ASIC design. The risks involved are also reduced as the majority of the design has have been previously tested and proved on past units.

LIBRARY MODULE DESIGN

Each module has a standard set structure; Digital input - Digital filter - Operational logic - Output drive.

The digital filters are designed to replicate a standard RC low pass filter. A fixed filter design is used for all timers but this can be modified to change its performance if required. Filters may be either preset or cleared, have any number of sampling stages, and have any debounce period. The standard debounce period is 50 mS but it may be as large as 500 mS for inputs such as ignition.

A master counter chain controls all the timing functions. The counter chain consists of a long series of 4 bit ripple counters and flip-flops. The counter chain divides down the incoming frequency - typically 8kHz - down to the slowest frequency required. The slowest frequency may well have a period of 1 minute in some designs (HRW timers), so a typical chain might have 20 stages. Various frequencies are picked off the counter chain and shipped round the chip to the different modules. Filters are normally run off the 64Hz frequency line, courtesy light delay timers off the 1 Hz line and so on.

The operational logic is connected to the filters and any required frequencies. The standard library blocks normally only require minor timing changes from the standard version to operate as defined in the customers specification.

The output drive can take various forms - from push/pull stages capable of driving a few mA to 24mA open drain zener clamped outputs.

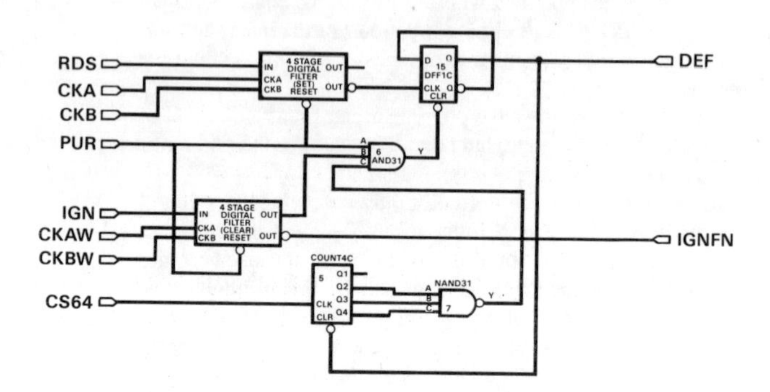

Fig 1 Heated front window 'library module'

When all the library modules have been combined it is often possible to rexamine the design and offer additional features to the customer. These enhancements may take the form of offering greater performance from the existing I/O, or increasing I/O to get a new desirable feature. One example might be performing a rear wipe when reverse is selected and front wipe is on. Complex logic feature such as this, that would require additional discrete components in a conventional timer, can be implemented within the ASIC with the addition of a few gates at very little extra cost to the customer. These additional features often prove a valuable selling point to the customer.

Another enhancement that can be offered is an in car diagnostic test. One of the problems that has plagued many automotive suppliers in the past is of units being returned that when checked had "no fault found". It was decided that a multifunction timer warranted a form of diagnostics to assist the car manufacturer. A diagnostic module has been designed as a standard library part. This module requires no additional I/O to operate and isolates the inputs from the outputs for independent testing. Once the unit is placed into diagnostic mode then the operation of any input causes the lighting alarm buzzer to operate. At the end of the input test, the unit then tests all the outputs by operating them in a set order. The unit then reverts to normal operation. This diagnostic facility allows the customer to operate a quick and easy test for use at the end of line or for checking out whether faults lie within the harness or unit on any returned car.

LIBRARY MODULE TESTING

Before any ASIC can go for masking it has to be fully tested. A testing strategy has been developed for testing ASIC timers. Each module is individually tested to prove that the functional design is correct. When all the library modules are combined a new set of test patterns have to be generated that will exercise the internal nodes of the design and that can be used by the vendor to test each chip. Most ASIC vendors stipulate that a minimum of 85 % coverage is required before the design will be submitted for lay out. This minimum of 85% coverage gives a high degree of confidence that every ASIC will operate. The test patterns have certain constraints placed upon them. There must not be more than 10000 unique patterns and the total number of clocks is limited. An excessive number of clocks will obviously slow down the ASIC production line and therefore have cost implications. The long counter chains that make up a timer must be tested fully and this immediately poses a problem. A HRW timer may well take 8 million clocks to time out and therefore additional test logic has to be implemented. The test strategy involves injecting the master oscillator frequency into the counter chain at carefully selected points. This allows the counter chains to be preset so that they have almost timed out. The ASIC is then returned to non-test mode and then only a few clocks are required to ripple through and finish the timing event. To access this test mode a spare pin may be used as a test pin.

The final test patterns must then be run with different delays to simulate the effect of interconnect delays and gate loading. Maximum ,minimum and nominal simulations are run and the outputs obtained from each must match exactly. Some vendors also stipulate that a unit simulation must be run. This assigns a standard 1 unit delay to every element of the design. A flip flop will therefore have the same delay as an inverter. This simulation must match the other simulations. Unit simulations are very good for discovering race hazards and marginal clock and clear pulse widths. Worst case simulations where maximum and minimum simulations are used together are not run as it is not considered that they offer a realistic simulation result.

At this point the design is ready to hand over to the vendor.

APPLICATION OF ASIC TIMERS

When is an ASIC the right solution for an automotive timer?

On purely economic terms, to make financial sense the volumes must be high enough - this does not normally pose a problem in automotive design where volumes are regularly in the 100's K. Tooling costs for aa ASIC start in the region of 10K pounds. Lower volume solutions would normally be met by a gate array where tooling costs are substantially lower but piece price is greater. Higher volume solutions being met by a standard cell where additional special features are available and the piece price is lower.

In terms of complexity there must be sufficient I/O and functionality to warrant the use of an ASIC. A single function timer with limited I/O would not fully utilise the ASIC or the package it was in and could probably be adequately designed using conventional components. Designs over 400 gates and of 10 or more I/O are probably suitable for an ASIC application. The combining of all the single functions into a multifunction timer, results in harness and packaging improvements. This reduces the cost and number of connections required and increases the reliability of the unit over its discrete counterparts.

The design must be primarily digital. Analog features are available on standard cells but the resulting design is then less elegant.

The design must be fixed sufficiently early to allow the ASIC to be designed, fabricated and tested. Changes to a single timer function can't be made easily once masked. However, if a time had to be changed within one of the library modules, then very little extra validation is required. A simple functional check will prove if the timing is correct - the hardware would be unchanged and therefore still validated. Leadtimes for a standard cell are approximately 3 months, after handing over a working database, though this figure is being regularly bettered. Gate arrays can be turned round in a few days using facilities such as E-beam machines, but the faster the turnaround the higher the charge.

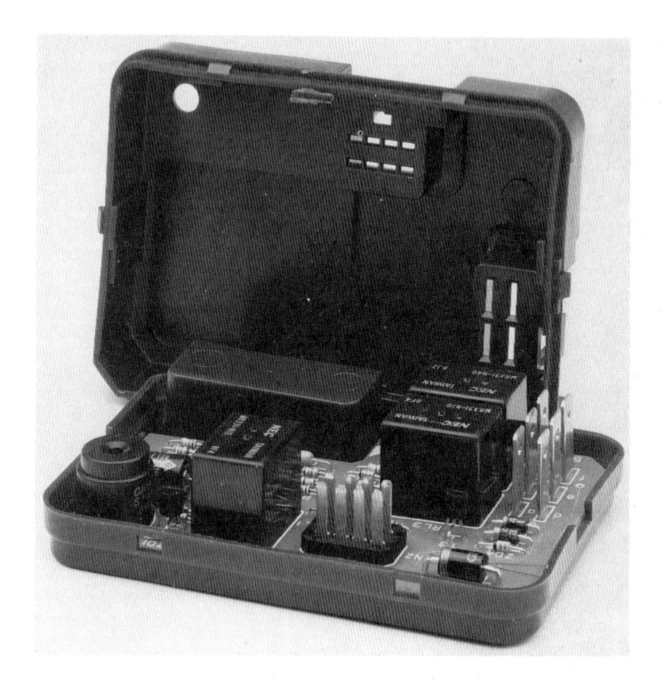

Fig 2 Three function vehicle timer

When board size is limited an ASIC produces a compact and tidy design. With relative freedom of pinout and package type on the ASIC, a multifunction timer can be made so that the whole design flows neatly on the PCB, thus optimizing the layout. This enables clean PCB designs and with intelligent designing the effects of humidity, EMC and transients can be reduced.

CONCLUSION

The modular ASIC approach allows multifunction timers to be produced quickly, easily and reliably using standard library modules. The use of simple interface hardware means that the final product will be an elegant combination of ASIC and discrete components. As the majority of the design will have been tested on past products, this will give the customer added confidence that the product will work first time and be on time.

C391/063

Adaptive noise cancellation for road vehicles

D C PERRY, BSc, MIEE
Lotus Engineering, Norwich, Norfolk
S J ELLIOTT, PhD, MIOA, MASA and **I M STOTHERS**, BSc
Institute of Sound and Vibration Research, University of Southampton
S J OXLEY, BSc
Texas Instruments, Bedford

SYNOPSIS A system for adaptive noise cancellation using multiple loudspeakers and microphones controlled by a digital signal processor to combat engine driven low frequency sound pressure 'booms' found in modern automobiles is described. Applications of the technique to control other resonant noise and vibration problem are outlined.

1 INTRODUCTION

The trend towards lighter weight vehicle construction tends to increase the problem of low frequency booms in a motor vehicle.

The in-line four cylinder four stroke engine is a particularly strong source of input for such booms. Both its dominant out-of-balance and firing frequency are at the same frequency of twice engine speed - that is, at second order of engine rotation. The typical four cylinder engine has an operating range from 900 r/min to 6000 r/min resulting in strong vibration inputs from 30 Hz to 200 Hz.

A motor vehicle can be schematically described as shown in Figure 1. The engine, coupled to its drivetrain, exhaust system and induction system become a distributed source of noise and vibration to a structure which is itself an interaction of mechanical boundaries and air cavities.

The drivetrain is likely to exhibit bending and torsional resonances in the frequency range of 30 Hz to 200 Hz. Likewise the exhaust system may well suffer from both cavity resonances and mechanical resonances in this same frequency range. These sources couple both structurally and through airborne radiation with the vehicle structure and its air cavities. The sources combine to provide inputs throughout this range, which can be very resonant in character.

The body structure and its air spaces are 'tuned' receivers of the source inputs. Acoustic modelling can predict the resonant frequencies and mode shapes of the airspaces, but the boundary conditions provided by the body are far from rigid.

Accurate mode shape predictions require a good knowledge of the boundary conditions (1). 'Nastran' modelling attempts to overcome this problem of coupling between air cavities and boundaries (2), but the mechanical boundaries and the interface connections to the sources are notoriously difficult to accurately model throughout the dynamic frequency range of 30 Hz to 200 Hz.

This modelling problem is compounded by the statistical variation of interior low frequency sound pressure which can be found across a sample of apparently identical vehicles. Second order variations of up to 12 dB have been measured. In fact, we have found that cars can also vary in second order boom sound pressure level from day to day. Such statistical variations are not totally understood, but differences in welding and panel stressing from body to body may account for some of the variation.

Figure 2 is a schematic representation of a vehicle interior and shows that the resultant sound pressure at an occupants position is a vector addition of a multitude of distributed boundary sources. This vector addition depends critically on the amplitude and phase relationship of these boundaries. Variations in detail from one vehicle body to the next may well affect this complicated relationship.

The problem facing today's noise and vibration engineer in the motor industry is that he can rarely find a single dominant source of a boom problem. The problem of statistical variation leads to the fact that a component improvement which helps a boom problem on one car can make the boom worse on the next, apparently identical, car. The real solution using conventional techniques is invariably one involving the reduction of a number of sources - often very difficult to achieve within the framework of practical changes and cost parameters.

The control of interior booms using an adaptively controlled system of loudspeakers and microphones able to react rapidly to the changing engine speed and to cope with variations in the vehicles acoustical characteristics would provide great advantages to an industry faced with the above problems.

Advances in modern electronic control systems, especially the availability of low cost single-chip Digital Signal Processor (D.S.P.)

devices such as the Texas Instruments TMS320 family, has rendered possible the suppression of such low frequency booms. This is achieved by the addition of 'secondary' loudspeaker sources within the vehicle interior to interfere destructively with the sound field produced by the 'primary' sources. A discrete number of microphone sensors is used to provide feedback control. The electronic control system described is the product of research and development at the 'Institute of Sound and Vibration Research' (3) of Southampton University and Lotus Engineering.

The inclusion of microphones into the vehicle also opens the door to inclusion of other adaptive acoustic interference suppression measures and convenience features into the vehicle, some of theses will be described.

2 THEORETICAL AND LABORATORY BACKGROUND

Computer and laboratory modelling has been carried out by I.S.V.R. (4) into the cancellation of the sound pressure in an enclosed space. A system of discrete 'secondary' loudspeaker sources and discrete sensor microphones is used to minimise the sound field resulting from a primary source. It has been shown that at a single low frequency reduction in total energy throughout the enclosed volume is possible provided that :

(a) The system is being excited at, or close to, a lightly damped acoustic resonance.

(b) The microphone sensors are placed close to pressure maxima of the primary acoustic field - that is, not close to the nodal planes of the primary field.

(c) The secondary loudspeaker sources can couple into the acoustic mode produced by the primary sources.

In practice, the number of microphone sensors should be greater than the number of secondary loudspeaker sources to balance the suppression of dominant modes without exciting residual modes.

It has been shown that it can be much harder to achieve successful cancellation at frequencies between dominant acoustic resonances because when there are several modes contributing significantly to the overall sound pressure level it may be difficult to position secondary sources to couple in successfully with all these modes. In fact, in this off resonance condition, the minimising of the sum of the squared microphone pressures may lead to an increase in the total global acoustic potential energy. In such conditions, it has been shown that the minimising of a more complicated cost function which not only includes microphone pressures, but also the strength of the secondary loudspeaker sources - the 'effort' term - may suppress this off resonance problem.

3 REQUIREMENTS OF A VEHICLE ADAPTIVE NOISE CANCELLATION SYSTEM

The following major criteria are necessary for a successful vehicle adaptive cancellation system :

(a) The system must use a practical number of loudspeakers and microphones.

(b) The system must use loudspeakers of practical dimensions.

(c) The system must achieve sound pressure reductions at the positions of <u>all</u> occupants.

(d) The system must respond very rapidly to any change in the vehicle sound pressure - that is the system must respond very rapidly to any sudden engine speed change or vehicle load change.

(e) The system must not cancel occupant speech or speech and music from the I.C.E. system.

(f) The cost of including the system into the vehicle must be able to be offset by the combination of reduction in vehicle development effort, decreased use of passive treatments , increased audibility of quiet passages of music, decreased driver fatigue and marketing advantage of adding this 'unique selling point'.

4 DEVELOPMENT OF A PRACTICAL SYSTEM

The algorithm used to adaptively control the low frequency booms was developed by I.S.V.R., the initial research being directed towards cancellation of booms in propeller driven aircraft; this is essentially a steady frequency problem. Developing the algorithm for use in motor vehicles involved refining its execution speed sufficiently to cope with the rapid engine speed changes encountered. Initial vehicle tests were made using the TMS32020, it being known that Texas Instruments would release D.S.P. devices with higher performance and more features. The upper limit chosen for the audio signals to be processed was set at approximately 500 Hz, since the modal density is becoming very high even at this frequency. The initial development versions of the computer were built using 12 bit Analogue to Digital Converters (A.D.C.) and 12 bit Digital to Analogue Converters (D.A.C.).

Figure 3 shows a schematic representation of the system as fitted to a medium size saloon car. The system shown incorporates six loudspeakers - one pair each on the dashboard, front doors and rear parcel shelf - and eight microphones fitted into the roof lining. This system would be referred to as a '6x8' system. It can be seen that the computer receives signals not only from the array of microphones but also an engine speed signal. This serves as a reference signal which the computer uses both to determine what frequency booms will be excited by the engine and to cancel only those booms which are coherent with the engine.

Using a '6x8' system measurements in excess of 15 dB reduction in booms have been made in all seat positions, covering a frequency range from 30 Hz to 200 Hz. Where a vehicle suffering from a limited number of booms is to be fitted, or where a limited number of seat positions need to be considered, it is possible to make reductions in both the number of loudspeakers and microphones. For example, a light commercial van or a two seater sports coupe only needs to provide cancellation in two seat positions, so a system comprising two loudspeakers and four microphones, a '2x4' system, could be satisfactory. Similarly, a medium size car suffering from only a limited boom range could be fitted with a system comprising four loudspeakers and eight microphones, a '4x8' system.

As previously mentioned, the optimum positions for the microphones and loudspeakers are to place them well away from nodes of the standing wave patterns. Determining the positions of these nodes can be done theoretically using a rigid boundary finite element model, and then confirming these predictions by measurements on a vehicle. Figure 4 shows a typical prediction for a light commercial vehicle. These predictions have been confirmed by means of an acoustical survey covering several hundred positions within the vehicle.

With the loudspeakers positioned such that they are able to couple in well with the acoustic modes the acoustic power required to achieve cancellation is minimised. This allows even loudspeakers having a cone diameter as small as .4.5 inches - 11.4 cm - to achieve excellent results at frequencies as low as 30 Hz when fitted into sealed enclosures with a volume of less than two litres. This good coupling into the passenger cabin also allows the generation of a strong bass signal from the I.C.E. system when sharing the same loudspeakers. Measurements taken in vehicles have shown typical drive powers to the loudspeakers of less than one watt. Thus with I.C.E. power amplifiers of typically 15 to 100 watts per channel there is ample headroom left for the music to be played without distortion. The reduction in overall sound level within the vehicle increases the dynamic range available for music reproduction, this being the difference between the vehicle sound level and the threshold of pain. This factor is especially important with the increasing use of C.D. and D.A.T. in vehicles with their dynamic ranges in excess of 90 dB.

Figure 5 shows in block diagram form how an adaptive noise cancellation system can be integrated with the I.C.E. system. The only significant changes to the I.C.E. system are the inclusion of a simple line level mixer, and to ensure that the power amplifiers are operative whenever the vehicle ignition is on or the I.C.E. is switched on.

Turning to system response time, the cancellation signal produced by the computer must be able to track rapid engine speed changes. Measurements have been taken of the system response time by simulating the engine boom using a driving loudspeaker fitted in the footwell and allowing the adaptive control system to cancel this boom. The phase of the signal to the driving loudspeaker is then changed by 180 degrees, so the adaptive system has to react to a worst case phase change. The response time for an optimised installation is typically 70 ms. To confirm the speed of response was adequate an installation has been made in a two seater sports car and measurements taken under full throttle acceleration in first gear.

5 TODAY'S ADAPTIVE NOISE CANCELLATION SYSTEM

Trials with the development versions of the Adaptive Noise Computer (A.N.C.) have involved fitting the system into over thirty different vehicles.

During these trials it has emerged that it may be possible to make use of cheaper A.D.C. and D.A.C. devices than the twelve bit items used during development. Consideration may also be giving to employing oversampling. This would allow use to be made of very much simpler anti-aliasing filters in the computer. The simplification could be of sufficient degree to enable the filters to be implemented using linear continuous time filters in place of the switched capacitor devices used in the development computers.

Taking these measures, along with the disposal of those items only required during the development phase, allows the A.N.C. to be implemented in approximately 30 Integrated Circuits (I.C.) and having a component cost of approximately 40 U.S. dollars in volume quantities.

6 TOMORROW'S ADAPTIVE NOISE CANCELLATION SYSTEMS

Further refinements in the software handling the microphone inputs will allow use of A.D.C. devices with even fewer bits, and schemes are being studied which remove the need for a complex, and costly, A.D.C. all together.

By working in close collaboration Lotus, I.S.V.R. and Texas Instruments can combine all the experience and disciplines from research through to production so as to provide forward projections based on implementations which are production feasible. This collaboration has already established that in less than twelve months the A.N.C. hardware can be reduced to only 4 I.C.s and its cost reduced by over 35 per cent. The detailed work necessary to turn this option into reality is already under way.

Looking further ahead, the hardware requirements are consistent with being able to integrate all the functions onto one piece of silicon. This device can be regarded as a variant of the standard D.S.P. with the addition of sufficient A.D.C. and D.A.C. channels to handle the needs of the adaptive control algorithms, not only for noise cancellation but for control of other noise and vibration problems.

The availability of this visible upgrade, and cost reduction, route is an essential factor in determining the production readiness and commercial viability of the automotive use of adaptive noise cancellation.

The reduction in interior booms, and resultant increase in available dynamic range, would create an environment in which the In-Car Entertainment (I.C.E.) system can be more easily appreciated, especially where a Compact Disc (C.D.) or Digital Audio Tape (D.A.T.) system is incorporated. The positioning of the loudspeakers for boom cancellation is also ideally suited to the requirements for the I.C.E. low frequency loudspeakers, this in combination with the very low power requirements for adaptive noise cancellation allows power amplifiers and loudspeakers to be shared.

The emerging trend to include digital signal processing of sources such as C.D. and D.A.T. leads to the inclusion of D.S.P. devices into the next generation of I.C.E. systems. For such systems the inclusion of adaptive noise cancellation could be simply the addition of software!

7 EXTENDING THE ADAPTIVE NOISE CANCELLATION SYSTEM PERFORMANCE

The current system uses less than fifty per cent of the available processing power of a TMS320C25, thus there is plenty of scope left for including further enhancements.

In some vehicles the reduction in second order are so great that the dominant engine related noise signal becomes that at four times engine rotational speed, i.e. fourth order. Using the present hardware a system combining simultaneous reductions in both second and fourth order booms has been constructed. With improvements in the processing power available using later generation members of the TMS320 family it would be a simple matter to extend the process to include further orders such as sixth or eighth order terms. The limitation then becomes the increased acoustic modal density at the higher frequencies corresponding to these orders. This problem can be solved by fitting small mid-range loudspeakers into the headrests of each seating position and routing the higher order cancellation signals to them. Obviously this increases the installation cost, but having individual loudspeakers carries with it the benefit of allowing each occupant in the vehicle to be able to listen to different I.C.E. signals also.

8 OTHER AUTOMOTIVE APPLICATIONS

Once a vehicle has been fitted with a system to cancel engine induced low frequency sound pressure booms, the remaining noise and vibration problems become more noticeable.

The same computer hardware can be applied to the cancellation of suspension induced booms, such as are experienced when a vehicle is driven over sharp bumps or potholes in the road surface. In place of the engine speed reference signal, each suspension system is fitted with accelerometers. These feed a signal to the A.N.C. which travels faster than the mechanical signal through the structure of the vehicle. This enables the A.N.C. to predict any booms which would occur and generate the appropriate wideband signal to cancel them.

By replacing the loudspeakers with actively controlled engine mounts, the same A.N.C. system can be used to cancel the unpleasant vibrations fed into the vehicle from the engine. This problem is especially noticeable on transverse engined front wheel drive vehicles, where the engine mounts are a compromise between being stiff enough to control engine 'shunt' and soft enough to provide vibration isolation. The ultimate engine mounting system must surely be one which combines the Lotus Active Suspension system to control the engine mounting low frequency vibrations below 30 Hz, with the A.N.C. system to cancel higher frequency vibrations fed through to the body. Since both of these two systems use Texas Instruments TMS320 family devices, the combination makes even more sense.

The inclusion of microphones into the vehicle as an integral part of its noise and vibration package opens up the possibilities for many enhancements to both the vehicle and its electronic accessories. Among the possibilities are :

(a) Providing dynamic equalisation of the signal from the I.C.E. system to ensure that each occupant receives a correctly levelled and balanced signal.

(b) Using the microphones for the 'hands-free' feature of cellular telephones. An enhanced A.N.C. system could even electrically cancel very high order engine and suspension booms in the signal fed to the telephone transmitter, possibly making the person receiving the call unaware that a car 'phone is being used!

(c) Providing speech enhancement from one occupant to the others, a form of in-car public address!

(d) By employing electrical cancellation, as already mentioned, a signal free enough from interference to provide reliable performance from a speech recognition system could be generated. This could simplify the operation of many ancillary items fitted to the vehicle, and thereby improve safety.

Adaptive noise cancellation not only finds a role within the passenger cabin though, its use can be extended to controlling noise levels in both the exhaust and induction systems. Both these areas are acoustically simpler to control since the situation is essentially that of a duct, which can only sustain longitudinal low frequency standing waves. Cancellation of these would only require a system comprising one sensor and one actuator, a '1x1' system. Intake cancellation could be achieved using a loudspeaker, but exhaust noise is best cancelled by using a modulated air stream injected close to the exhaust manifold. Cancelling exhaust noise at the tailpipe still allows the whole exhaust pipe to be a resonant source of noise both inside and outside the vehicle.

It is even possible to envisage an intake system containing two loudspeaker-type devices, one to cancel the exterior low frequency noise, and one to provide the correct standing wave within the inlet tracts to ensure that a high pressure area exists when the inlet valves open - a form of 'electronic supercharger'. This would employ a minor modification to the algorithm to maximise the acoustic energy, rather than minimise as is achieved during cancellation.

9 CONCLUSIONS

The availability of low cost D.S.P. devices has enabled the idea of adaptive noise cancellation, which was first suggested as early as 1934, to be turned into a system which is both small enough and fast enough to be fitted into motor vehicles.

Also important is that availability of a practical system gives the Noise, Vibration and Harshness (N.V.H.) engineers in the motor industry a tool which, for the first time, enables them to cancel low frequency booms regardless of their source. This can be used to significantly reduce development times, and achieve low weight vehicles.

The application of electronic feedback control systems has already made great advances in fuel economy and exhaust emissions, now these benefits can be extended to provide a quieter and more comfortable vehicle for its occupants, and to reduce noise levels outside the vehicle.

10 REFERENCES

(1) NEFSKE, D.J. & HOWELL, L.J. Automobile Interior Noise Reduction Using Finite Element Methods. SAE 1978, paper no. 780365.

(2) WOLF, J.A. Jr., NEFSKE, D.J. & HOWELL, L.J. Structural-Acoustic Finite Element Analysis of the Automobile Passenger Compartment. SAE 1976, paper no. 760184.

(3) ELLIOTT, S.J. et al. A Multiple Error LMS Algorithm and its Application to the Active Control of Sound and Vibration. IEEE Transactions 1987, Vol. ASSP 35, pages 1423-1434.

(4) NELSON, P.A. et al. The Active Minimisation of Harmonic Enclosed Sound Fields. Journal of Sound and Vibration 1987, Vol. 117, pages 1-58.

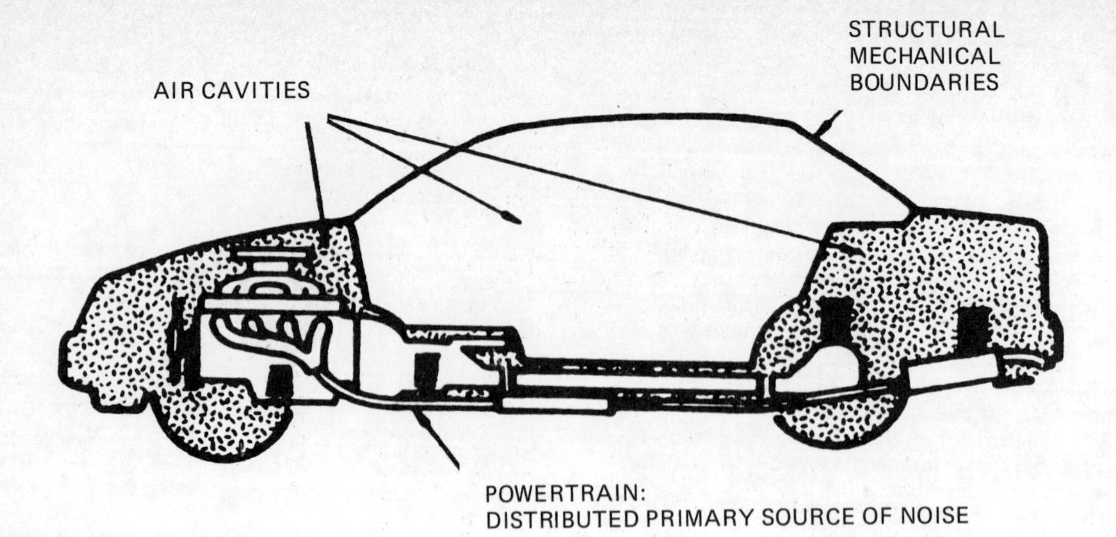

Fig 1 Schematic representation of vehicle noise and vibration sources, coupled to an interactive structure of air cavities and mechanical boundaries

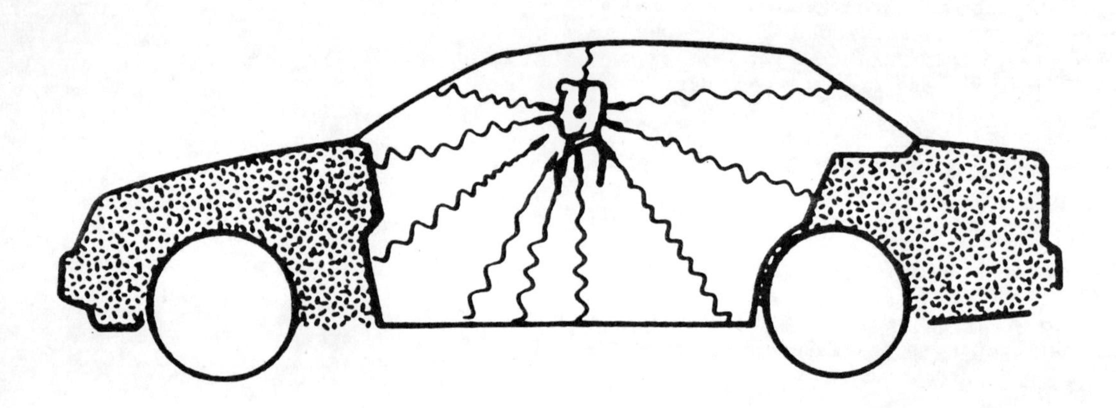

Fig 2 Schematic representation showing that resultant sound pressure at driver's head position is a superposition of the distributed radiation from all sources

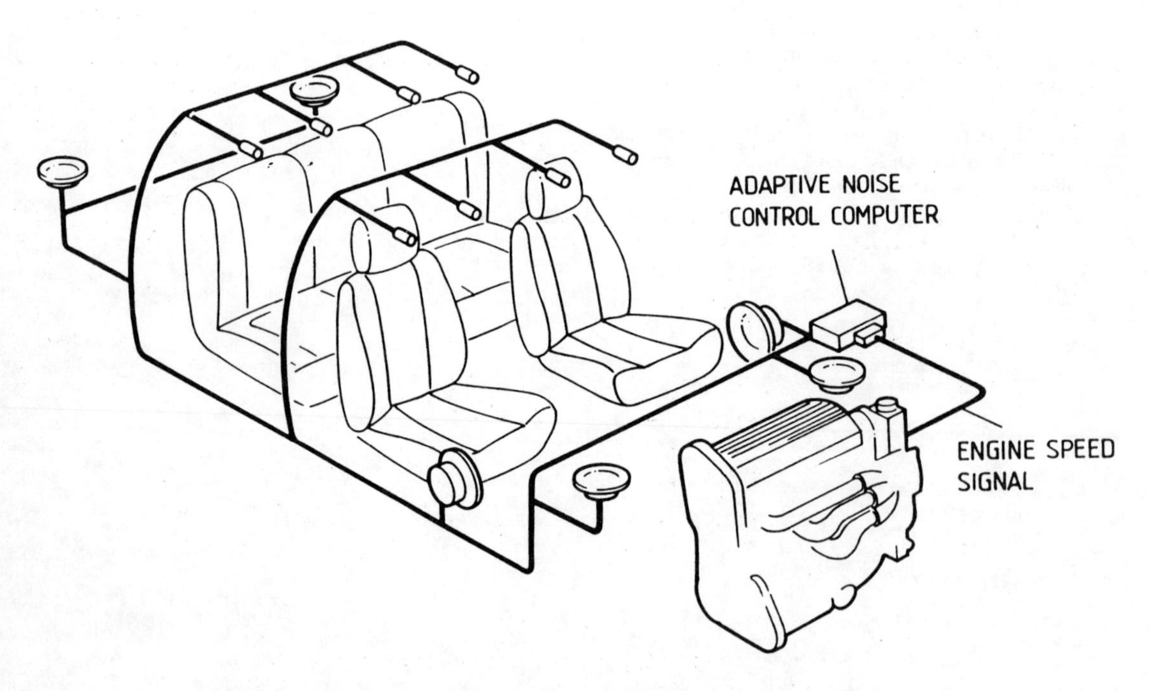

Fig 3 Schematic representation of a '6x8' adaptive noise cancellation system fitted to a medium size saloon car

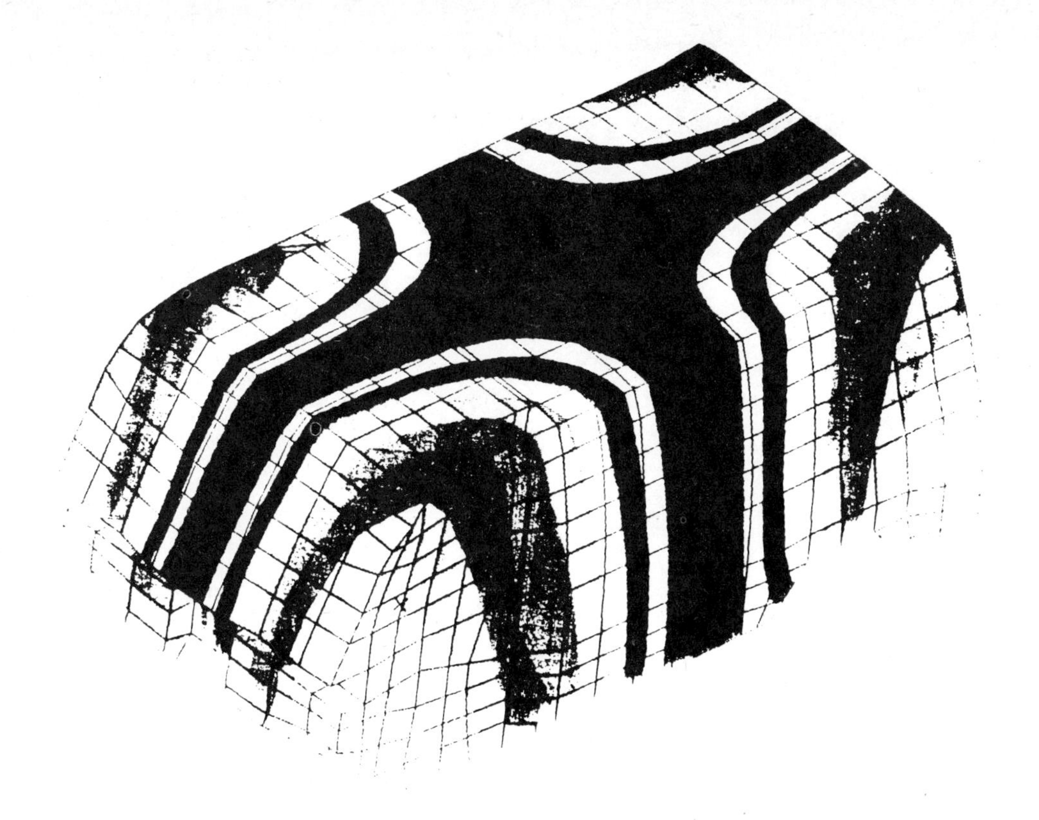

Fig 4 Typical predicted mode shape for a light commercial vehicle

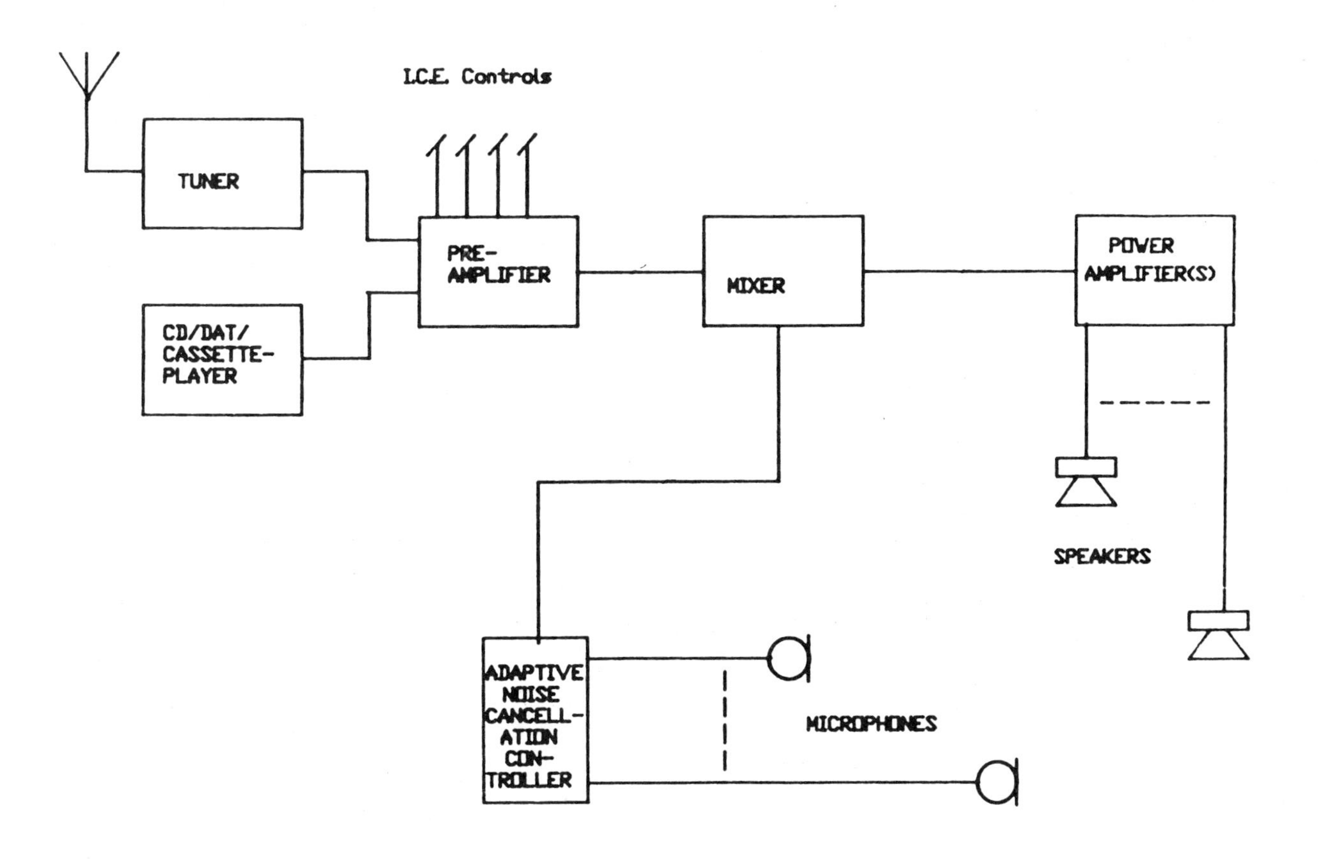

Fig 5 Block diagram of integrated adaptive noise cancellation and in-car entertainment system

C391/084

High performance timer eases the load on automotive system designers

V B GOLER, BSEE and **G MILLER**, BSEE
Motorola Inc, Austin, Texas, USA

SYNOPSIS Timer structure evolution for automotive control is described. The two fundamental competing timer approaches; counter vs. capture/compare are discussed. The evolutionary development of the prevailing approach, that of the capture/compare is traced in various timer implementations, ending with the latest advance in timer structures; the Time Processor Unit of the Motorola 68332.

1 INTRODUCTION

Automotive system designers have increasingly looked to electronics to replace existing mechanical systems in an effort to provide both more and better control. This drive toward increased controllability through the use of electronics has been mandated by stricter government limits on permitted emissions from new vehicles, and by government imposed fuel fleet requirements. To meet these requirements automotive electronic control components have evolved, becoming more sophisticated. It has been largely through the development of more highly integrated microcontrollers, that has provided automotive system designers with the components from which to meet increasingly stringent automotive control requirements.

As more systems have been brought on chip and those already present enhanced, it has been the evolution of the timer subsystem resident on microcontrollers that has had the greatest impact on overall automotive system control performance. In the following paragraphs the evolution of microcontroller timer subsystems will be described and the resulting automotive system control advancements shown. A historical evolutionary perspective of timer submodules is discussed starting with the two fundamental competing timer approaches. The first silicon implementation of the prevailing timer approach was realized on the Motorola 67002. The enhancements made to the prevailing timer approach is traced from this first realization, progressing to the Motorola 6801/68HC11A8, and ending with the Intel 8061/8096. The latest advancement in timer submodules that of the Motorola 68332 is discussed with emphasis on the applicability of its features to provide enhanced automotive system control. Finally, the other submodules of the 68332 are briefly described with emphasis on how they complement the timer submodule to provide an enhanced microcontroller for automotive system control.

2 COUNTER VS. CAPTURE/COMPARE TIMER

In the early 1980's two fundamental approaches were taken in the design of timer subsystems. These two approaches have been given the names counter and capture/compare because of the use of this hardware to perform timing functions. In the counter approach, input measurements are performed by utilizing a known frequency time base to increment a counter over the period measured. The value of the counter accumulated over the measured interval multiplied by the period of the counter time base, yields the time duration of the measured period. Output events are scheduled by loading a down counter with the number of counts corresponding to an interval of time as calculated by determining the difference from the scheduled output event time and a reference time. In the second approach capture register(s) and compare register(s) are referenced to a free running time base. This free running time base can be likened to a clock with a periodicity of two raised to the power of the timer bit length. When an input event occurs the value of the free running timer (clock) is stored in a capture register. An output compare is scheduled simply by determining the time (clock time) when the pin transition is to occur and storing that value in the compare register. The timer advances, and when equal to the value in the compare register the output pin transition occurs. The capture/compare concept is often referred to as the "time of day" approach.

Several problems are encountered by the automotive system designer in trying to use a timer subsystem based on the counter approach. One problem encountered in scheduling an output event is the time that it takes to load the counter must be taken into account with respect to the value loaded into the counter. This makes exact scheduling difficult. However, of greater concern is the inability to easily relate output events to input events. Such is the case when trying to schedule fuel or spark synchronized to an engine reference angle. The counter approach has largely been abandoned in favor of the capture/compare or "time of day" approach. In the "time of day" approach an output compare can be scheduled by simply determining the clock time when the event is to occur, as opposed to determining the time left. Setting up the output compare becomes less critical since the transition time only has to be scheduled before it is to occur. Because engine position can be synchronized to a time base via an input capture, and engine speed calculated, scheduling of an output compare for determining spark firing or fuel pulse initiation can be easily done.

3 TIMER MODULE 67002

The timer submodule of the 67002 was the first to utilize the "time of day" approach. This timer submodule contains one 8 bit TIMER COUNT REGISTER (TCR) which increments every 2 microseconds, two 8 bit CAPTURE registers and one 8 bit COMPARE register. Interrupts are generated every 512 microseconds whenever the TCR rolls over, whenever the TCR increments to equal the value of the compare register (maximum of once every 512 microseconds), and when a specified transition occurs on an input capture pin initiating the value of the TCR to be saved in the capture register.

As originally conceived the 67002 consisted of a two chip set, a microcontroller and an I/O chip. The I/O chip contained the analog portion of a dual slope A/D converter with a 10 channel multiplexer, and 10 output state latches which were independently set or cleared. Multiplexer channel control, A/D conversion initiation, and output latch control were communicated by sending an 8 bit control word over a 1 Mhz synchronous serial link to the I/O chip. By loading a value into the compare register, and taking into account the shift delay in sending the control word to the I/O chip, events could be controlled with a resolution of 2 microseconds. Once the time of the event and the command word were loaded, the microcontroller was free to perform other tasks. Because the 67002 could control up to 10 outputs on a timed basis, the 67002 was useful in controlling both single and multi-point sequential fuel injectors, and in controlling spark advance/retard. Voltage converted pulse widths were fed from the I/O chip to the 67002 which utilized one input capture register in conjunction with the CPU to calculate the pulse width which corresponded to a known voltage. The other input capture register was used to measure the time between reference angles, which was used to calculate engine speed which in turn was used to predict spark advance and retard.

For automotive system designers the 67002 microcontroller had serious limitations for advanced applications. Since the control word sent from the microcontroller to the I/O chip could only control one output per transfer, timing accuracy was lost when multiple outputs were to be changed simultaneously. The interrupt latency, service time, and control word transfer time of the 67002 microcontroller became the limiting factor controlling the frequency at which the outputs could be changed. Assuming an operating frequency of 1 MHZ, the minimum time two outputs could change was approximately 40 microseconds. In addition because the TCR was only 8 bits, interrupt setup time plus interrupt service time was constrained to be less than 512 microseconds, otherwise the integrity of the measurement would be lost, since software counters were used to keep track of longer timer intervals. Because of the interrupt handler time constraint, writing the control software was greatly complicated.

4 TIMER MODULE 6801/68HC11A8

The 6801 timer submodule extended the bit lengths of the registers over that of the 67002 timer submodule. This timer submodule, shown in figure 1 at the upper left, contained one 16 bit TIMER COUNT REGISTER (TCR) which was incremented every 1 microsecond thereby offering improved resolution over the 67002 timer. However the 6801 timer contained only one 16 bit CAPTURE register, and one 16 bit COMPARE register. Unlike the 67002 whose compare register could singularly control 10 output state latches, the compare register of the 6801 timer controls but one output state latch. Because the TCR bit length was extended from 8 bits to 16 bits the interrupt handler time was not nearly so constrained with always having to update software counters to maintain timing integrity, which eased the burden of writing the control software. However this timer suffered from the same problems as the 67002 timer when multiple outputs were to be changed simultaneously, only more so. Since the 6801 microcontroller is a slower machine than the 67002 microcontroller, the minimum time two outputs could change increased to approximately 60 microseconds.

The 68HC11A8 timer, shown in figure 1 at the upper right, offered some improvement over that of the 6801 timer, increasing the number of capture registers to three, the number of compare registers to five, added additional control capability for compare register 1 to affect the output latches of the other four compare registers, and added an eight bit hardware event counter. All of this additional hardware did serve to provide the automotive system designer improved performance in controlling spark advance/retard, and in single and multipoint simultaneous (all injectors turned on at the same time) fuel injection. The time to change two events on two different output latches could now be made zero by utilizing two compare registers, if the events could be scheduled far enough in advance. Having achieved this, automotive system designers now wanted better controllability, to change strategy at or near the time to the scheduled event, and at a faster rate. The goal of the automotive system designer is to always utilize the most recent information available to schedule events. The time to respond to an interrupt and service the interrupt then become the limiting factors.

5 TIMER MODULE 8061/8096

The 8061 timer, shown in figure 1 at the bottom, expanded upon the "time of day" approach by implementing a FIFO (first in first out) on the input side and a CAM (content addressable memory) on the output side. The input FIFO is three bytes wide by 12 deep. When a transition occurs on any of the 8 input pins not only is the associated TCR time recorded but each TCR time is "tagged" with an eight bit identifier word which is used to determine on which input the transition occurred, and the transition edge. This approach allows pulse widths as small as one TCR count to be measured. On the output side, there is another three byte by 12 deep stack. Into the CAM is loaded the 16-bit event time, and the 8-bit tag byte identifying one of 10 outputs to be affected, and the output latch state. For every TCR count, each entry in the queue is examined, and if a compare for equal found, the associated tag byte directs the output state for one of the output state latches. The output CAM allows a pulse width as narrow as one TCR count to be generated. The 8096 timer has the same timer structure as the 8061 timer, but contains fewer timer pins.

The 8061 timer offers improvements over the 68HC11A8 timer both in terms of the number of timer pins, and in the ability to program multiple event times affecting a single pin in advance. However, like the 68HC11A8 timer, the time required to change a previously scheduled event is still limited by the time required to respond to an interrupt.

6 TIMER MODULE 68332

In order to achieve even better control, automotive system designers want, and are demanding the flexibility to wait as late as possible to a scheduled event time, before possibly changing the scheduled event time, in order that the newest most recent data be utilized. The number-one constraint of microcontrollers in automotive applications is the inability to perform high frequency timing functions in which the latest available data is utilized. High frequency timing has been primarily limited by the CPU overhead associated with servicing the timer system along with the other peripherals. In addition, the lack of flexibility associated with the timer pins has limited microcontroller versatility. (dedicated timer pin e.g. input capture) The timer module of the 68332 takes the time of day approach a quantum leap beyond all previous timers. The Time Processor Unit (TPU) of the 68332 optimizes high-speed timing, and offers unparalleled flexibility in timer pin configuration. The 68332 microcontroller implements a processor dedicated to timing tasks, freeing the CPU from much of the overhead associated with performing timing tasks.

The TPU, shown in figure 2, is a peripheral device of the MC68300 family, that manages time by performing simple as well as complex timing tasks. Viewed as a special-purpose microcomputer, the TPU performs operations on only one kind of operand: time. The TPU, a semi-autonomous device, is configured with control and data information by the CPU module, but operates independently.

Since the TPU is a microcomputer which operates independently of the CPU, very complex timing tasks can be executed by the TPU with minimal CPU intervention. CPU overhead associated with servicing the timer and other peripherals no longer affects automotive controllability. Consequently, design margin calculations become much simpler, being confined to the TPU worst-case scenario, which is easily calculated. More importantly, system latency which affects controllability is reduced with the TPU, by approximately an order of magnitude over conventional microcontrollers.

System constraints due to dedicated functionality of timer pins is not a concern with the TPU, since all pins have identical functionality and perform a wide variety of timing tasks. Providing flexibility in pin functionality eases the burden of design changes, made to accommodate new or different needs.

The TPU interfaces to the external world through 16 input/output pins, each of which is associated with a timer channel. A timer channel consists of a 16-bit capture register, a 16-bit match register, and a 16-bit greater-than or equal-to comparator. Each channel can be synchronized to one or both of the two 16-bit free-running TCRs, TCR1 and TCR2. TCR1 is clocked by a 2- bit pre-scaler, which is clocked by the internal system clock. TCR2 is clocked by a 2-bit pre-scaler, which is clocked by the internal system clock or an external clock. The resolution capability provided to each pin by channel hardware is the system clock divided by 4 (240 nanoseconds at 16.67 MHz). Providing a greater-than or equal-to comparator for each pin ensures the integrity of TPU timing tasks, and removes this burden from the system designer.

The TPU, being a service request driven machine (and not interrupt driven), has four request sources associated with each channel. The CPU module can initiate service of a channel, and has priority over the other sources. An input pin transition, which causes the capture register to be loaded with the selected TCR, can initiate service. An equal-to or greater-than compare of the match register and a selected TCR, which may result in an output pin transition, can initiate service. Finally, a channel itself can link to another channel and initiate service, enabling channels to operate in concert.

Servicing of pins is orchestrated by a real-time task scheduler, which allocates service time to each channel according to priority. Each channel is assigned a priority of low, medium, or high. The scheduler divides the TPU processor's time into seven time slots. Out of each seven slots, four are allocated to high-priority channels, two to mid-priority channels, and one to low-priority channels. Servicing of the channels is done by a dedicated RISC-like processor. The processor executes instructions, contained in a ROM control store, at a rate of one every two clocks, or 8 million instructions/second at 16 MHz. The factory-programmed ROM control store contains a library of time functions, which may be allocated to any pin(s). RAM parameter registers associated with each channel (6-8) are used as a shared data space between the TPU and CPU, with special usage for each time function.

A time function is a program in the control store for controlling TPU facilities to effect a time-related task. The TPU will be available in several factory-programmed versions, containing up to 16 time functions designed for specific real-time applications. The initial version contains time functions ranging from simple digital input/output to complex automotive functions. Time functions included facilitates:

- Angle-based engine control
- Stepper motor control
- Pulse width modulation
- Frequency measurement
- High-time accumulation
- Frequency divide/multiply
- Pulse accumulator
- Output compare
- Input capture
- Digital I/O

Realizing that the TPU could not solve the myriad of timing problems with a set of general-purpose time functions, new time functions can also be defined by users. Emulation capability is provided to facilitate the creation of new time functions.

Studying the programmer's model for a channel, shown in figure 3, helps in understanding how the TPU operates. The function select field programs which time function is assigned to the channel e.g. $9 indicates pulse width modulation. The priority field assigns one of three priority levels for servicing. The host sequence field selects one of four modes of operation for a time function. The host service request field indicates one of three types of a host service request issued by the CPU. All time functions have one host request type for channel initialization, but other types vary from function to function. The interrupt enable and status fields enable and indicate the occurrence of a channel interrupt. The time function assigned to a channel specifies interrupt occurrence.

As an example of a time function, and an illustration of how the TPU enhances automotive controllability, three time functions that facilitate angle-based engine control will be discussed. Two functions, period measurement with additional transition detection (PMA) and period measurement with missing transition detection (PMM), detect special teeth on the flywheel, crankshaft, etc, and track engine position and speed. Special teeth generally exist in systems comprising only one variable reluctance transducer (sensor), to indicate a reference point from which ignition firing points and fuel-injection pulses are triggered. A missing flywheel tooth can serve as the reference point; however, in some applications an extra tooth, several extra teeth, or even multiple missing teeth may define the reference point. Systems comprising two sensors, one indicating position and the second contributing engine-cycle information, may also be controlled by the TPU. In either system, TCR2 is clocked by the engine position sensor signal and represents course engine position.

The position synchronized pulse function (PSP) utilizes the position and speed information calculated by PMA/PMM to generate angle-based output pulses. Since flywheel teeth generally do not give adequate position resolution, extrapolation, based on historical speed data, is used by PSP to resolve outputs to a more accurate angular position.

PMA and PMM can accommodate a wide variety of angle-based systems, with any number of special teeth, since important attributes are programmable via the programmer's model of a channel shown in figure 4. These functions accept the digital input from the position sensor and check for the special tooth condition, which can be programmed dynamically in RATIO:

(1) PMA detects the additional tooth condition if the current measured tooth period is less than the last measured tooth period multiplied by RATIO.

(2) PMM detects the missing tooth condition if the current measured tooth period exceeds the last measured tooth period multiplied by RATIO.

These functions can operate in one of two modes. PMA, operating in count mode, can count the number of additional teeth programmed in MAX_ADDITIONAL before clearing TCR2 i.e. setting TCR2 to a starting value, indicating top dead center, or some other position. Operating in bank mode, PMA can clear TCR2 upon the occurrence of an additional tooth when BANK_SIGNAL is non- zero. BANK_SIGNAL can be set to a non-zero value by the CPU or by another channel executing one of the other TPU functions. PMM has similar modes of operation.

PMA and PMM also provide a degree of noise immunity through two checks:

(1) When each special tooth is detected, an exact number of regular teeth, programmed in NUM_OF_TEETH, must have occurred since the last special tooth, and

(2) When each regular tooth is detected, the number of regular teeth counted by TCR2 is checked. This number must not exceed the

If either of these checks fail, or if the calculated period, stored in PERIOD, exceeds 16 bits, the CPU is interrupted and notified of the error condition in ERR.

PSP performs the necessary extrapolating operation for outputs at programmed ignition firing points and fuel-injection pulses. A programmer's model for a PSP channel is shown in figure 5. A rising output transition can be programmed for a precise engine angle, specified as a tooth number in ANGLE1 and a fraction of the previous tooth period in RATIO1. A falling output transition can be programmed for a precise engine angle, specified as a tooth number in ANGLE2 and fraction of period in RATIO2, or it may be programmed to occur at a time period following the rising transition, as specified in HIGH_TIME. The former mode of operation is for engine controls which are a function of engine position only, referred to as angle-angle mode, and the latter is for controls which are a function of engine position and time, referred to as angle-time mode. Once programmed, the TPU will continually generate outputs at the programmed ignition firing points or fuel injection pulses. Hence, the system designer can easily control engine timing with minimal CPU interaction. The accuracy of the TPU also enhances controllability; in general, the TPU can provide better than 0.1 degrees resolution for systems with a large number of teeth.

Our example will comprise one sensor and one flywheel tooth every 10 degrees, less one tooth, resulting in a total of 35 teeth. Updating MAX_MISSING on channel 0, programmed for PMM in count mode, to $01 will result in TCR2 being reset upon each occurrence of the missing tooth. MAX_MISSING indicates the total number of missing transitions the TPU will count before resetting TCR2 to $FFFF. Consequently, the teeth will be represented by a TCR2 value of zero to thirty-four. The current engine position, or tooth number, is read by the CPU in TCR2_VALUE, which is updated by the TPU. Programming four channels for PSP, channels 1 and 2 for angle-angle mode and channels 2 and 3 for angle-time mode, typical outputs could be as shown in figure 6. PERIOD_ADDRESS on these channels is programmed with the address of PERIOD on channel 0 so that the speed data calculated by PMM can be used to extrapolate outputs.

Many of the other TPU time functions may also be used for engine control, relieving the CPU of costly interrupt servicing for routine tasks. Functional programmability of TPU pins also provides the needed flexibility for automotive engine control.

In addition to the control performance provided by the Time Processor Unit module, the 68332 contains four other highly integrated modules. The other modules include: a MC68000 family CPU; a serial module which contains both asynchronous and queued high speed synchronous serial sub-modules; 2K bytes of static RAM with standby capability; and a System Integration Module containing programmable chip selects, system clock, periodic interrupt and system protection features. Each of these modules coupled with the TPU module is designed to provide the automotive system designer with the highest performance microcontroller available.

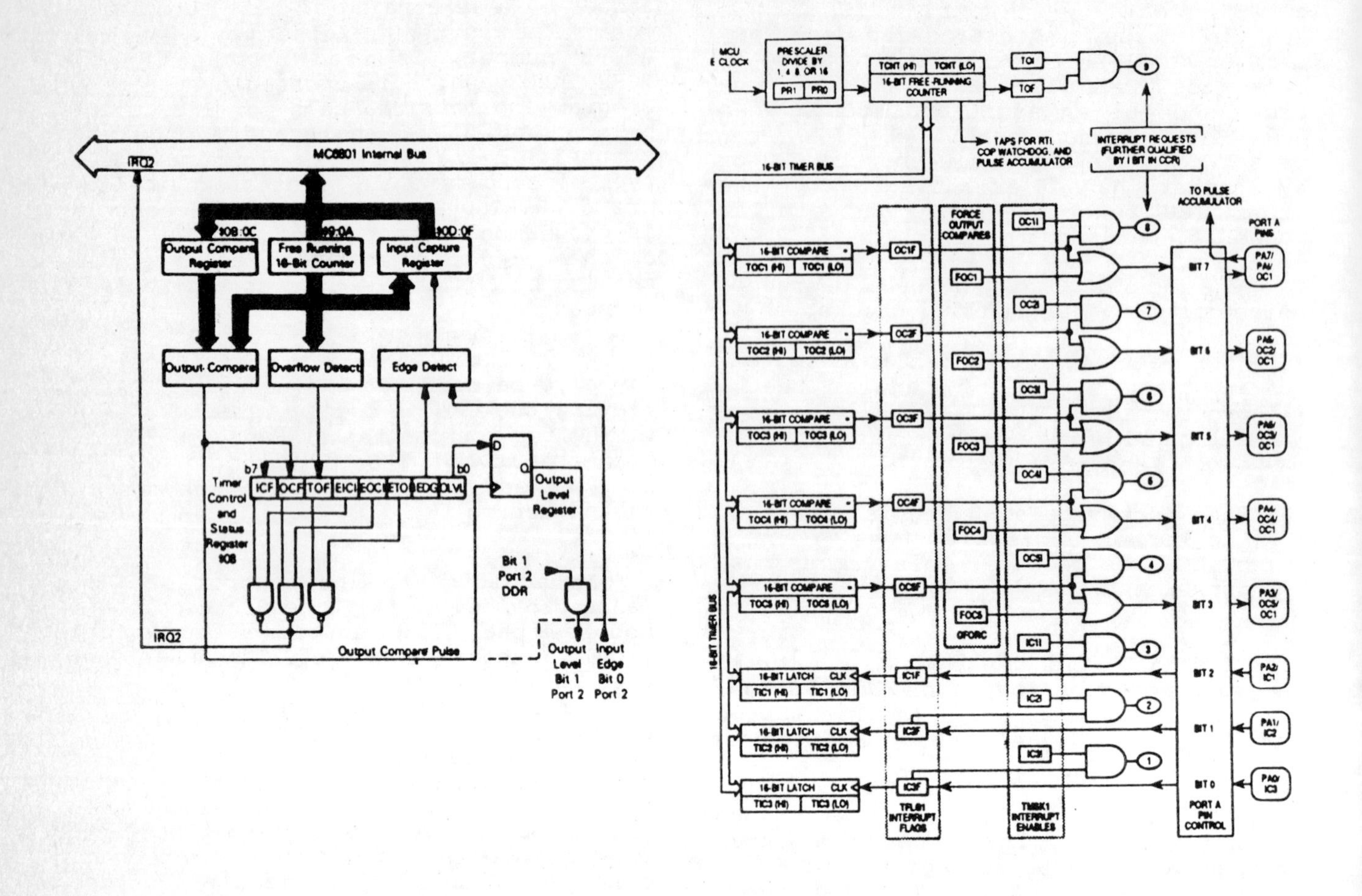

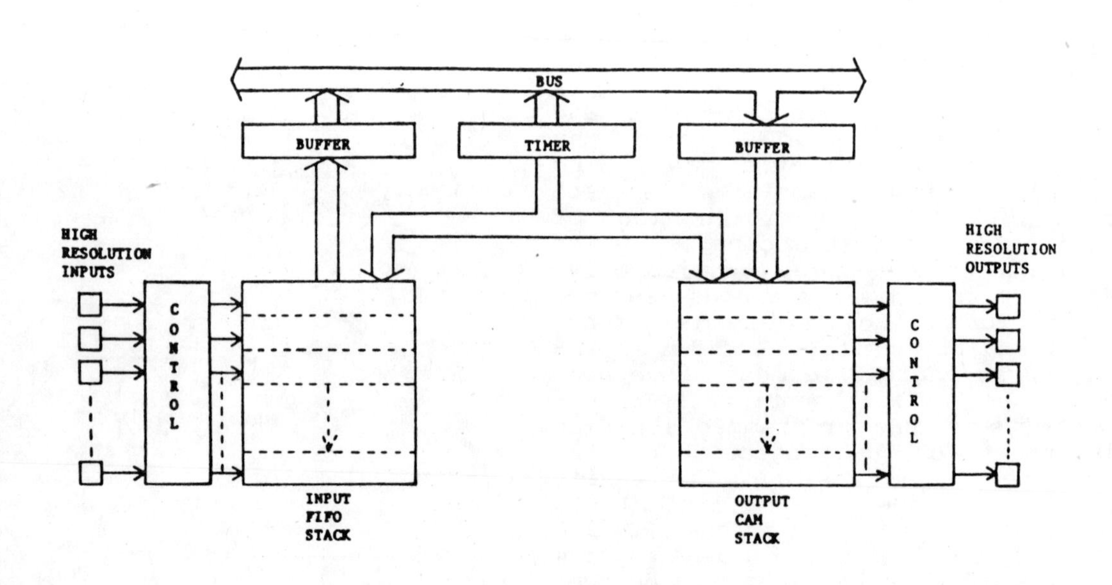

Fig 1 Motorola 6801, 68HC11A8 and Intel 8061 timer structure

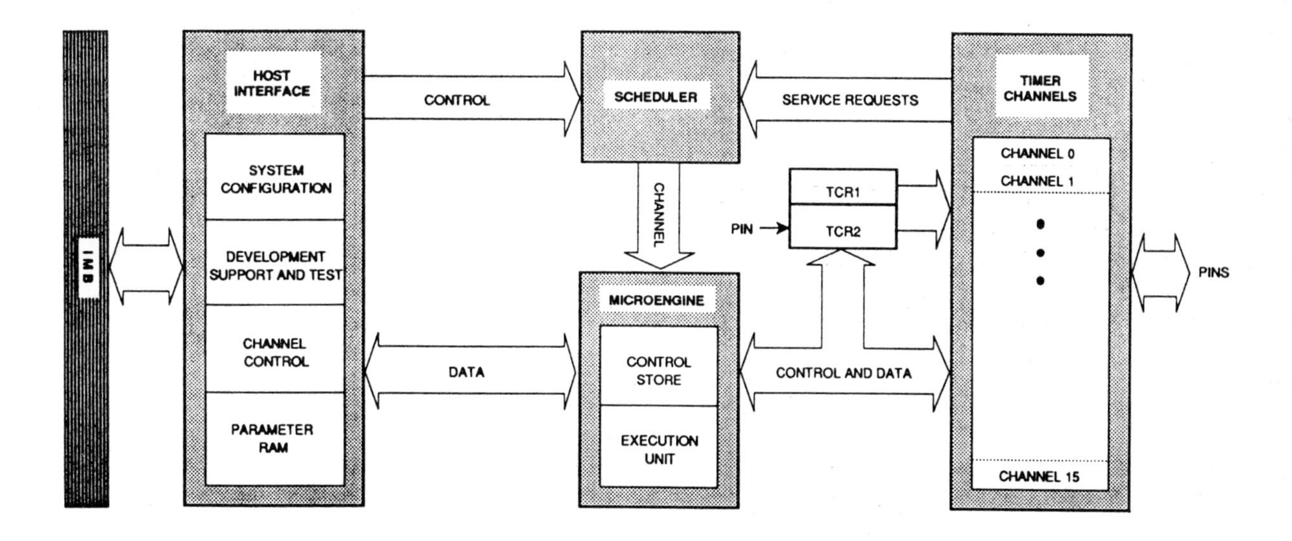

Fig 2 TPU simplified block diagram

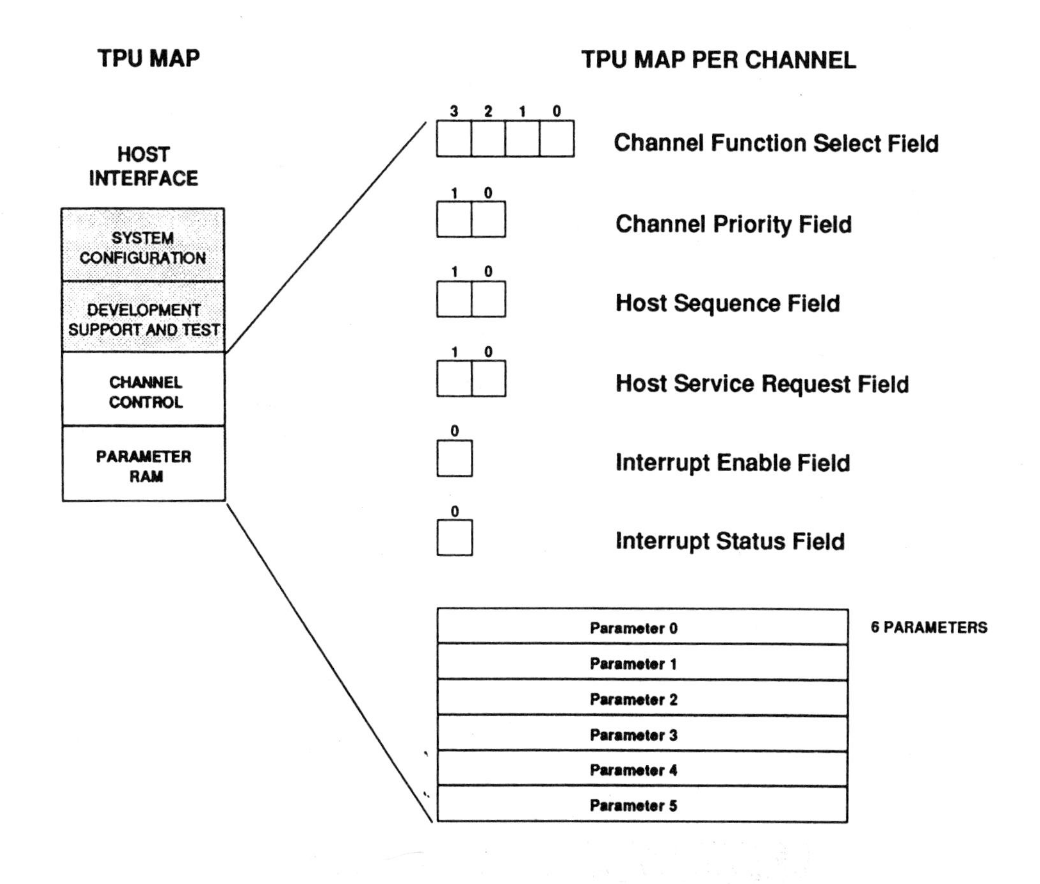

Fig 3 Programmer's register map for each channel

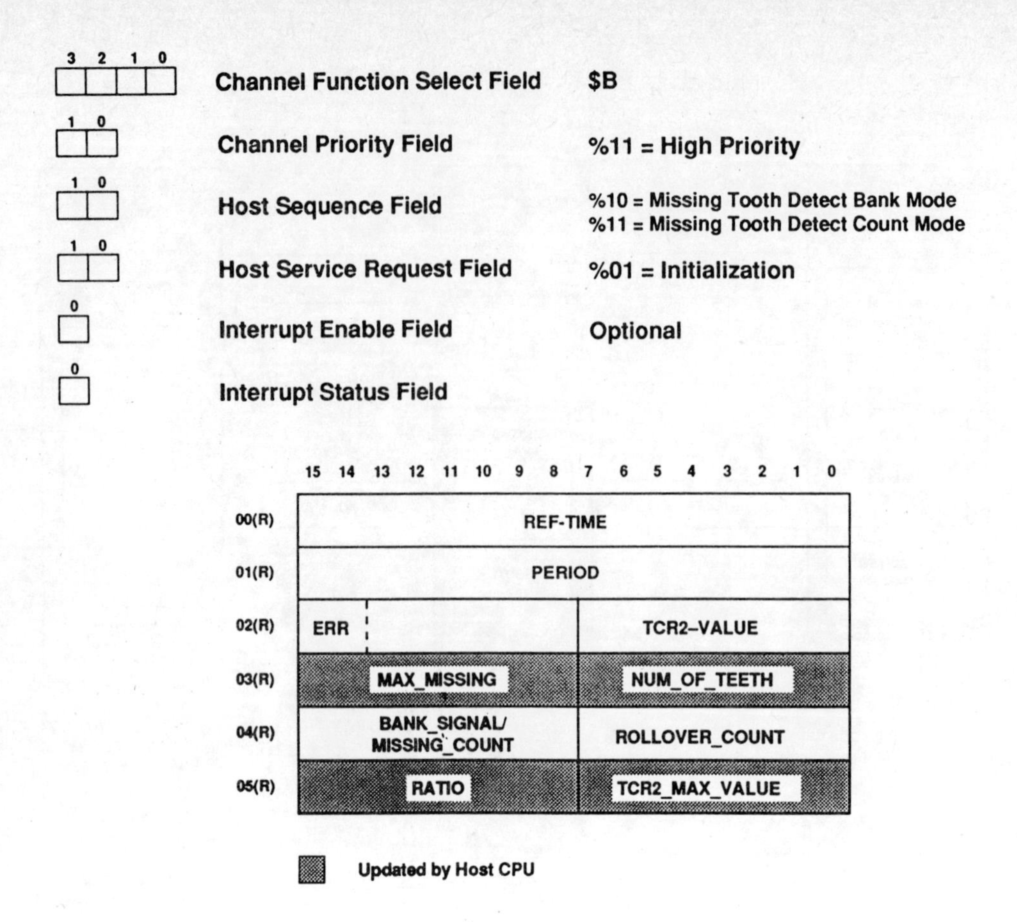

Fig 4 PMM time function

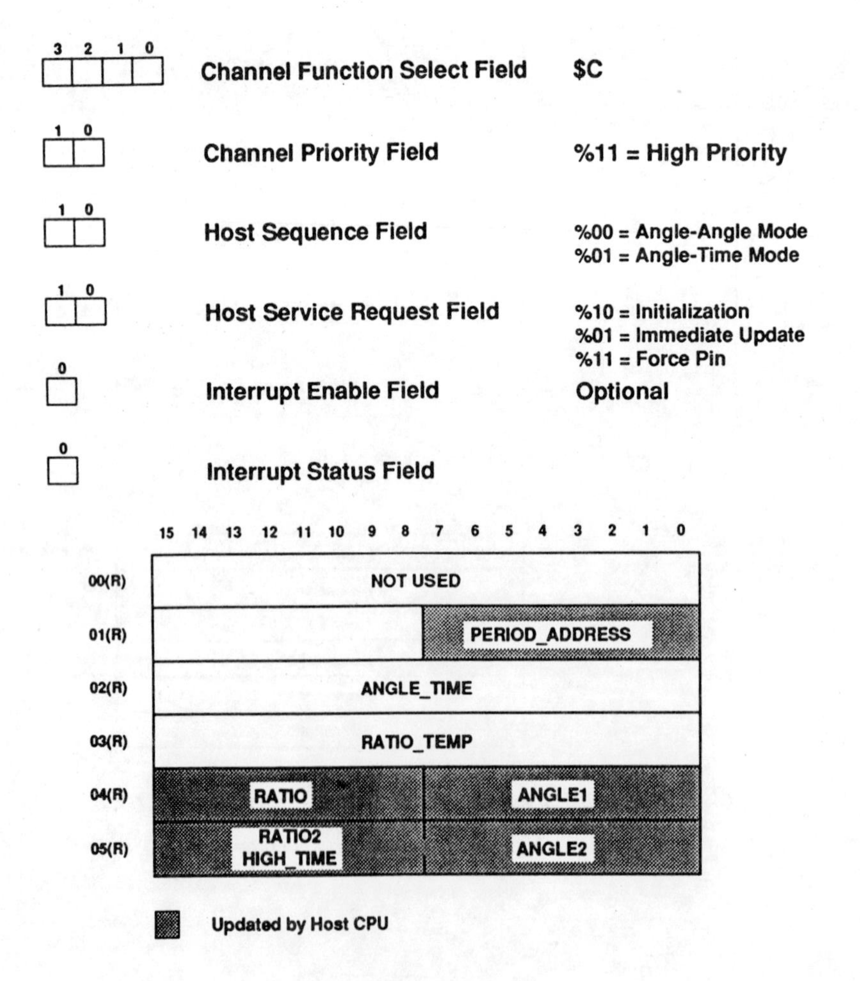

Fig 5 PSP time function

MISSING TOOTH — ONE TOOTH EVERY 10°, LESS ONE TOOTH RESULTS IN A TOTAL OF 35 TEETH

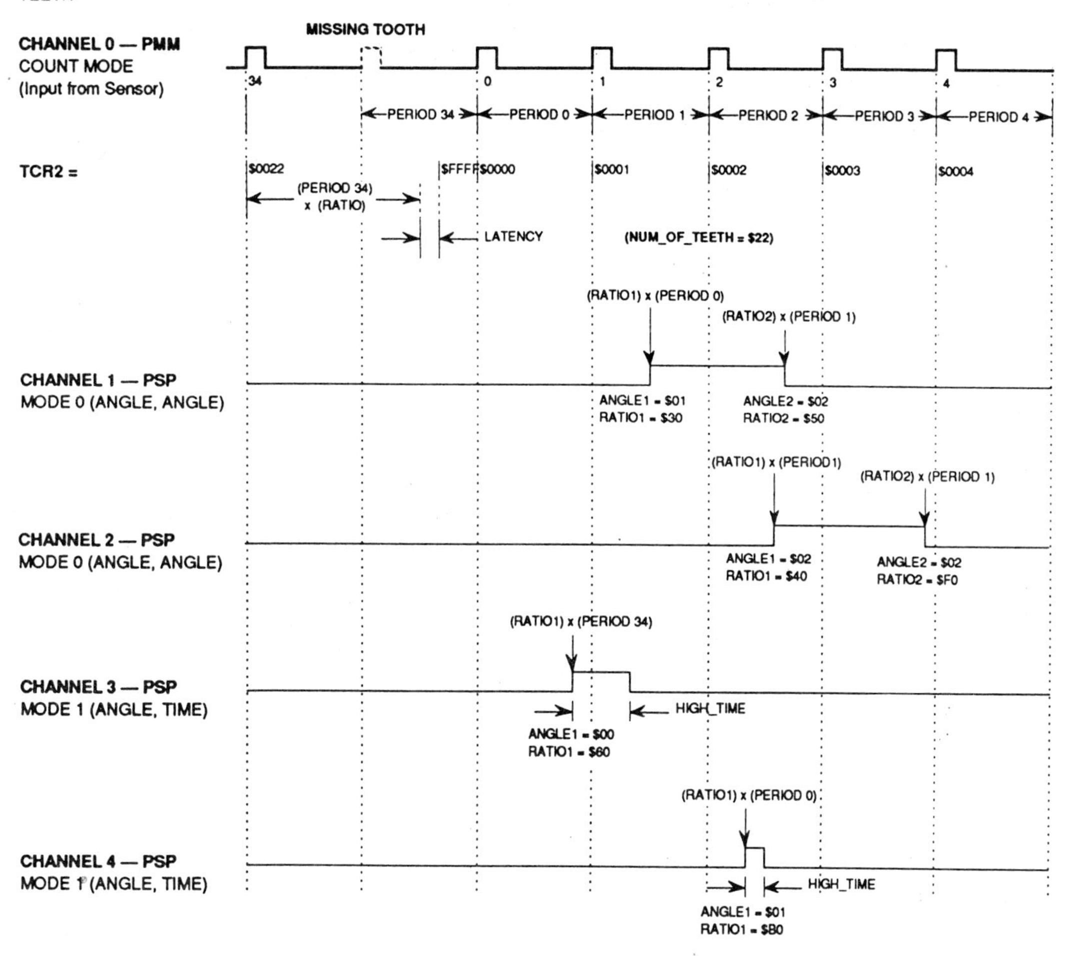

Fig 6 Angle-based engine control — PMM and PSP

C391/021

On aspects of automotive noise and its effect on the performance of pitch determination algorithms for speech

M R VARLEY, BSc, R J SIMPSON, BTech, MSc, PhD, CEng, FIEE and T J TERRELL, MSc, PhD, CEng, FIEE
School of Electrical and Electronic Engineering, Lancashire Polytechnic, Preston

SYNOPSIS Two pitch determination algorithms for speech are presented based on different analysis techniques. Results showing the effects of noise from vehicles on the performance of these algorithms are presented and discussed.

1. INTRODUCTION

There are several potential uses for speech technology in the automotive industry. One important potential application is for hands-free dialling for car phones (1), this is particularly relevant since legislation against making telephone calls whilst driving, either exists or is to be introduced in many countries. Another area of interest is speech recognition and identification for security and control applications. Additionally, on the assembly line, small vocabulary speech recognition systems may be used to improve testing procedures, whereby operators communicate with a system via speech using two-way radios (2). In such applications, it is often beneficial for an estimate of the pitch of the speech signal to be available, as many analysis and recognition techniques are enhanced when a reliable pitch estimate is incorporated (3,4).

In most applications, it is necessary to derive the pitch from the speech signal itself, obtained using a microphone, rather than using a separate transducer to directly obtain pitch estimates (5). For 'clean' speech recorded in a relatively noise-free environment, several techniques are available for pitch determination. However, when the laboratory environment is replaced by a practical situation, for example a moving vehicle where considerable noise may exist, the algorithms can become much less reliable. This paper describes two pitch determination algorithms (PDAs) (6), and assesses their performance for speech embedded in automotive noise.

2. PITCH DETERMINATION ALGORITHMS AND THEIR IMPLEMENTATION

Pitch determination algorithms may be broadly classified into two groups: short-term analysis PDAs, which process a short section of speech during which the pitch is assumed to be constant, and time-domain PDAs, which process the signal in the time domain (7). In this section, two PDAs are introduced, one based on a short-term analysis technique, and the other operating in the time domain. The algorithms used in this investigation are based on established pitch estimation techniques which have been extensively modified by the authors in order to improve performance. The algorithms have been implemented using the computer programming language 'C' and evaluated using an IBM PC-XT, in conjunction with the Interactive Laboratory System (ILS) PC digital signal processing software versions 5.0 and 6.0.

In both algorithms, an important modification is incorporated which limits the maximum rate of change of pitch frequency to 1%/ms, corresponding to the maximum rate of change of pitch that a voice can normally achieve. Many gross pitch errors are removed by introducing this criterion (6).

2.1 Modified SIFT algorithm

The SIFT (simplified inverse filter tracking) algorithm (8) for pitch estimation, is based on autocorrelation analysis of the residual signal obtained by linear predictive coding (LPC) analysis. A general block diagram for the SIFT algorithm is shown in Fig. 1(a). The speech is initially sampled at 10kHz, then digitally low-pass filtered, using a 3-pole, 2dB ripple

Chebyshev filter with a cut-off frequency of 800Hz, and downsampled by 5:1 to give a sampling frequency of 2kHz. This is effectively the same as sampling the analogue signal at 2kHz, but the 10kHz rate is used so as to retain consistency with the algorithm described in section 2.2.

A fourth order inverse filter, modelling the vocal tract, is used to yield the residual signal, which is taken as an approximation to the glottal excitation signal. The inverse filter design is produced by the autocorrelation method of LPC analysis on non-overlapping 32ms frames of speech data. For a pitch frequency of 100Hz, just over three pitch periods are included in a 32ms frame, and it may be assumed that pitch changes are minimal within that frame. For each frame of residual signal, a scale factor is applied so as to minimize quantization errors prior to signal storage.

The short-time autocorrelation sequence,

$$r(n) = \sum_{i=0}^{N-1-n} e(i)\, e(i+n)$$

where N = number of samples in an analysis frame, of each frame of the LPC residual exhibits a strong peak at a lag equal to the pitch period for voiced speech, whilst no such peak exists for a frame of unvoiced speech. Interpolation using a ratio of 4:1 is applied about this peak in order to achieve a higher degree of accuracy for the pitch period estimate, giving a resolution of 0.125ms for the pitch period. Voiced/unvoiced (V/UV) decisions are implemented by thresholding the interpolated peak in the autocorrelation sequence and applying the decision algorithm specified in (8).

2.2 Modified secondary feature algorithm

This algorithm is based on a time-domain pitch estimator originally proposed by Tucker and Bates (9) in which, after a preprocessing stage, the speech signal is represented by positive and negative pulse trains and these are then matched to estimate the pitch. A block diagram of the arrangement is shown in Fig. 1(b).

The preprocessor in the modified algorithm is a centre clipper (9,10) whose levels are updated every 10ms. Clipping levels are determined by examining the preceding and subsequent 10ms segments of speech, and setting the clipping level to 80% of the smaller of the peak values in these segments. Since positive and negative signal values are processed independently, the clipping levels are not necessarily of equal magnitude. All samples of a pulse are passed through the centre clipper providing at least one sample exceeds the appropriate clipping level in magnitude. This modification reduces the number of single-sample pulses which may otherwise be erroneously matched.

The output from the centre clipper is in the form of two pulse trains, one positive and one negative. Since these are derived from the same speech signal, and therefore cannot overlap, it would be possible to store both pulse trains as one signal, but for ease of programming and processing they are kept separate. Each pulse is characterized by three parameters: pulse width, energy and ratio of maximum amplitude to square root of energy. The pulse trains are tracked simultaneously and when a pulse is encountered, its parameters are compared with those of the previous pulse of like sign. If corresponding parameters of the two pulses are sufficiently similar, defined by a specified tolerance value, then the attempted match is successful and the pitch period is taken as the time between the two peaks. Current and previous pulse parameters are compared in this way for a maximum time span of 25ms, corresponding to a minimum pitch frequency of 40Hz, or until a match is found.

3. PDA PERFORMANCE ASSESSMENT

Speech and noise data were recorded in the passenger's position in a freight Rover van (Sherpa 300 series, 2.5 litre diesel), and in a Montego 1.6L estate car, on the test track at the Leyland Technical Centre, Leyland, Lancs. The signals were recorded using a Nagra analogue tape recorder, and subsequently digitized using a 12-bit A/D converter, sampling at 10kHz under ILS control.

In both vehicles, samples of noise alone, and male speech in noise, were taken at various selected speeds. Recordings were taken using a close-speaking microphone and also a microphone situated close to the vehicle dashboard, intended to simulate a dashboard mounted microphone. For each data set, the utterance 'one, two, three, four, five, six, seven, eight, nine, oh' was spoken over a duration of approximately six seconds. Each data file for analysis therefore consists of about 60000 sampled data values containing the full utterance.

Figures 2 - 6 show selected representative pitch contours/estimates produced by the algorithms when presented with various data sets. Fig. 2 shows the output from each algorithm with clean speech, recorded in the car with the engine off, as the input signal. Sections where there are no pitch estimates ideally correspond to segments of the signal where there is either no speech, or the speech is unvoiced, i.e. the vocal cords are not vibrating, such as in the fricative 's'. Although a few errors are present in all the pitch contours for clean speech, there are significantly more for the female speech results from the modified SIFT algorithm, shown in Fig. 2(b). Female speech typically has a higher pitch frequency than male speech and is therefore more difficult to analyse. In this case, the errors are largely due to the second peak of the autocorrelation sequence being selected as representing the pitch period, causing a halving of the estimated pitch frequency. Visual comparison of pitch contours for noisy speech with those of Fig. 2 provide a means of assessing the performance of the PDAs. In all cases, the results presented are for signals obtained using the close-speaking microphone, since those obtained when the microphone was located near the dashboard produced significantly more errors.

Figures 3 and 4 show selected results for speech signals recorded in the car. It can be seen that the modified SIFT algorithm produces acceptable results for both male and female speech at the three speeds used. The modified secondary feature algorithm gives reasonable results for male speech at all speeds, whereas the pitch estimates for female speech are mostly erroneous.

Selected results for speech signals recorded in the van are presented in Figs. 5 and 6. It is apparent that the van noise causes considerably more errors in estimated pitch values than the car noise, to the extent that in some instances the algorithms fail to produce any reasonable pitch estimates. This is to be expected, since the speech is subjectively more difficult to interpret when corrupted by comparatively higher noise levels in the van.

4. CONCLUDING REMARKS

It is apparent from the tests carried out, that PDA performance is affected in all cases by the location of the microphone. Results indicate that a speech recognition system designed to work in a moving vehicle would require a microphone situated close to the speaker.

Additionally, it is clear that both PDAs used in this investigation perform significantly better in the car than in the van. It is suggested that improved sound-proofing of the van, to reduce engine and road noise audible to the driver and passengers, would also result in an improvement in PDA performance.

Although the investigations have been carried out using an IBM PC-XT, and consequently do not operate in real time, it should be noted that specialized hardware is available which enables a compact, real time implementation of such algorithms. Work is continuing on the implementation of the PDAs using a single-chip CMOS digital signal processor.

5. ACKNOWLEDGEMENTS

The authors wish to thank members of the Speech Research Group at GEC Research, Hirst Research Centre, Wembley, and members of the Advanced Technology Group at the Leyland Technical Centre, Leyland, Lancs. for their encouragement and collaboration with this research work.

6. REFERENCES

(1) BLOMBERG, M., ELENIUS, K., LUNDSTROM, B. & NEOVIUS, L. Speech Recognizer for Voice Control of Mobile Telephone, *Proc. European Conference on Speech Technology*, vol.2, pp210-213, September 1987.

(2) CASSFORD, G.E. Applications of Speech Technology in the Automotive Industry, *IEE Colloquium on Speech Processing*, Digest no. 1988/11, pp10/1-10/4, January 1988.

(3) RABINER, L.R. & SCHAFER, R.W. Digital Processing of Speech Signals, pp319-321, Prentice-Hall, Englewood Cliffs, 1978.

(4) WHITAKER, L.C. & PEARCE, D.J.B. Larynx Synchronous Formant Analysis, *Proc. European Conference on Speech Technology*, vol.1, pp323-326, September 1987.

(5) HESS, W. & INDEFREY, H. Accurate Pitch Determination of Speech Signals by Means of a Laryngograph, *Proc. IEEE International Conference on Acoustics, Speech and Signal Processing*, 1984.

(6) VARLEY, M.R., SIMPSON, R.J. & TERRELL, T.J. On the Performance of Pitch Extraction Algorithms for Speech with Acoustic Noise, *Proc. Speech '88, 7th FASE Symposium*, book 2, pp621-628, August 1988.

(7) HESS, W. Pitch Determination of Speech Signals: Algorithms and Devices, p5, Springer-Verlag, Berlin, 1983.

(8) MARKEL, J.D. The SIFT Algorithm for Fundamental Frequency Estimation, *IEEE Transactions on Audio and Electroacoustics*, vol. AU-20, no. 5, pp367-377, December 1972.

(9) TUCKER, W.H. & BATES, R.H.T. A Pitch Estimation Algorithm for Speech and Music, *IEEE Transactions on Acoustics, Speech and Signal Processing*, vol. ASSP-26, no. 6, pp597-604, December 1978.

(10) DUBNOWSKI, J.J., SCHAFER, R.W. & RABINER, L.R. Real-Time Digital Hardware Pitch Detector, *IEEE Transactions on Acoustics, Speech and Signal Processing*, vol. ASSP-24, no. 1, pp2-8, February 1976.

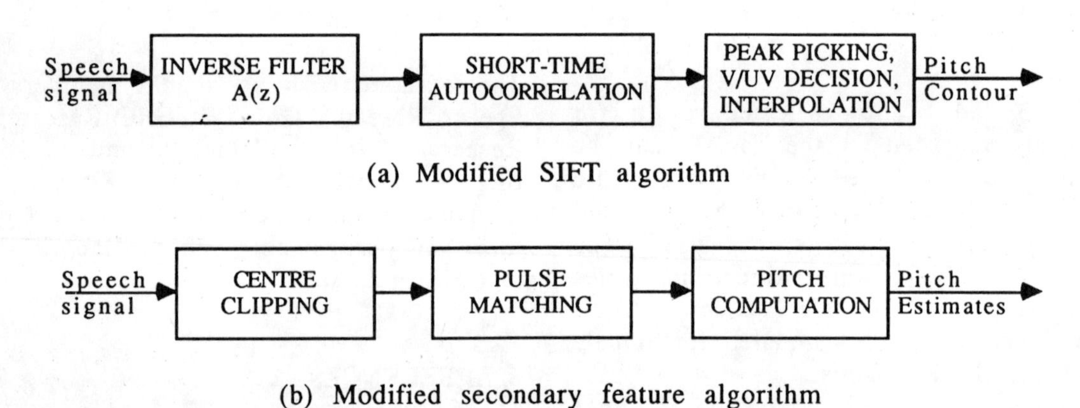

Fig 1 Block diagrams of the pitch determination algorithms

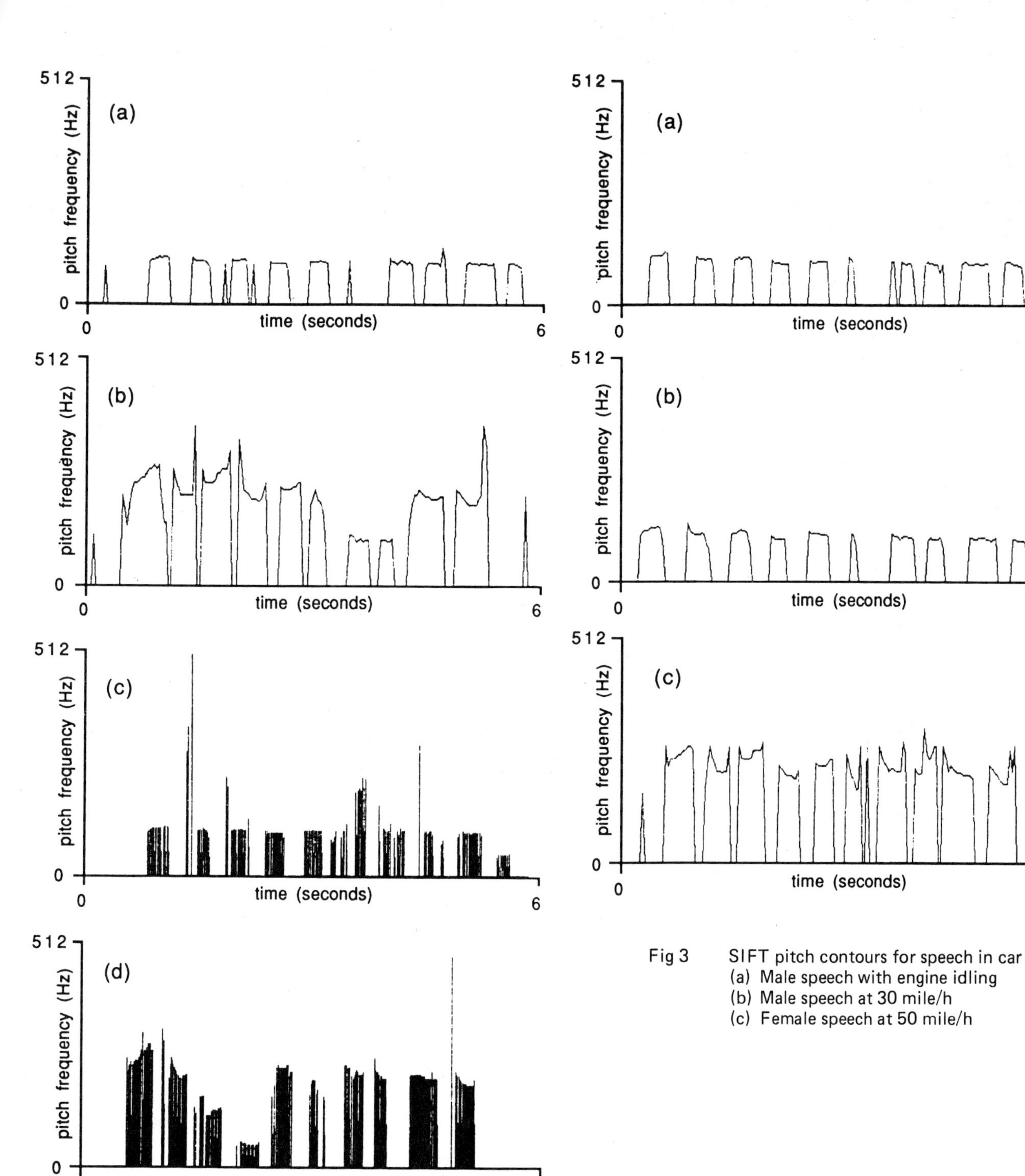

Fig 2 Pitch contours/estimates for relatively clean speech
(a) SIFT algorithm output for male speech
(b) SIFT algorithm output for female speech
(c) Secondary feature algorithm output for male speech
(d) Secondary feature algorithm output for female speech

Fig 3 SIFT pitch contours for speech in car
(a) Male speech with engine idling
(b) Male speech at 30 mile/h
(c) Female speech at 50 mile/h

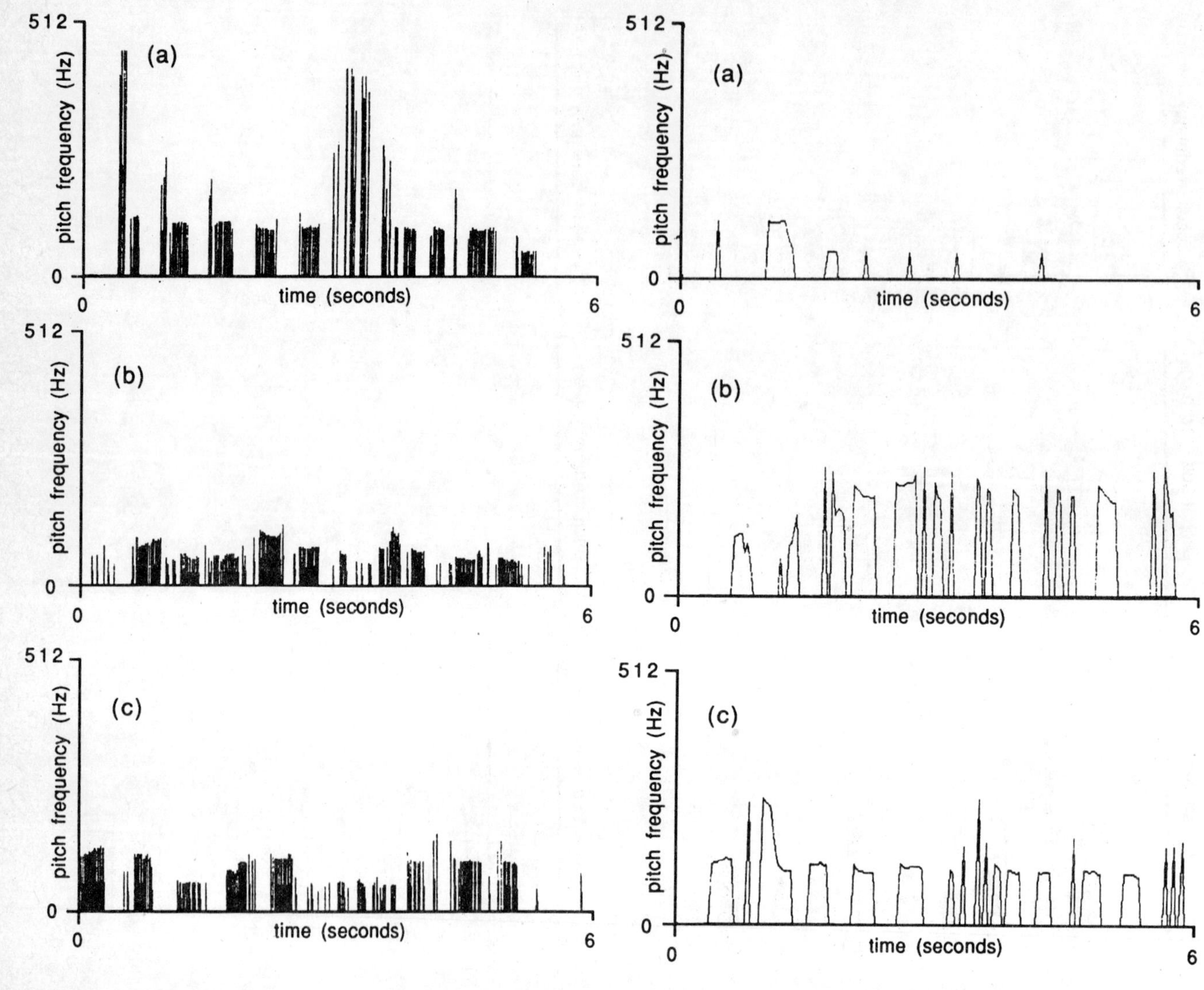

Fig 4 Secondary feature algorithm pitch estimates for speech in car
(a) Male speech with engine idling
(b) Female speech at 30 mile/h
(c) Male speech at 50 mile/h

Fig 5 SIFT pitch contours for speech in van
(a) Male speech with engine idling
(b) Female speech at 30 mile/h
(c) Male speech at 50 mile/h

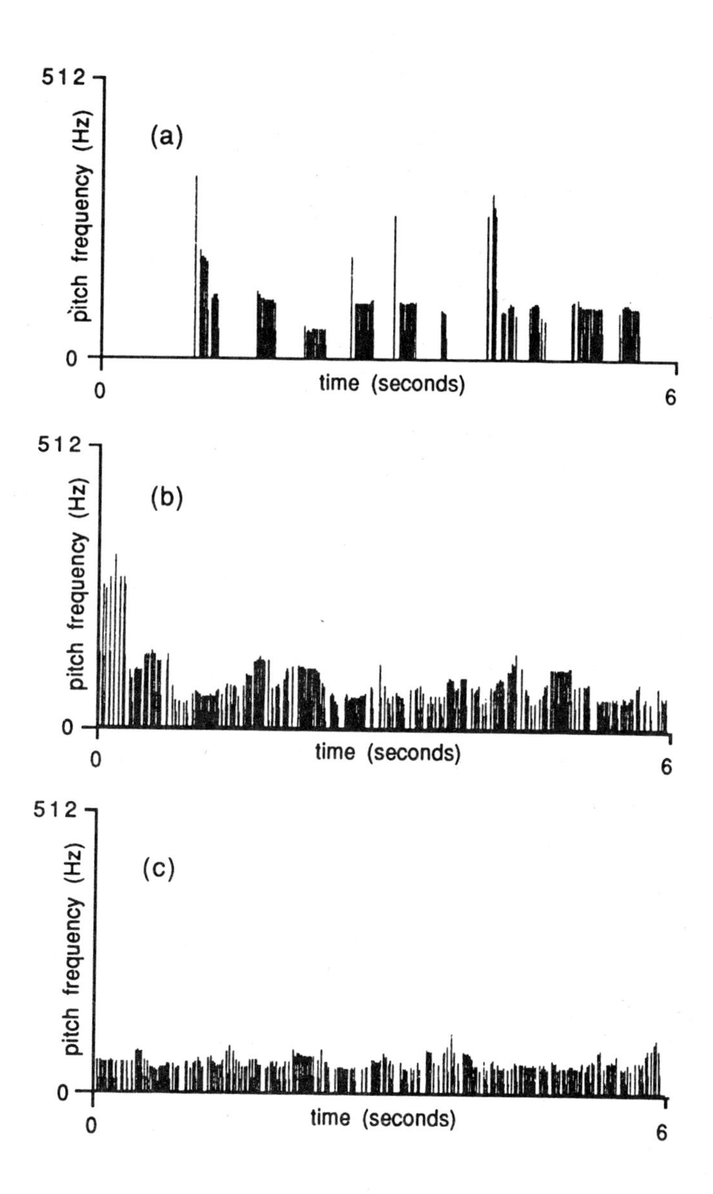

Fig 6 Secondary feature algorithm pitch estimates for speech in van
(a) Male speech with engine idling
(b) Male speech at 30 mile/h
(c) Female speech at 50 mile/h

C391/019

The implications of engine management systems in vehicle security

T J KERSHAW, CEng, MIEE and **D WRIGHT**, BSc, AMIEE
Rover UK, Lighthorne, Warwick

SYNOPSIS; The potential use of Engine Management Systems as part of a comprehensive vehicle Security System is discussed. The paper includes a description of the Engine Management Security system used in the 1988 Rover Security Concept Car. Follow up work is also described.

1. INTRODUCTION

Car related crime has become big business. Because of its seriousness we feel we should share some of the things we are doing now in order to achieve the 'thief proof' car of the future. It is probable that some of the ideas discussed here will not be adopted, but some just might be.

Car related crime has increased from 846,349 instances in 1985 to 1,048,153 instances in 1987 in the UK alone, and looks like rising much further. Car thefts, that is theft of the vehicle rather than just its contents are split into two main types :-

1.1 The Joy Rider

An opportunist thief, out to steal a car for the thrill of being on the 'wrong side of the law'. This type of thief is usually discouraged by any form of vehicle alarm, often an 'after market' system, simply because it is easier to find another car without an alarm than it is to disarm the vehicles alarm system.

1.2 The Professional Thief

This type of thief, in many instances steals vehicles to order (mainly luxury top of the range models) in order to resell, often in a disguised form. The professional thief is not so easily deterred.

To help overcome the thief, and anticipate our customers needs we are looking at new ways in which to 'design in' vehicle security.

2. ENGINE MANAGEMENT SECURITY.

The increasing application of high technology electronics in modern vehicles, especially in the Engine Management Systems area, brings with it some useful spin offs for Vehicle Security Systems. Most Engine Management Systems are now microprocessor based , and have the potential to b used as an extremely effective part of a Vehicle Security System . They have the ability to disable the engines total function - ALL fuel injectors AND ignition sparks can be made inoperative, and cannot be easily 'hot wired' or linked out in order to start the vehicle.

Because many of the new generation of Engine Management Systems have diagnostic serial transmit/receive links between the 'outside world' and the ECU's (Electronic Control Units) microprocessor, they lend themselves well to coded vehicle security functions for little or no extra cost.

An ideal place to store security related information specific to the vehicle is in EEPROM (Electrically Erasable and Programmable Read Only Memory) within an Engine Management ECU. This memory is normally used for areas such as tune, positional information, and fault data retention, but can also be used for storing additional data such as the vehicles VIN (Vehicle Identification Number) , and vehicle registration etc.

The ECU can be set up to store data in a write once only mode, for such information as the build date, and the factory set VIN code, or a multiple rewrite mode for such information as day to day driver defined passwords.

Access to stored data can be by single or multiple password codes depending on the level of security required for the data. Some data can be accessed and set by the manufacturer, some by the dealer, and some by the driver. This method of security password access can also be used to set performance and speed limit characteristics of a vehicle, since it is the Engine Management System that controls these functions.

Incorporating vehicle specific information within the Engine Management System allows easy electronic traceability. The system can be interrogated at any time, and automatically at service via the dealer vehicle diagnostic equipment. The technology also exists to perform this task remotely as we will discuss later.

Rover developed an Engine Management Security System for the Security Concept car which was exhibited at the Home Office conference on Crime Prevention in December 1988. A Rover 800 Sterling was selected as a basis for the car. Figure 1 shows the Rover Security Concept Car. The vehicles Engine Management System ECU was modified to include security functions. A current production Motorola MC68HC11 microprocessor was used to control the main power feeds within the ECU.

3. THE CONCEPT CAR ENGINE MANAGEMENT SECURITY SYSTEM.

Figure 2 shows the block diagram of the basic Engine Management System used on the concept vehicle.

The system consists of an Alarm unit, which has inputs from door and steering column Infra-red receivers, and the Engine Management ECU. The Engine Management ECU controls the Fuel, Ignition, and Starter systems of the vehicles engine.

Assuming the vehicle security system is in the 'locked' mode, the procedure to unlock the system is as follows; A serial infra-red door unlock code is transmitted from the vehicle ignition key (often called a 'PLIP') to receivers situated in the door handle assembly on the front driver's and passenger's doors. Figure 3 shows the ignition key with its built in infra-red transmitter, and the infra-red receiver built into the vehicle door lock. Code is fed from the receiver to the vehicles Alarm ECU.

When the correct door unlock code is received by the Alarm ECU the alarm system de-activates and the central door mechanism unlocks to allow driver access to the vehicle. At this stage the Engine Management System is still in a 'locked state'.

In order to drive the vehicle away the ignition key is inserted in the column lock and turned to unlock the mechanical lock as normal. The PLIP is then operated again. A receiver within the face of the column lock bezel sends the serial code again to the Alarm ECU. This causes the Alarm ECU to transmit the Engine Management Security Code to the Engine Management ECU.

For display purposes the Security Concept Car uses an ASCII version of *ROVER for the Engine Management Security Code. This code is fed into the Engine Management ECU's microprocessor via its serial communications interface. The software within the Engine Management ECU is configured to look for the first character, in this case "*" and until it identifies this character it continues to read in characters from the link . When a "*" is read the system looks for the next character in the code sequence "R". If the character is correct the system continues to read in more characters until the password is read in. If however any of the characters in the incoming sequence are not correct the code reading is reset and the Engine Management System again looks for the first character in the code sequence. This prevents the system being unlocked by running through all codes combinations with an external code generator.

When the correct code is accepted by the Engine Management System the normal engine functions and the starter motor are enabled. The engine can then be cranked and started in the normal way by turning the ignition key in the column lock. The Engine Management Security System is reset to the 'locked' mode by switching off the vehicle's ignition circuit.

Each time the vehicle is to be restarted the column lock/PLIP security system has to be used. To lock the car the doors are closed and the PLIP is operated outside the car.

This was the basic Engine Management Security System decided upon in order to meet the vehicles build date.

Further work to improve the Engine Management Security System has however continued in the form of a driver defined password system. In this particular application a vehicle Trip Computer is used as a key pad and display in order to allow the input into the system of a driver defined security password.

Figure 4 shows a block diagram of the expanded Engine Management Security System

Figure 5 shows the trip computer keyboard

The driver defined security coding system is linked in by a serial bus.

With this particular system the driver enters a four digit PIN (Personal Identification Number) on the trip computer key pad. The trip computer sends this code to the Engine Management ECU via the serial communications link. In order to 'unlock' the Engine Management Security System the codes from both the trip computer and the vehicle alarm system must be correct.

There are 10,000 possible combinations of a four digit code and it is unlikely that a prospective thief could enter the correct PIN, but as a further security measure the driver is allowed three attempts to correctly enter the PIN. If the PIN is entered incorrectly three time in a row the Engine Management ECU 'locks out' and the message 'REFER TO DEALER' is displayed. The system can be reset by powering down the Trip Computer and the Engine Management System with the column ignition switch. This allows the driver to return to a vehicle that has been tampered with, and still be able to input the correct PIN and drive the vehicle away.

In the case of a forgotten PIN the Engine Management System can only be unlocked by authorised dealers that have the correct diagnostic equipment and system passwords.

To maintain a high level of security it is likely that the number of dealers with security access would be limited. Information stored in the ECU can be accessed via the Engine Management Systems Diagnostic Connector.

Once the driver has entered the correct security code in to the Trip Computer the code can be changed, or the driver defined option disabled. The disabling option is to allow for short stop/start situations (buying a newspaper etc.) where entering the driver code would prove inconvenient.

The PIN changing facility makes the system more secure and allows the vehicle to be delivered with a default code; thus it is not necessary to dispatch a separate notification of the PIN to the driver as in the current British banking system cash till cards.

4. SYSTEM PRACTICALITY AND EXPANSION POSSIBILITIES.

In order for the system to work correctly the serial security code in the alarm ECU and the required code stored in the Engine Management System must be the same. This means that the alarm and Engine Management ECU must be a matched pair, or that one of the ECU's must have the capability to learn the other's code. This can be achieved by a track side computer which reads the alarm code and stores it in a 'write once' only location of the Engine Management Systems EEPROM memory. The Engine Management System can also be set up to automatically read in the serial code from the alarm ECU when it is initially powered up on the vehicle (at the end of the track). This removes the need for a trackside computer, but has a potential problem of ECU's being powered up with incorrect codes and is not favoured.

To complement the driver defined coding system provided by the Trip Computer, or as a replacement for the ignition key, a further level of security could be provided by a 'Smart Card'. This is a personal memory card, similar to a credit card with a built in microprocessor and memory. Such a system is however expensive in hardware (card reader etc) and will not be viable until either costs are reduced, or other vehicle personal memory features are required by the customer.

5. REMOTE VEHICLE SECURITY.

A novel method of vehicle security that can be employed, although it smacks a bit of Big Brother, is to remotely access the vehicles Security System. This can be achieved by employing a communications link to the vehicle. Such a link could be via a vehicles mobile telephone system. It is possible to obtain stored identity and vehicle service condition information of a specific vehicle. If considered acceptable, it could be made possible to remotely change selected data within the car. In this way a car could be disabled or have its drivability characteristics altered remotely.

It is also possible to determine the approximate location of the vehicle.

Mobile telephones operate in a cellular system and a vehicle could be located within a certain cell area. Problems with this type of system do however exist. Large areas of Britain (for example Scotland and Wales) are not yet covered by the cellular telephone system. Cell sizes also vary. Central London for example has many small cells covering the area, making the chances of pin pointing the vehicles location much more likely. Vehicles located in larger cells would be more difficult to find. Information as to exact position within the cell is not available on the current mobile telephone system. With the announcement of British Telecom's new GSM (Groupe Special Mobile) cellular mobile telephone system, positional location of a greater accuracy is now a possibility. Designed for digital data transmission rather than the present analogue transmission system it is configured to give radial distance information from the cell transmitter to the receiver. Figure 6 illustrates this point.

A vehicle could transmit its own identification code periodically, or on command, to a central data base which monitors and updates vehicle location. Although a doubtful application for the general public, a remotely accessible vehicle security system could prove attractive to fleet operators and insurance companies.

6. A DISTRIBUTED VEHICLE SECURITY SYSTEM

In order to meet the increasingly stringent Emissions regulations and performance/quality requirements, Vehicle Management Systems need to monitor more and more parameters. The number of ECU's per vehicle is also increasing. We now have Electronic control of Fuelling, Ignition, Gearbox, Cruise Control, Cooling, and Braking Systems. Many vehicle ECUs require the same information, and repeat similar calculations. A possible solution to the duplication of connectors, sensors and resource on the vehicle is to network the Vehicle Management Systems together.

An ideal network system to achieve this appears to be the class 'C' CAN (Controller Area Network) system.

A vehicle equipped with a versatile serial bus communication system between its ECUs has a wonderful security opportunity. Every ECU on the vehicle network can become a security ECU distributing the security system across the vehicle.

Figure 7 shows a block diagram of a CAN based vehicle security network.

After power up no ECU would function correctly until they all recognise each other. To defeat such a system every ECU on the network would have to be replaced at the same time. Even if this was done the stolen vehicle would almost certainly be detected at the next service if it were connected to a central database via the dealer's diagnostics equipment. The vehicle identification passwords would be incorrect or duplicated.

A full description of the CAN system and its protocol are beyond the scope of this paper and for further information the reader should consult the references section.

7. VEHICLE SECURITY - WHO PAYS ?

As with most things in life vehicle security costs money, and the more secure systems become the more inconvenient they become to use.

Whether the customers of the future will demand the features outlined in this paper is not yet known, but someone will have to pay for them. One possible solution to this dilemma is to use insurance premiums as an incentive, after all it is the customer that pays for car theft today via insurance premiums and the loss of no claims bonuses. If premiums were weighted to penalise vehicles that can be easily stolen then we may have the justification we need to manufacture the 'Thief Proof' car.

Who knows one day we may see the sales slogan, "If you can steal this car from the showrooms you can keep it".

8. ACKNOWLEDGMENTS.

The authors would like to thank all their colleagues within Rover who helped with the material for this paper, and the Concept Car. Special thanks go to Fred Coultas, John Fowler, Bill Shirley, and John Tidbury who were responsible for most of the work on the car, Dave DuMont who was responsible for the monumental task of the vehicle wiring, and to Chris Nutt of Nieman Klaxon who developed the vehicle's Infra-red locking and Alarm ECUs. Thank you.

9. REFERENCES

1. INTEL Automotive Handbook ISBN 1-55512-030-X for CAN.

2. CAR CRIME Report of the Working Group for the Home Office Standing Conference on Crime Prevention 1988.

3. A GUIDE TO TOTAL ACCESS COMMUNICATION SYSTEMS-DTI 1985.

Fig 1 Rover security concept car

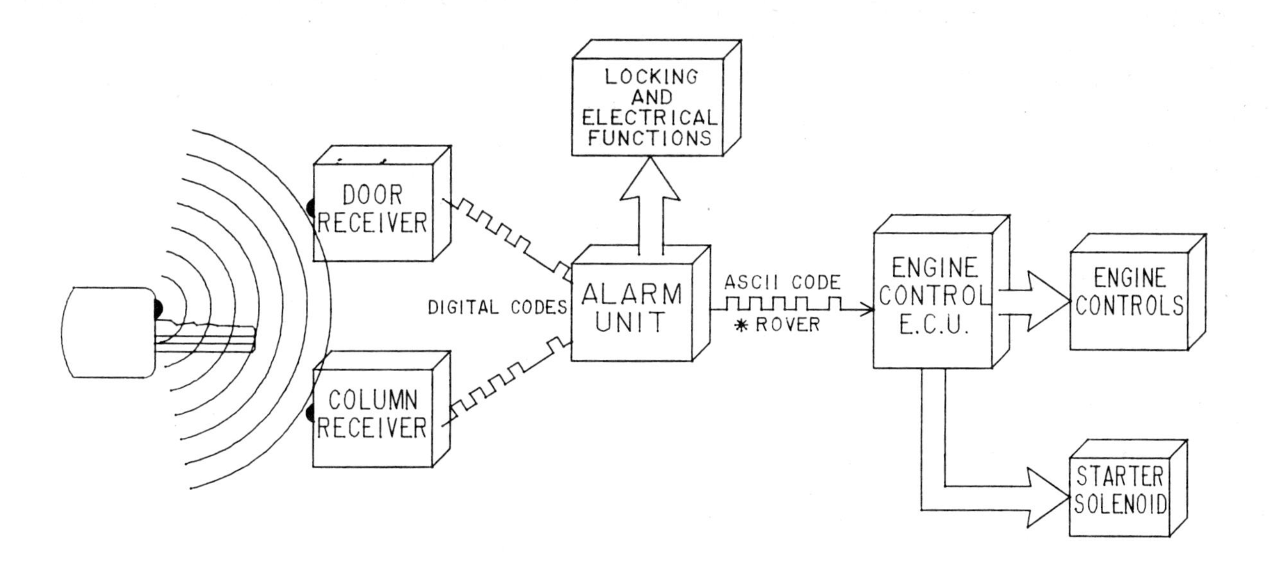

Fig 2 Block diagram of the basic engine management system used on the Rover security concept car

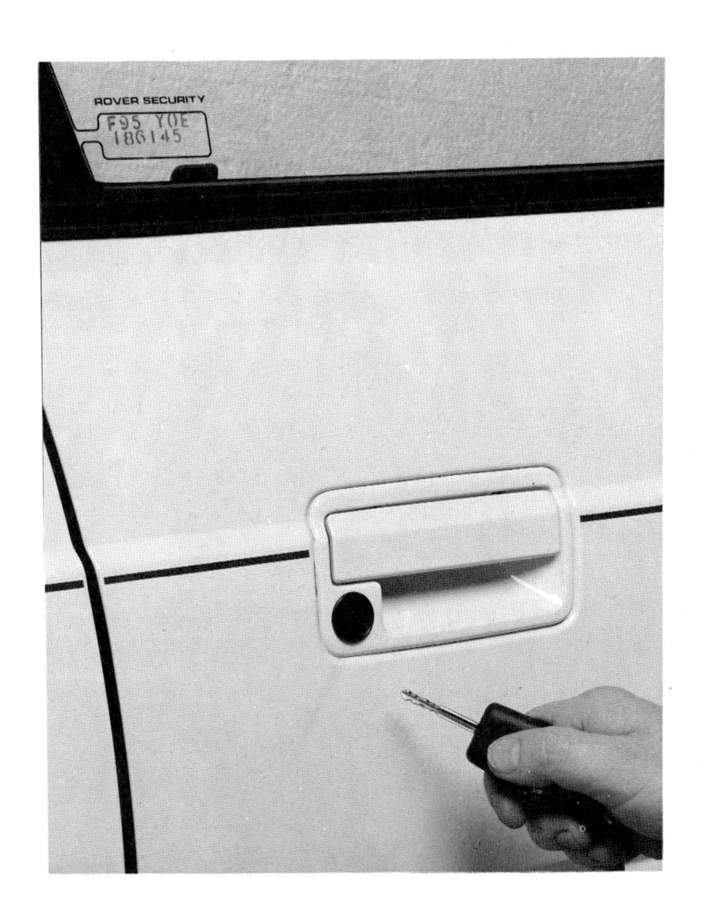

Fig 3 Infra-red 'PLIP' ignition key

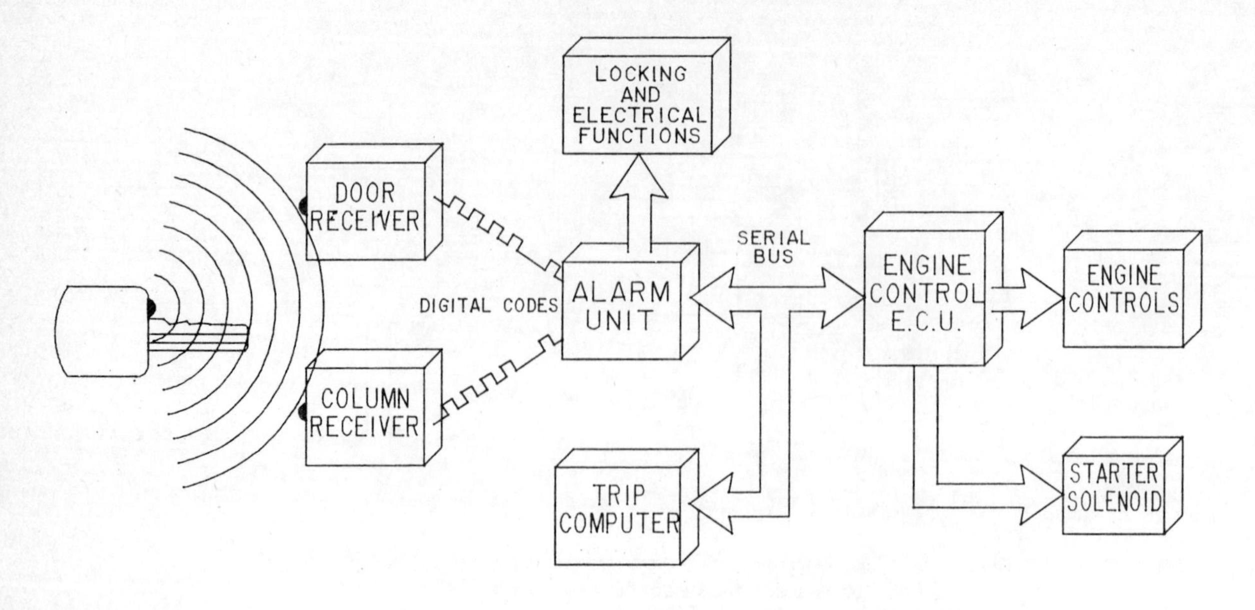

Fig 4 Block diagram of the expanded engine management security system incorporating a trip computer linked via a serial communication bus

Fig 5 Trip computer keyboard

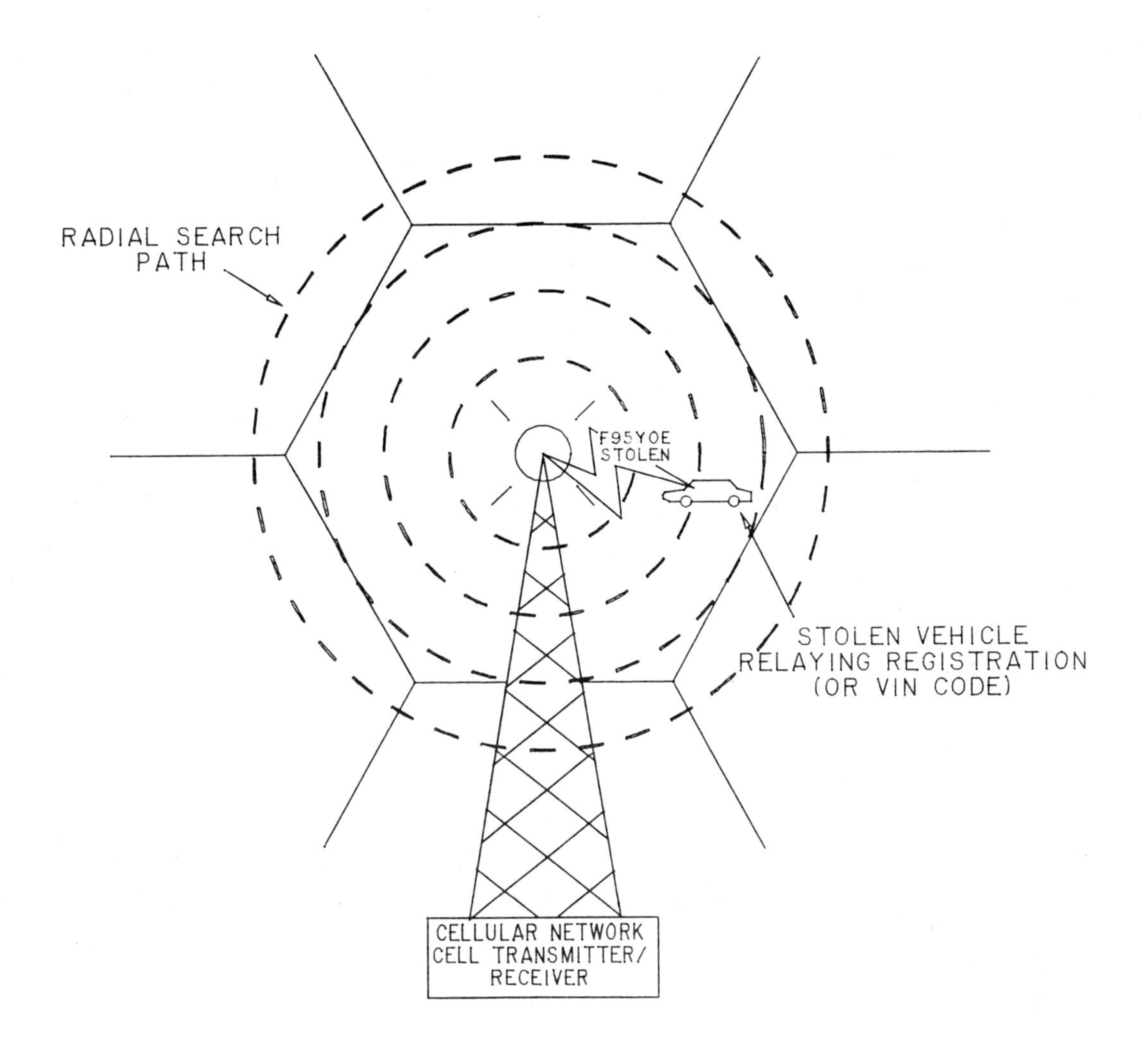

Fig 6 Illustration of the use of the GSM telephone system in vehicle location

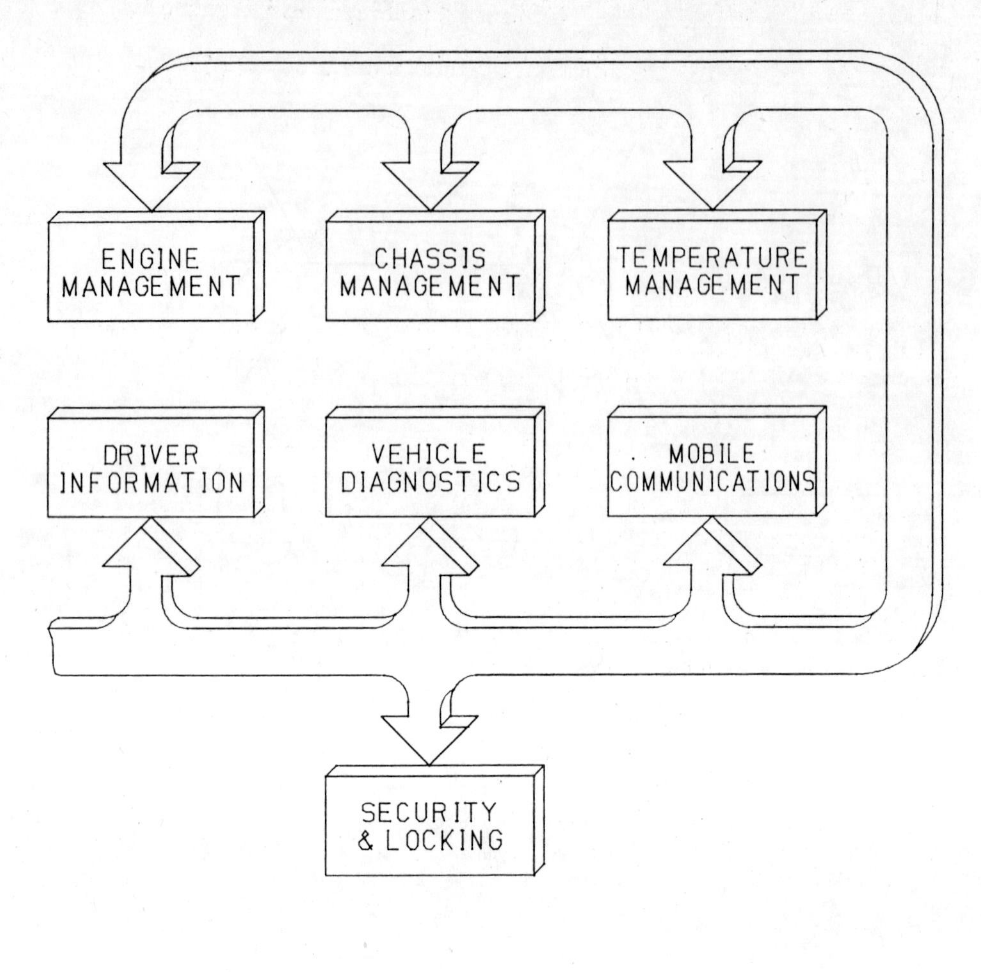

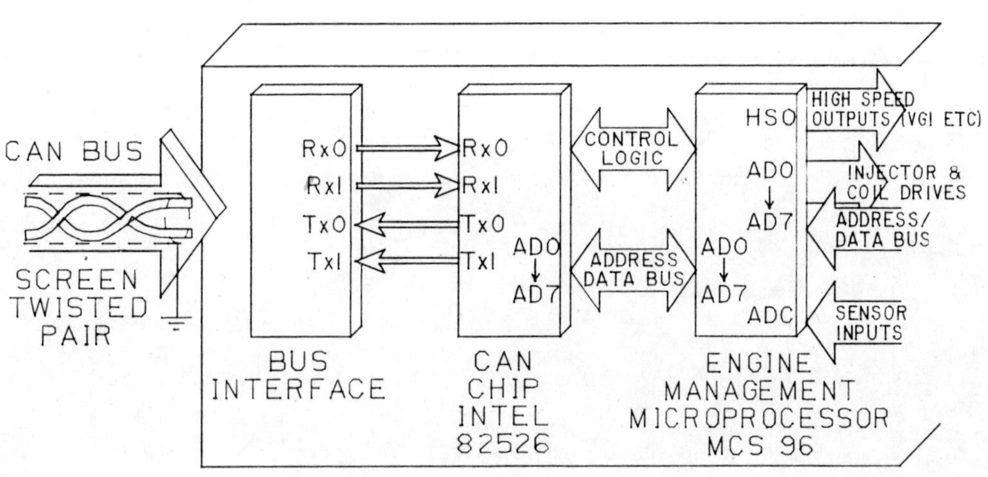

Fig 7 Diagram of a controller area network vehicle security system

C391/051

A totally integrated electronic control system for the engine and driveline

Y OHYAMA, Dr, JSME, SAEJ, SAE
Hitachi Limited, Hitachi-City, Japan

SYNOPSIS A system is described, which implements efficient co-ordination of functions of the engine and driveline control subsystems. The principle of the system is based on control strategies, which involve engine combustion and driveline dynamic characteristics.

1. INTRODUCTION

Progress in electronics is providing countless technological improvements in vehicular performance and functions. Production applications of new technology such as artificial intelligence and communications require a control strategy for efficient co-ordination of functions of the engine, driveline, brake, steering, and suspension control subsystems.(1)-(4)

This paper presents an overview of a totally integrated electronic control system for the engine and driveline which gives efficient coordinated actions between these subsystems. The principles of the system are based on control strategies, which involve engine combustion and driveline dynamic characteristics.

The engine control subsystem has a torque servo with adaptive control, which uses the signals related directly to combustion. The driveline control subsystem has a controller which is based on two-degrees-of-freedom control strategies.

Efficient co-ordination of the functions of the engine and driveline control subsystems in accordance with vehicle performance can be achieved and a marked improvement in driveability and fuel economy can be attained.

2. SYSTEM OUTLINE

To realize increasing demands for driveability and fuel economy, various electronic subsystems are applied to give a totally integrated vehicle electronic control system.(4) These include engine control subsystem and driveline control subsystem, as shown in Figure 1.

The driveline is composed of the transmission which has a highly adaptive shifting capability. G_{c1} is the control unit of the engine control subsystem and G_{c2} is the unit of the driveline control subsystem.

Two observers are included in the subsystems. One of them, A is for the vehicle body. Observer A estimates the output torque of the driveline, using signals from the G sensor(4) which measures acceleration. In the conditions of zero acceleration which could correspond to any steady vehicle speed and

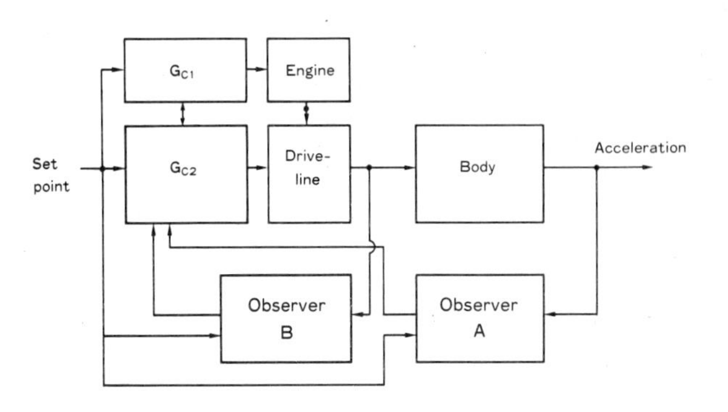

Fig 1 Engine and driveline control subsystems

driveline torque, the output torque is estimated using signals from the engine. Observer B estimates the output torque of the driveline using signals from the driveline and the engine. The set point which refers to the torque in accordance with the driver's demand, is sent to G_{c1} and G_{c2} from the driver information control unit.(4) The engine torque and transmission shift are adjusted by the output of the units G_{c1} and G_{c2} so as to accommodate the real vehicle acceleration to the driver's demand.

The estimated values of observers A and B are sent to the units which adjust the air, fuel quantity, spark advance of the engine, and the transmission shift accordingly. Since engine torque and transmission torque are controlled in real time, there is adequate vehicle acceleration to meet the driver's demands.

3. FUNCTION

The engine is controlled on a cylinder-by-cylinder basis to improve fuel economy and minimize vibration, over the full range of torque and speed. The engine control subsystem has a torque servo with adaptive control during cold start and warming, which uses the signals related directly to combustion. An adequate combustion velocity under all operating conditions can be maintained to improve fuel economy and driveability.

For conventional vehicles, the throttle valve of the engine is controlled directly by the acceleration pedal. For the system shown in

Figure 1, the throttle is controlled by the signal from the automatic drive controller which has functions based on fuzzy control and stability augmentation strategies.[4]

The engine is closely coupled with the driveline containing the transmission which has a highly adaptive shifting capability. The driveline and engine are electronically controlled as a unit in response to the driver's demands, using the network shown in Figure 1.

4. ENGINE CONTROL SUBSYSTEM

Figure 2 shows a block diagram of the engine control subsystem, which has a torque servo with adaptive control during cold start and warming. The set point refers to the torque determined by the driver information control unit in accordance with the driver's demand.[4] The torque is compared with the output of H_1 (the real torque). Fuel, air quantity and spark advance for the engine are controlled by the unit G_{c1}, to minimize the error.

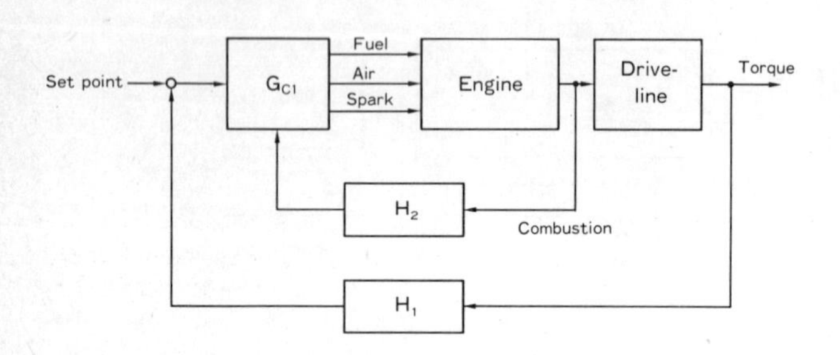

Fig 2 Torque servo with adaptive control during cold start and warming

The unit G_{c1} also receives the outputs of H_2, which are direct signals related to combustion (for example, the signal from the in-cylinder pressure sensor, or the combustion flame sensor[5]).

Figure 3 shows an example of experimental results of the mass transfer of fuel film on intake or cylinder model vs. temperature at a variety of heat transfer rates. In Figure 3, air velocity is 10 m/s, film velocity v_f is 0.1 and 1 m/s, the wall temperature is 100°C. The mass transfer decreases when the heat transfer is low and the temperature at which the mass and heat transfer balance is low. The fuel evaporation rate, which depends on the mass transfer is limited.

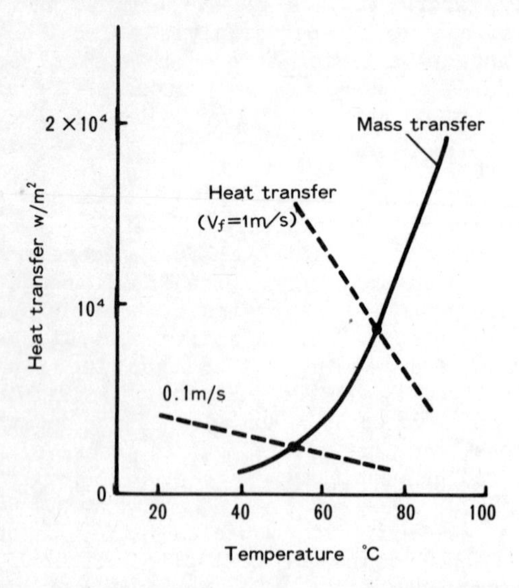

Fig 3 Heat and mass transfer of fuel film

In the case that a sufficient evaporation time cannot be obtained, part of the fuel clings to the cylinder wall, and leads to diffused combustion. If diffused combustion occurs, it will cause an increase in the intensity of radiation near the 700 nm wavelength and distortion in the signal of the combustion flame sensor[5], as shown in Figure 4. Experimental results using a 1.8ℓ 4 cylinders engine show that hydrocarbon emissions with diffused combustion are over 7,000 ppm, higher than that of 3,500 ppm without diffused combustion.

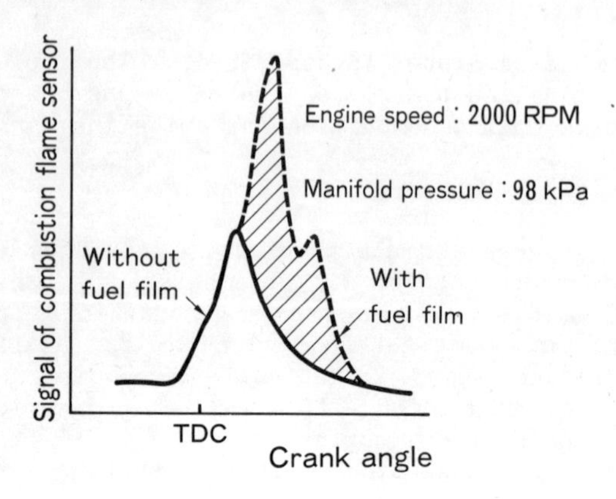

Fig 4 Signal of combustion flame sensor

The unit G_{c1} controls the fuel quantity to maintain the desired pre-mixed combustion level during cold start and warming of the engine, using the signal. For example, experimental results show the diffused combustion disappears when the intake manifold pressure of the engine decreases from 98 kPa to 63 kPa and the fuel quantity decreases 30 percent under the condition of engine speed of 2,000 rpm and temperature of 0°C. In the same engine conditions, engine torque with diffused combustion decreases about 3 percent, compared to that without diffused combustion.

The maximum fuel quantity is limited, so that there is no adherence of fuel to cylinder walls. Good mixture formation without a liquid film in the cylinders can be attained. Consequently, a marked improvement in the fuel economy and exhaust emission reduction during cold start and warming is realized. The unit G_{c1} controls not only the fuel quantity but also the fuel injection timing, not to misfire.

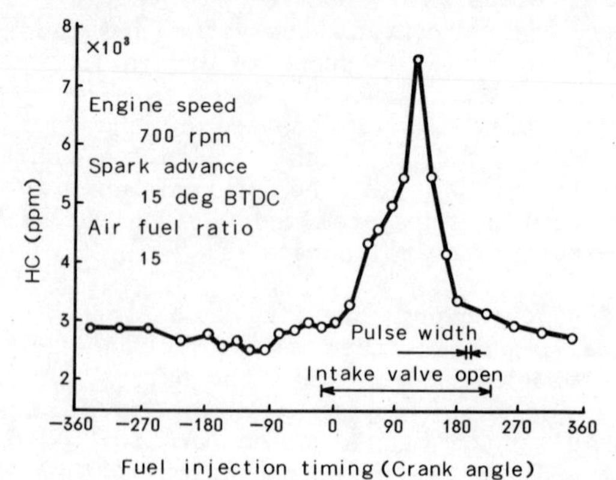

Fig 4b HC versus fuel injection timing

The signal of the combustion flame sensor is zero if perfect misfire occurs. The sensor can detect misfire more clearly than pressure sensors. Experiments show that HC emissions in idling conditions depend on the fuel injection timing as shown in Figure 4(b). If misfire occurs, it will cause an increase in HC emissions and the distortion in the signal of the sensor. The fuel injection timing is controlled to maintain HC minimum, using the signal. The sensor is useful not only during cold start and warm up, but also after warming.

The direct combustion control results in significant improvements over current systems, which measure secondary parameters such as the air-fuel ratio of the exhaust gas.

The maximum fuel quantity depends on the mass transfer, as shown in Figure 3. Then, the maximum torque is decreased about 20 percent during cold start and warming. But the torque of drivelines is compensated by adjusting the transmission shift. Simulation results based on the model shown in Figure 2 show that the desired pre-mixed combustion level can be maintained without subsequent loss of driveability and with 3 percent improvement in the fuel economy during the test of mean speed of 40 km/h.

5. DRIVELINE CONTROL SUBSYSTEM

Response improvement is important for requirements in driveability. This can be attained by using the two-degrees-of-freedom control strategy, as shown in Figure 5.

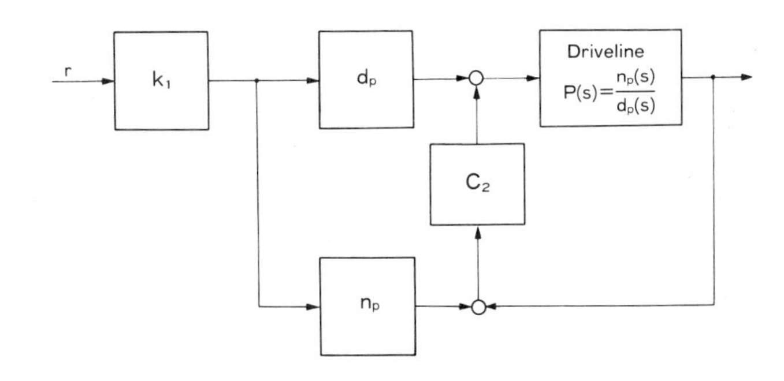

Fig 5 Two-degrees-of-freedom control strategy

Set point response characteristics and feedback characteristics can be selected independently, by using the design procedure as follows.

The set point refers to the torque determined by the driver information control unit[4] in accordance with the driver's demand. The procedure is carried out as follows, when the transfer function of the driveline is $P(s) = n_p(s)/d_p(s)$. Figure 6 shows an example of the driveline dynamic model.

(1) k_1 is determined so as to attain desirable response characteristics against the change of the set point of the driveline torque.

(2) If the desirable model transfer function is G_M, k_1 is set as G_M/n_p.

(3) The response characteristics G_{yr} are described by

$$G_{yr} = n_p \cdot k_1 = G_M \quad (1)$$

This shows that model-following control can be executed.

(4) The gain k_2 of C_2 is designed to be so high that sensitivities against noise and parameter changes of the driveline become low.

(5) For $P(s) = 1/(s+1)$, if k_2 is selected as the following equation,

$$k_2 = \frac{(s+1)}{\rho s + 1} \quad (2)$$

the sensitivity function Fs is described by

$$Fs = \frac{\rho s}{\rho s + 1} \quad (3)$$

(6) If ρ is small, then Fs is small and the sensitivity can be kept low without subsequent loss of the response characteristics.

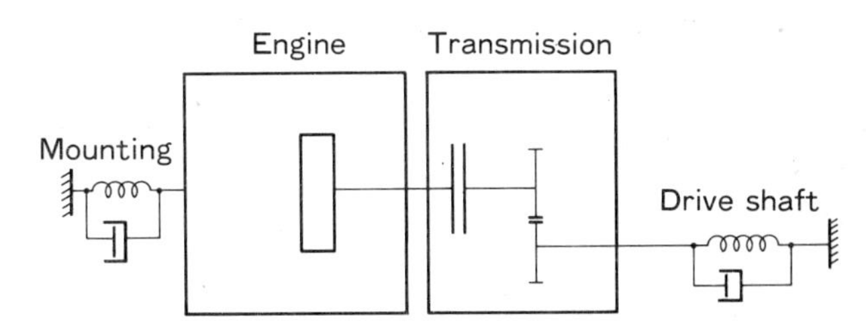

Fig 6 Driveline dynamic model

The vehicle under acceleration and deceleration conditions with the two-degrees-of-freedom control strategy responds more quickly and smoothly than it would without them.

Efficient co-ordination of the functions of the engine and driveline control subsystems in accordance with vehicle performance can be achieved. Figure 7 shows experimental results of vehicle performance characteristics, which were conducted using a vehicle with inertia weight of 1,440 kg, manual transmission, an engine of 2.0ℓ, 6 cylinder. A curve ⑤ shows acceleration when the throttle valve is closed with 1st gear. A curve ③ shows acceleration when the throttle valve is reopened in the same conditions. A curve ③ shows acceleration in the same reopening condition with the control. A smooth curve ③a for acceleration can be attained with the fuel, air quantity and spark advance control.

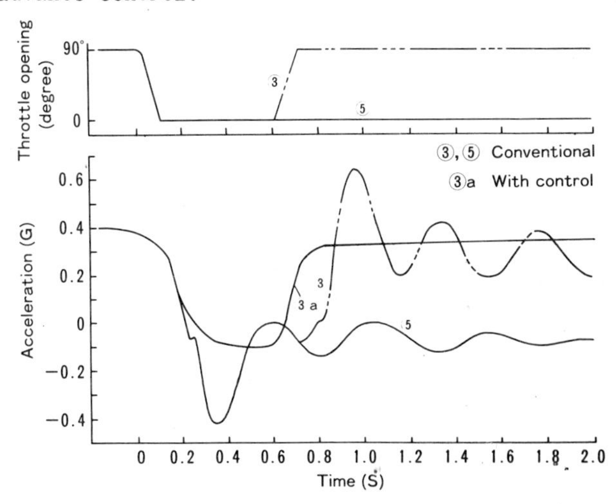

Fig 7 Vehicle dynamic characteristics

Effects of elasticity of the drive shaft as shown in Figure 6 on unstable characteristics can be reduced.

6. SUMMARY

This paper presented an overview of a totally integrated electronic control system for the engine and driveline, based on efficient co-ordination of engine and driveline subsystems.

(1) Functions of the system to give better driveability and fuel economy were clarified.

(2) The principles of the system were based on direct combustion control and two-degrees-of-freedom control strategies.

(3) The desired pre-mixed combustion level could be maintained by using direct combustion control, without subsequent loss of driveability.

ACKNOWLEDGEMENTS

The author thanks Y. Nishimura, T. Minowa and N. Kurihara for their contributions and support to the design and development work.

REFERENCES

(1) J.G. Rivard, "Automotive Electronics in the Year 2000", SAE Paper 861027, 1986

(2) S.A. Haider, J.A. Griffin, "Powertrain Torque Management", SAE Paper 8700381, 1987

(3) A. Numazawa, "Overview and Future Plan of Automotive Electronic System", SAE Paper 851060, 1986

(4) Y. Ohyama, "A Totally Integrated Vehicle Electronic Control System", SAE Paper 881772, 1988

(5) T. Nogi, "Mixture Formation of Fuel Injection Systems in Gasoline Engines", SAE Paper 880558, 1988

C391/044

Thermodynamic simulation of a turbo-charged spark ignition engine for electronic control development

A M FOSS, R P G HEATH and **P HEYWORTH**
Cambridge Control Limited, Cambridge
J A COOK and **J McLEAN**
Ford Motor Company, Basildon, Essex

Synopsis

A nonlinear dynamic model of an intercooled, turbocharged spark ignited engine is developed for the purpose of evaluating microprocessor-based wastegate and ignition control laws. The engine system is thermodynamically modelled and matched to engine data. The microprocessor representation contains the actual algorithms used to implement the control law.

Notation

A_c	Effective compressor area
cp	Specific heat (constant pressure)
c_w	Wastegate valve time-constant reciprocal
fuel	Fuel flow
gratio	Gear ratio
$I\_$	Inertia
l_c	Effective compressor length
load	Load
$N\_$	Speed
$P\_$	Pressure
$PR\_$	Pressure ratio
R	Gas constant
$R\_$	Resistance coefficient
spark	Spark angle
swepv	Engine displacement
$T\_$	Temperature
$TR\_$	Temperature ratio
$U\_$	Power
voleff	Volumetric efficiency
$V\_$	Volume
$W\_$	Mass flow
X_h	Throttle position
X_w	Wastegate position
X_wd	Wastegate position demand
γ	Ratio of specific heats (cp/cv)
ρ	Density of air
$\_o$	ambient
$\_c$	compressor
$\_i$	intercooler
$\_h$	throttle
$\_m$	inlet manifold
$\_e$	engine
$\_x$	exhaust manifold
$\_t$	turbine
$\_w$	wastegate
$\_n$	nozzle
$\_s$	shaft

1 Introduction

Turbocharging is a common method of increasing automotive engine power and torque output by harnessing exhaust gas energy to increase the pressure and density of the air charge delivered to the cylinders. The mechanism by which the charge pressure is controlled is regulation of the amount of exhaust flow introduced to the turbine. This is accomplished via manipulation of a wastegate by-pass valve in the exhaust stream.

The Ford Motor Company are developing a high performance, intercooled, turbocharged engine for the speciality vehicle market. Precise electronic control of turbocharger wastegate function and engine ignition timing over a broad operating regime is required to maximise performance and minimise detonation and associated power and durability degradation. Specifically, the control problem is to regulate engine inlet manifold pressure to a desired value as a function of engine speed and throttle position, while avoiding engine knock. Of course the unstable flow characteristics associated with compressor surge must be avoided while providing good transient throttle response performance without excessive turbocharger lag. These objectives must be satisfied as components age over the life of the vehicle, at various altitudes and temperatures, and with fuels of varying quality and octane rating.

In order to evaluate the control strategy over all operating conditions and life-cycle of the engine, a nonlinear, dynamic simulation model was developed. This was a physical equations based model consisting of the engine, turbocharger and microprocessor-based controller. The simulation captures the relevant dynamics of the powerplant, provides reasonably accurate prediction of steady-state behaviour and incorporates the control law in a form that is representative of the actual software implementation.

2 Modelling Approach

The intent of the system model is to evaluate the control algorithm. The system model consists of two distinct subsystems: the thermodynamic description of the turbocharged engine and the algorithmic description of the microprocessor-based controller. This structural organisation is illustrated in Figure 1. The digital controller representation contains all the necessary inputs from and outputs to the engine/turbocharger model and faithfully implements the microprocessor control program in order to exercise the exact software under consideration.

The turbocharged engine model is developed as a lumped parameter, continuous system with the assumptions of isentropic flow of a perfect gas with constant specific heats, and stationary air-to-fuel ratio (A/F). The system is represented as a collection of components and control volumes, the characteristics of which may be developed by the application of the conservation relationships for mass, energy and momentum along with the employment of regression equations to describe complex processes. Similar models have been developed for control analysis of naturally aspirated reciprocating engines [1,2,3,4]. The application of this modelling technique to rotating machinery is common in the aerospace industry [5]. The interrelationship of the various system components is illustrated in the block diagram of Figure 2. The key static elements of the model are the component mass flow characteristics and the engine power generation functions. The dynamic elements include the turbocharger, intercooler, engine inlet and exhaust manifolds, wastegate actuator and drivetrain load. The forms of the functional relationships associated with the static elements were developed by evaluation of engine-dynamometer steady-state test results. The dynamic parameters are functions of engine geometry and gas properties and can be validated by dynamic engine system tests.

The intent of the turbocharger simulation model is to test the actual software to be implemented in the vehicle microprocessor in a controlled environment. With this in mind, the controller representation was constructed as a FORTRAN subroutine called from the main program at simulation time intervals characteristic of the looptime of the microprocessor. The high level language code exactly reproduces the algorithm contained in the vehicle microprocessor allowing evaluation not only of the turbocharger control law but of the implementation of that law and the detection of unforeseen interactions or side effects due to control regime entrance or exit strategies. Table 1 lists the inputs and outputs from the simulated microprocessor.

Subsequent sections will address the required system tests, the matching of test data to the functional representation of the system components, the integration of the individual components into the overall system model and an illustration of the controller/system interaction.

3 Model Description and Static Matching

In order to statically match the thermodynamic model of the turbocharged engine, data were obtained from an instrumented engine operating on a dynanometer at 160 different points under conditions of constant speed in the range from 2000 to 6000 r/min, constant ignition timing in the range of 10 to 35 degrees before top dead center (BTDC) and for various throttle settings and wastegate positions spanning the full range of capability. The engine was operated at a constant representative A/F of 12.5. Measurements included temperatures and pressures for each volume element, engine and turbocharger speeds, dynamometer load, and air and fuel mass flow rates. A complete list of the recorded parameters is included in Table 2. The following paragraphs detail matching the engine model to the static data. The model was matched element by element with whole engine steady-state conditions used as a cross check.

Table 1: Microprocessor model inputs and outputs

Inputs	Outputs
Ambient Temperature	Spark Advance
Compressor Temperature	Fuel Demand
Engine Speed	Wastegate Boost
Inlet Manifold Pressure	
Throttle Setting	
Knock Information	

Table 2: Recorded engine parameters during testing

Engine Settings	Pressures	Temperatures	Other Measured Parameters
Throttle	Ambient	Ambient	Dyno Load
Engine Speed	Intercooler	Compressor	Air/Fuel Ratio
Wastegate Boost	Inlet Man.	Inlet Man.	Turbo Shaft
Spark Advance	Exhaust Man.	Exhaust Man.	Fuel Flow Rate
	Nozzle	Nozzle	
	Wastegate	ECT	

3.1 Compressor

The rate of change of mass flow across the compressor is developed from the momentum equation for the effective control volume between the compressor inlet and the intercooler with consideration for the compressor pressure ratio which is related to the turbocharger shaft corrected speed and flow.

$$\frac{dW_c}{dt} = \frac{A_c}{l_c} \times (PR_c \cdot P_o - P_i) \quad (1)$$

The pressure rise through the compressor is accompanied by a temperature increase related to the pressure ratio and efficiency of the compressor. These characteristics are supplied by the turbocharger manufacturer in the form of a compressor performance map relating pressure ratio and efficiency to mass flow rate at various shaft speeds. Excellent correlation was noted between the characteristics supplied by the manufacturer and the measured compressor data as illustrated in Figure 3.

3.2 Intercooler

The intercooler is an air-to-air heat exchanger which enhances engine power generation and aids in detonation control by reducing the temperature of the inducted air charge. The increased mass of air in the cylinders allows more fuel to be supplied for a constant A/F and consequently greater power generation. This process is represented by the calculation of a static temperature ratio across the intercooler and, in concert with the compressor discharge temperature, a resultant manifold temperature. The temperature ratio across the intercooler is a function of the heat exchanger effectiveness or thermal ratio which is defined as the ratio of actual heat transfer to maximum possible heat transfer. The thermal ratio is a function of the cooling medium available, heat exchanger geometry and cooling medium flow rate [6]. For a typical automotive application of an air-to-air heat exchanger mounted in the engine cooling air stream, high effectiveness values may be achieved [7]. For the dynamic model, a constant thermal ratio of 0.7 was assumed based on measured intercooler temperature ratios distributed over the range 0.5 to 0.9. Hence, the temperature of the inlet manifold was calculated from the temperature at the outlet of the compressor and ambient conditions:

$$T_m = T_c - 0.7 \times (T_c + T_o) \tag{2}$$

The lumped temperature of the heat exchanger is assumed to be the average of the temperatures before and after the component. The pressure rate dynamics of the intercooler are described in terms of the intercooler volume, gas constant, ratio of specific heats and ingress and egress mass flow rates in association with the lumped intercooler temperature.

$$\frac{dP_i}{dt} = (\gamma \times R \times \frac{T_i}{V_i}) \times (W_c - W_h) \tag{3}$$

3.3 Throttle

It was hypothesized, and confirmed from measured data, that the corrected flow through the throttle was a function of throttle count and pressure ratio. Thus, a tabular function of this form was fitted to the measured data. Best resolution was obtained by expressing the function in the form

$$W_h \sqrt{T_m}/P_m = f(X_h, P_m/P_i)$$

3.4 Inlet Manifold

The dynamic equation for the inlet manifold relates the rate of change of pressure to the differential mass flow rates and manifold temperature.

$$\frac{dP_m}{dt} = (\gamma \times R \times \frac{T_m}{V_m}) \times (W_h - W_e) \tag{4}$$

Conservation of momentum is satisfied by assuming that a uniform pressure exists in the intake manifold between the throttle body and the intake valves. This assumption, although valid for a low frequency dynamic representation, precludes simulation of high frequency acoustic propagation [8,9].

3.5 Engine

The necessary static relations for the engine submodel are the pumping mass flow, exhaust temperature and power generation functions. The engine mass flow is a function of engine speed, charge density and volumetric efficiency. Volumetric efficiency can be expressed as a regressed function of engine speed and manifold pressure.

$$W_e = N_e \times swepv \times voleff \times \rho \tag{5}$$

The total energy release is proportional to the engine mass flow assuming constant combustion efficiency and air-to-fuel ratio.

Functions were fitted to the proportions of available power going into mechanical and gas stream power. The power entering the gas stream was modelled as a function of the total power and spark angle, and the fraction going into mechanical power was modelled as a function of manifold pressure and spark angle.

$$\begin{aligned} U_{gas} &= 0.28455 \cdot U_{tot} - 0.001 \cdot spark \cdot U_{tot} - 1962 \\ U_e/U_{tot} &= 0.1993 + 0.0021 \cdot spark + 1.2478e^{-4P_m} \end{aligned}$$

The exhaust temperature relationship is related to the gas stream energy by

$$T_x = T_m + \frac{U_{gas}}{W_e \times cp} \tag{6}$$

The engine dynamic relationship expresses the rate of change of engine speed as a function of the developed torque (expressed as the ratio of generated power to engine speed), the engine load and the total inertia of the engine and vehicle.

$$\frac{dN_e}{dt} = \frac{\frac{U_e}{N_e} - load}{I_{tot}} \tag{7}$$

Engine detonation is a complex function of combustion chamber geometry, operating condition, ignition timing and fuel properties resulting in an extremely rapid in-cylinder pressure rise and audible 'knock'. A knock condition is normally sensed using a broad band piezoelectric accelerometer mounted on the engine block to sense vibration associated with detonation. The accelerometer signal, if greater than a calibrated threshold, generates a 'flag' which is used in the ignition timing and wastegate control algorithms. This 'flag' can be used to retard ignition timing and/or open the turbocharger wastegate until knock-free performance is attained. No attempt was made to model either the combustion process or the detonation feedback sensor characteristics. Rather, an external knock input flag was included in the controller representation in order to independently exercise the appropriate algorithm.

3.6 Exhaust System and Turbine

The relevant exhaust system components consist of the exhaust manifold, turbine, wastegate and turbine nozzle.

The mass flows which define the pressure rate dynamics of the exhaust manifold consist of the ingress total engine mass flow rate of air and fuel and the egress flow rate proportioned between the flow through the turbine and flow through the wastegate.

$$\frac{dP_x}{dt} = (\gamma \times R \times \frac{T_x}{V_x}) \times (W_e + fuel - W_t - W_w) \quad (8)$$

The mass flow characteristic of the radial outflow turbine is approximated over a range of turbocharger speeds by a single function relating turbine pressure ratio to corrected flow

$$W_t\sqrt{T_x}/P_x = f(P_x/P_n)$$

The temperature ratio across the turbine is a function of turbine efficiency which varies primarily with pressure ratio and, less strongly, with speed. These relations are developed from manufacturer's data (validated by engine dynamometer measurement), tabulated as break-point functions and accessed by linear interpolation.

The wastegate actuation pressure is controlled by a two port actuator communicating with the high and low pressure sides of the compressor and a boost signal from the microprocessor. This pressure acts against a spring to open the wastegate.

As the turbocharger speed increases, the differential pressure applied to the actuator acts to overcome the spring force and open the wastegate; as the pressure differential decreases, the spring loaded actuator closes the wastegate, returning the system to the higher speed. The differential pressure signal is duty cycle modulated to control boost by bleeding the high pressure signal back to the compressor inlet based on the feedback signal from manifold pressure. At small values of duty cycle, no signal is bled and the turbocharger speed (and ultimately the intake charge pressure) is controlled about a point set by the spring. Higher values of duty cycle act to close the wastegate and increase the boost pressure.

The flow through the wastegate is calculated from the pressure difference across the valve and the wastegate position which is a linear and hardware calibratable function of wastegate actuation pressure.

The difficulty of measuring wastegate flow directly in the actual engine operating environment necessitated an oblique approach of estimating wastegate flow to determine this functionality. The turbine characteristic for the condition of fully closed wastegate was used to estimate the turbine flow at any operating condition and the wastegate flow was then assumed to be the difference between the actual flow and the estimated value. The analysis demonstrated that wastegate flow begins at an actuation pressure of about 0.21 bar and increases linearly.

Dynamics are imposed on the wastegate displacement by the assumption of predominately first order behaviour with respect to desired position. The associated time constant is estimated from the physical characteristics of the valve and verified by dynamic tests.

$$\frac{dX_w}{dt} = c_w \times (X_w d - X_w) \quad (9)$$

The dynamics of the exhaust nozzle are defined by the turbine and wastegate mass rates, the turbine temperature ratio and the nozzle characteristic flow as a function of pressure ratio across the nozzle.

$$T_n = T_x \times TR_t \quad (10)$$

$$\frac{dP_n}{dt} = (\gamma \times R \times \frac{T_n}{V_n}) \times (W_t + W_w - W_n) \quad (11)$$

3.7 Turbocharger Shaft

The rotational dynamics of the turbocharger shaft are expressed in terms of the turbocharger polar moment of inertia, angular acceleration, and the difference between the power applied to the turbine and expended by the compressor.

$$U_c = W_c \times (T_c - T_o) \times cp \quad (12)$$

$$U_t = W_t \times (T_x - T_n) \times cp \quad (13)$$

$$\frac{dN_s}{dt} = \frac{U_t - U_c}{N_s \times I_s} \quad (14)$$

3.8 Component Integration

Subsequent to the development of the component functional relationships, the individual model elements must be integrated into the overall structure defined by Figure 2 and compared to the system performance observed on the engine-dynamometer and the dynamic characteristics illustrated by vehicle testing. The predicted system steady-state performance is compared to actual engine data in Figure 4.

4 Dynamic Matching

Subsequent to validation of the steady-state characteristics of the model, system dynamic matching was addressed. This process involved two categories of experimental evaluation: high frequency dynamics associated with system pressures, mass flows and the turbocharger speed were matched using engine dynamometer tests. Low frequency dynamics associated with the engine output shaft were matched using in-vehicle tests.

4.1 High Frequency Engine Matching

The principal dynamic tests used to validate the system model were performed on the engine dynamometer operating under constant speed control. These transient tests consisted of applying fast throttle ramps and step inputs of spark angle and wastegate duty cycle at several steady-state operating points. System pressure response was measured at the compressor outlet, intake manifold plenum, and exhaust manifold at the turbine inlet and exit. Turbocharger shaft speed and engine torque responses were also monitored. The dynamometer tests were repeated on the open-loop simulation model with engine speed held constant. Figure 5 shows the response of the model to a step in wastegate duty cycle.

4.2 Low Frequency Engine Matching

Low frequency engine matching is concerned with the validation of effective vehicle and drivetrain inertias to match observed engine speed dynamics. Prior to transient response matching, it was necessary to develop a vehicle model with algorithms for the load imposed on the engine by the vehicle and the effective powertrain inertia.

For a specific vehicle, imposed engine load is primarily a function of velocity, gear ratio and various empirical coefficients associated with aerodynamic drag and tyre rolling resistance. For the vehicle model, the imposed load was resolved as a function of gear ratio and engine speed, and was derived from steady-motion vehicle measurements:

$$load = (R_{eng} + R_{veh} \cdot gratio) \times N_e^2 \qquad (15)$$

Subsequent to establishing the load torque relationship, the low frequency system dynamics were matched. This was accomplished by applying (to the degree possible in-vehicle) step throttle inputs and measuring engine speed responses at numerous vehicle speeds throughout the operating range, from 20 to 120 km/h in five different gear selections. Replictation of these tests on the engine-vehicle model provided accurate engine speed dynamic representation. Figure 6 shows the response of the engine-vehicle model to a step change in throttle setting.

5 Controller and Engine Integration

The evaluation of the implemented control algorithm with respect to the problem of inlet manifold pressure regulation can be carried out by inspection of simulated time response data such as those presented in Figure 7 for a step input applied to the throttle of the integrated engine-vehicle-controller system. Such results will be used for strategy development as well as for system calibration and hardware development projects.

6 Summary

The approach described in this paper provides a versatile simulation model of a high performance turbocharged engine for control development that is accurate over a wide operating range. The method of model building results in component submodels incorporating appropriate nonlinear characteristics matched to actual engine data. The modular structure imposed on the system allows straightforward modifications to simulate different engines or to examine the effects of hardware changes. The characterisation of the controller is a realistic representation of the software implementation allowing the actual control algorithms to be exercised.

References

[1] Powell, B.K. A Dynamic Model For Automotive Engine Control Analysis. Proc. 18th IEEE Conference Decision and Control, 1979, pp.120-126.

[2] Delosh, R.G. , Brewer, K.J. , Buch, L.H. , Ferguson, T.W.F. and Tobler, W.E. Dynamic Computer Simulation of a Vehicle with Electronic Engine Control. SAE Paper 810447, 1981.

[3] Dobner, D.J. A Mathematical Model for Development of Dynamic Engine Control. SAE Paper 800054, 1980.

[4] Wu, H. , Aquino, C.F. and Chou, G.L. A 1.6 Litre Engine and Intake Manifold Dynamic Model. ASME Paper 83-WA/DSL-39, 1984.

[5] Onions, R.A. and Foss, A.M. Improvements in the Dynamic Simulation of Gas Turbines. AGARD CP324-27, 1982.

[6] Kays, W and London, A.L. *Compact Heat Exchangers*. 1964, McGraw-Hill.

[7] Watson, N. and Janota, M.S. *Turbocharging the Internal Combustion Engine*. 1984, MacMillan Publishers Ltd.

[8] Yuen, W.W. *A Mathematical Engine Model Including the Effects of Engine Emissions*. 1982, Dept. of Mechanical and Environmental Engineering, University of California, Santa Barbara, CA.

[9] Servati, H.B. Investigation of the Behaviour of Fuel in the Intake Manifold and Its Relation to SI Engines. PhD Thesis, 1984, University of California, Santa Barbara, CA.

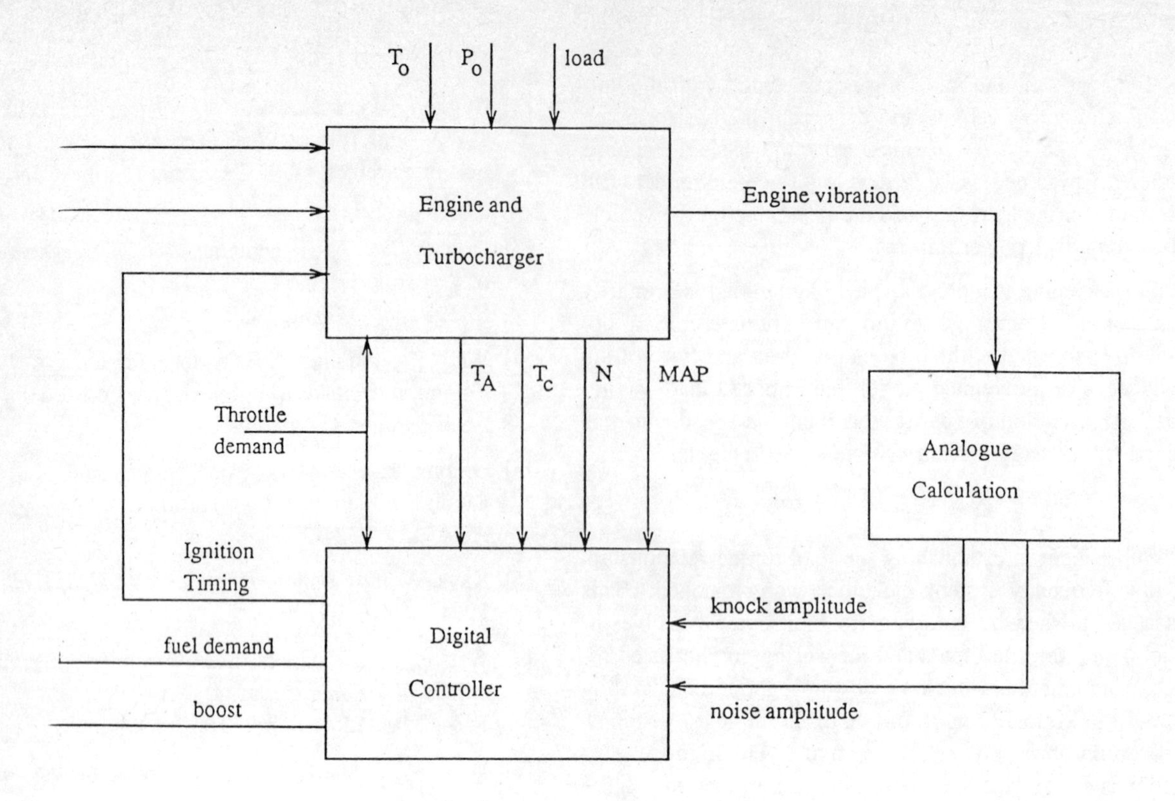

Fig 1 Overall model structure

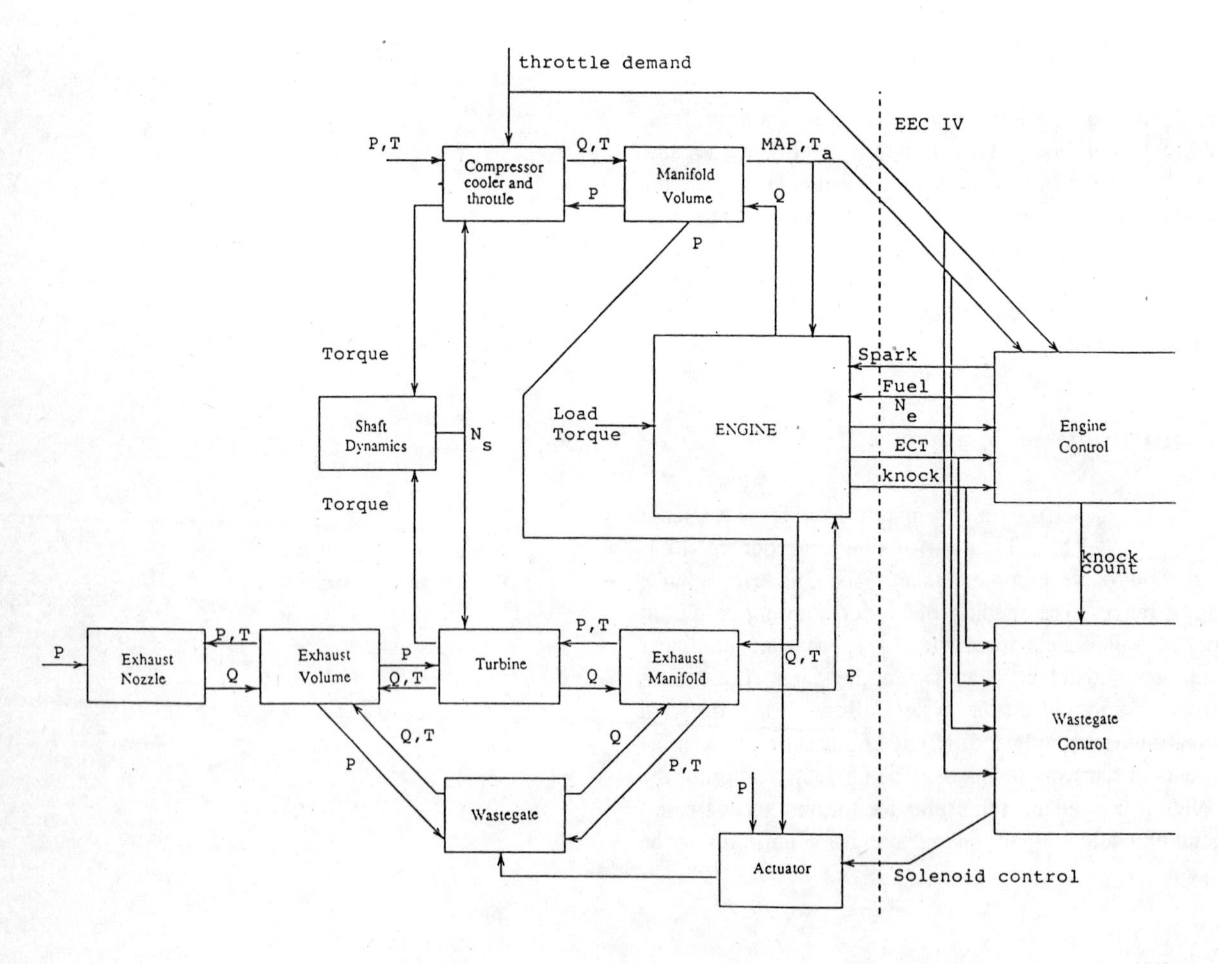

Fig 2 Information flow diagram

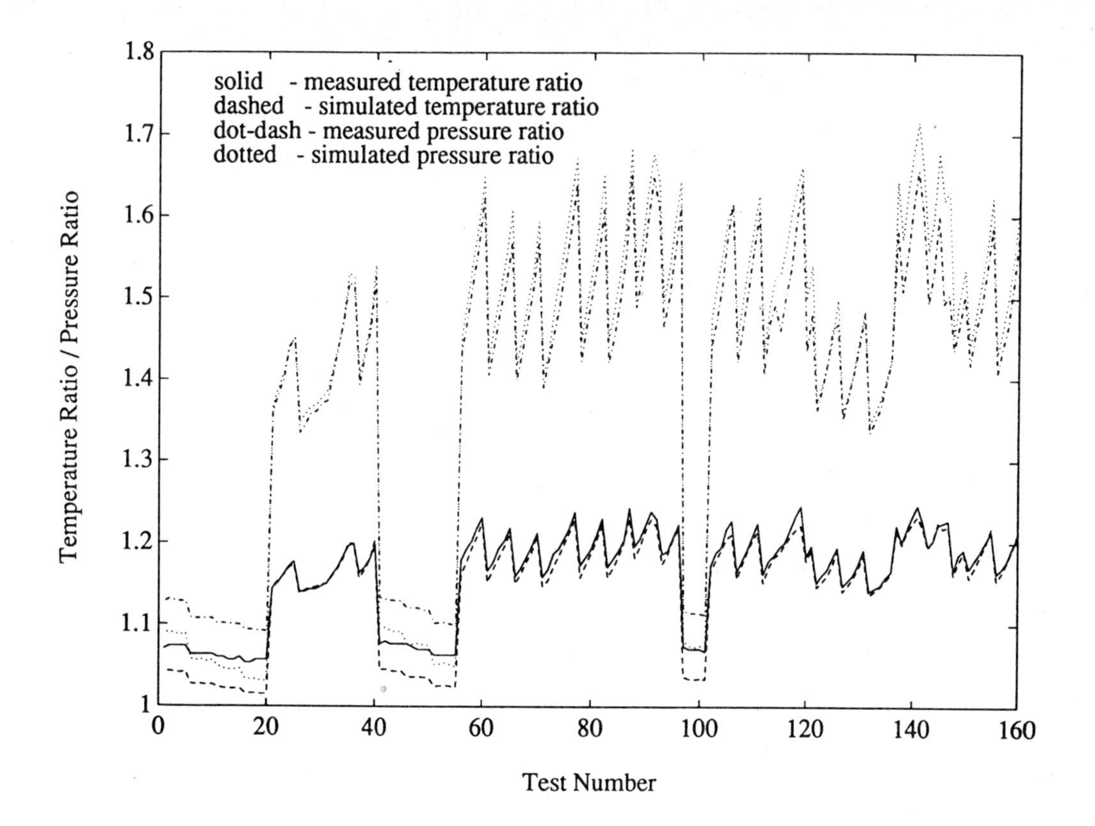

Fig 3 Compressor pressure and temperature ratios comparison

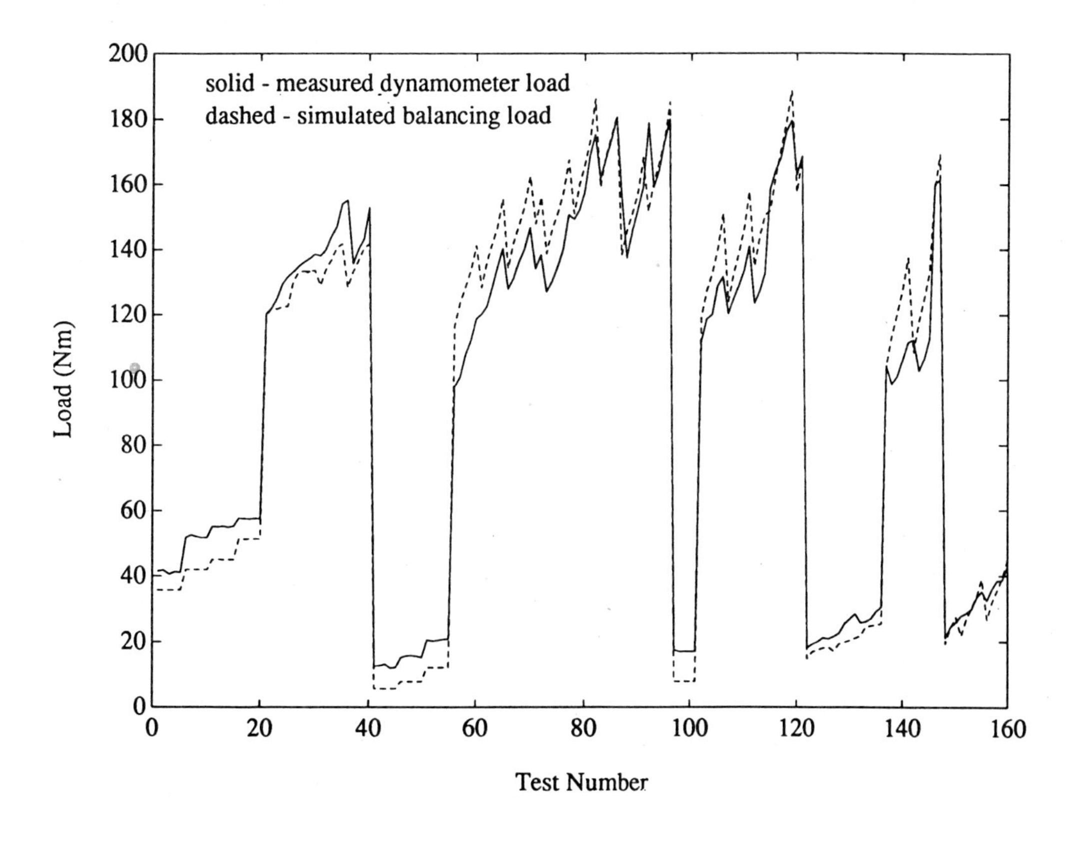

Fig 4 Comparison of measured load with simulated load

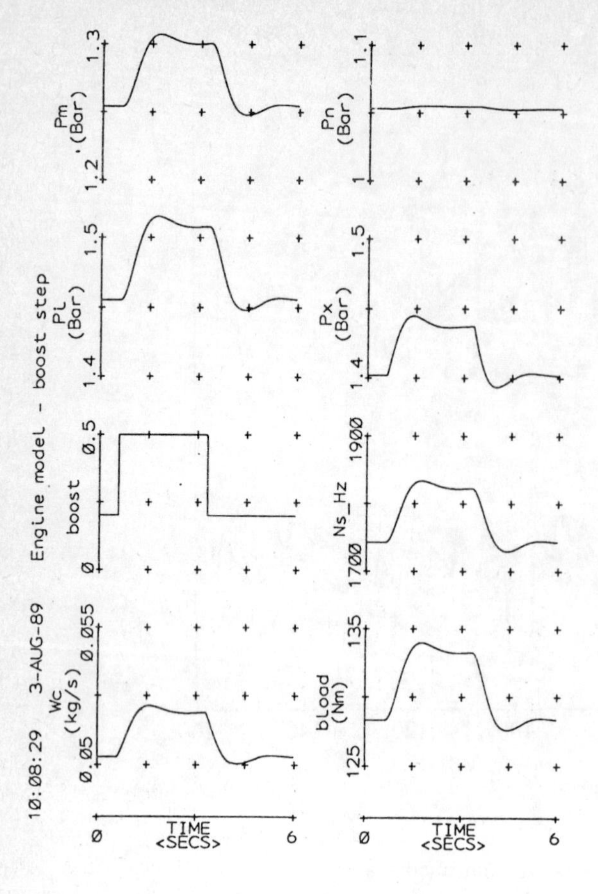

Fig 5 Engine model – boost step

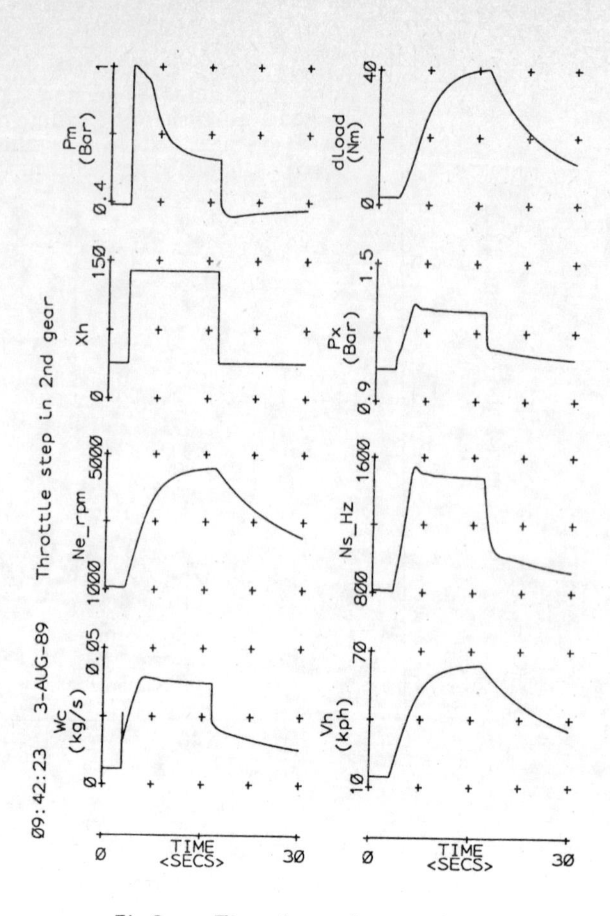

Fig 6 Throttle step in second gear

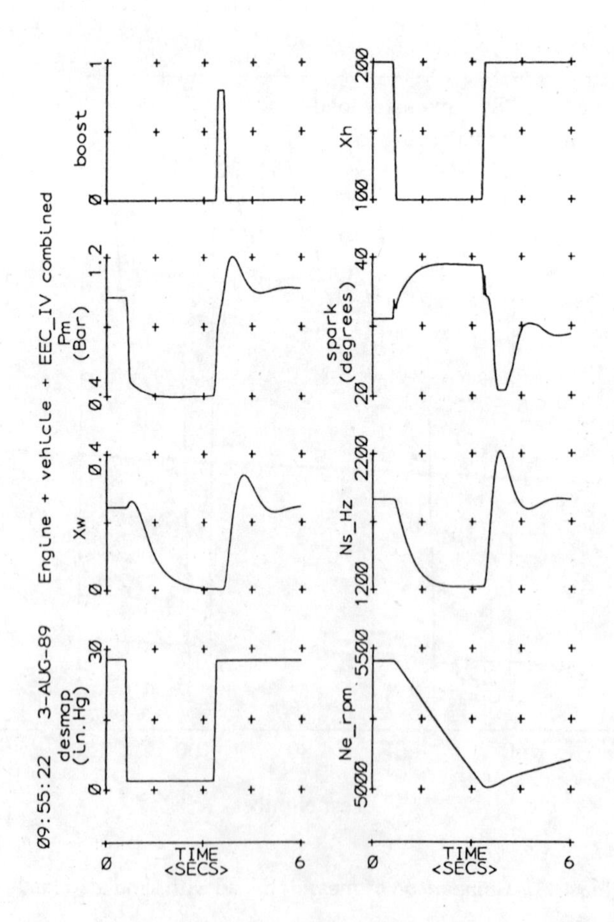

Fig 7 Engine, vehicle and EEC IV models combined

C391/025

System concepts for serial data communication in cars

U WENKEBACH and **B RECKELS**
Philips Bauelente GmbH, Hamburg, West Germany

Abstract:

In the past, discussions concerning automotive networks have been dedicated mostly to the protocol and the components used. However, it is necessary, to spend some effort in an investigation on the automobile as a complete electronic system in itself.

A system concept for an in-vehicle multiplex wiring network using the CAN-protocol is presented, covering the topics:

- o requirements on the in-vehicle network regarding robustness against faults
- o concept for the behaviour of the total system as the result of the behaviour of all electronic subsystems (network members)
- o required hardware for different approaches.

Introduction:

In-vehicle networks in cars offer several advantages for the customer and for the manufacturer. They enable car manufacturers to use a multiplex wiring system, which increases the performance of a car in terms of comfort, reliability and safety. Furthermore, distributed intelligent control systems for drivetrain management can now easily communicate with each other. This may lead to reduction of fuel consumption and an overall improvement of the drivetrain performance. Beyond this, in-vehicle networks will become mandatory in the near future- think of the California Air Resources Board proposals, which prescribe a serial data outlet in cars in California in the near future.

As a result, several car manufacturers as well as their suppliers investigate in-vehicle networks. However, the discussions concerning automotive networks have been dedicated mostly to the protocol and the components used.

But before a reliable in-vehicle network can be designed, the requirements on the car as a total system have to be defined precisely. After this has been done, the requirements on the in-vehicle network can be described. If an aspect during the design phase is omitted, the car and the built-in network may have an unpredictable behaviour. This must be absolutely avoided by a precise definition of all possible system states.

In the following, a system concept for an in-vehicle network such as a multiplex-wiring system is presented.

Functional Requirements on a Multiplex Wiring System in a Car

Two basic groups of requirements on a multiplex-wiring system (MUX) in a car exist: electrical requirements (e.g. noise immunity) and functional requirements. Some of the functional requirements are listed below:

- o The system has to be fully functional, when the car is active under normal operation (i.e. someone is inside the car),
- o degraded function must be ensured in case of a disturbance, which affects the data communication. If the disturbance disappears, the affected network nodes must recover into an appropriate state,
- o a low power mode function has to be provided when the car is inactive, say parked, so as not to drain the battery.

The system concept presented in this paper uses a busprotocol. Since there are some relations between a system concept and the protocol used, the requirements on the logical busprotocol for this system concept are:

- o Multi-master facility: every module connected to the network may transmit a message. The order is fixed by an arbitration algorithm.

- o Broadcast philosophy / shared memory concept: an undisturbed message transmitted via the network is acknowledged by every module connected to the network. There is no dedicated module, which has to acknowledge. Every receiving module may use the incoming data.

- o Extensive measures against disturbances (e.g. CRC). This guarantees as well, that a properly received message was actually transmitted by another module/s.

These requirements are fulfilled with the CAN-protocol ('Controller Area Network'), which was used for the implementation of this system concept. As a result, all timing values provided in this paper were calculated using the actual CAN-protocol as available from various vendors. The system concept itself, however, may be realized with any other protocol as well, which meets the requirements mentioned above.

When a protocol such as described above is used for multiplex wiring tasks, it is an urgent need, that every module is able to determine its own status (e.g. with or without network communication failure) and to react appropriately. In other words, the transmitter of a message must not keep track of the fact, if his distinct message has reached its receiver, as long as no network error was signalled. Every receiver should "know", what to do, if no communication is possible via the network. This fact speeds up the data transfer and is commonly known as "shared memory concept", where the shared memory is represented by the network itself.

System concept for an in-vehicle network: Overview

In order to develop a system concept for an in-vehicle network, the following main steps were carried out:

- o Identification of all main states, a car and the built-in network may occupy,

- o description of the main states,

- o identification of the relationship between the main states,

- o description of the transitions between the main states.

Figure 1 shows the principle scenario for an in-vehicle network we had in mind, when developing our system concept. The application for the network is a 'Class A' multiplex wiring system. The ISO- classification for this type of network is 'low speed body wiring and control functions such as switching of exterior lamps'.

Connected to the multiplex wiring bus are front modules, a dashboard module and one or even more door modules and rear modules.

A car has to consume very low power when it is parked. Therefore, a multiplex wiring system must be able to enter a so called sleep- mode. LA, local activation, is the name of a signal from a switch in the door or at the steering wheel. This signal is used to reactivate the network, when the car was parked.

Class 'A' applications are most challenging in terms of complexity of their functions. Other applications like those in engine management require less system states, because they normally do not require a sleep mode. Our work also applies for class 'C' applications, because they are a subset of class 'A' applications in terms of complexity of network functions.

To obtain a simple, yet precise graphic representation of the complex situations, we chose the technique of state diagrams with detailed description of the transitions.

Figure 2 introduces all five main states, a car and its MUX- system can occupy: Very often people only see the apparently most important states, which a car and its in-vehicle network can occupy:

- o the active mode, where the car is in use and all systems have to be fully functional and

- o the sleep mode, when the car is not in use -say parked- and all electric systems are required to consume very low power so as not to drain the battery.

If however any failure occurs, the network may indicate incorrect behaviour to the user when only these two main states are realized. What happens, as an example, if a network node such as the front module is totally disconnected from the bus by a connector failure or damaged buswires? If this happens in the evening and the driver cannot switch on the headlamps, the car may become unusable.

This bad situation can be prevented with a third state, the

- o default mode. This mode is entered by a network node, as long as no communication is possible via the network. Herein is defined, which tasks then have to be carried out, in our example the headlamps are always switched on, so that the car does not become unusable in the dark.

The next two states are used for process synchronization: It is of utmost importance, that the network in total has a well defined and known condition at any time. Since nodes in a network may have different microcontrollers with different application tasks, the change of states for the whole network must be coordinated. This is done in the wake-up procedure and in the fall-asleep procedure.

One of the main objectives of a system concept is to define precisely, when a certain state will be entered and how it will be left, i.e. the description of the transitions between the main states. Due to the fact, that there is no inherent redundancy in a MUX system, two problems are encountered, when the following questions have to be solved:

- o When a transient network failure begins, a module will enter its default mode DM. How can this module gain information about the rest of the MUX system, when the failure is gone, i.e. which state should the module occupy after the failure?

- o When a module is totally disconnected from the network, it must enter its default mode DM to support a reasonable behaviour of the car in total. Because the module is connected to the continuous supply voltage of the car, this state will be occupied as long as the disconnection lasts. This will cause problems: E.g. the default mode for the main head lamps should be 'on'. This may drain the battery, when the car is parked, because the disconnected module does not leave the default mode.

Especially the second case shows, that a global information about the state of the MUX system must be provided to any module. For safety reasons, this information is not allowed to be transmitted only via the network. This leads to two slightly different system concepts, which differ mainly in the behaviour in case of disconnected nodes as mentioned above:

- o System 'A': without a global activity signal,
- o system 'B': with a global activity signal.

Due to the global activity signal, system 'B' has a superior fault tolerance than system 'A'.

In the following, both approaches will be discussed.

System 'A' Operation

Figure 3 shows all main states and their transitions of system 'A'. A description of the system behaviour regarding states and transitions is given in the following chapters.

Transition: Sleep mode to Wake-up Procedure

From sleep mode SM a wake-up signal WUE forces either the network or a single module to enter the wake-up procedure WUP. This signal may be:

- o a certain condition on the bus lines (a 'dominant' bus condition). This dominant bus condition can be generated by one or more modules connected to the network or a severe disturbance on the bus lines,
- o a local activation signal LA generated at one module e.g. by the door lock switch. This signal is sensed by the door module, which in turn issues a dominant bus condition in order to wake-up all other modules.

After entering the state WUP, every module starts up and performs some initialization tasks, but preserves the static behaviour it had during sleep mode e.g. 'park light on'. This is important in case of an incidental wake-up. After a certain time, any module will start to send a special telegram via the network, the wake-up message. This enables all modules to perform an individual check, whether communication via the network is possible for them or not.

The time from a wake-up event until the total system is in the wake-up procedure calculates roughly (the internal calculation time in the software is omitted in all calculations):

T_system_wake_up <= 2 * T_osc_start + 3 * T_wum [1]

with

T_osc_start	average start-up time for the local quartz oscillator,
T_wum	duration of the wake-up message.

For a typical quartz oscillator and a bitrate of 10 kbit/s this formula gets

T_system_wake_up <= 2 * 10 ms + 3 * 5 ms =35 ms.

In the case, that data transmission takes place, but cannot be received without errors by a node, this distinct node has a bus error (BERS). The node will enter the default mode DM.

Transition: Wake-up Procedure to Active Mode

After all modules have sent their wake-up message, one of the modules equipped with a local activation LA facility must start to send an active mode message (AMM). This will lead the modules and therefore the MUX system to the active mode. The time, within the message AMM must occur is

T_active_mode_confirmation =
T_active_mode_message + T_delay [2]

with

T_active_mode_message	length of the active mode message itself
T_delay	safety margin to compensate for various delays

With 10 kbit/s this value is approx.

T_active_mode_confirmation =
56 bittimes + 500 us = 6 ms.

Is this message omitted, all modules have the information, that the wake-up was due to noise. They will enter the fall-asleep procedure, if no bus error is being detected.

Active Mode

Being in the active mode, every module equipped with the local activation facility LA sends regularly (e.g. every few seconds) a dedicated message, the active-mode message AMM. In order to reduce the bus load, it is desirable, that only one node issues the active mode message AMM at one time. This can be easily achieved by a timer in the intelligent part of the module (e.g. an on-chip timer). The timer is controlled by a simple algorithm implemented in software:

- o Active mode message AMM and time limit reached: send active mode message AMM
- o active mode message AMM received: reload timer with timeconstant

The regular circulation of a real message like the active mode message AMM has several important advantages:

- o Since an actual received or transmitted message crosses all seven layers of the ISO/OSI layer model, a single node can really decide, whether communication via the network is possible or not.
- o Only the combination of two messages (the active mode message AMM and the sleep mode message SMM) will bring a node into sleep mode. If this scheme is disturbed due to noise, this is recognized as a failure and the disturbed module will react appropriately. This guarantees a predictable behaviour of the system also in this case of disturbance.
- o A node, which has entered the default mode DM due to a temporary failure has information, in which state the other nodes (i.e. the system) are, when communication via the network is possible again.

In the case, that data transmission takes place, but cannot be received without errors by a node, this distinct node has a bus error (BERS). The node will enter the default mode DM.

Transition: Active Mode to Fall-asleep Procedure

When the local activation signal LA becomes inactive (e.g. the user of the car locks the door from outside) the MUX- system must enter the low power mode. The node, which has sensed the change-of-state of the local activation signal LA issues immediately a dedicated message, the sleep mode message SMM. The system (i.e. the sum of all other nodes) responds to this message:

- o If another module senses its local activation signal LA still active, an active mode message AMM is issued to indicate, that sleep mode is not allowed at this moment. The node, which has issued the sleep mode message stays active and acts from now on like a module without the local activation facility.
- o If no other module disagrees with the sleep mode, the system enters the fall-asleep procedure after a certain time T_smr.

The timing considerations for this time T_smr are:

T_smr >= T_longest_message +
T_longest_error_frame +
T_amm + T_smm + T_delay [3]

where

T_longest message	duration of the longest message possible,
T_longest error frame	duration of the longest error frame possible,
T_amm	duration of the active mode message AMM,
T_smm	duration of the sleep mode message SMM,
T_delay	duration of various delays.

With 10 kbit/s this value is

T_smr >= 22 ms

Fall-asleep Procedure

In the fall-asleep procedure each module carries out some housekeeping in order to prepare for the low power mode. Besides other application specific tasks, the modules perform mainly the cancellation of any message, which would have been send. After this, the sleep mode will be entered. If during the fall-asleep procedure a wake-up signal is sensed (e.g. a local activation signal becomes active), each module enters the wake-up procedure directly.

Default Mode

Purpose of the default mode DM is to guarantee an explicitly defined and reasonable behaviour of the car as a total, even when the network communication is not possible due to a transient or continuous failure.

In general, this mode will be entered, when communication via the network is not possible and a reasonable behaviour of the functions of the car controlled by the disturbed node must be maintained. Functions of the controlled devices connected to the node in the default mode are to be defined by the application under safety aspects.

Entry to this mode is always initiated by the information BERS, bus error signalled. This information is transferred to the intelligent part of a module, when the busline related controller (e.g. a CAN-controller) has no undisturbed access to and from the network. Within a network using CAN-protocol products, this information is reflected by the 'bus-off' signal of the CAN-controller.

Another information defines the state, when the BERS information is absent -communication should be possible- but no messages are received. This may be due to a complex failure or simply, because the rest of the system is already in the sleep-mode. This information is named MDNO, message detectable, but not occurred.

Being in the default mode, the distinct node stops transmission in order to prevent possible network failures, enters the application specific default actions and awaits an undisturbed message received via the network.

If the bus error BERS is absent again, the node can decide, which state the system occupies by means of a decision table shown in figure 4 and figure 5. If the module is not equipped with a local activation facility and the BERS information is no longer present, the module should receive one of the circulating messages as e.g. the active mode message AMM. Then the state of the rest of the system is easy to predict. If no message is received, this module must stay in the default mode, because it can not decide, whether the rest of the system is sleeping or a more complex network failure is still present, which is not detectable by the buscontroller. In contrast, a module equipped with the local activation facility LA will enter the wake-up procedure, if no BERS is present and the LA signal is active as depicted in figure 5.

Here the main problem of system 'A' becomes visible: if a network error is present (BERS is true) and no further information is provided, it is not possible for a module to come from the default mode into the sleep mode via the fall-asleep-procedure. This is simply due to the lack of redundancy which is inherent in system 'A' and which lead to the development of system 'B'.

System 'B' Operation

In order to solve the problems discussed above, an information about the status of the rest of the network is required, whenever communication is not possible. This information may be obtained from a single wire connected to all safety relevant modules. This wire is named global activation, GA, and is sketched in figure 6. GA is controlled by the ignition switch, which is already available in almost every car (terminal 15).

Figure 7 shows this more fault tolerant solution. Generally speaking, system 'B' is a super set of system 'A'. The main difference and advantage compared to system 'A' is, that the transition from the default mode DM to the fall asleep procedure FAP and finally to the sleep mode SM is now possible.

This is also reflected in the following decision tables in figure 8 and figure 9. Figure 8 describes a module without the local activation signal LA but with the global activation signal GA: If a communication via the network is not possible (BERS is true) and the car is active (i.e. the engine is running and/or the steering wheel is unlocked) the module has entered its default mode DM.

Despite of the fact, that no communication via the network is possible, the disturbed module has the information, that the car is still in use due to the global activation signal GA. When the global activation signal GA becomes inactive, the regular sleep mode will be entered via the fall asleep procedure FAP.

Being in sleep mode, the disturbed module will become active again, when GA becomes active. This is included in the wake-up event WUE definition in system 'B'. If there is still a network failure, the default mode DM will be entered again.

Figure 9 shows the decision table for a module equipped with the local activation signal LA and the global activation signal GA. It can be seen, that both LA and GA must be inactive during a network failure situation, before the fall asleep procedure FAP can be entered.

Required Hardware for System 'A' and System 'B'

Figure 10 sketches the connection between the modules of an in-vehicle network in a car, when PHILIPS- CAN products are used. The physical bus consists of two wires, which connect all modules in a bus topology. Every module employs a transceiver circuit, a buscontroller unit and a microcontroller.

Figure 11 shows the schematics for an intelligent busnode in more detail. Connected to the bus wires (twisted pair recommended) is the PHILIPS Multiplex Transceiver PMT. This device incorporates a 5V power supply and reset logic as well as a transceiver, which transforms logical bit values to their physical bit representation on the bus. The CAN protocol is realized by a PHILIPS 82C200 CAN protocol controller, which is connected to an automotive dedicated microcontroller, the PHILIPS 83C552. The microcontroller controls the application, which may be connected to the various ports of the 83C552. On this microcontroller, either system 'A' or system 'B' software may be implemented.

The wires to the local activation signal LA (e.g. a door lock switch) and the global activation signal GA (e.g. terminal 15) are drawn in dotted lines. This indicates, that this module circuitry may serve for all combinations of system 'A' and system 'B'. The advanced or safety relevant modules perform system 'B' type software and are connected together to a global activation signal GA. The more simple modules, which are not safety relevant or have no critical default operation perform the more simple system 'A' type software.

Summary

Presented in this paper were two proposals for the design of the logical behaviour of a multiplex wiring system in a car. The implementation employs three dedicated messages to control the status of the network:

- o the wake-up message,
- o the active-mode message, which is issued periodically and the
- o sleep-mode message.

Since these messages cross all 7 ISO/OSI layers up to the application layer, a true test of all layers is performed periodically. One of the main results is, that the network system reacts predictably in any case in a fixed and predictable time. In fact, these messages are sort of overhead, since they do not contribute directly to the data exchange. However, this overhead is comparatively small: in our implementation it is less than 1% of the network capacity.

The works made visible, that a different thinking is required during the design phase of electronics for cars, when in-vehicle networks are used. The design work is not finished, when the modules are physically connected with buswires. If functions are safety relevant such as the switching of the headlamps, additional measures such as the global activation signal GA must be provided and processed by the software. This costs money mainly in the development phase of the network system and a small additional wiring effort to those modules during production.

Future efforts will be focused toward fast and efficient implementation of the two proposals on several microcontrollers used in the automotive industry.

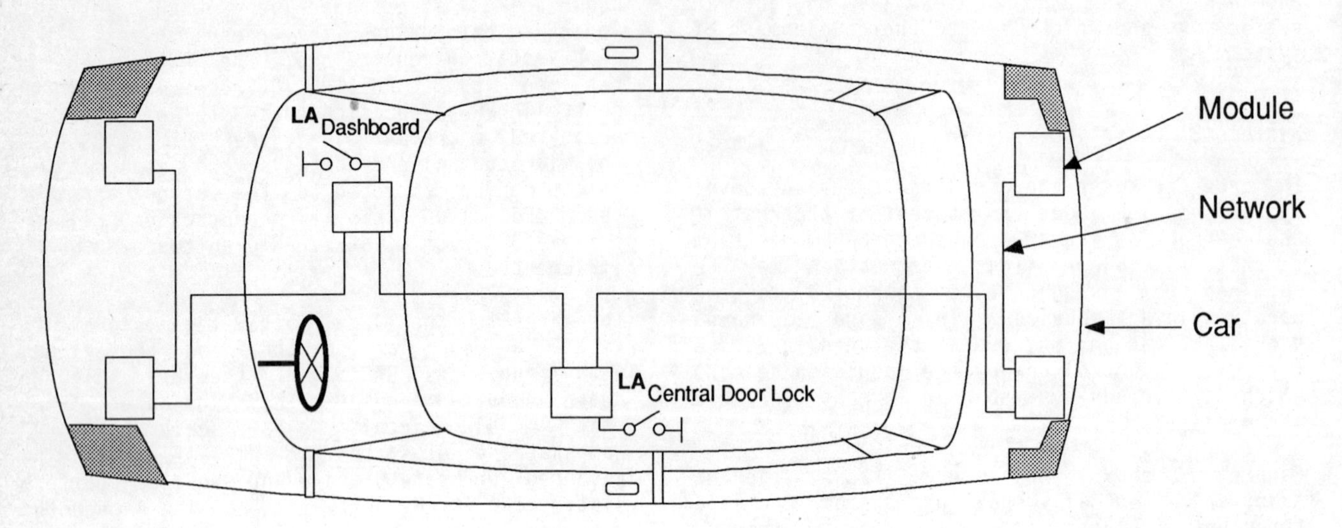

Fig 1 MUX system in a car

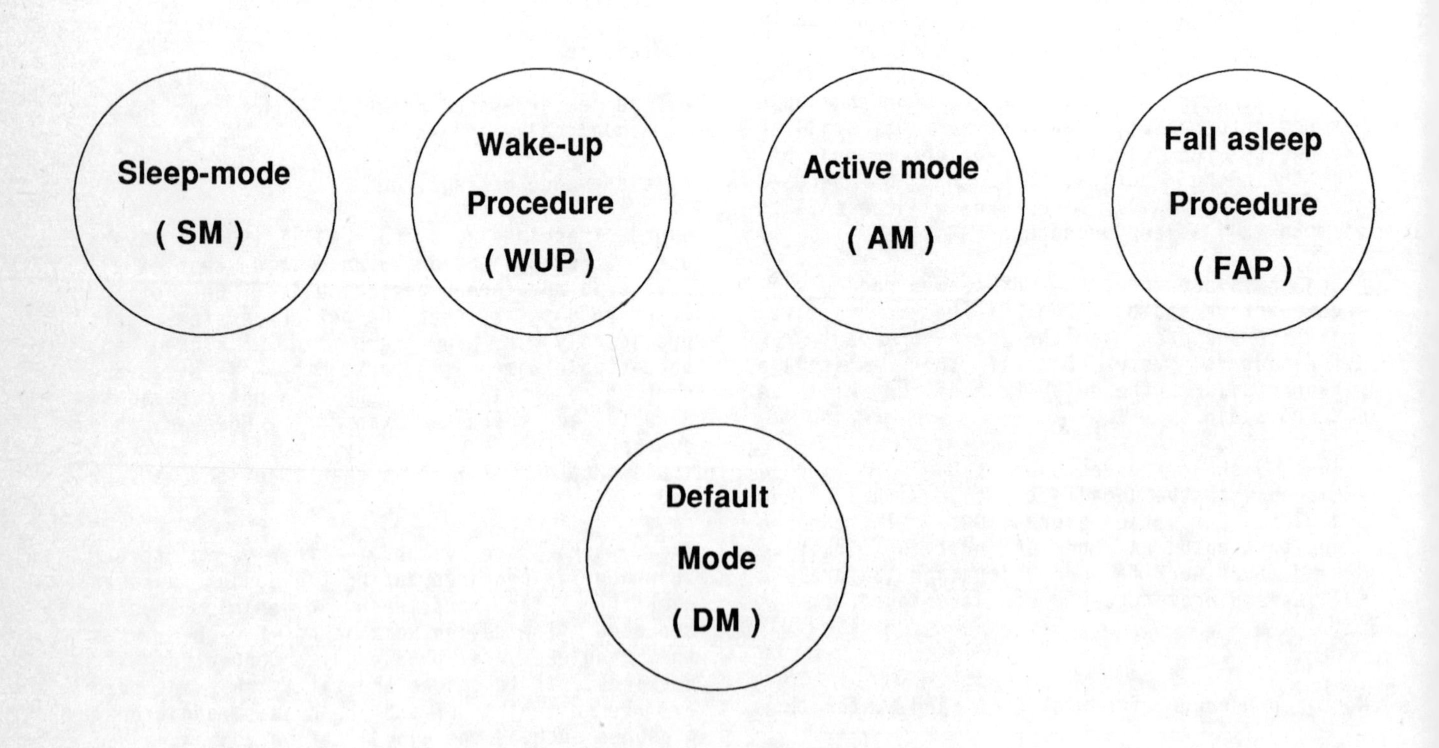

Fig 2 General view of all system states

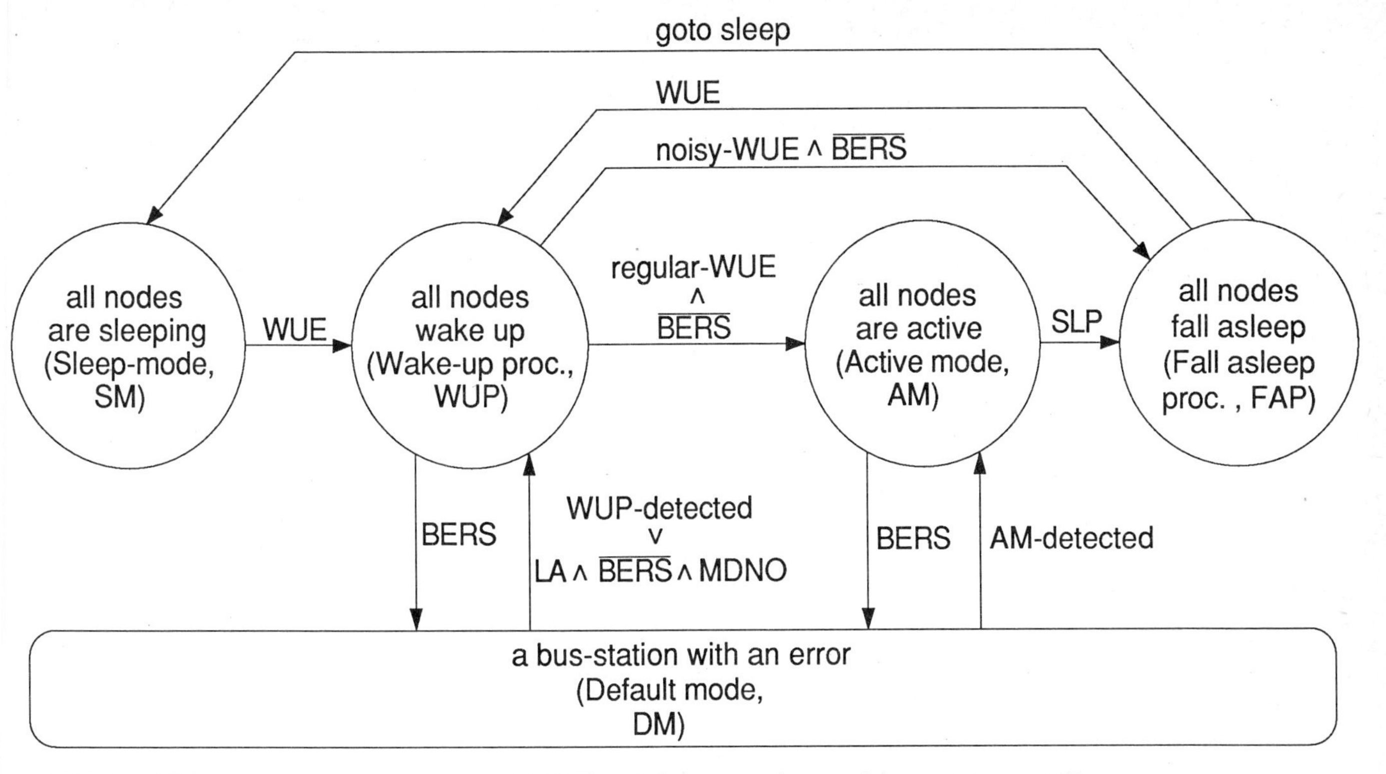

WUE : "Wake-up event"
BERS : "Bus-error signalled"
LA : "Local activation signal"
MDNO : "Message detectable, not occured"
SLP : "Sleep mode to be entered"

Fig 3 System 'A' operation

BERS	$\overline{\text{BERS}}$ ∧ $\overline{\text{WUM-detected}}$ ∧ $\overline{\text{AMM-detected}}$ ∧ $\overline{\text{SMM-detected}}$	$\overline{\text{BERS}}$ ∧ WUM-detected	$\overline{\text{BERS}}$ ∧ AMM-detected	$\overline{\text{BERS}}$ ∧ SMM-detected
Default Mode	Default Mode	Wake-up Procedure	Active Mode	Active Mode

Fig 4 System 'A', no LA-signal

	BERS	$\overline{\text{BERS}}$ ∧ $\overline{\text{WUM-detected}}$ ∧ $\overline{\text{AMM-detected}}$ ∧ $\overline{\text{SMM-detected}}$	$\overline{\text{BERS}}$ ∧ WUM-detected	$\overline{\text{BERS}}$ ∧ AMM-detected	$\overline{\text{BERS}}$ ∧ SMM-detected
$\overline{\text{LA}}$	Default Mode	Default Mode	Wake-up Procedure	Active Mode	Active Mode
LA	Default Mode	Wake-up Procedure	Wake-up Procedure	Active Mode	Active Mode

Fig 5 System 'A', with LA-signal

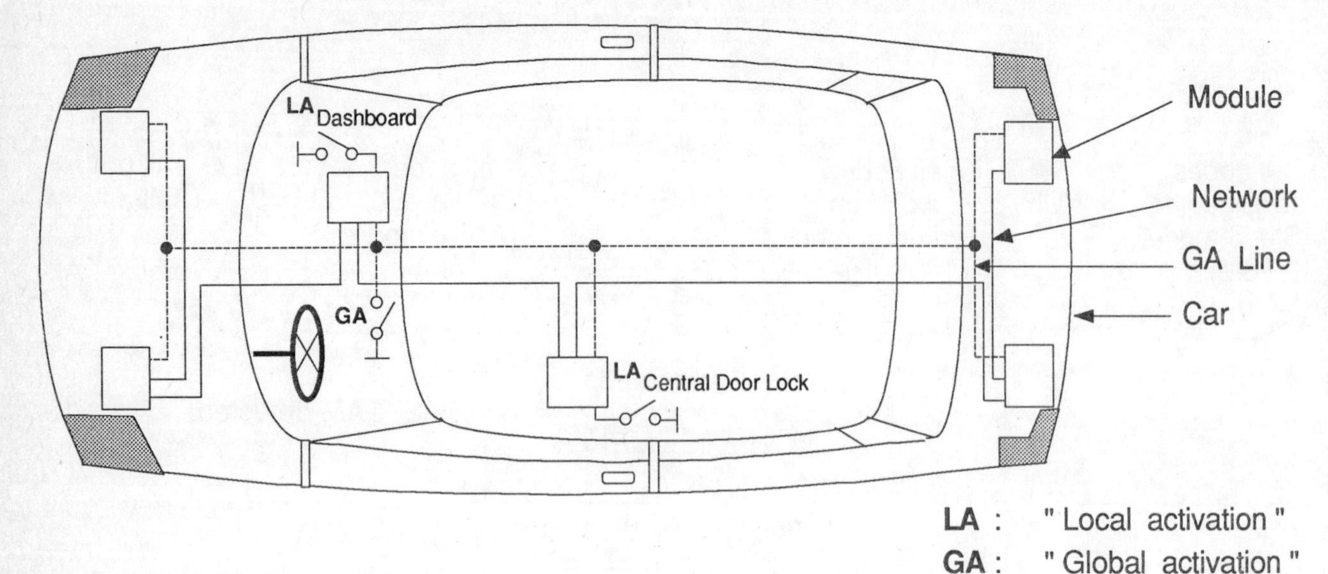

Fig 6 MUX system in a car

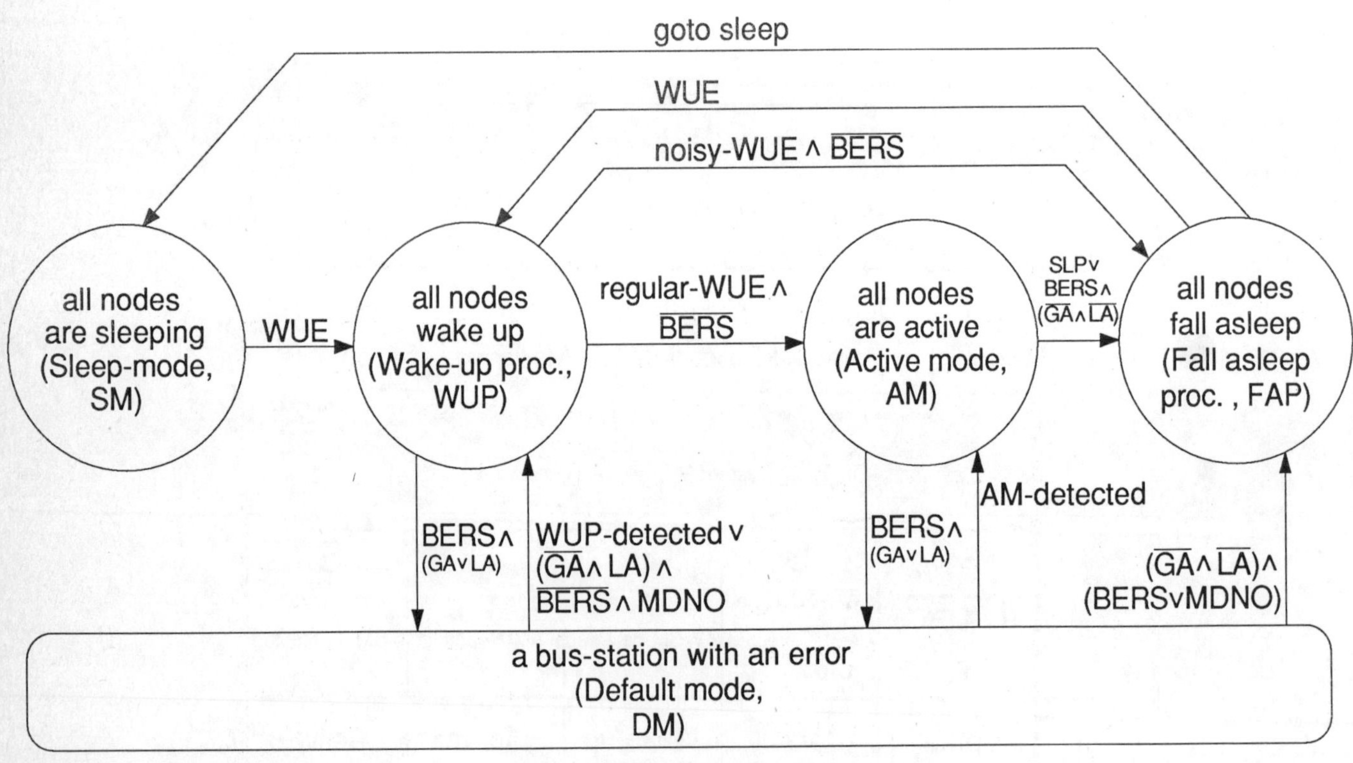

Fig 7 System 'B' operation

	BERS	$\overline{\text{BERS}}$			
		$\wedge \overline{\text{WUM-detected}} \wedge \overline{\text{AMM-detected}} \wedge \overline{\text{SMM-detected}}$	$\wedge$ WUM-detected	$\wedge$ AMM-detected	$\wedge$ SMM-detected
GA	Default Mode	Default Mode	Wake-up Procedure	Active Mode	Active Mode
$\overline{\text{GA}}$	Fall asleep Procedure	Fall asleep Procedure	Wake-up Procedure	Active Mode	Active Mode

Fig 8 System 'B', no LA-signal

	BERS	$\overline{\text{BERS}}$			
		$\wedge \overline{\text{WUM-detected}} \wedge \overline{\text{AMM-detected}} \wedge \overline{\text{SMM-detected}}$	$\wedge$ WUM-detected	$\wedge$ AMM-detected	$\wedge$ SMM-detected
$\overline{\text{LA}} \wedge \text{GA}$	Default Mode	Default Mode	Wake-up Procedure	Active Mode	Active Mode
$\text{LA} \wedge \text{GA}$	Default Mode	Wake-up Procedure	Wake-up Procedure	Active Mode	Active Mode
$\text{LA} \wedge \overline{\text{GA}}$	Default Mode	Wake-up Procedure	Wake-up Procedure	Active Mode	Active Mode
$\overline{\text{LA}} \wedge \overline{\text{GA}}$	Fall asleep Procedure	Fall asleep Procedure	Wake-up Procedure	Active Mode	Active Mode

Fig 9 System 'B', with LA-signal

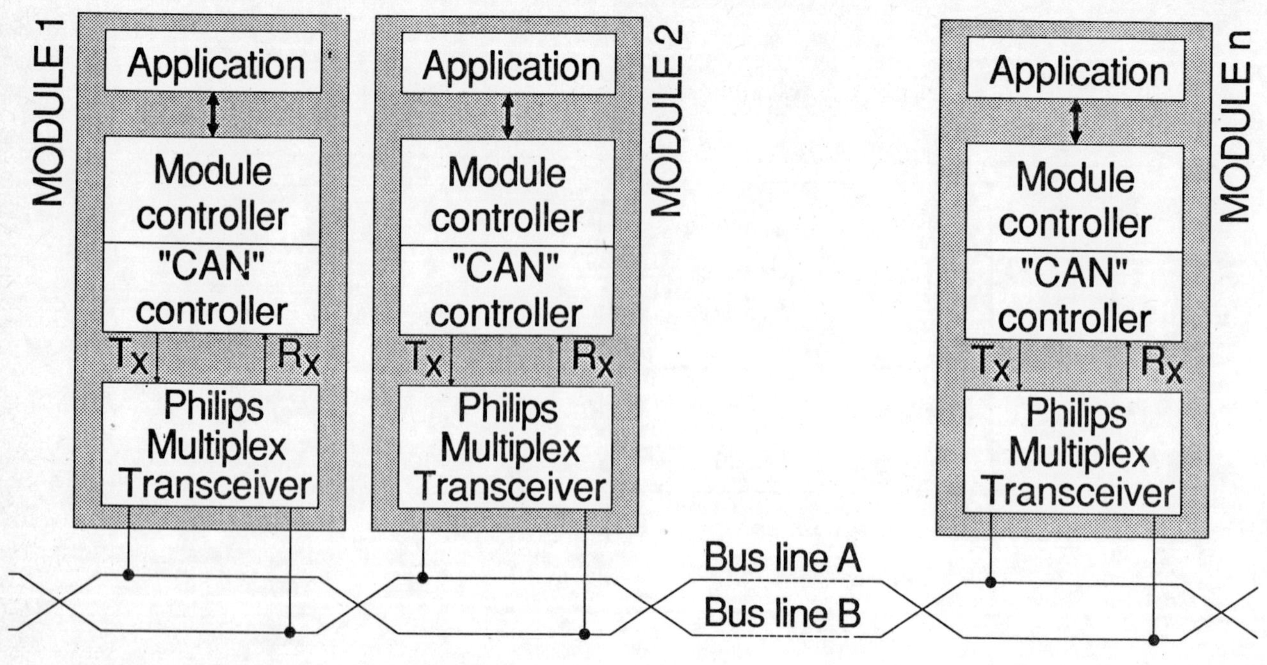

Fig 10 An in-vehicle network

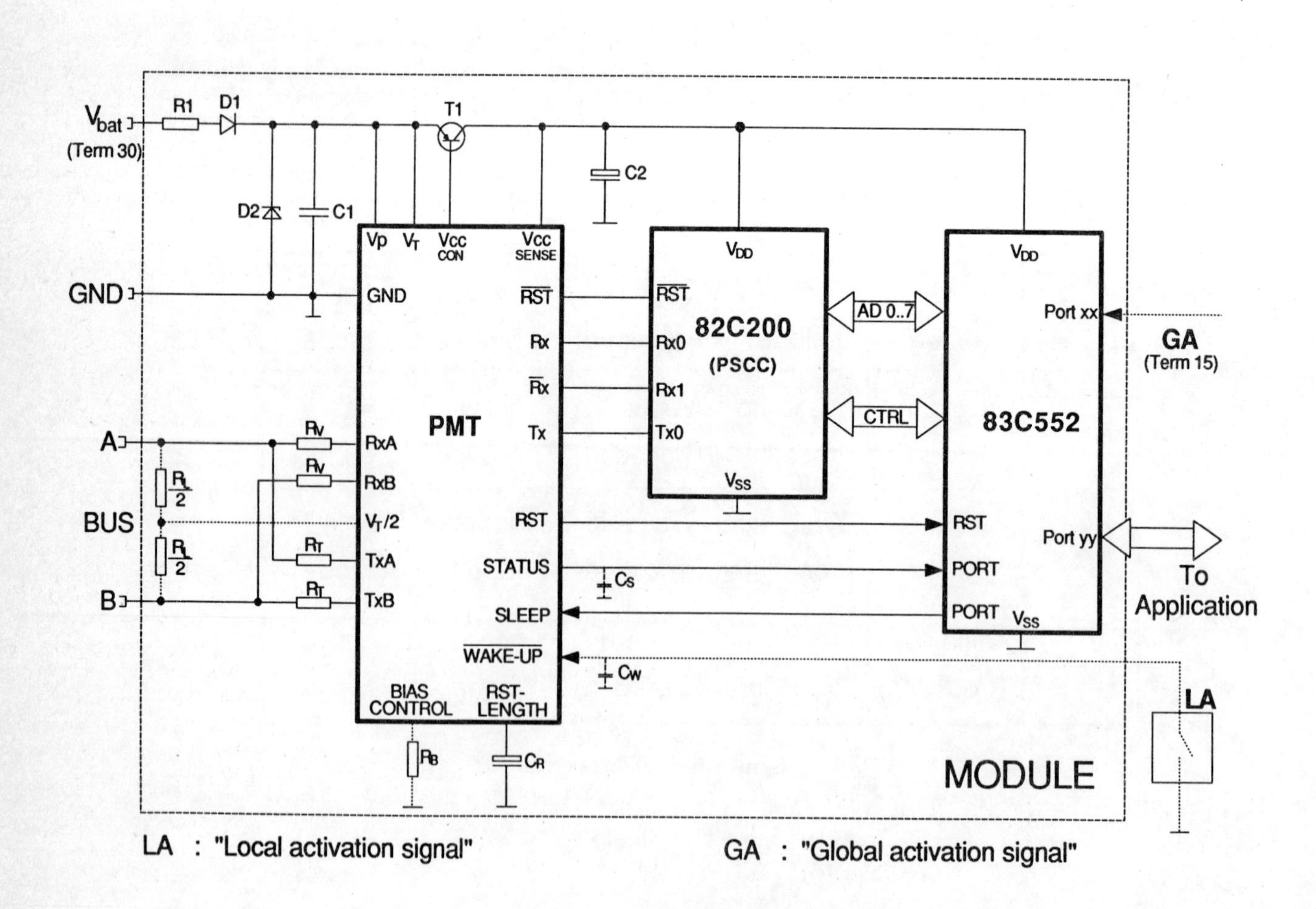

Fig 11 A single network node

C391/035

Advanced engineering measurement and information systems of future vehicle wiring systems—multiplex

F HEINTZ, Dipl-Ing and **W BREMER**, Dipl-Ing
Robert Bosch GmbH, Ettlingen, West Germany

SYNOPSIS Due to the possibilities of modern electronics the conventional vehicle wiring system can be completely redesigned in a new serial structure. The Robert BOSCH company has developed CAN (Controller Area Network) for serial data communication. This protocol is already used in control units with high speed data transmission and is being put forward worldwide as a proposed standard by the International Standardization Organisation (ISO). For slower Multiplex systems a special CAN unit (basicCAN) has been developed with considerably simpler hardware structure but with an identical protocol. A comparison between both systems shows that serial structures can be used in future with more and more advantages.

1 GENERAL

Whilst at the beginning of vehicle development, the wiring in the vehicle was a virtually unknown topic, even less a problem, the wiring harness today is an extensive, extremely jumbled network with leads measuring several kilometers in length. Unless countermeasures are taken, the extent and complexity of the vehicle wiring system will rise drastically over the next few years due to the constant increase in new electrical and electronic systems and assemblies. At present it is assumed that the on-board electrical network will increase in complexity by a factor of 2 every 10 years. (1) This is already leading to great difficulties, not to mention those to come in the future, in the area of the wiring system, e.g. in its implementation and re-equipment, and is also causing problems of space in the bodywork, wasted time in fault detection and diminishing reliability if current wiring structures are retained.

Electronic components available today - microprocessors, memories, power semiconductors etc. - offer interesting starting points for a new, simplified organization of vehicle wiring. Such novel wiring systems have become known under the name of "Multiplex". Similar system prototypes in vehicles were presented as early as the seventies by the Robert BOSCH company and several other companies. (3), (6), (9), (10)

Besides the generally accepted advantages of Multiplex systems (wiring harness standardization, reduction in wiring, simplified removal, ease of repair, increased vehicle reliability due to monitoring equipment contained within the system (7)), there are considerable improvements in important areas of vehicle wiring:

- Reduction in the documentation, drawings and costs involved in manufacture
- Redundancy of many small electronic systems
- Improved variance for different countries and customer equipment
- Improved testing of vehicle and individual components during manufacture and servicing
- Simplification of operating elements and switching logic.

Such aspects together with the constant drop in price of electronic components, particularly power electronics, leads one to expect that Multiplex systems may be produced in the future with advantages in cost and reliability.

2 SERIAL BUS SYSTEMS IN MOTOR VEHICLES

International committees (ISO, SAE) have long been working intensively to normalize and standardize serial bus systems. Depending on the intended application, buses with different rates of data transmission are to be used which, according to a suggestion from SAE, are designated Class A, Class B, Class C and Class D (Class A: approx. 1 kHz, Class B: up to approx. 100 kHz, Class C: up to 1 MHz, Class D: above 1 MHz data transfer rate). However, this classification is purely arbitrary and says nothing about the actual system configuration and the technical equipment used. Fig 1 shows as an example a diagnosis bus already standardized by ISO in a vehicle as well as a Class C system for control unit linkage and a Class B Multiplex system.

2.1 ISO Diagnosis

Via the diagnosis interface the condition of all control units with a corresponding interface can be diagnozed with the aid of simple equipment (lamps, ohmmeter) or with complex testers depending on the level of equipment available in the workshops. Normally no data traffic occurs on this bus lead. Only after a suitable stimulus of the control units to be tested do these send a test protocol with all diagnosis data. The transfer speed is variable and can be adjusted in dialog between the vehicle and the tester. The requirements for the diagnosis interface and its configuration are internationally standardized in the ISO Standard ISO/DIS 9141. (11)

2.2 Control unit linkage (Class C)

Serial buses with transfer rates up to 1 MBaud are used e.g. for control unit linkage (real-time control systems) and in the field of communication, as illustrated in Fig 2. Control units for ABS (Anti-Lock Braking Systems), ETC (Electronic Traction Control), gearbox control, engine management etc. exchange important sensor information, computation results and control signals at a high transfer rate (up to 1 MBaud). For this BOSCH has suggested a special transfer concept, the so-called CAN protocol (Controller Area Network) (4)(5)(8). In conjunction with the Intel company a stand-alone module (Type 82C900) was firstly developed for the CAN protocol, which is already available on the market. Other semiconductor firms will also be offering functionally graduated microprocessors with integrated CAN interface in the near future.

It can be seen in Fig 2 how communication systems, e.g. the car radio, telephone and future navigation systems are connected to one another via a CAN bus.

2.3 Multiplex system (Class B)

The data rate lies within the range of 10 kBaud up to a maximum of about 50 kBaud. Besides controlling the different loads (bulbs, motors etc.), this relatively slow bus serves to interconnect sensors, switches and smaller electronic systems if this is possible within the constraints of the transfer speed. Such systems are usually referred to as Multiplex systems.

In addition, internal and external diagnosis data can be distributed and processed via this bus. For reasons of uniformity, it is highly advisable to use the same protocol for a Multiplex system as for the control unit linkage. The lower data rate required means that considerably lower cost CAN modules can be used; these are known as basicCAN and will be soon available as stand-alone modules as well as integrated in low-priced microprocessors.

2.4 Bus network links

The diverse bus systems with their different transfer rates form independent areas operating without further external data communication. In several cases (e.g. for diagnosis purposes and for general driver information) it is necessary to exchange data between buses of different data rates. Suitable gateways between those buses should be provided, e.g. in a central display and control unit, as indicated in Fig. 2. Such a central display and control unit could also contain the interface of the external diagnosis, which, in this way, would possess a diagnosis link to all serial buses and the units connected to them.

3 MULTIPLEX SYSTEM

Fig. 3 shows some basic features and some of the differences between current wiring systems and a Multiplex system. Previously both the load selection and the power distribution were performed by the operating elements in the area of the instrument panel. The individual loads are supplied via one cable with the cross section required for the necessary total current. In modern vehicles the number of leads around the instrument panel has risen to far over 100 in passenger cars and commercial vehicles. The same applies e.g. for climate control systems and around the doors and seats particularly in luxury vehicles. The leads run via several levels of connection (fuse boxes, several frame plugs depending on location, relays) with several hundred individual contacts. Multiplex replaces this structure by a serial data transmission system with just one or a maximum of two information leads. The parallel signals from the operating elements are converted in this example into a serial data telegram by a participant station integrated in the instrument panel and then transmitted to the other participant stations. These are situated either directly beside the loads or load modules, or else they constitute a fully integrated unit together with the loads. There, the arriving serial telegrams are converted again into parallel commands and carried out e.g. via suitable "intelligent" power switches. These power switches must not only possess the ability to switch the individual loads on and off, but they must also include an overcurrent protection and transmit status information to the driver. This overcurrent protection is necessary since the fusing of each individual load is no longer possible. Only fire-break fuses (or over-current releases) for a complete load unit can be provided (see bottom of Fig. 3). In one typical area of vehicle wiring (cockpit + lamp units) alone, up to 50 % of the conventional contacts previously required can be saved. If one also considers the monitoring of loads, as it is standard in several vehicles today (failure monitor), the number of saved contacts increases even more dramatically.

3.1 Power switches

The power switches required in Multiplex systems must be able to switch ohmic and inductive loads, particularly the various lamps and direct-current motors. These loads are normally connected to ground on one side so that so-called high-side switches are generally required. They must contain a minimum of "intelligence" (control switching, protection, reporting the switching condition) in order to be used properly. Within the International Standardization Organisation (ISO/TC22/SC3/WG1), an expert group is working on the standardization of requirements for such IPS (Intelligent Power Switches), and particularly on the specification of switch characteristics, of short-circuit protection, assignment of connections and voltage drop ranges, in order to ensure the necessary compatibility between different products. Besides numerous power semiconductor manufacturers presently trying to produce completely integrated IPSs, several firms are working on a hybrid realization with a MOS power element and a separate controller chip. Such solutions have the advantage that they can be better suited to vehicle requirements, and particularly, comprehensive interference-suppression (limiting diodes, capacitors) can be directly realized in the hybrid. The result is a very compact component with only 5 external connections; Table 1 contains its most important data.

Table 1 Example of data for a hybrid IPS

Rated Voltage	12 Volts
Rated current	≤ 10 A (ohmic load) ≤ 5 A (inductive load)
Oper. temperature	- 40°C .. + 125°C
Voltage drop at 25°C at 125°C	(at rated current) ≤ 0.4 volts ≤ 0.8 volts
Quiescent current	≤ 100 μA
Current limiting	up to 70 A max. 10 μs up to 20 A max. 600 ms > 10 A after 0.5 s

A certain difficulty arises in detecting whether high currents are due to a short circuit or to the inrush-current occurring when bulbs and motors are switched on. (These currents may partly lie up to 15 times higher than the rated currents). When using such semiconductor switches as bulb drivers, it must also be taken into account that, due to luminosity requirements the voltage drop across the switch should not exceed 0.4 V. (Similar recommendations state a maximum of 0.6 V difference between battery and bulb connection (2).) These problems become much less serious in on-board networks of 24 V (the current is halved, the power loss is quartered). Besides simple "intelligent" switches for activating bulbs, half bridges are required, as are full bridges for reversible motors (these may be e.g. a combination of half bridges). Considering the typical current ranges in vehicles, it can be seen that approximately those versions listed in Table 2 will be required in future:

Table 2 Type spectrum for IPS

VERSION	RATED CURRENT RANGES			
	3 A	10 A	20 A	30 A
Simple switch	x	x	x	x
Half bridge		(x)	x	x
Full bridge			x	x

Considering 24 V networks in commercial vehicles, approximately 15 different types are required (this number may be reduced in future if the 24 V network is also installed in passenger cars). At present, semiconductor switches for currents above 30 A are not yet available so that relays will continue to provide the economical solution.

3.2 BasicCAN, Physical Layer

For data transmission both between the electronic control units and in Multiplex systems the CAN protocol can be recommended for reasons of uniformity and the advantages arising from this. For Multiplex with its low transfer rates (50 kBaud up to maximum 100 kBaud), the chip-size of the module can be reduced considerably. Compared with CAN, the entire part with message processing and acceptance filtering can be omitted. These tasks are performed by the microprocessor of the participant station. This station can be connected by both a capacitive and a resistive coupling, the so-called physical layer, to a (parallel or twisted) two-wire bus lead. The connections are designed for minimum leakage currents so that the system can run on, at least in part, with the ignition switched off (necessary for parking lamps, hazard warning lamps etc.). Furthermore, short-circuits or interruptions in one of the two bus leads do not cause a system failure. Data transfer continues uninterrupted on a single lead. A bus monitor indicates this failure. What is even more important, the entire circuit is protected against typical on-board network interference voltages. The left-hand side of Fig 4 illustrates the suggestion for a capacitive bus connection in which the bus leads are connected to + 5 volts and ground via high-ohmic resistors. The bus signals reach the receiver inputs via a network of resistors, a control transistor and appropriate linking capacitors. This electronic circuit has the advantage that variations in direct voltage and potential displacements on the bus leads do not cause any malfunction. Due to the low number of external components it can be constructed in small and compact form as a SMD module. On the right of this diagram is the suggestion for a resistive bus connection as is particularly suitable for integration so that the external components do not become too large. Besides the CAN interface, a special detection and switching logic circuit is integrated into the processor which distinguishes between short-circuits in the bus leads and different potentials and automatically switches to the lead remaining intact. The driver transistors of the CAN module are able to supply up to approx. 100 connected stations at a transfer rate of 50 kBaud over a line length of around 130 m (at 100 kBaud this is still 50 m). This is more than sufficient for Multiplex applications. If necessary the system can also be operated in the vehicle merely with a single data lead, which reduces the number of contacts. However, the redundancy of a two-wire lead is lost. The inevitably delayed execution of a command occuring in serial transfer, is negligible even at the low transfer rate of 20 kBaud. Table 3 shows the reaction time to a so-called "stress operation". Depending on the priorities selected, this results in delays which are absolutely uncritical.

Table 3 Worst case delays

Simultaneous control of all assemblies via CAN 20 kBaud. Priority: 2000 highest, 0 lowest.

Assembly	Priority	Execution (ms)
Windscreen washer	4	49.8
Windscreen wiper	99	19.0
Stop lamp, right	13	7.1
Stop lamp, left	14	8.1
Direction indicators	102	14.2
Low beam	104	9.5
High beam	103	11.8
Headlamp flasher	1250	4.7
Foglamps	101	16.6
Horn	1300	2.4
Window lifter	7	45.0
Temperature sensor	30	21.3
Fan	5	47.4

Each individual signal was allocated its own priority for the worst case estimation, which in practice is not essential. The switching commands merely consist of ON/OFF information which only requires 1 bit each and so several other items of switching information can be compiled into a telegram and transmitted as a single object. So several signals can be transmitted simultaneously with only 2.4 ms delay.

3.3 Participant station

On the right of Fig. 5 are the details of a complete participant station as is required in the Multiplex system. The station is of hybrid construction with a microprocessor chip (with basicCAN and - if necessary - with A/D or D/A converters), a power switch hybrid (4 to 7 switches in appropriately graduated current ranges as required) and with the protection switching for the sensor and operating switch inputs. In the future, participant stations will certainly also be supplied integrated within the loads, e.g. complete front and rear lamp units or complete engine assemblies which can be directly connected to the Multiplex cable. On the left of Fig. 5 is an "intelligent" seat adjustment motor, the housing of which contains a complete participant station. A potentiometric position information is available for the seat-memory function. Such an "intelligent servomotor" possesses only 4 external connections, i.e. the two voltage supply leads and the two Multiplex leads.

3.4 Network topology, examples

The entire system can be wired in any star or circular structure. However, for reasons of safety, particularly the energy supply will be in star form with the positive side effect that the lead cross sections are not too large and unmanageable. The individual leads are protected against fire hazard with safety fuses or overcurrent releases. In case of particularly stringent requirements on data transfer safety, the structure of the information lead can also be star-shaped, although in this case a bus monitor must be provided. Fig. 6 illustrates the design for a full Multiplex system in a luxury vehicle. Two distribution points are present from which the total of 13 interactive stations in the vehicle are driven.

Multiplex wiring systems can be used with particular advantages in commercial vehicles, especially omnibuses. It is possible, with only 10 participant stations in a common omnibus to reduce the wiring harness with more than 100 leads going from the front main instrument panel to the rear distribution station to a standardized Multiplex-cable of only 14 leads, as we could demonstrate in a test car. The failure monitoring of the entire system and of the connected instruments can be integrated into the system in an extremely simple way. The necessary monitoring information is automatic and virtually cost-free, since the status of the connected load is determined continuously and can be transmitted e.g. via the intelligent power switch (bidirectional bus structure). (7), (10)

4 SYSTEM EVALUATION, ADVANTAGES, PROBLEMS

The main advantages of Multiplex systems can be found in the following areas:

* advantages in operating elements and displays
* advantages in wiring
* advantages in storage and manufacture
* saving in individual control units
* advantages in different vehicle versions

The opportunities for simplification in the area of the operating elements, for example, are shown particularly impressively. Typical switches mostly have several switching levels and up to 15 external contacts (partly designed for very high currents), which can be redesigned in almost all cases to a maximum of 4 contacts in a simple circuit structure with minimal currents flowing. The following Table 4 illustrates the total possible savings in contacts and wire cross sections for 5 typical switches.

Table 4 Savings in operating elements

	No. of cable connections		Total cable cross section [mm²]	
SWITCH	Stand.	MPX	Stand.	MPX
Light	15	4	15,5	2,0
Fog-Light	5	4	3,5	2,0
Rear Fog-Warning Light	4	4	2,75	2,0
Hazard-Warning	10	4	11,0	2,0
Heated Rear Window	6	5	3,0	2,0
Total	40	21	35,75	10,5

The advantages in the field of the operating elements and displays can be summarized as follows:

* Greater reliability by means of less moving parts and greatly reduced contact stress, longer service life
* More simple testing and modification by transferring the logistics into the software
* Lower storage costs by means of standardized wiring harness components
* Reduction in price by means of:
  - simplification of switches (mechanical, electrical)
  - larger quantities (standard elements)
  - less space required for installation
  - advantages in weight (mounting of switches)
* Pre-assembly of the instrument panel is possible
* New ergonomic design freedom in the area of the instrument panel by means of simpler and smaller operating elements and more flexible wiring
* Simple control of graphic instrument panel displays

In the area of the wiring, the most important advantages may be summarized as follows:

* Saving in weight by using wiring harness with a low number of connections
* Uniform plug connections (number and variety)
* Less connecting procedures (assembly, manufacture)
* More flexible laying of cables (bending radii)
* Smaller cable lead-throughs and simpler sealing
* Simplification in assembly in serial manufacture and servicing due to:
  - uniform wiring harnesses
  - possibility to pre-assemble modules (doors, instrument panel)
  - simple testing of pre-assembled modules
  - simplified functional control and error elimination by means of self-diagnosis

Particularly the possibility to pre-assemble modules and the simple, largely automatic testability of individual modules and of the entire vehicle Multiplex wiring harness with its connected components should be emphasized here again. Using suitable testers - simple test units or standard PCs - all components connected to the serial CAN bus can be activated and tested. Visual inspections (window lifters, wipers, lighting) are also possible, as is the checking of sensor functions (fuel tank level, oil pressure, temperatures). Similarly, the emergency functions can be checked by fault simulation. The following cost-reducing advantages present themselves in the areas of storage, manufacturing preparation and control:

* No different versions of wiring harness (logistics performed by the software), modular system
* Reduction and simplification of manufacture documentation in design, testing and work preparation
* Changes of function when changing models can be performed inexpensively (software)
* Different national versions do not require different versions of wiring harness and electrical system (standard hardware is operated with different software)
* Fully automatic testability of the entire electrical wiring (servomotors, switches and sensors) in vehicle manufacture at the end of the production line
* Lower costs by reducing finishing operations
* Reduction in warranty requirements
* Reduced assembly time due to simplified wiring harnesses and concentration of components in the control units

An important point of view, often overlooked when considering Multiplex, is the possibility to replace virtually all small control units in the vehicle by such a system. In their place, numerous more elegant solutions better suited to the requirements can be found due to the microprocessors in the individual participant stations. (Examples are: pulse-width modulated motor controls, speed-dependent windscreen wiper control, trip recorder functions, map control, improvement in control and regulation processes by fully utilizing the entire sensor information available, service support thanks to fault registration - particularly important for sporadically occurring faults, such as e.g. wobbly contacts). Some examples for small control units which could be completely dispensed with due to Multiplex are listed below:

* Check control
* Bulb warning unit
* Control unit for delayed interior lighting
* Central information unit
* Control unit for seat-belt warning lamp
* Rear windscreen wiper control relay
* Control unit for central locking system
* Anti-theft warning system and radio monitoring
* Control unit for rear windscreen heating
* Hazard warning lamp relay
* Control unit for wash-wipe interval
* Control unit for kick-down function
* Control unit for auxiliary washer system
* Control unit for headlamp washer system
* Control unit for seat adjustment and seat memory

To provide a sensible answer for the question of the costs of a Multiplex system, particularly in comparison to conventional wiring, all the advantages listed above must be systematically assessed and an assignment of possible cost savings must be derived. An often performed simple cost comparison between only the hardware components - on one hand the conventional wiring harness, on the other hand Multiplex and electronic components - inevitably leads to a false picture of the situation and a wrong assessment of this new technology. A detailed analysis considering all cost-relevant factors shows that, even from today's point of view, Multiplex can lead to clear cost savings at least in some areas of passenger cars and in the entire field of commercial vehicles.

REFERENCES

(1) EHLERS, K. WILLE, R. Das Automobil und seine Steckverbindungen. VW-Zeitschrift Impulse.

(2) DIN-Norm 72600 Elektrische Spannungen an Glühlampen in Kraftfahrzeugen und deren Anhängern

(3) VDO-Schrift Querschnitt 1: Signalbus

(4) KIENCKE, U. DAIS, S. Application Specific Microcontroller for Multiplex Wiring SAE Technical Paper 870515

(5) KIENCKE, U. DAIS, S. LITSCHEL, M. Automotive Serial Controller Area Network SAE Technical Paper 860391

(6) DIODATO, C. PERISSINOTTO, B. TREVISANI, R. Electric Wiring System - An Electronic Management for Busses SAE Technical Paper 850449

(7) HEINTZ, F. BREMER, W. Möglichkeiten der fahrzeugfesten Überwachung mit Kabelbaum-Multiplex. BMFT 78, S.244ff

(8) BOTZENHARDT, W. LITSCHEL, M. UNRUH, J. Bussystem für Kfz-Steuergeräte. VDI-Berichte Nr.612 S.459ff.

(9) DILLENBURG, R. HEINTZ, F. ZABLER, E. Multiplexsystem als Kabelbaumersatz im Kraftfahrzeug. BOSCH Technische Berichte Band 5 (1975) Heft 2, S.91ff.

(10) BREMER, W. HEINTZ, F. Was bieten die Karosserie-Zentral-Elektronik mit Kabelbaum-Multiplex und die On-Board-Diagnose; BOSCH, Kundenbrief Nr.14, Mai 1977

(11) HEINTZ, F. Normung einer Diagnoseschnittstelle. Internationaler Norm-Entwurf ISO/DIS 9141. ATZ Automobiltechnische Zeitschrift 89 (1987)

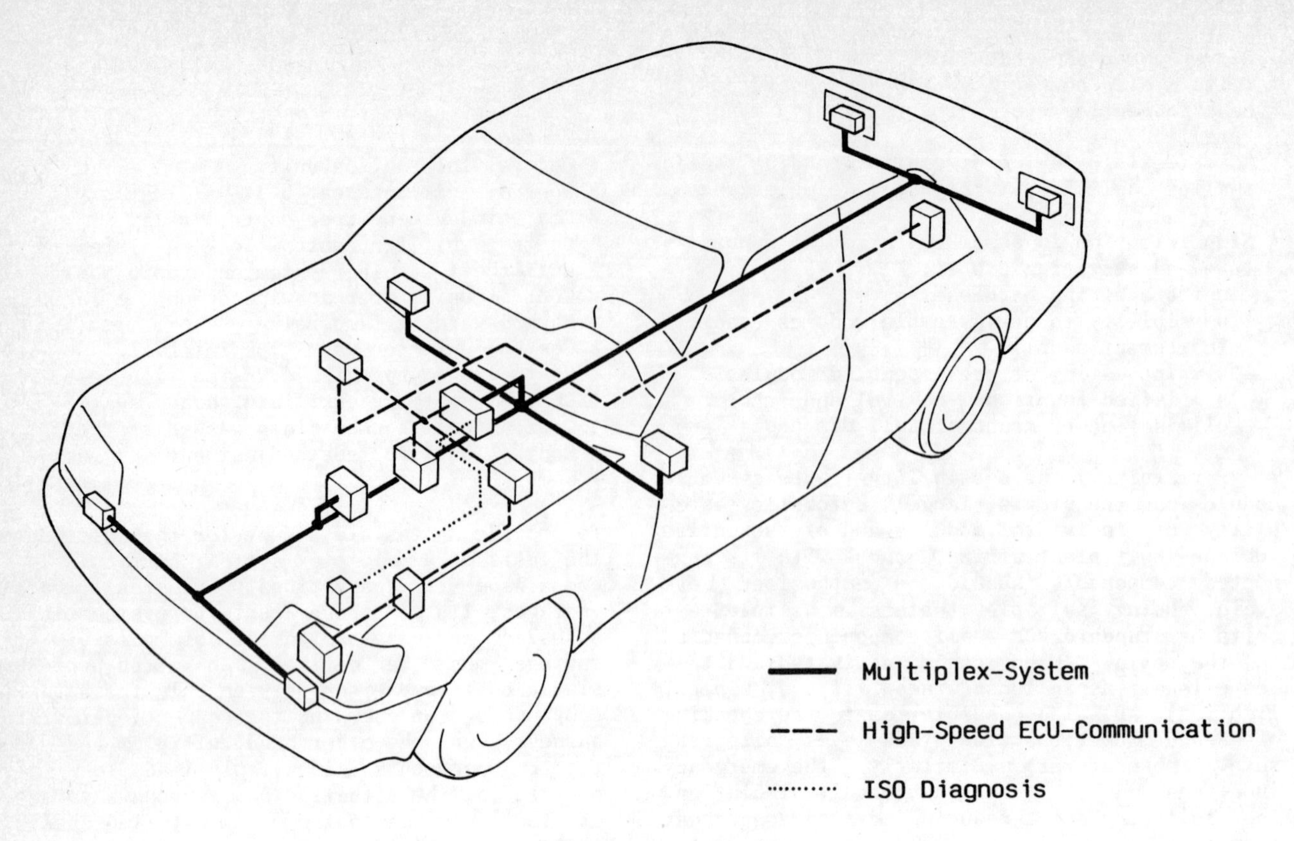

Fig 1 Different serial buses in a vehicle

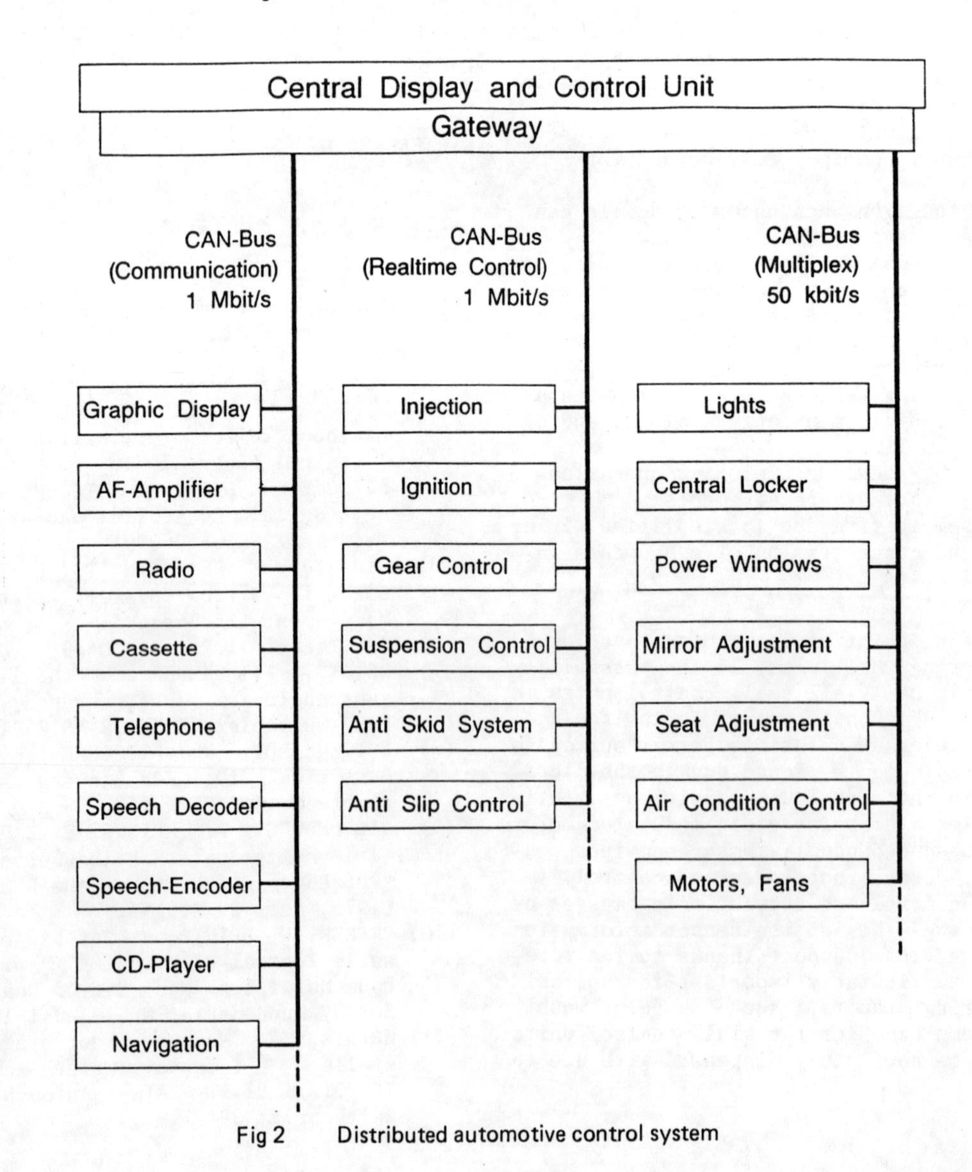

Fig 2 Distributed automotive control system

CONVENTIONAL WIRING

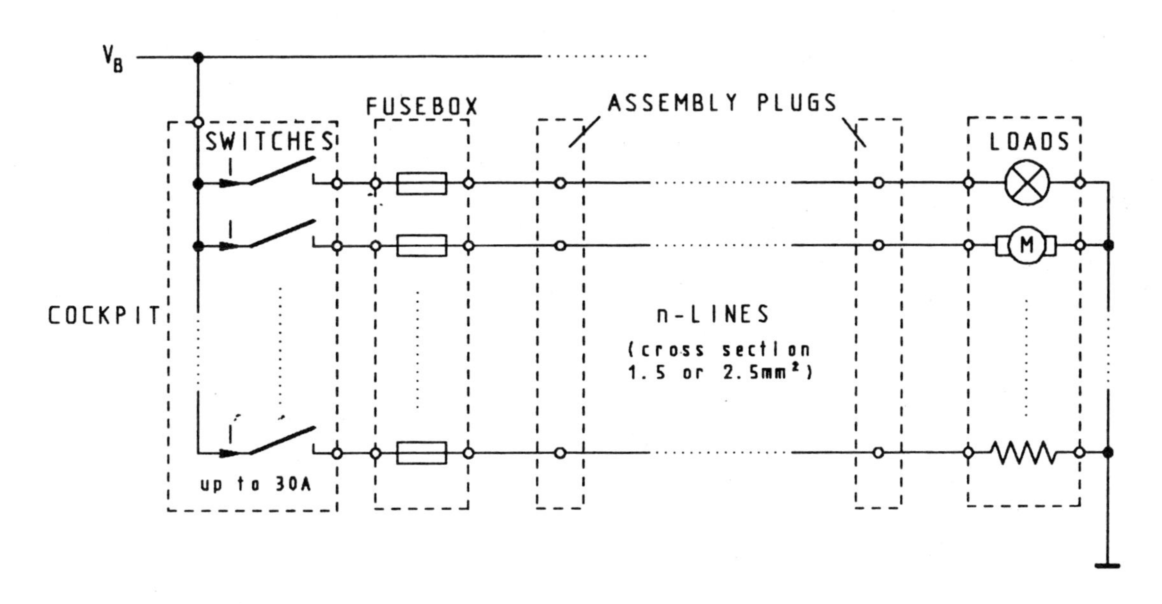

MULTIPLEX WIRING

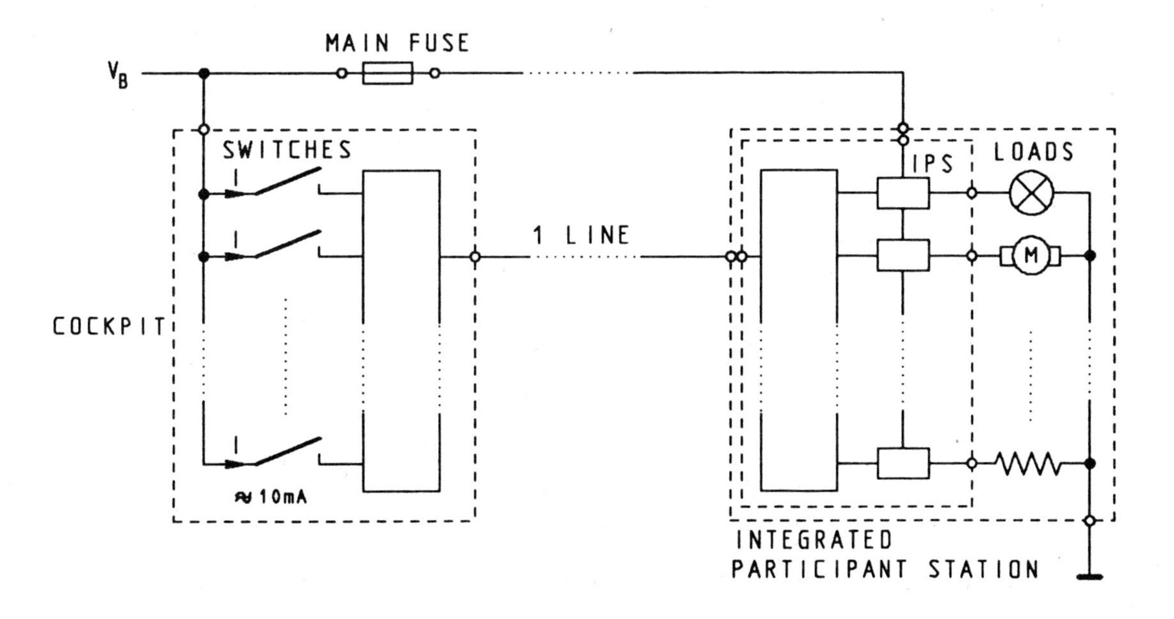

Fig 3 Comparison between conventional and multiplex wiring

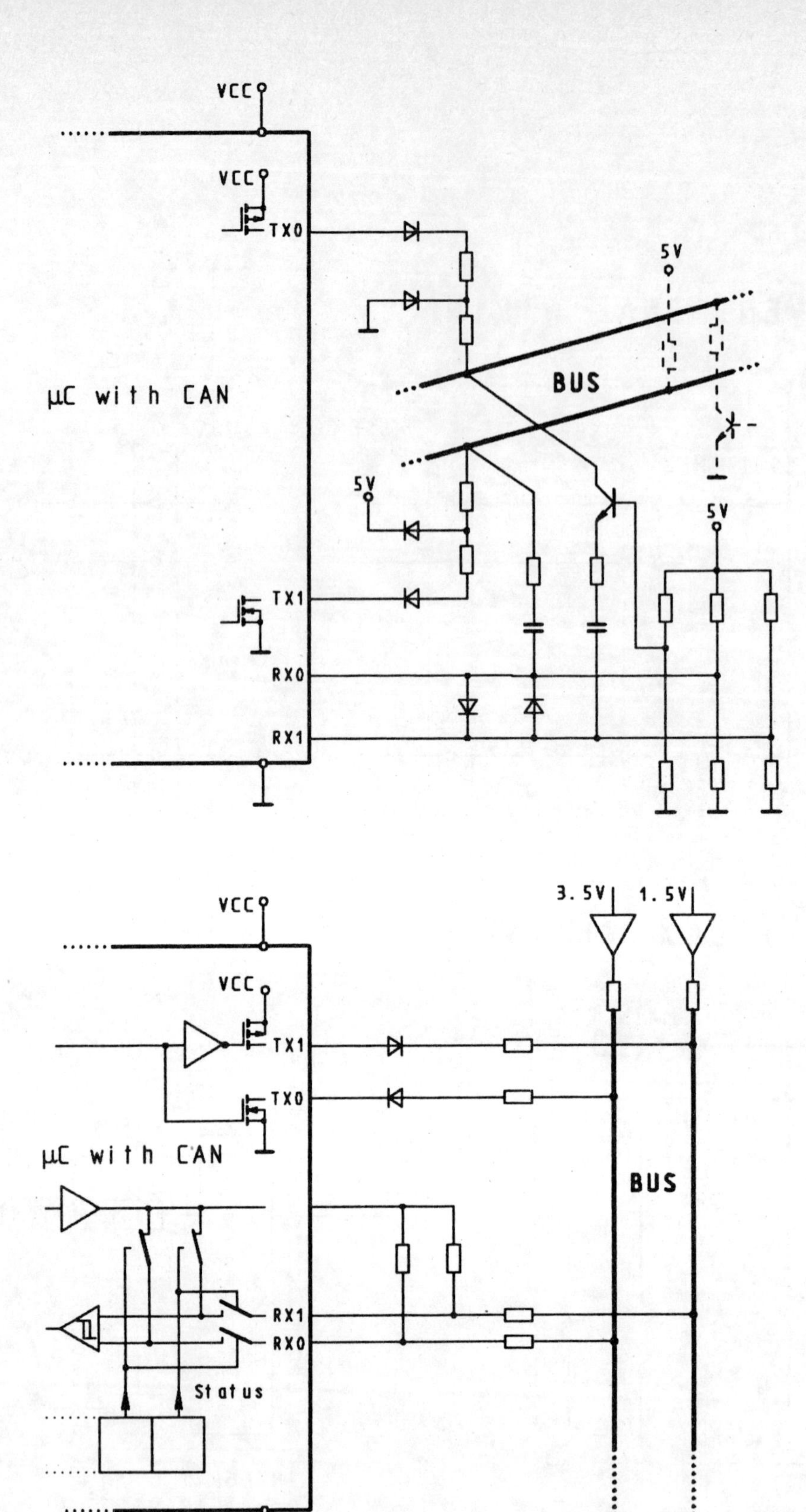

Fig 4 CAN physical layer

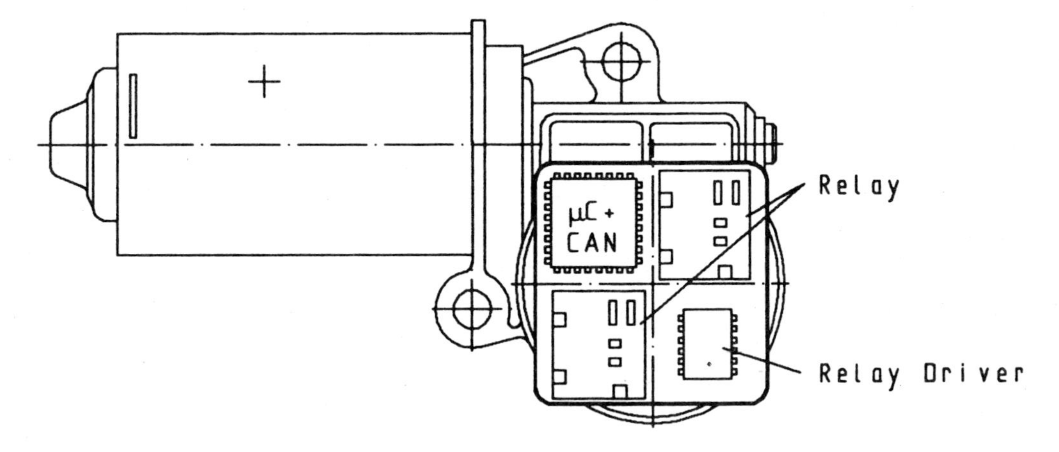

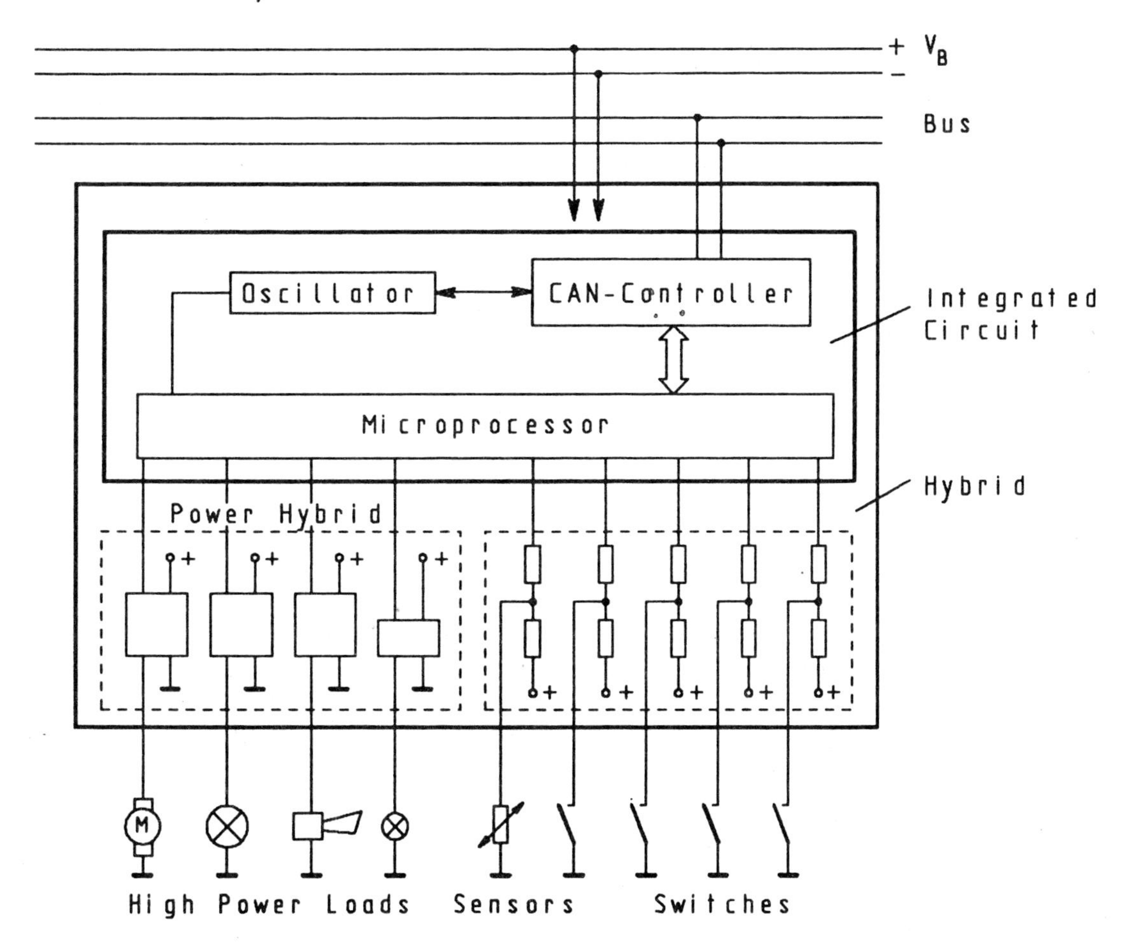

Fig 5 Participant station

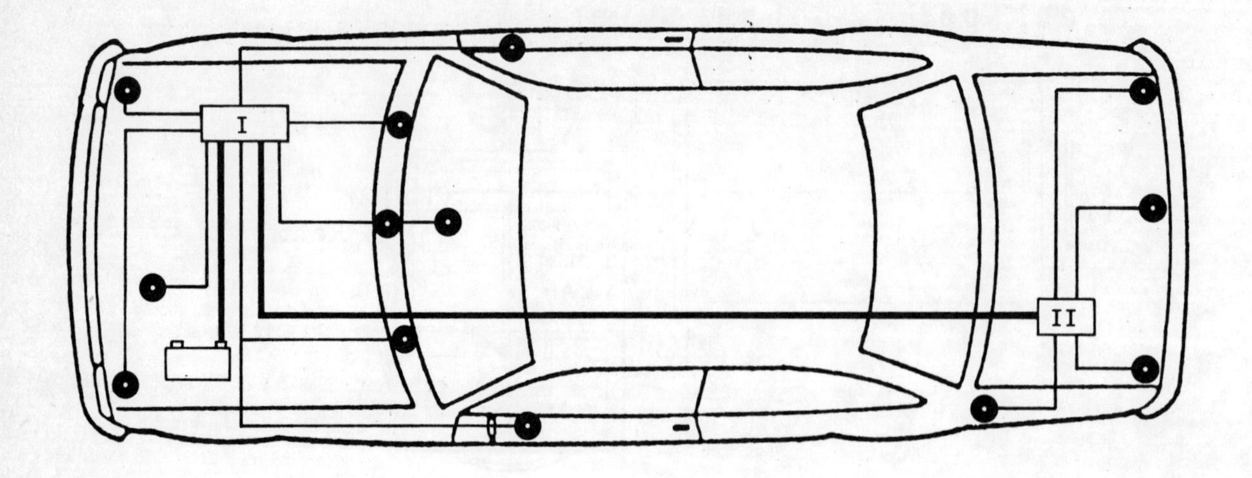

Fig 6 Multiplex wiring in a vehicle

C391/012

An automated test and programming system for electronic control modules

L RETTNER, Dipl-Ing(FH)
Siemens AG, Nuremburg, West Germany

ECOS-i - an automated Test- and Programmingsystem for Electronic Control Modules

ECOS-i utilizes the characteristics of ECU controlled systems to check the functioning of the complete circuits: ECU, sensors, actuators and cable connection.

Programming of ECUs at the Assembly Line:

The amount of ECU variants can be reduced drastically by programming the correct set of firmware parameters into the ECU after it has been built into the vehicle. This feature has to be supported by the ECU firmware.

Quality requirements in the automotive industry have increased. In order to keep repair costs down it is necessary to find faults and repair them before delivery. The repair costs in the plant are about 15 % of the costs in the garage.
As a result of the revolutionary developments in the field of electronics and microprocessor-technology more and more "intelligent" electronic control units are being built into cars.

These ECUs are also used for components which are essential for the functioning of the car.

For example:
- engine management
- transmission control
- anti-skid system
- 4 wheel drive
- on board computer
- instrument cluster

The introduction of these ECUs has increased the driver's comfort, performance of the car and efficiency of the engines, but also the complexity of sensors, actuators and harnesses.

It is therefore very important to check the car thoroughly before delivery. Siemens has developed the menu-driven programmable test and programming system ECOS-i for this purpose:

<u>Test System configuration</u>
(picture 1)

The specific demands of the factory environment had to be taken into account in selecting the test station components.

Special demands were:
- high speed test run because of high production rates
- fast response to operators' inputs
- adjustment to existing vehicle identification systems
- test-station components must be rugged

1. <u>Identification system</u>

Since the vehicles are produced to customers' specifications the variety of specifications is correspondingly large. The test program has to be adjusted to the vehicle specification. The information about the vehicle - the vehicle data record - has to be fed into the test system.
Examples:

- Barcode reading of production number
  The production number is the key to read the appropriate vehicle data record.
- computer connection to production scheduling system
  A production scheduling system transmits the vehicle data record to the test system at the right time.
- automatic identification system
  A data carrier is fixed to the vehicle and contains the vehicle data record which is read out of the data carrier by a reading station connected to the test system.

2. <u>Dialogue System</u>

The dialogue system consists of an input/output device; either a wireless or wire-connected handheld terminal or a monitor with IR-remote control. It is used for operator guidance, for example "change gear", "accelerate" and to feed back information into the system, for example "start test", "acknowledge".

3. <u>Result Printer</u>

At the end of the test the result OK/NOK and, if wanted, certain set points and actual values are printed out.

4. Check Out System Interface (serial connection to the vehicle)

The connection to the vehicle is established by the COSI which converts the messages from and to the test computer into a form which the ECU can understand. The COSI is transparent regarding the information (physical parameters, message format, baudrate).

5. Digital Outputs

Digital outputs enable the testsystem to control test station components. It is possible, for example to switch the brakes of the rollers in the roller test bed on or off, depending on the engine temperature read out from the ECU.

6. Test Station Computer

The test station computer runs the test programm, controls all the test station components and transfers the test results to the host computer for statistical evaluation.

7. Host Computer

The test parameters are entered and maintained via interactive menus on the host computer. The host computer supplies the test computer with parameters via copy, archive and transfer menus.

The test results coming from the test station are received and stored in different statistical evaluation files. Actual value distributions can be printed out and error bursts can be printed out.

It is also possible to route the test station print-outs from one test station to the host and from there to another test station to produce one complete print-out at the end of the test-line.

8. Test Program

The "system operator" doesn't need computer-specific knowledge to assign system parameters, since every input is menu driven.
A test program consists of
- sequences
- test steps

The sequences define the order of tests.
A sequence has the following parameters:

sequence number:
defines the order of processing
ECU-number:
points to a list which contains ECU-specific data; for example ECU-address
LOR: (logic operation record)
determines whether this test sequence will be processed by comparing with the vehicle data record.
condition flag:
The flag number and the status of the flag (0 or 1) determine whether or not the sequence is to be processed. The status of the flags can be affected by the result of previous tests.
result flag: (see condition flag)
text:
This text appears on the dialogue system during processing of the test sequence.
timeout:
Defines timeout in the event of a test condition not being fulfilled.
test step number:
Up to five test steps per test sequence can be processed.
options:
Result is not intended for statistical evaluation.
Test abort if test is NOK.
digital output:
A digital output can be set for one second.

A test step has the following parameters.
Test step number:
reference for the sequence
condition flag: (see sequence)
text: (see sequence)
test type: The test type number causes one of the following test types to execute:

a) status comparison of different bits in a certain ECU-byte
b) tolerance check of an ADC-channel
c) read error memory of ECU
d) tolerance check with period count (-check)
e) ECU-identification number check
f) delete error memory

9. Connection of the vehicle, start of communication (pictures 2 and 3)

The ECUs are equipped with a serial two-way data link. The data links of the different ECUs are all connected to the same pin on the diagnostic socket. The test-system is connected to this diagnostic socket.

In order to access a specific ECU to start a test programm, all devices connected to this diagnostics pin must be initiated by the same procedure.

The initiation procedure is started with an idle line, after which the ECU specific address is transmitted to the ECU (transmission rate 5 bd). Once the ECU recognises its own address, it responds with a syncronisation byte (55 H) at a specific transmission rate. This byte is used by the COSI to measure the baud rate and adjust to it.

From now on the data transfer will be carried out at this transmission rate.
The ECU sends two key bytes, the second key byte is complemented by the test system and returned to the ECU. This concludes the communication initiation procedure. Data is now transformed with a procedure which can vary for each ECU. The communication is terminated either by a message or by a wire break signal on the diagnostic link.

10. Example of message structure

In the system presently implemented, a distinction is made between two different message types:

a) request message with relevant reply message, for example:

- test system requests "read RAM"
- ECU answers by outputting the contents of the RAM

b) command message with relevant ready message, for example:

- testsystem commands "delete error memory"
- ECU answers "ready"

request messages:	responses:
read ECU identification	output ECU identification
read snapshot	output snapshot
read RAM	output RAM contents
read ROM/EPROM/EEPROM	output ROM/EPROM/EEPROM contents

command messages:	acknowledge messages:
delete error memory	ready
write RAM	ready
write EPROM/EEPROM	ready
end diagnostic mode	ready

11. Programming of ECUs

ECUs can be programmed with the function "write EEPROM".

In order to reduce the variety of ECUs, which depend on vehicle type, engine type and special fittings, only one ECU-type is built into the car. The vehicle is connected to the test system at a programming station (similar in design to a test-station but without a dialogue system).

After identification of the car, the computer selects the correct data record for this ECU and transfers the data with the function "write EEPROM" to the ECU. After that, the function "read EEPROM" is used to check that the data has been correctly transferred.

"End of line" programming is already in use for on board computers. Programming for engine management units is being planned.

Conclusion

The test- and programming system ECOS-i can be used for vehicle specific testing of the ECU functions with sensors, switches and actuators built into the vehicle.

The test-parameter can be adjusted to the vehicle and test-station characteristics by the car manufacturer, via interactive menus.

Statistical analysis can be used to quickly identify errors and to ensure constant quality. The variety of ECUs can be reduced drastically with the programming function,and technical changes can be brought into production much more quickly.

appendix:
ECU: electronic control unit
ECOS-i:electric check out system for ECUs
COSI: check out system interface

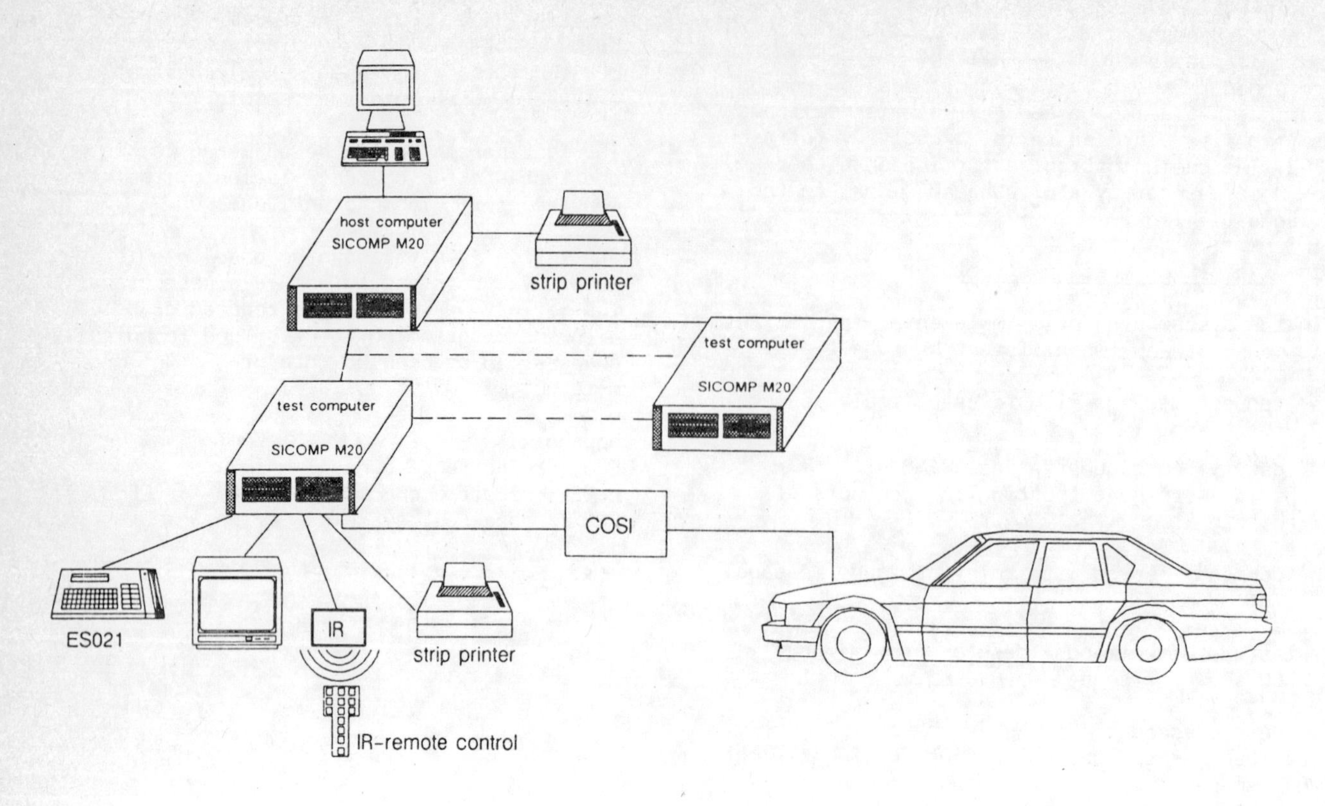
host computer
SICOMP M20
strip printer
test computer
SICOMP M20
test computer
SICOMP M20
COSI
ES021
IR
strip printer
IR-remote control

Fig 1

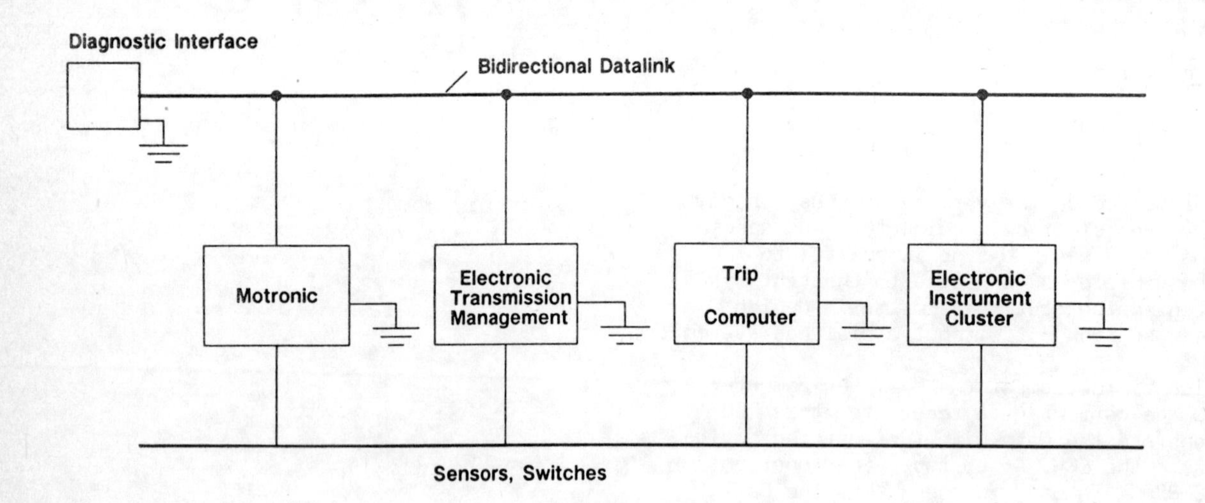
Diagnostic Interface
Bidirectional Datalink
Motronic
Electronic Transmission Management
Trip Computer
Electronic Instrument Cluster
Sensors, Switches

Fig 2

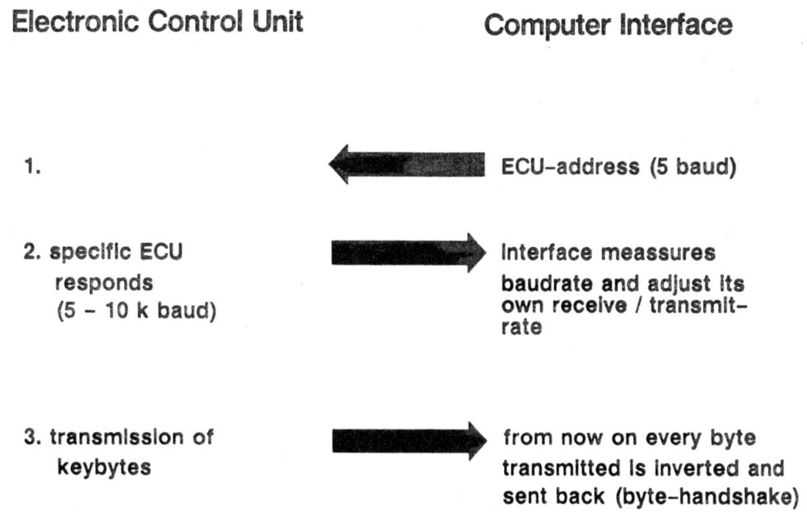

Electronic Control Unit		Computer Interface
1.	←	ECU-address (5 baud)
2. specific ECU responds (5 - 10 k baud)	→	Interface meassures baudrate and adjust its own receive / transmit-rate
3. transmission of keybytes	→	from now on every byte transmitted is inverted and sent back (byte-handshake)

Fig 3

C391/077

Innovative concept for high-speed I/O processor dedicated to engine control system

F AUSSEDAT
Texas Instruments France, Velizy-Villacoublay, France
G MAUREL
Texas Instruments France, Villeneuve Loubet, France

SYNOPSIS : On account of the increase of complexity of the engine control systems, microcontrollers of Electronic Control Units need to be more sophisticated above all regarding the process of timing functions which is the heart of the sytem.
Conventional microcontrollers use timers with capture and compare registers to achieve the control functions required. However, the CPU needs to interact to a large extend with timers and dedicated registers. This interaction reduces dramatically the processing power available for the other functions.
TEXAS INSTRUMENTS made a major step forward by introducing the modular TMS370 microcontroller Family which allow "custom" microcontrollers with dedicated engine control modules.
The High Speed Control and Acquisition (HSCA) module is one of these modules. By combining input capture and output compare features, event counter and timers which are all software definable within a flexible RAM based organisation, time, event or event plus time related tasks are able to be set up in advance of their action and then release the CPU to operate independently of these tasks.
Since all the HSCA control information is held within the register RAM, the CPU has easy and fast access to these parameters (i.e, duty cycle update of PWM outputs automatically driven by the HSCA module).
The paper describes in detail how signals in engine control system such as sprak time event, dwell time generation, dwell time feedback, fuel injection time, PWM signal outputs, knock window generation can be automatically driven by the HSCA and this completely independently of the CPU.
The paper presents a static ignition and multi-points sequential fuel injection system using a TMS 370 with the HSCA module.

TMS 370 FAMILY OVERVIEW

The TMS 370 Family consists of VLSI 1.6 uCMOS modular microcontroller which integrate advanced functions such as program EEPROM, data EEPROM, program EPROM, program contact ROM and a library of peripheral modules such as A/D converter, Serial Peripheral Interface,Serial Communication Interface, HSCA, Timers, custom modules ...

The TMS 370 CPU is a register to register oriented 8-bit CPU where access time to the register file is a single system clock cycle. It instruction set contained in a control ROM, can be optimized to the applications (dedicated intruction set).

As the TMS370 family was developed with a modular design methodology, the right configuration can be specify for each kind of engine control system (size of ROM, RAM, EEPROM, choice of standard modules, integration of custom modules ...), and can be immediatly emulated using the configured TMS370 XDS in circuit emulator.

HSCA INTRODUCTION

The High Speed Control and Acquisition (HSCA) is an automotive oriented peripheral module which shares the internal register file with the TMS 370 CPU.

Since all the HSCA control information are held within the Register RAM, the CPU has easy and fast direct access to these parameters. Therefore, the RAM access is shared between the 32-bit HSCA bus and the 8-bit CPU bus.

The HSCA has six input capture pins with dedicated 32-bit capture registers implanted inside the register file. An 8-bit event counter is associated to input capture features.

Up to 16 output pins can be driven by output compare functions. The HSCA provides a 20-bit default timer with 16-bit virtual timers with independent programmable overflow.

A software configurable watchdog is integrated thus a simple Serial Communication Interface (diagnostic port) with independant software set up of baud rate for receive and transmit lines.

18 to 24 independant interrupt vectors with 2 software selectable priority levels are provided associated to inputs, outputs, SCI and timers functions.

HSCA ARCHITECTURE

The HSCA is a RAM based module and as such, occupies an area of the internal register file. The RAM size used by the HSCA is determined by the functions selected by the programmer. Seen from the HSCA, the RAM is organised by 32-bit words.

At the top of the RAM (figure 1) an image of the 20-bit default timer, a copy of the flag bits for capture pins and an image of the 8-bit event counter are always held. There after six 32-bit dedicated capture registers are implemented, which are followed by a circular 32-bit captures storage buffer. Below is the area used to store HSCA commands and definitions (again by 32 bit words).

Both the circular buffer and the COMMANDS/DEFINITIONS area lengths are software defined by the programmer.

All the RAM below the end of the command area are general purpose registers. However, it should be noted that as all the functions above are also within the register file, the data in these locations may be directly manipulated with normal register based instructions.

The HSCA module can be considered as having 2 distinct, but related sections. The first one deals with all inputs related functions (timer capture, event counter etc...), the second one is the COMMANDS/DEFINITIONS area covering timer definitions, output related actions (setting or clearing a pin) outputting PWM signals etc...

INPUT ARCHITECTURE (figure 2)

Six input pins (CP1-CP6) are available for data capture on which rising edge, falling edge or both can be selected to capture 20-bit default timer value and the 8-bit event counter value. For each input, a hardware interrupt can be generated on the selected edge and each input has a dedicated 32-bit capture register into the register file. In addition, CP6 pin has a software selectable choice to work as event counter pin only (to increment the HSCA 8-bit event counter) or event pin and timer capture pin (in circular buffer).

The circular buffer can store 32-bit capture values (20-bit default timer value, 8-bit event counter value and pin capture identification) triggered directly by the input pins (CP5-CP6), and/or 16-bit timer values trigged by Commands in the COMMANDS/DEFINITIONS area.

In the Engine Control system described in this paper, (figure 7) the crankstaft reference is entered on a TMS370 harware interrupt pin (INT3). This information is used as a reference for the sequential fuel injection.

The Engine speed signal, provided by a toothed wheel, 60 teeth minus 2, drives the CP6 pin in the mode event counter pin only (the two missing teeth of the toothed wheel allow to determine the position of the Top Dead Center of cylinders). The event counter which is cleared each Engine cycle (every two engine rounds) by a command in the COMMANDS/DEFINITIONS area allow to number each tooth of the toothed wheel.

In addition, commands always in the COMMANDS/DEFINITION area ("double event compare" command) might generate interrupt requests on a programmed number of tooth. That means, as each

REGISTER FILE

<— 32-bit —>

8-bit Event Counter	20-bit default timer	
storage Event Counter	CP1 20-bit capture register	CAPTURES
storage Event Counter	CP2 20-bit capture register	
storage Event Counter	CP3 20-bit capture register	
storage Event Counter	CP4 20-bit capture register	
storage Event Counter	CP5 20-bit capture register	
storage Event Counter	CP6 20-bit capture register	
Circular Buffer (CP5,CP6)		
Offset Timer Definition		COMMANS/DEFINITIONS AREA
Cylinder 1 Dwell Start		
Cylinder 1 Spark Time		
Cylinder 2 Dwell Start		
Cylinder 2 Spark Time		
Cylinder 3 Dwell Start		
Cylinder 3 Spark Time		
Cylinder 4 Dwell Start		
Cylinder 4 Spark Time		
Cylinders 1,4 Pre-spark Interrupts		
Cylinders 3,2 Pre-spark Interrupts		
Open Knock Window		
Close Knock Window		
Injector 1 Timer Definition		
Set/Reset Injector 1 Output		
Injector 2 Timer Definition		
Set/Reset Injector 2 Output		
Injector 3 Timer Definition		
Set/Reset Injector 3 Output		
Injector 4 Timer Definition		
Set/Reset Injector 4 Output		
TDC signal generation		
Timer Definition		
PWM signal 1 generation		
Timer Definition		
PWM signal 2 generation		
Timer Definition		
PWM signal 3 generation		
SCI baud rate definition		
Others commands		
General purpose Register File		

Fig 1 Register file organization with engine control commands in the COMMANDS/DEFINITIONS area

tooth has a number and corresponds to an Engine angle, software tasks can be executed in synchronisation with the Engine Cycle to a precise engine angle (missing teeth checking, ignition advance angle computation, dwell time computation, fuel injection time computation etc...).

Dwell feedback signal is entered on CP2 which allows to capture the default timer then to check the dwell value.

COMMANDS/DEFINITIONS AREA

All commands and definition entries are 32-bit words, stored in the register file following the circular buffer (figure 1). The Engine Control commands such as Spark time, dwell time, fuel injection time, knock window generation, PWM signals etc... are programmed in this area. Those commands are executed sequentially, starting at the top of the area until the end and repeated each resolution (the resolution is determined by the programmable HSCA time base). Below, a summary of available definitions and commands :

Definitions :

. Define a virtual timer (set up a 16-bit virtual timer cleared when the programmed maximum value is reached).(frequency set-up).

. Define a baud rate timer (set up a 16-bit virtual timer acting as baud rate generator).

. Define an offset timer/event counter (set up a 16-bit virtual timer cleared when an event occurs on input pin CP6).

. Define an offset timer (set up a 16-bit virtual timer cleared when a programmed event value matches the event counter).

Commands :

. Standard timer compare command (actions occur when compare value matches the referred timer).

. Double event compare command (actions occur when event compare value matches the event counter).

. Conditional event/timer compare command : (actions occur when the two compared values -one time/One event- matches timer and event counter).

The continuation of the paper describes all these definitions and commands with the help of engine control command examples which highlight the powerful of HSCA module for engine control system.

IGNITION CONTROL (figure 3)

The new distributorless ignition systems require two or four output pins to drive two or four ignition coils (semi-static or static ignition system). HSCA module can drive automatically these ouputs by using programmed commands into the COMMANDS/DEFINITIONS area. For each ignition output pin, one command set the output to a given engine angle previously calculated (dwell time) and another command resets the same pin at the ignition advance angle (spark time). The used command in this case, is the "conditional command" which can set or reset a pin when its timer compare value is matched with the referred timer and its event counter compare value is matched with the event counter (figure 3). The event counter compare value (EV1, figure 3) corresponds to a number of tooth so to an engine angle and the timer compare value (T1, figure 3) corresponds to an angular interpolation between the tooth number EV1 and the tooth number EV1 + 1.

But before to program these commands, the programmer has to write an "offset timer/event counter" command (DEF 1, figure 3) which allows to set up a 16-bit virtual timer incremented every resolution and cleared each time an event occurs on input pin CP6 (teeth from engine speed signal). Before the clear action, the 16-bit virtual timer can be stored in the circular buffer. This allows to capture the tooth signal period in the circular buffer (very useful to check the missing teeth).

Once the dwell angle and the ignition advance angle have been processed, the CPU updates the event counter compare values (EV1, EV2) and the timer compare values (TC1, TC2) into the two ignition commands (CD1, CD2, figure 3). So, when the event counter reaches the EV1 compare value of the CD1 command and the offset timer (cleared by each tooth) reaches the TC1 timer compare value of the CD1 command, the output pin is set (Dwell time).

In the same way, when the two compared values (time, event) of CD2 command are matched by the offset timer and event counter, the output pin is reset (spark time).

In order to get a better angular interpolation precision, an interrupt can be generate the tooth (EV2-1) before the calculated ignition tooth (EV2). This allows to know exactly the period of the teeth signal (captured in the circular buffer) just before the spark time in order to get the right timer compare value TC2 (angular interpolation). This can be done by the "double event compare" command (CD3, figure 3).

We can notice, in the case when the engine parameters don't change from an engine round to an other , the compare values in the commands don't need to be updated and the HSCA drives the ignition coil outputs automatically without CPU intervention.

INJECTION CONTROL

A four cylinder sequential fuel injection system requires four distinct injectors so four independant outputs. HSCA module can drive automatically the start of fuel injection and the fuel injection time for each injector outputs by using specific offset timers.

So for each output, an offset timer is defined in the COMMANDS/DEFINITIONS area. The timer is cleared when the event counter (incremented by the tooth signal) reaches the event compare value (EV4, DEF2, figure 4) in the offset timer definition. This allows to program the start of the fuel injection to a precise engine angle. The fuel injection time is driven by the standard timer compare command (CD4, figure 4) of which the timer compare value corresponds to the fuel injection time. So, when the previous offset timer is cleared (event counter = EV4 in DEF2) the timer compare command (CD4) set the selected injector output, and when the offset timer reaches the programmed timer compare value, the same command (CD4) reset the same selected injector output.

So for a four cylinders sequential fuel injection system, four offset timers and four standard timer compare commands are required and the part of the CPU is only to update the start of the fuel injection and the fuel injection time when it's necessary.

PWM SIGNAL GENERATION

In Engine Control System, some actuators require PWM signals to be driven (EGR, Idle speed, fuel pump etc...). As the number of 16-bit virtual timers is programmable, the HSCA can provide as much as PWM signals with independant frequencies that the Engine Control System requires.

To generate a PWM signal, the "virtual timer definition" is used and set up a 16-bit virtual timer (DEF3, figure 5) incremented every resolution and cleared when the maximum value (Time out value) is reached. This definition fixes the PWM signal frequency.

To fix the duty cycle of the PWM signal, a "standard timer compare" command is used, where the compare value (TC1) corresponds to the duty cycle. So, when the 16-bit virtual timer (T1) is cleared, a selected output pins is set or reset and when the compare value (TC1) matches the timer, the same pin is reset (or set).

Interrupts can be enabled when the timer comparison occurs or/and when the virtual timer is cleared.

KNOCK WINDOW GENERATION (figure 6)

To extract information containing the knock, a measuring window (knock window) synchronized to the crankshaft is necessary. This window is generally angularly constant but can be modified according to the Engine Control Strategy.

HSCA can generate automically the knock window thanks to "conditional -time event- compare" commands which can set then reset the output at very precise engine angles (these commands have been explained in ignition control paragraph).

TOP DEAD CENTER (TDC) SIGNAL

For diagnosis and adjusting, it's very usefull to output a signal synchronized with the TDC. As all the teeth of the toothed wheel have a number, the number of the teeth corresponding to the TDC has been programmed in the "double event compare" command and the number of the opposite teeth of the wheel in the same command in order to set an output pin on the first event compare and reset the same pin on the second event compare. So, in this example, the output is toggled each semi engine round.

CONCLUSION

In the case where the engine is in a temporary stationnary state (engine parameters are temporary stable) all the outputs described above (ignition coil outputs, injector outputs, PWM signals, knock window generation,etc ...) are driven by the HSCA module without any CPU intervention (CPU may be in idle mode). This highlights the HSCA powerful which releases CPU of all real time I/O tasks. This allows to the CPU to be entirely dedicated to the processing tasks.

In addition, as described in the COMMANDS/DEFINITIONS area (figure 1), all engine control Commands are written directly in this area. This shows the easiness to program engine control tasks. For example, to add a PWM signal output, the programmer has just to add a "virtual timer" and a "standard compare" command in the COMMANDS/DEFINITIONS area.

The TMS 370 modular approach (design and test), the automotive oriented peripheral module (HSCA, EEPROM, ADC, custom modules), the high immunity to noisy automotive environment (ESD and latch up I/O protections, extended temperature range), the system integrity (EEPROM write protection, privilege mode, watchdog, oscillator monitor) make the TMS 370 microcontrollers fit perfectly with Engine Control systems.

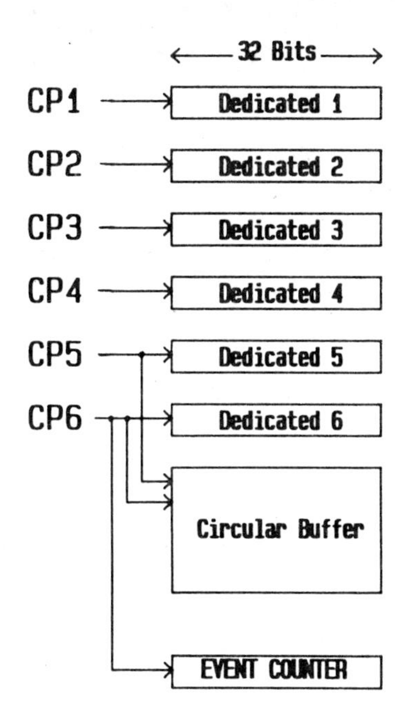

Fig 2 HSCA input pin architecture

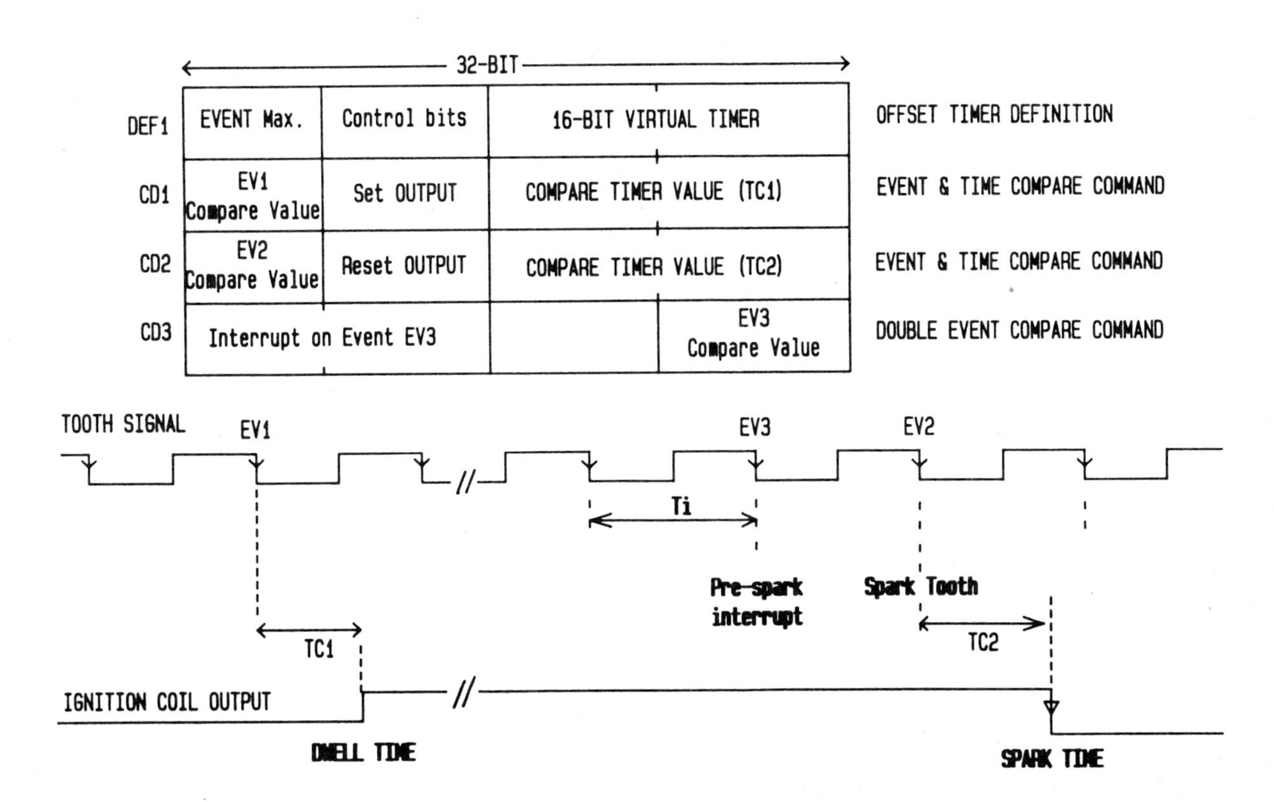

Fig 3 Ignition coil commands

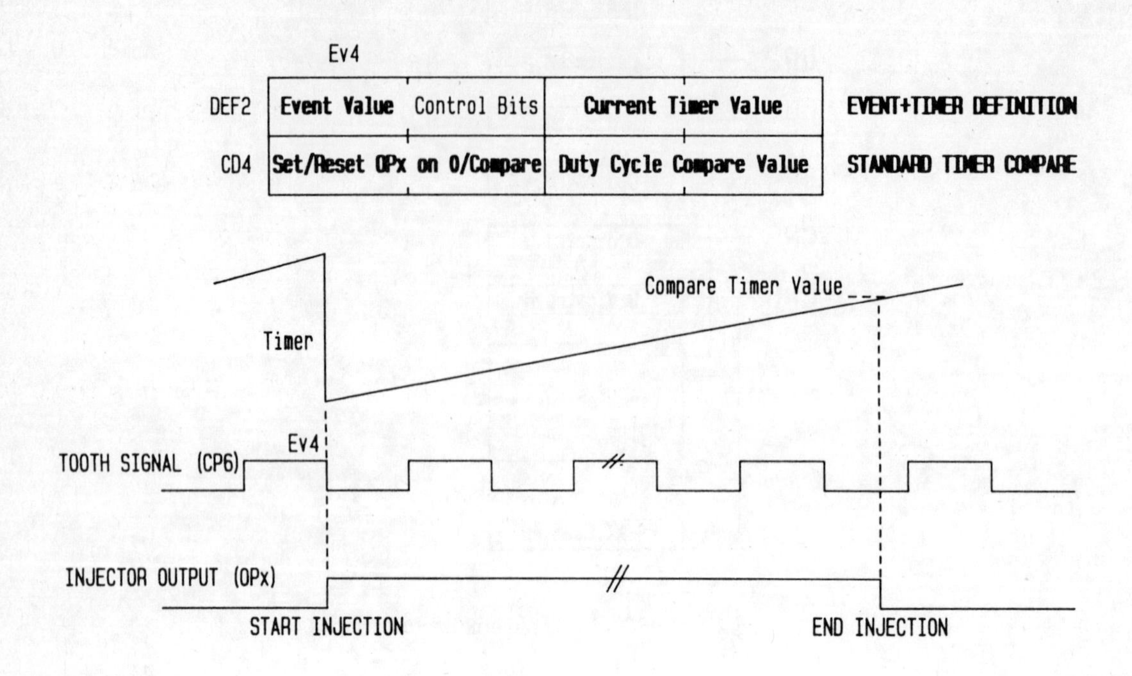

Fig 4 Injector commands

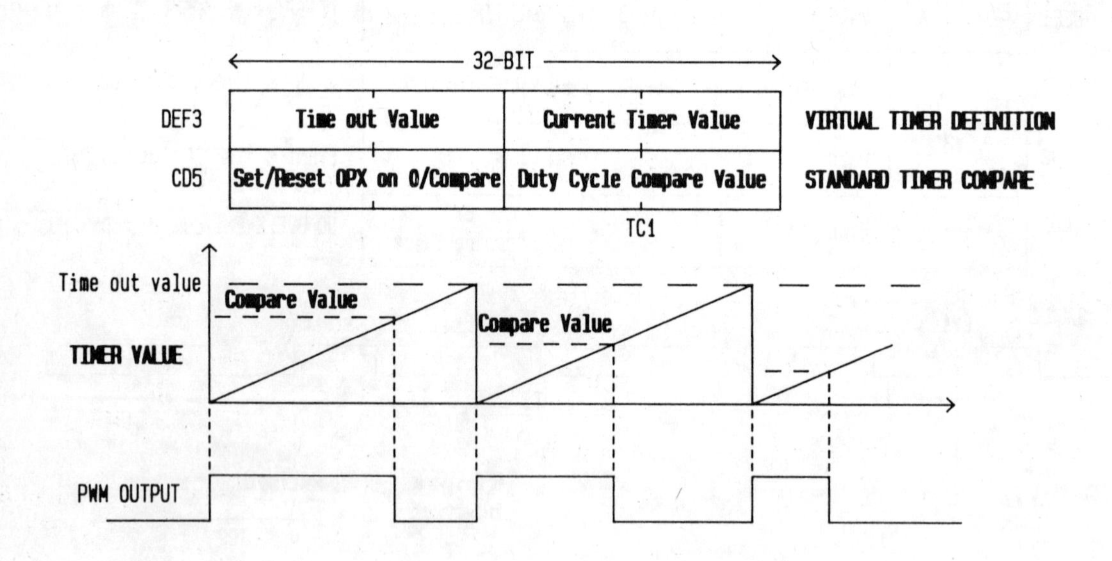

Fig 5 PWM signal generation

CD5	Event1= (Set Knock Tooth)	Set OP7	Angle Adjustement Time (T1)	EVENT & TIME COMMAND TIME
CD6	Event2= (Reset Knock Tooth)	Reset OP7	Angle Adjustement Time (T2)	EVENT & TIME COMPARE COMMAND

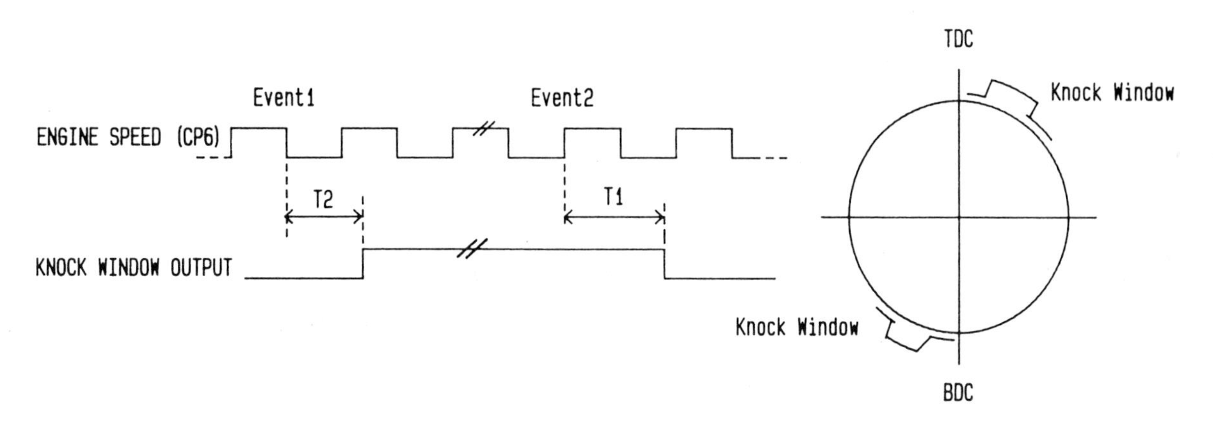

Fig 6 Knock window generation

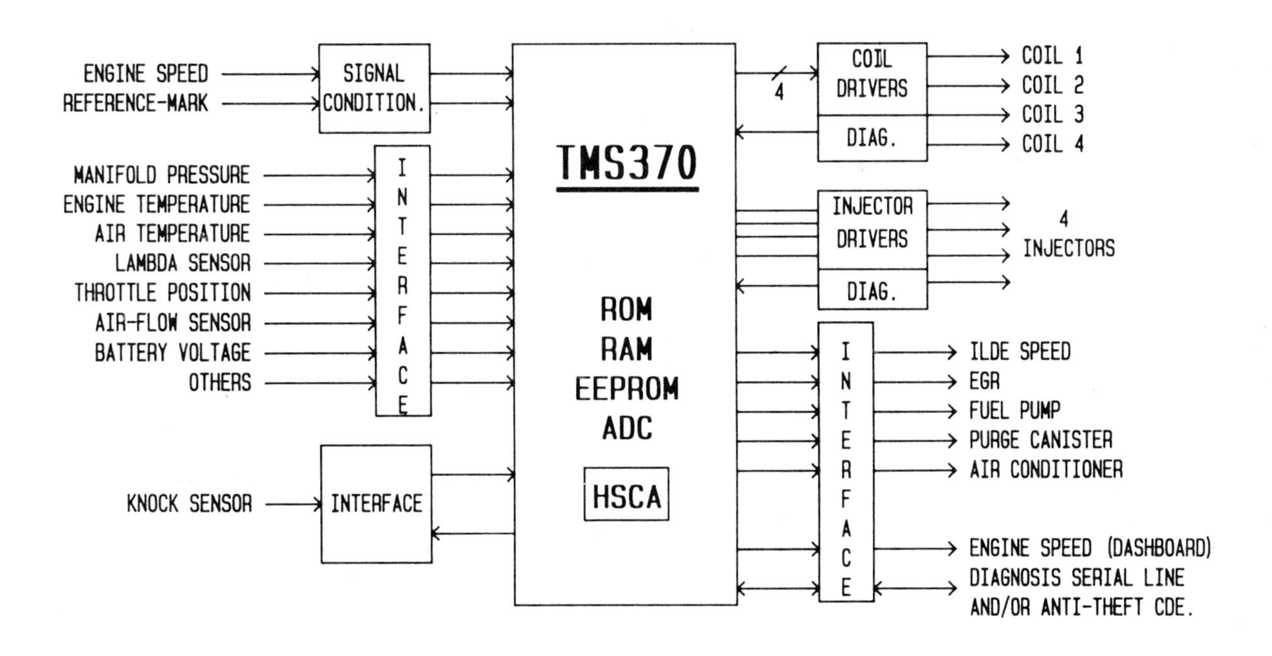

Fig 7 Engine control system block diagram

C391/040

Optimized microcontroller input/output (I/O) for electronic injection and ignition control

S M McINTYRE, BSCS
Intel Corporation, Chandler, Arizona, USA

ABSTRACT

The amount of LSI logic to control and monitor engine performance is constantly being integrated. Injection control modules range in functionality from the simple fuel metering to sequential fuel injection with distributorless ignition. This paper will review the type of I/O processing power required for engine control today and look at what will be required of I/O processing of tomorrow.

INTRODUCTION

Controlling engines electronically is becoming more and more common due to stiffer regulation of emissions. Controlling the fuel intake into the throttle body, or even into individual cylinders, is becoming a *required* function. In today's engine control units (ECUs) fuel metering is a standard function. This task is either throttle body injection (TBI), multiport injection (MPI), or sequential fuel injection (SFI). Each of these require CPU controlled outputs based on engine information such as camshaft position, oxygen content in the exhaust gases, throttle position, air density, etc. The placement of these outputs and the ability to move edges with high resolution may make the difference between a poor-running engine and a high-output power machine.

Although ignition control is not a standard function in today's ECUs, it will become more and more popular as the cost/performance curve dictates. In today's engines this function can be in stand-alone modules, handled by a distributor based on an ECU-multiplexed output, or integrated within the ECU but handled through another CPU. The placement of these control signals are also based on camshaft position and engine RPMs.

The key to both engine control signals is the camshaft position signal.

CAMSHAFT POSITION SIGNALS

There are many ways to get engine position and engine RPMs. One is to use a two-signal input in the ECU, one for crankshaft and one for the camshaft. Another is a single camshaft signal with more "teeth" and represent the first cylinder position as a missing tooth. An example of the waveform this alternative would produce is shown in figure 1.

Since the camshaft (CAM) rotates only once for each 720 engine cycle degrees, the CAM is a better place to pick up both engine RPMs and engine position. The more teeth the CAM wheel has, the better the resolution the ECU has on the placement of both the spark and injection signals.

Most microcontrollers have devised a method to detect when a rising or falling edge occurs on a processor input, typically called *input capture.* Input capture records the "real time" (in processor units) when an edge is detected. From this information the microcontroller programmer can determine engine speeds or engine position in degrees (0-720). Usually, though, there are some resolution errors because of the microcontroller's ability to detect that incoming edge.

TODAYS PROCESSOR INPUT and OUTPUT

As it was stated, input captures would determine the time when a CAM edge is found. Based on this information, an output signal is generated to drive the fuel injectors or spark plug coils. Figure 2 shows how an output signal is generated.

Notice that when the *nth* CAM signal is found, the fuel injector or spark output is turned on. This positive edge output triggers a interrupt to the processor as well as turn on the injector or spark. Next the microcontroller interrupt service routine sets up the duration of the output pulse based on other collected information. The microcontroller output capable of such a task is called *compare outputs* or *high-speed outputs.* This feature can transition outputs, trigger interrupt events, or in some cases reset timers and start Analog-Digital conversions. All based on a pre-programmed timer/counter value. However, the outputs are only based on one timer/counter and are not based on both the CAM position and internal processor's "real time". The interrupt- driven compare outputs take away valuable CPU time in order to perform their function.

A good example of *compare outputs* in silicon is the 8096 HSO units. It can accomplish all of the above metioned functions including toggle more than one pin using only one memory location, but each edge is still based on one timer/counter base.

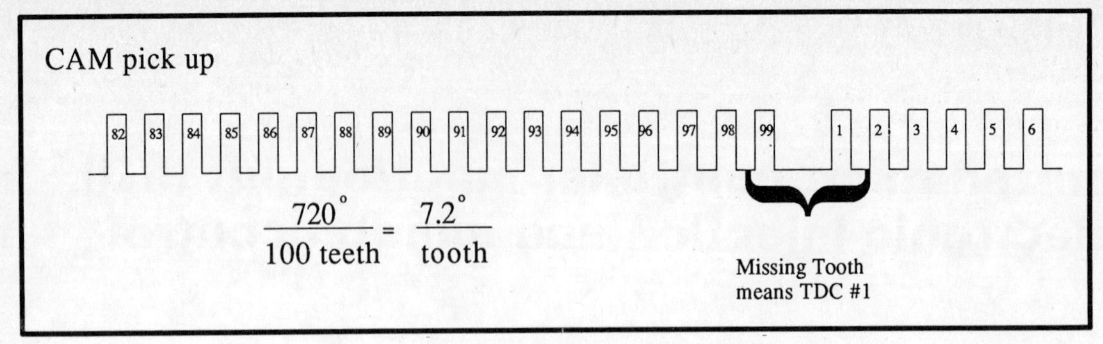

Fig 1 Camshaft signal pick-up to the ECU

TOMORROWs INPUT and OUTPUT FUNCTIONs
In the above example of input and output signals the following improvements could be made for tomorrows I/O on microcontrollers:

1. The input signal could be programmed to automatically increment an internal counter in order to keep track of the current tooth (engine position in degrees - one tooth = 7.2 degrees).

2. The same input signal can be *captured* in order to put the engine position (in increments of 7.2 degrees) information into processor "real time" terms. This will allow easy processor access and calculations for output signals.

3. The output signal could be "automatic". A RAM location is set up with the OFFSET (from the CAM position in "real-time" units) and an additional RAM location is set up for the DURATION. This would allow maximum pulse placement flexibility with low or no CPU overhead. Resolution of this signal needs to be in the one microsecond range, therefore rising-edge placement (OFFSET) needs to be to equal zero.

A co-processing loop would determine the correct OFFSET and DURATION values for each cylinder and placed in the RAM while determining the exact CAM speed and position. A look-up table approach to these OFFSET/DURATION values could be performed, but if the I/O is "automatic" it might save enough time to perform the mathematical calculations to accomplish better engine control.

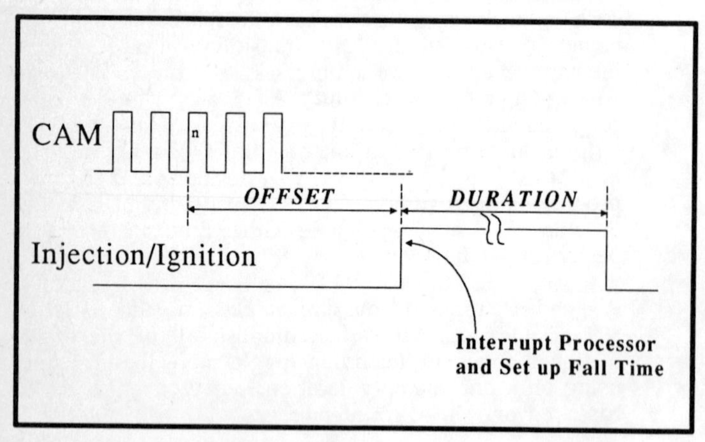

Fig 2 Spark and injection output signals based on CAM input signal

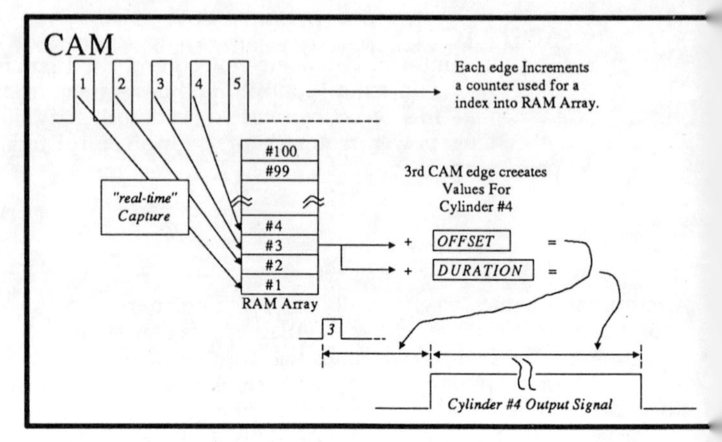

Fig 3 Automatic input and output engine control signals

A WORKING EXAMPLE
Suppose the CAM position sensor produced a signal such as the one in Figure 1, and that the output signals for the spark and injectors are based on this signal as shown in Figure 2. If every rising edge of the CAM signal were "time stamped" into a RAM location and indexed via the edge number for easy access, an array of 100 numbers would be generated (one for each CAM rising edge. The array pointer would be reset when the "missing tooth" was found).

This table of "real-time" values could be used to generate the output signals. OFFSET and DURATION values would be added to these numbers and placed in their respective control registers. Figure 3 illustrates such a task.

Notice that the output signal placement is totally flexible based on "real-time" (internal microcontroller time) resolution. If this internal timer had the ability to be scaled, resolution would be up the individual programmer.

These inputs and output tasks would happen without microcontroller intervention and CPU processing would not interfere with or be interfere by I/O processing.

One implementation os such I/O in silicon is that of *Event Processor Array (EPA)* and *Peripheral Transaction Server (PTS)*. These two feature combine to form the previously mentioned "automatic input and output processing" ideal for next generation engine controlling.

CONCLUSION

The main input and output signals in ECUs are based on generic microcontroller I/O features (Capture/Compare). The I/O used to control tomorrows engines will need to be "fine tuned" to meet a new set of unique requirements.

This paper reviewed only the I/O processing: a dedicated I/O processor with the ability to manipulate injection and ignition signals with the resolution needed for tomorrows engines. Ideally this I/O processor would be integrated into the microcontroller better utilizing the single ECUs' overall size, high-temperature requirement, and integrated function.

Having this I/O processor behind a powerful 16-bit CPU would make engine control easier to integrate, and in turn more cost effective.

C391/062

An advanced racing ignition system

T MEARS, BSc, AMIEE
Lucas Automotive, Birmingham
S J OXLEY, BSc
Texas Instruments, Bedford

SYNOPSIS This paper describes the rationale and development of a high performance racing ignition controller based on a Digital Signal Processor. Applying new techniques such as these in the high pressure racing environment allows companies such as Lucas to develop strategies for production engine management systems in the 1990's. The vastly increased processing power available allows designers to begin to consider control techniques previously considered impractical for low cost production systems.

1 INTRODUCTION

The current generation of high performance racing engines have been developed to such a degree that a 15000 r/min V12 engine is now a reality. Such an engine, by its very nature, requires full electronic control of both fuelling and ignition in order to extract the maximum performance. Traditional electronic engine management systems (EMS) are unable to provide accurate control for such an engine - the main barrier being processing speed. Lucas have applied a single chip Digital Signal Processor (DSP) from the Texas Instruments TMS320 family to achieve distributorless mapped ignition for high performance racing engines. In future, mapped sequential fuelling will be added with the DSP controlling a slave processor. The alternative to using a DSP was to implement the system as a multiprocessor configuration which is both inelegant and difficult to develop and maintain as a reliable system.

2 SYSTEM REQUIREMENTS

The system being described is required to be able to control a Capacitor Discharge Ignition (CDI) system on a variety of engines up to a V12, 15000 r/min Formula 1 version. Additionally, the system must be able to be tailored to a variety of engine geometries and firing orders.

The dominant factor for such engines is the operating speed of the system - the V12 engine referred to above, with a 40 degree V-angle and 10 degree timing markers, requires processing of degree markers that are a mere 111uS apart at full speed. It is obvious that conventional microcomputers with minimum instruction cycle times of 2 to 4uS (complex instructions such as multiply may take 10 times this period to execute) could not be used to implement a single processor system.

The speed requirement is the reason for using CDI on racing engines - conventional inductive coil based systems would be unable to build up sufficient energy in the time between sparks at full engine speed.

In order to obtain the maximum performance from an engine, the following time critical operations must be performed accurately in real-time:-

1 Record the period between adjacent teeth on a timing wheel mounted on the engine - typically at 10 degree. intervals.

2 Recognise and maintain synchronisation with a missing tooth on the timing wheel and an independent TDC marker.

3 Trigger the CD circuit at an angle defined by a three-dimensional map (16 by 64 points - throttle angle against speed). Full interpolation is provided between the discrete points on the map with modifying functions applied for temperature, boost, pressure, etc.

Points 1 and 2 require precise measurement of the tooth intervals without latencies caused by interrupt actions which can give an uncertainty at least as long as the longest instruction. The output function, point 3, requires rapid mathematical processing to allow the ignition timing to be based on the most up-to-date information as possible. It then requires an output to be driven at a precise time after a specified tooth number.

In the short term a separate fuelling controller is being used, with the ignition controller passing speed and synchronisation information to allow mapped sequential injection to be achieved.

In order to meet these requirements without using a processor with the speed of a DSP, a multiprocessor system would be mandatory. Multiprocessing creates many additional problems in terms of synchronisation, data sharing and overall maintainability. There are systems available with up to 10 processors in one controller - a nightmare to develop and use in the high pressure racing world.

3 ATTRIBUTES OF THE DSP BASED MICROCONTROLLER

The device at the heart of the ignition system is the TMS320E14 from Texas Instruments. This device takes the first generation CMOS DSP core from the industry standard TMS320 family and adds the functions found in more complex microcontrollers.

Firstly let us define a DSP (1) - it is generally accepted that such a device must be a single chip with on-chip memory (RAM/ROM) and a single cycle hardware multiplier. In its original form the DSP was intended for real-time digital processing of analogue signals. In essence it was designed to perform filter functions which can be treated discretely as a sum of products. The same mathematical functions are required for many digital control systems used today - see Fig 1, PID implementation. What at first may appear rather odd instructions, are in fact functions that normally take several instructions to implement in conventional microcontrollers, i.e. LTD.

LTD - loads Register T with data from memory

- adds Register P contents into the Accumulator

- data in memory is copied to next higher address

This type of instruction is very useful for map interpolations to derive values between map sites.

The fact that all instructions execute in a single cycle means that with a 25MHz crystal each instruction takes 160nS.

However, the reason that the DSP has not been used in automotive systems to any great extent is due to its previous requirement for several support chips to handle I/O and timing functions. In the TMS320E14 an event manager has been added that provides input capture and output compare facilities in hardware - this ensures that critical time related functions occur independently of the CPU, thus avoiding the associated latencies. Additionally there are 16 I/O lines which may be manipulated independently under software control. An on-chip serial port and Watchdog timer complete the additions that have turned the DSP into a microcontroller - see Fig. 2, TMS320E14 block diagram.

4 SYSTEM IMPLEMENTATION

The TMS320E14 is an EPROM device with 4K words of EPROM and 256 words of RAM. Whilst the controller could easily be implemented without additional memory, the capability to address a further 4K words of off-chip memory has been exploited. The off-chip memory comprises 2K words of EPROM, used for map storage and 2K words of non-volatile RAM for diagnostic and telemetry functions.

Outputs used to drive the CD circuits are driven from 4 of the 6 output compare registers - the system being able to multiplex in software these 4 signals on to the 12 outputs required.

A block diagram of the system is given in Fig 3 which serves to highlight the integration of I/O functions on to the DSP chip, thus minimising the requirement for support circuitry.

The programme is a conventional real time control implementation in which time critical responses are performed in the foreground (interrupt driven) routine, and non-time critical calculations and management tasks are performed in the background routine.

The primary foreground task is an interrupt routine triggered by the signal from a 36 tooth wheel with ten degree tooth spacing. The flywheel has one missing tooth situated at T.D.C. on the reference cylinder. Since on a V8 engine, there are two crankshaft revolutions for a complete firing of each cylinder, it is not enough to simply detect the missing tooth to synchronise the engine. A second signal derived from a half engine speed sensor situated on the camshaft is used to indicate the cycle.

The software is designed to operate in the range of 51 to 16000 r/min on engine configurations up to and including V12. At high speeds the frequency of interrupts from the crankshaft sensor is given by:-

Engine speed = 16000 r/min

= (16000 X 6) degrees/s

Time for 10 degrees = 104uS

Hence frequency = 9600Hz

At this speed, it is the phenomenal processing power of the TMS320E14 that enables control to be achieved. Running at 16MHz, single cycle execution is 250nS, enabling 416 instructions to be executed in the tooth period. This enables both the interrupt task and a significant proportion of the background task to be completed in one tooth period. For example, on a V8 engine, there are approximately 9 tooth periods between sparks, the processor is easily capable of cylinder by cylinder update of the advance angle.

The background task uses engine speed and throttle angle to address the main ignition map. This map has 16 load (throttle position) and 64 speed sites making a 1K map. Only 8 bits are needed, but since the DSP is word oriented, each memory location contains two contiguous map values. This means that the 16 by 64 site main ignition map actually uses 512 words of memory. The hardware uses an external 8 bit A/D converter to measure the throttle angle. This raw value is filtered using the equation:

Filtered position =

$$\frac{3 \text{ X previous filtered position}) + \text{measured position}}{4}$$

The hardware multiplier plus very simple divide mechanism enables extremely fast and reliable implementations of the above type of algorithm. Since the microprocessor does not have a right shift instruction, the author tends, where possible, to avoid using left shifts to do division because the load accumulator with shift instruction is sign extended. This, where the dividend has a one in bit 15, requires masking of the extended bits. It is far simpler to use the subtract with carry instruction in this application. In general terms, processing speed is high enough to use slower algorithms in order to conserve memory.

The filtered throttle position is used to derive the load site and load interpolation steps. These are both numbers in the range 0 to 15, and fix precisely the load sites accessed on the ignition map. Load breakpoint preshaping is programmable, enabling the load breakpoints to be grouped closer together in an area of the map in which close throttle preshaping is required. Usually, the breakpoints are grouped closer together where the throttle first begins to open. The breakpoints are spaced wider as the throttle is opened further.

Engine speed is measured by timing the tooth period and filtering in a similar way to the throttle position. This parameter is global to the background task and the speed site is calculated as a number between 0 and 64. For a speed range of 16000 r/min, the speed breakpoints are fixed at 250 r/min, but for a reduced range, the breakpoint separation is programmable. Basically, the time for 250 r/min is divided by the tooth period to produce the speed site. The subtract with carry divide instruction is very useful here, because the remainder from the division is conveniently located in the high part of the accumulator. This is then used to calculate the speed interpolation steps.

Load and speed sites together with the interpolation steps are fed into the main ignition interpolation routine, which uses the four surrounding map sites to the engine operating position to calculate the interpolated ignition advance map value. This routine contains eight multiply and 4 additions, as well as data manipulation, and executes in 63 cycles, which is 15.75uS. For comparison this is 15 times faster than the same algorithm on the Motorola 6805 running at 4MHz.

The background task also handles diagnostics and telemetry via a serial communications routine, and measurement, filtering, and preshaping on the following ignition modifiers.

1) Air temperature
2) Coolant temperature
3) Barometric pressure
4) Overall trim
5) Boost pressure
6) Air humidity

The foreground task performs time critical control tasks, including the conversion of the ignition angle into a timer value which is loaded into the output compare structure. The primary tasks carried out in the input capture interrupt routine are as follows.

Synchronisation is initiated and maintained using missing tooth detection. On receipt of the tooth interrupt, the period between this tooth and the previous tooth is read from the input capture FIFO. If the missing tooth has either not been initially detected, or is expected then the missing tooth detection algorithm is implemented. A successful detection is valid if

Tooth period $<$ 5/8 X previous period.

After successful synchronisation, the tooth is identified, and calculations for the engine cycle are performed. Basically, the first tooth after TDC is called tooth 0, the next tooth 1 etc, up to tooth 9, on a V8 engine, when the cycle repeats itself for the next cylinder.

On tooth zero, the tooth to fire count is calculated, and decremented on each successive interrupt, until the firing tooth is arrived at. The tooth to fire counter is calculated from the advance angle by dividing it by fifty.

The remainder from this division is used to calculate the advance degrees. When the tooth to fire count has decremented to 1, the time for ten degrees at this point is used to calculate the timer value, by multiplying the advance degrees by the period timer, and dividing this result by ten.

When the tooth to fire count is zero, the angle timer value is loaded into the appropriate compare register, and the action register is enabled, and the correct channel selected. When the timer matches the compare register, compare output will go high, triggering the CD circuit, and sparking.

In conclusion, the input capture interrupts are used for mathematical manipulation, loading timer values, counting teeth, and selecting the correct channel for the relevant cylinder. The overhead of these tasks is easily managed by the TMS320E14 at very high speed, whereas other conventional microprocessors simply cannot perform them in time. Hence, a single DSP can be used in place of a multiprocessor system.

5 FUTURE DEVELOPMENTS

Having achieved the ignition control performance required by current racing engines, Lucas are working on expanding the system to full engine management. With the current DSP microcontroller there are insufficient output lines to control 12 injectors as well as 12 channels of CDI. Consequently, it is the I/O limitation rather than CPU power that requires a slave processor to handle the fuel injector outputs. The intention is to use a TMS370 8 bit microcontroller to drive the injectors sequentially under direct control of the DSP controller. The majority of the fuelling calculations will take place in the current ignition controller with the slave processor being passed the appropriate injection timing information.

Whilst we have concentrated on the racing applications in this paper, Lucas have used this programme to measure the effectiveness of the DSP for Automotive engine control. New control strategies are being developed to enhance the performance of engine control for passenger cars in order to both increase efficiency and decrease emissions.

One of these strategies is adaptive ignition control whereby the control system applies small perturbations to the engine's running condition to determine the optimum torque/speed point. Lucas have great experience of such a technique and expect it to be applied in future production systems (2).

Another technique yet to be exploited in production is that of cylinder pressure sensing (3). In this case a pressure waveform is used to provide closed loop control of the engine. At present the main barrier to this technique is the availability of a robust, cost effective sensor. It is well known that there are several developments under way including ones internal to Lucas. However, the pressure signal inside a cylinder is of a complex form that requires much filtering and processing. DSP's have already been used in research applications to extract the information contained in this complex system - speed, combustion quality, engine health, etc. Again it is the real-time digital filtering ability of the DSP that is its strength for this function. The controller described appears to have sufficient spare capacity to be able to handle a cylinder pressure sensor - the story is common, the electronics are available and cost effective, but it is the sensor technology that is lacking.

6 CONCLUSIONS

It has been effectively demonstrated that a microcontroller with DSP functions included can provide the core for a high performance ignition controller. The efficiency of the instruction set coupled with its speed of operation would allow engine management to be carried out on a single chip - the limiting factor is the amount of timer driven I/O available on the current device. The merits of having a very fast processing core can be summarised as:-

1) Control data updated closer to the time it is used.

2) No tradeoff of control functions against engine speed.

3) Opportunity to include new control techniques in single processor system, i.e. cylinder pressure sensing.

The TMS320E14 is the first step at availing the DSP functionality in a microcontroller device - it is expected that the lessons learnt from this and other automotive control projects will further enhance its capability as new devices are brought to production.

This system should not be viewed solely as a faster version of current systems, but rather one which may be used to effectively apply the more complex strategies required of engine management systems in the 1990's (4) - and in a reliable and cost effective manner.

REFERENCES

(1) LIN, K. Trends of digital signal processing in automotive. International Congress on Transportation Electronics, Dearborn, 17-18 October, 1988.

(2) HOLMES, M. and COCKERHAM, K. Adaptive ignition control. 6th International Conference on Automotive Electronics, London, 12-15 October, 1987.

(3) ANASTASIA, C.M. and PESTANA, G.W. A cylinder pressure sensor for closed loop engine control. SAE International Congress and Exposition, Detroit, 23-27 February, 1987.

(4) HATA, Y. and ASANO, M. New trends in electronic engine control - to the next stage. SAE International Congress and Exposition, Detroit, 24-28 February, 1986.

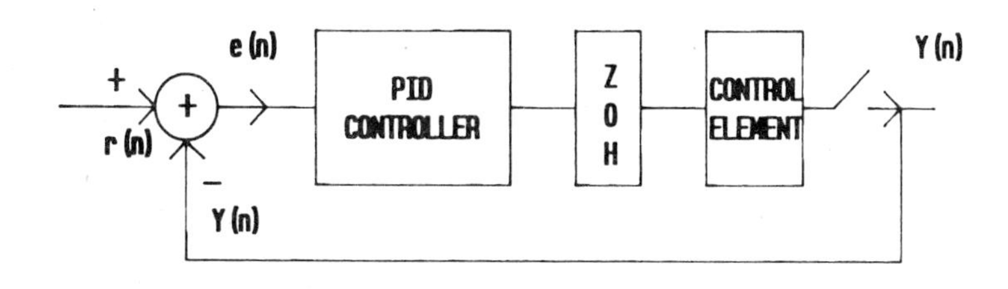

y(t) = Kp*e(t) + Ki e dt + Kd*de/dt

e(t) = error signal. Kp, Ki & Kd = PID constants.

Converting into discrete form (using rectangular approx.):

y(n) = y(n-1) + K0*e(n) + K1*e(n-1) + K2*e(n-2)

K0 = Kp + Kd/T + Ki*T, K1 = -Kp -2Kd/T, K2 = Kd/T

Where T = sampling interval.

CODE IMPLEMETATION

```
IN     E0,PA0    GET NEW SAMPLE
MPYK   0         CLEAR P REG
LAC    YN        ACC=y(n-1)
LT     E2
MPY    K2
LTD    E1        ACC=y(n-1)+K2e(n-2)
MPY    K1
LTD    E0        ACC=y(n-1)+K1e(n-1)+
MPY    K0                       K2e(n-2)
APAC             ACC=y(n-1)+K0e(n)+
SACH   YN               K1e(n-1)+K2e(n-2)
OUT    YN,PA1
```

EXECUTION TIME = 2.2uS @ 25MHz

Fig 1 PID control algorithm

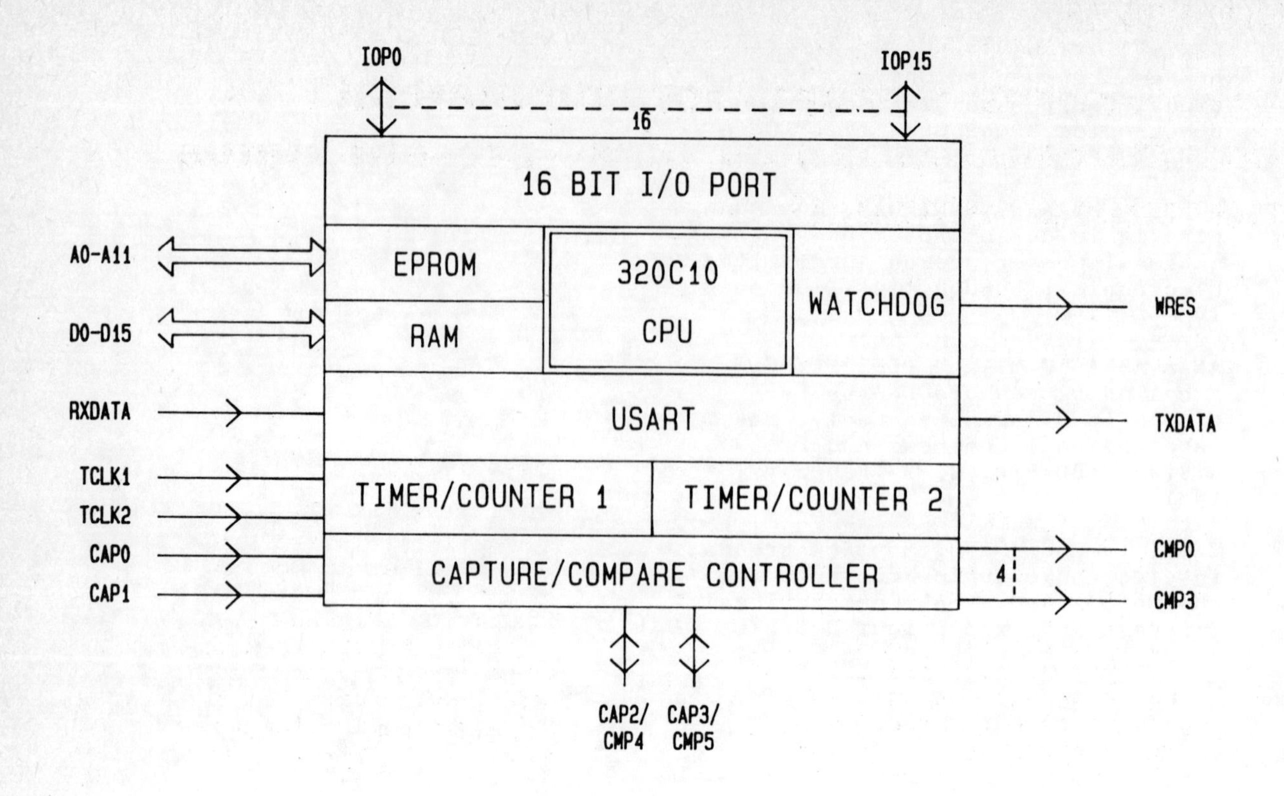

Fig 2 TMS320E14 hardware organisation

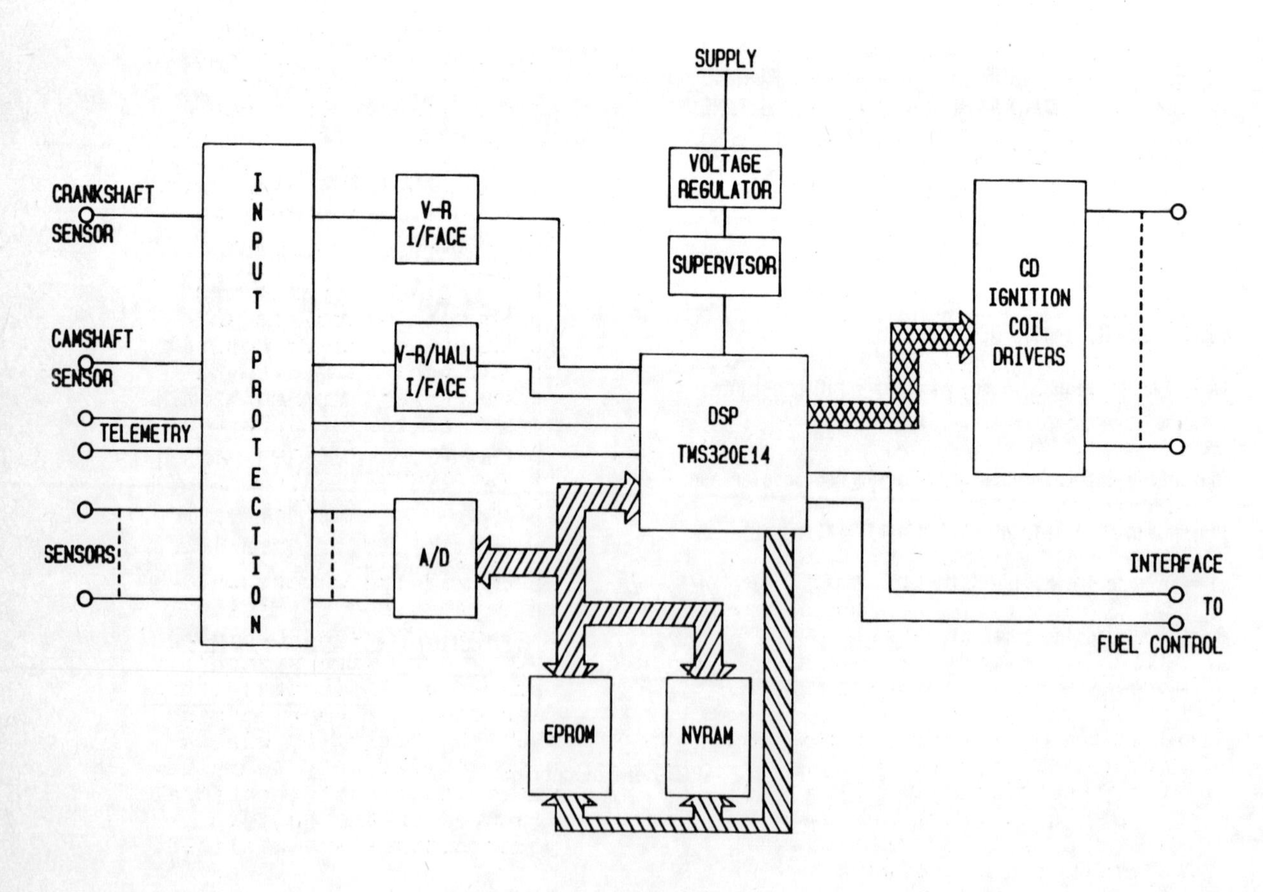

Fig 3 Lucas racing CDI system

C391/078

Drive-by-wire systems for commercial vehicles and passenger cars—present status and future perspective

E S MAUSNER
VDO Adolf Schindling AG, Frankfurt, West Germany

SYNOPSIS

In recent years development and introduction of electronically controlled systems in commercial vehicles and passenger cars have tremendously improved the function and performance of automobiles. New engine management systems, such as electronic drive-by-wire control, have been developed and the first generation has proven its reliability and performance in field use.

This report summarizes the current overview and future perspective in electronic drive-by-wire control systems. The emphasis will be placed on integrated functions as idle speed control, cruise control, speed limit control and traction slip control. In spite of thorough environmental- and life tests and, as it is impossible to reduce the failure rate of any single component to zero, the system must therefore be designed such that, in case of failures the vehicle can still be safely and properly driven. For this reason a new actuator with integrated set point sender, failesafe features and an electronic unit with two micro-controllers and ASICs have been developed. This new and innovated system opens the door to a new generation of reliable, electronic drive-by-wire systems. Further functional improvement will be achieved by system integration with electronic fuel injection, electronic ignition and an anti-lock brake system with automatic slip control via a controller area network.

1. INTRODUCTION

Cruise control and road-speed limiter units have been produced in volume production by VDO Adolf Schindling AG since 1975. According to the requirements, those systems can be installed in parallel to the existing accelerator pedal power travel.

Based on the experience of volume production, it may seem feasible to replace the linkage or the Bowden cable of the accelerator pedal power control with a "drive-by-wire system". This would mean that the accelerator pedal would have a linkage to a set point sender. This sender supplies a voltage signal proportional the accelerator position to the electronic controller. The circuit comprises processing and logic functions and its output energizes an actuator, which is mounted on the vehicle engine. A short linkage transmits the motion of the actuator shaft to the control lever of the injection pump or to the throttle valve.

Often the question is posed, if it is worthwhile to replace the traditional linkage or Bowden cable by a "drive-by-wire system" under cost aspects. In this case one must bear in mind that the "drive-by-wire system" will only be chosen where additional functional features in the engine power control are utilized.

In case that only one function, for example roadspeed

limiting or cruise control, is desired, one may have cost advantages if the unit is developed just for this very application.

If, however further functions are desired, the flexibility of a "drive-by-wire system" for commercial vehicles as well as for passenger cars will have evident cost advantages in comparison with a system with seperate engine control units.

Moreover a drive-by-wire system almost becomes a necessity when you consider the need for the various functions in modern engine management systems.

With regard to the quick change of the technical requirements and still more limited capacity in research and development a standard for the controller hardware with high flexibility in software functions has been defined. To create the most economic system we had to do market research and in-depth system studies evaluating functions, costs and flexibility to meet the future need and wishes of our customers. The increasing volume of VDO's drive-by-wire systems shows the success.

2. DRIVE-BY-WIRE FOR COMMERCIAL VEHICLES

The drive-by-wire system which is used in todays commercial vehicles incorporates the following functions:

- electric transmission of the accelerator position to the control lever of the diesel injection pump
- system self-calibration, i. e. automatic adaptation of the actuator travel range to the travel of the individual pump lever
- road speed limiting
- rpm limiting
- engine speed stabilization independent of load
- safety functions and power stage for a warning lamp
- serial interface to an anti-lock braking system.

In addition figure 2 shows the possible extension of functions in the designed drive-by-wire system in comparison with an integrated electronic diesel control unit.

For every function an individual adaption to the various engines or trucks respectively is necessary and it is obvious, that the whole complex functions can only be managed if a powerful advanced microcontroller with appropriate memory size is used. The software is written in an modular form and has been verified in extensive tests.

Only the beginning of injection and the end of injection cannot be controlled by a drive-by-wire system. This is due to the mechanical diesel pump which does not provide an internal access.

Irrespective of the overall system expenditure the drive-by-wire system has special advantages in flexibility of adaption, installation and, last but not least, in the replacement of single components in case of failures.

As required by the specification and the type of vehicle, it is possible to activate all the different features available by design.

Three groups of functions define the character of the system:

- basic functions
- convenience functions
- safty functions.

Basic functions

The main basic function is the direct transmission of the set point sender signal via the electronic control unit to the accelerator motor. The set point sender also provides the pedal forces required to ensure precise power control and driving convenience. The characteristic curve of the pedal forces can be tailored to the particular application and installation position. For safety reasons, two springs are used to generate the pedal return force.

Convenience functions

Starting acceleration control (Fig. 3)

This function is useful to avoid jerky starting accelerations of busses, which may be embarassing to standing bus passengers. Reduction of fuel consumption and less wear on the vehicle are additionel advantages.

If the bus begins to accelerate or is in slow driving condition below 30 km/h, the set point sender signal will not be transmitted directly but the motion of the actuator lever follows a progressive acceleration curve.

Above this range of limitation the atuator motor will follow the accelerator pedal motion via the ECU in a linear relation.

Hand throttle

A special ECU input is provided to connect a hand throttle potentiometer which allows to increase the idle speed continously so that the engine rpm will be increased in warm-up operation after the cold start.

Tempostat = automatic cruise control

This function keeps the speed constant when the driver activates the cruise control.

The electronic controller receives the different signals furnished (actual speed, driver's input, actuator position), evaluates them by comparison and - in case of deviation of the actual cruising speed from the set value - will supply a differential voltage to the actuator until the disparity is eliminated.

A defined threshold avoids activation of the cruise control function at speeds lower than approximately 40 km/h.

The operator's control is designed as a 4-function stalk switch with the following functions:

1. Accelerate and set

   It enables acceleration or setting of a speed by tapping the switch. Of course, it is also possible to accelerate in the conventional manner be pressing the accelerator pedal. In this case the selected cruising speed can be stored by tapping the switch.

2. Decelerate and set

   With this switch, the power control of the vehicle is returned to idle where it remains until the vehicle has been retarded down to the desired speed.

3. Disengage

   The tempostat regulates smoothly down with the actuator going into the idle position. The driver resumes normal control by depressing the accelerator pedal.

4. Resumption

   When this function is activated at any speed over 40 km/h, the system will automatically resume cruise control on the level of the last speed input.

The system is switched off when either the brake pedal or clutch pedal is depressed or when speed falls below 40 km/h. Also the controller automatically rotates to idle position, when the vehicle is decelarated at a rate of more than 2 m/sec^2 and the brake light switch is unserviceable due to malfunction. Moreover the function is switched off if other safety functions in the E-GAS controller give an alert in case of malfunction.

Safety functions

Road speed limiting

Recently, various countries have stipulated stringent design specifications with respect to maximum permissible speed of commercial vehicles. France, for instance, has specified maximum speeds of 90 km/h and 80 km/h, respectively, depending on cargo, for all commercial vehicles over 10 t gross weight. However, to permit economical operation of commercial vehicles, they are equipped with "long" driving axle ratios, allowing them to attain maximum speeds far beyond those specified by legislation. The automatic speed limiter denies the driver the use of the vehicles maximum speed potential, while the objective of the economical operation is fully retained. The automatic speed limiter monitors the vehicle speed and prevents overspeeding even with the accelerator pedal fully depressed. The maximum permissible speed is a design feature of the controller stored in correspondence to legal provisions, while a fine adjustment capability is provided to match the system to the individual vehicle model specification.

If, despite of down-regulation, the vehicle exceeds the permissible maximum speed when going downhill, a telltale or a buzzer will come on. An additional function output is provided to activate an auxiliary brake (retarder) under this conditions.

Fig. 4 shows, the measurements which fulfill the British standard BS AU 217. Curve 1 represents the limited curve and curve 2 was measured with a vehicle of 260 KW enginepower. Inputs for fine adjustment allow individual adaption to every vehicle model specification so that it is ensured to meet the described standard under all conditions.

Automotive Load reduction

The driver initiates different operating commands e. g. with the set point sender, the stalk switch for cruise control, hand throttle etc.

These commands are converted from the "E-GAS" electronic controller to adequate actuator movements.

Because of road and/or vehicle conditions dangerous situations might occur for the passenger and for the vehicle if they are not timely sensed or properly assessed.

For that reason additional auxiliary information, supplied from sensors or other electronic units, is incorporated in the E-GAS software.

In articulated busses the E-GAS system will reduce engine power in 4 steps, related to the swivel angel between the two vehicle sections, to avoid that the vehicle jack knifes itself into the curve center.

The "E-GAS" electronic unit is supplied continuously with the swivel angle. In case the vehicle speed exceeds the speed allowed for a defined swivel angle and the swivel angle exceeds specific values stored in the electronic unit, the engine power is reduced automatically irrespective to the drivers operating action, until the driving condition is stabilized.

In reverse gear, the "E-GAS" system will also reduce the engine power automatically if the swivel angle overrides critical values and in addition the speed is limited up to 10 km/h.

Traction slip control interface

One of the most important reasons for the need of a E-GAS system is the traction slip control function. Different anti-lock control units have been used by the vehicle manufactures and these units also control the traction slip. The software of the traction slip control function generates electric pulses which are command signals for the E-GAS system to react as required. Because of the different units and their signal information two interfaces are used. Therefore it is possible to control the actuator movement speed or the actuator position. As the E-GAS system provides these both features, the vehicle manufacturer is able to choose a tailored system for his requirements .

To cope with the complexity of model variations, the system was designed after thorough system analysis and defined for high flexibility, especially considering small quantities. This concept ensures that development expenditure would be kept at a minimum for every new modified version. Actually approx. 60 to 70 % of the available features are utilized.

The important nature of the described functions in the vehicle give rise to the exacting requirements on the effective components. Functional safety and high life endurance under extremely severe conditions are absolutely indispensable.

The correlation between the "E-GAS" operational life time and the corresponding vehicle mileage has been established from many recorded road tests.

Fig. 5 shows a schematic diagram of load cycles in comparing the accelerator pedal angles under different driving conditions and an adequate life test.

Such measurements and test results are essential for the development and design of the components. They have to be available before production is started.

In addition, perodical tests during production will ensure the high quality standards are maintained.

The system is protected against EMI influence and no disturbance is observed in field strengths $\leq$ 100 v/m tested under stripline conditions.

EMI-tests for vehicle installation will be performed in cooperation with the vehicle manufacturer.

3. E-GAS for passenger cars

3.1 Existing system in current production

A unit similar to the E-GAS for commercial vehicles is in production for passenger cars.

Fig. 6 shows the three main components:

- set point sender unit
- actuator unit
- electronic control unit.

Utilized functions are:

- remote control
- map control
- cruise control
- traction slip control
- safety functions.

As with all E-GAS systems, the correct relation between set point sender position and actuator position is assured by potentiometers and switches. Malfunctions will automatically cause the system to switch over to an emergency drive mode and a mechanical emergency linkage ensures driveability.

Up to now there have been no complaints with the units in production since 1987.

3.2 Future development

Based on experience and requirements up to now, a new concept will be developed, with special emphasis on

- integration of set point sender, actuator and throttle
- idle speed control
- cruise control
- traction slip control
- engine drag force control
- emergency drive mode
- advanced safety functions.

Fig. 7 shows a prototype E-GAS featuring actuator integrated into a throttle body, which was used to investigate for studying the essential criteria to obtain stability and dynamic behaviour of an idle speed control in comparison with a common bypass idle speed valve.

With respect to durability special attention had to be payed to fouling deposits at the throttle valve gap, which might result in binding or sticking.

Contrary to all expectations it was found that, as a result of the specific nature of the throttle surface and of the surplus of targue such failures did not occur.

Variations in the airflow that may result from deposits, can be automatically recognized and compensated by the electronic unit.

Fig. 8 is a schematic block diagram of the idle speed function implemented in the software. The electronic controller provides a defined set engine speed to be maintained even under the influence of load changes. The set engine idle speed, varying with the individual application, may be defined as low as just above the stalling limit of the engine. With the additional evaluation of a temperature signal, the elevated engine idle speed required during warm-up may be lowered continuously by to the normal set value. The software features composite PID-action. On the basis of the ignition pulses, the controller receives the instantaneous engine speed input, compares it to the set speed value and in case of disparity, supplies a signal to the actuator which, in turn, regulates the air supply to the idling engine. Because of the delay occuring in the process of combustion, it is necessary in special cases to superimpose predetermined constants to the normal values, so that the idle speed is also stabilized under dynamic conditions (load change, activation of air condition, unit etc.)

As a result of the tests performed it was established that this kind of idle speed control gives results as good as or even better than the systems known up till now. Low air leakage and a wider dynamic range are prominent features of the new system.

In a further development step the set point sender has now been integrated into the trottle body (see Fig. 9).

In this approach the conventional linkage between the accelerater and the throttle body is fully retained. This ensures on the one hand, that in case of failure, throttle control can be maintained in the conventional manner. However, on the other hand, the integrated electric motor connected to the throttle via a magnetic clutch, can execute the E-GAS control commands.

Malfunction will automatically open the clutch and the throttle is free to move.

The throttle valve is connected to the lever on the outside of the thottle body not by a rigid shaft, but by a spring. The force of this spring is sufficient to overcome the forces of the vacuum at the throttle valve and the mechanical friction.

If traction slip control is activated, the throttle will close against the force of this coupling spring. Due to the predominant force of the return springs and the hysteresis in the gas pedal linkage, the occuring counter-force cannot be sensed by the driver in the accelerator.

Fig. 10 shows, that the electronic controller unit is designed with redundancy to make sure that all signals received from the set point sender, from the actuator and other sensors can be checked for identity and plausibility. Moreover both microcontrollers compare calculated values and output power signals. Integrated circuits, which are important, for safety are redundant components. And a custom integrated I/O-circuit is developed which incorporates to have additional hardware timers to detect both vehicle speed and engine speed via a redundant manner and to provide communication of the 16-bit microcontrollers. In accordance to the electronic interfaces of the other motormanagement systems, controlling for example ignition, injection, transmission and anti-lock brake system, a new innovated E-GAS system was created.

In a motormanagement system, where the ECUs are combined via microcontrolled networks, new functions can be realized, which improve the economics, reduce the exhaust emissions, increase safety and, last but not least, the comfort.

A joint system concept with other motor management systems using, intelligent interfaces for fast data transfer will enable compliance with future requirements, especially with respect to anticipated legislation.

4. Conclusion

The described E-GAS systems for commercial vehicles and passenger cars are electronically controlled motormanagement systems. These systems control the power of diesel or gasoline engines. Using these systems all engine power controlling functions can be realized. For the ECUs 16-bit microcontrollers have been chosen and which enable self-learning control algorithmes as used for self-adjustment and such features as idle speed control.

By integrating the set point sender and the actuator into the throttle body and by redundant system design availability, reliability and safety are improved further.

Reference:

Schlick, H.
Elektronisches Gaspedal,
VDO-Bericht Nr. 418, 1981

Pfalzgraf, M.
Zentrales Fahrzeugmanagement
für Nutzfahrzeuge

Ruschek, G.
(E-GAS) Automobil Industrie,
Vogel Verlag
Würzburg, 28 Jahrgang,
Heft 2/Juni 1983

Mausner, E.
Schneider, E.
Eine intelligente Leerlauffüllregelung mit adaptivem Verhalten
VDI Berichte Nr. 612, 1986

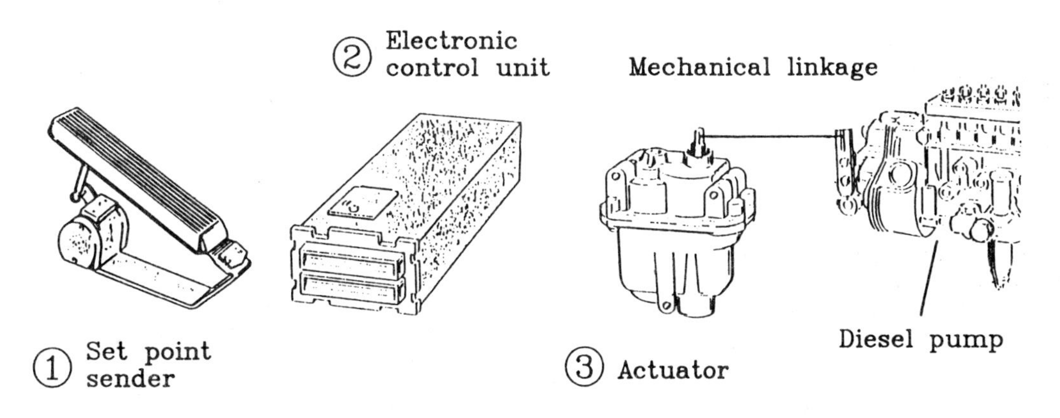

Fig 1 E-GAS first generation commercial vehicle

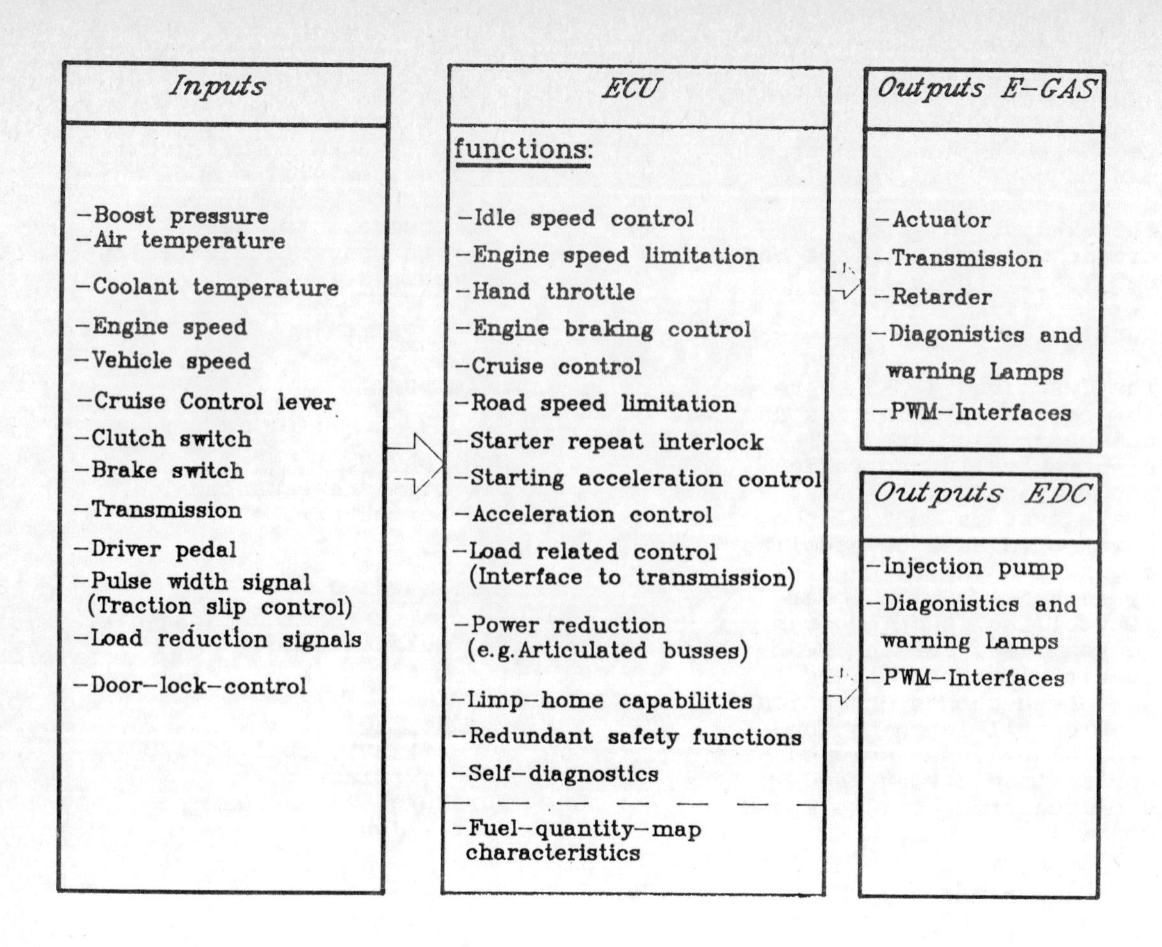

Fig 2 E-GAS and EDC functions commercial vehicle

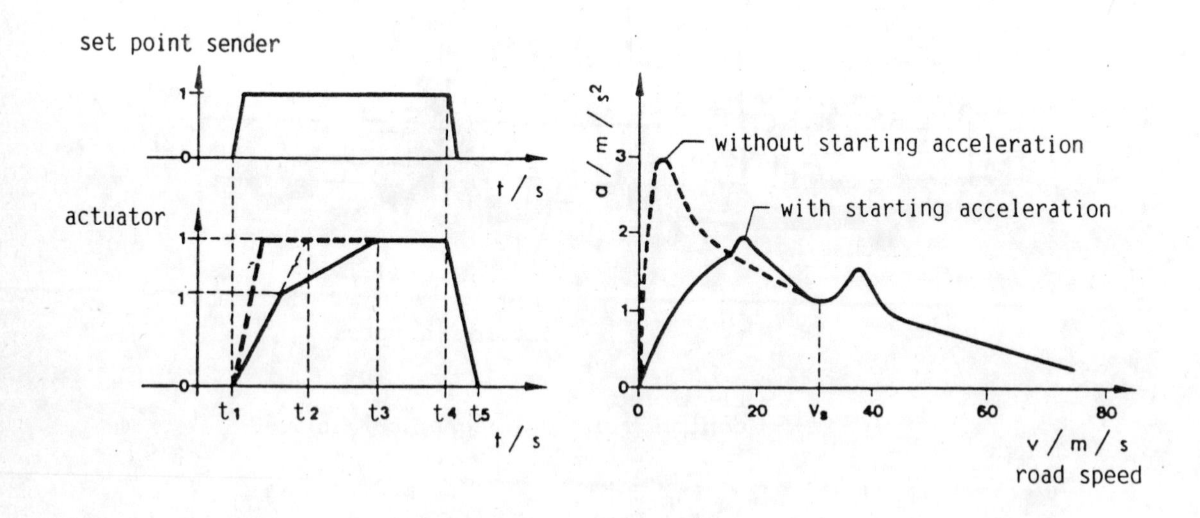

Fig 3 E-GAS first generation commercial vehicle — starting acceleration control

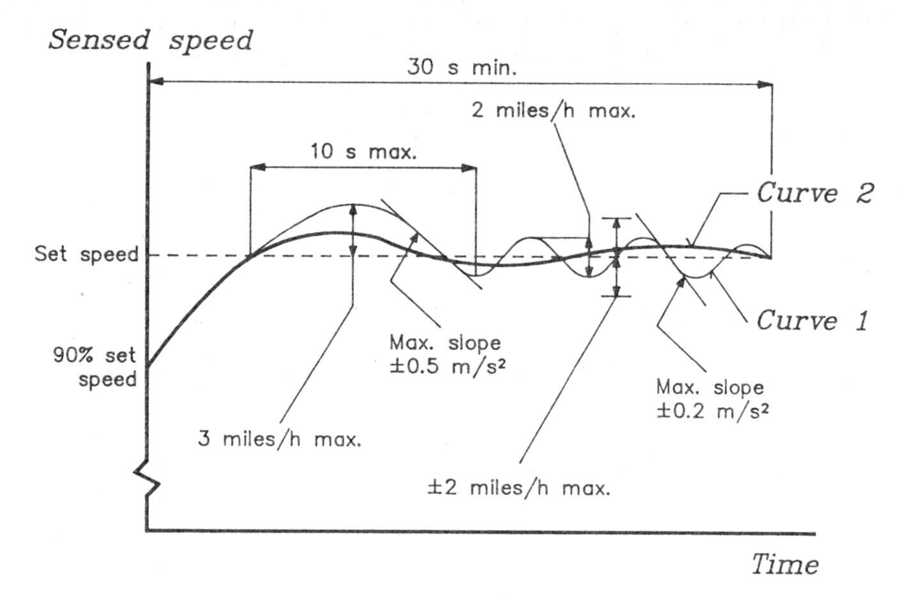

Fig 4 Road speed limiter (BS AU 217) — tolerance of limiter response characteristic

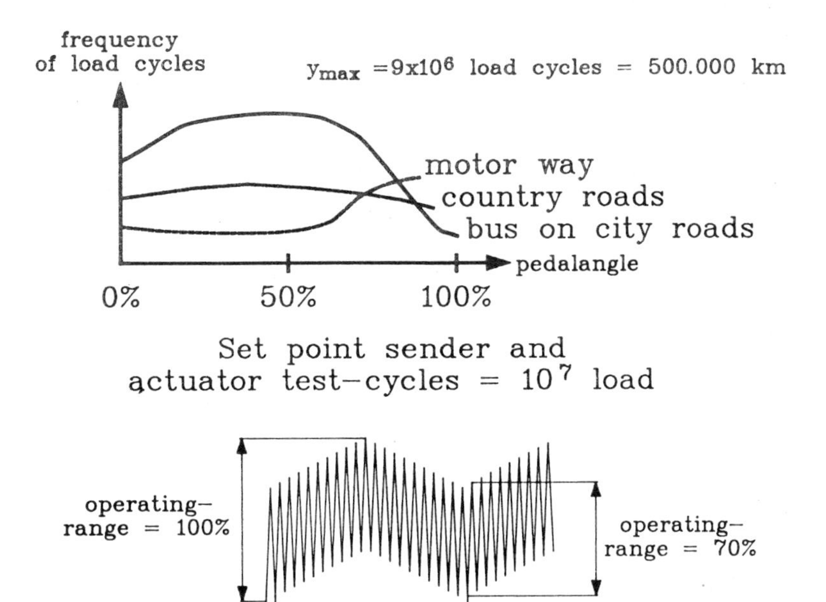

Fig 5 E-GAS first generation commercial vehicle — durability tests based on different road tests

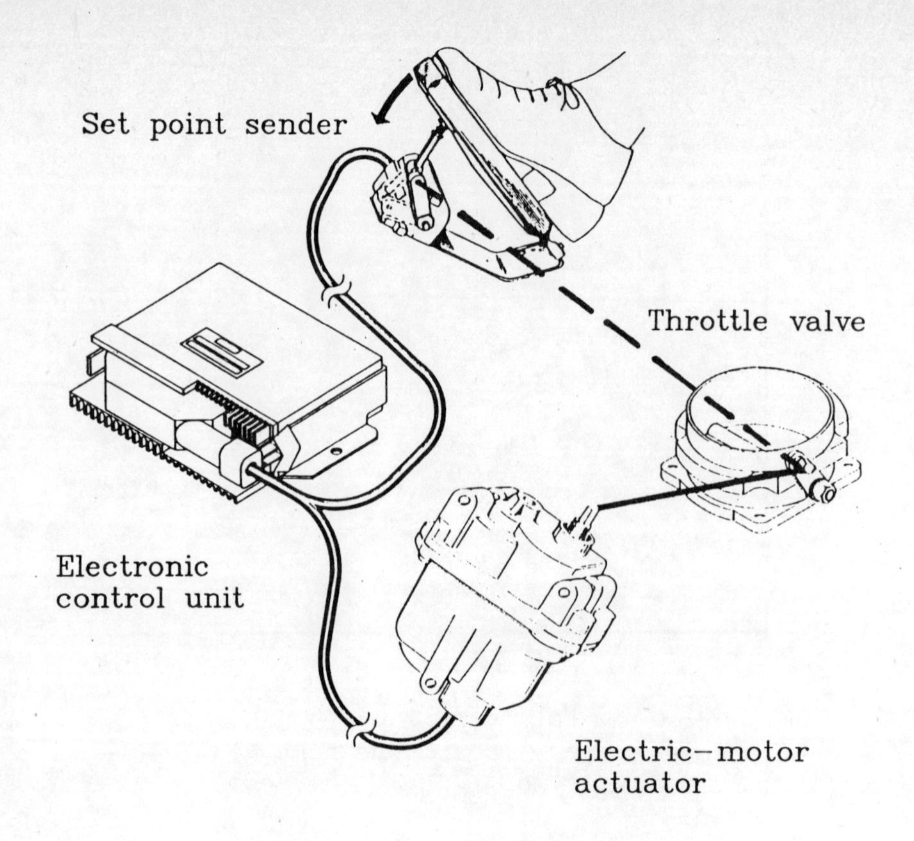

Fig 6 E-GAS first generation passenger car

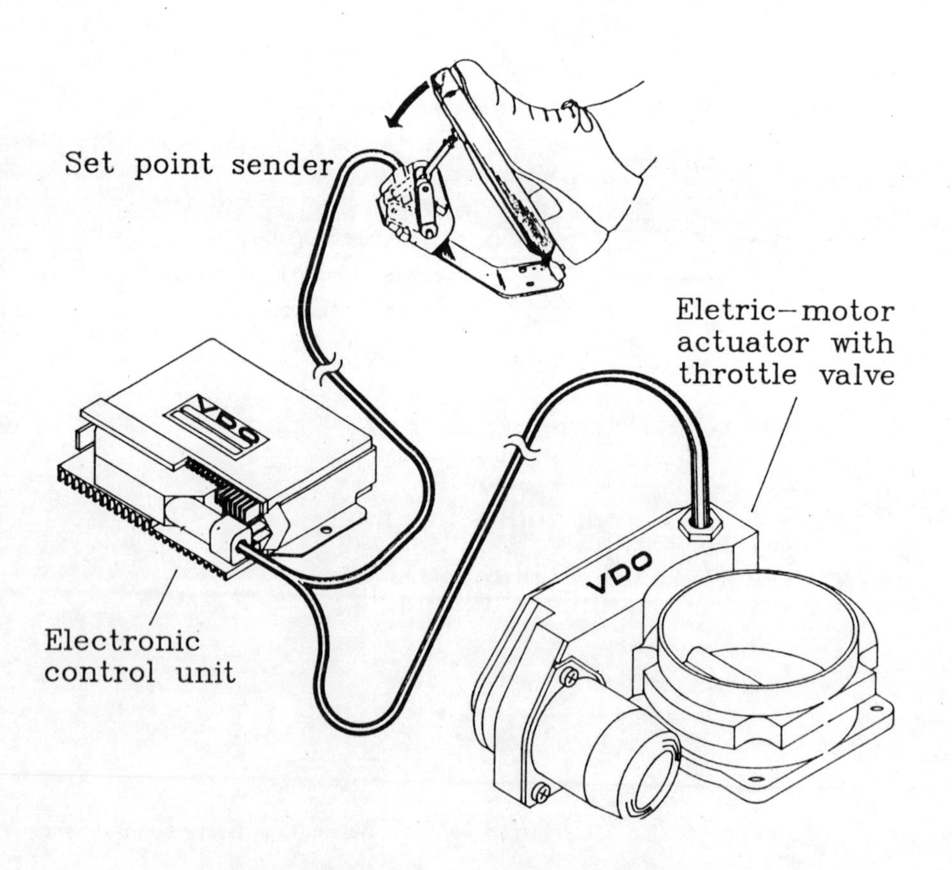

Fig 7 E-GAS second generation passenger car

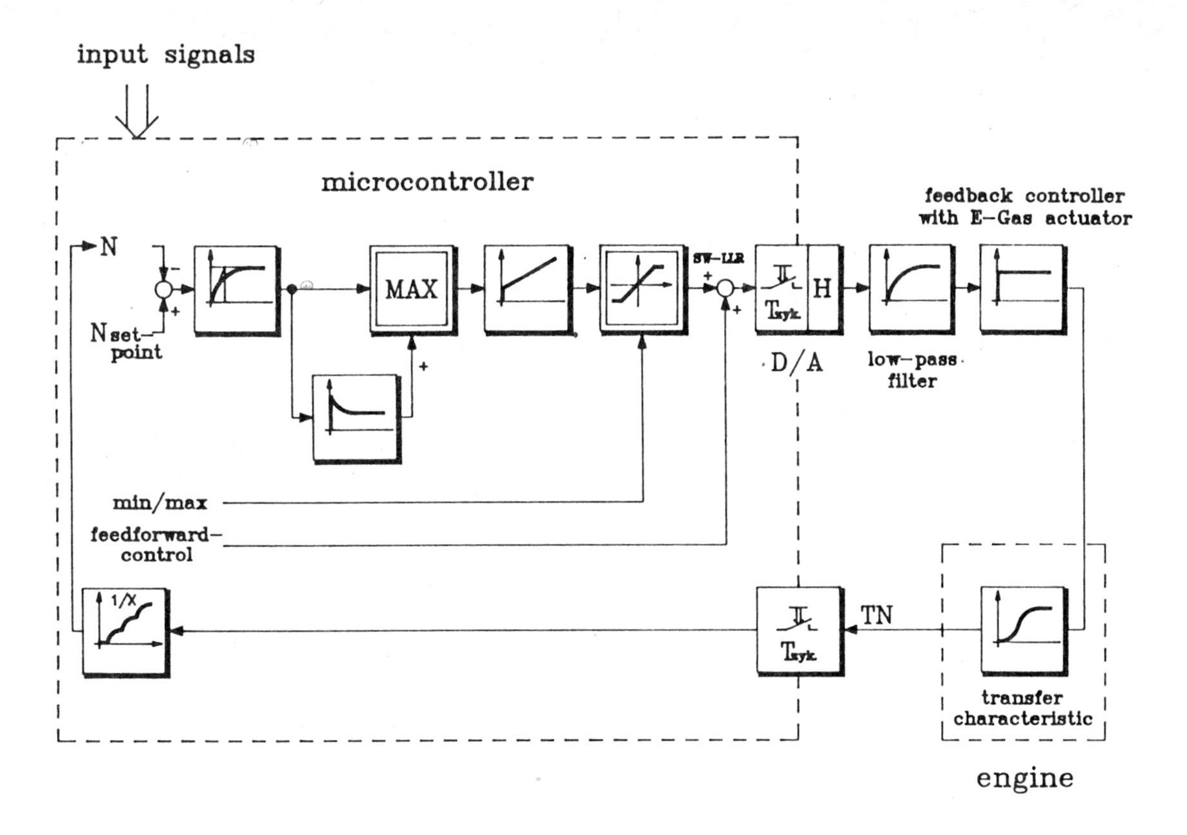

Fig 8 E-GAS third generation passenger car – block diagram of digital idle speed control

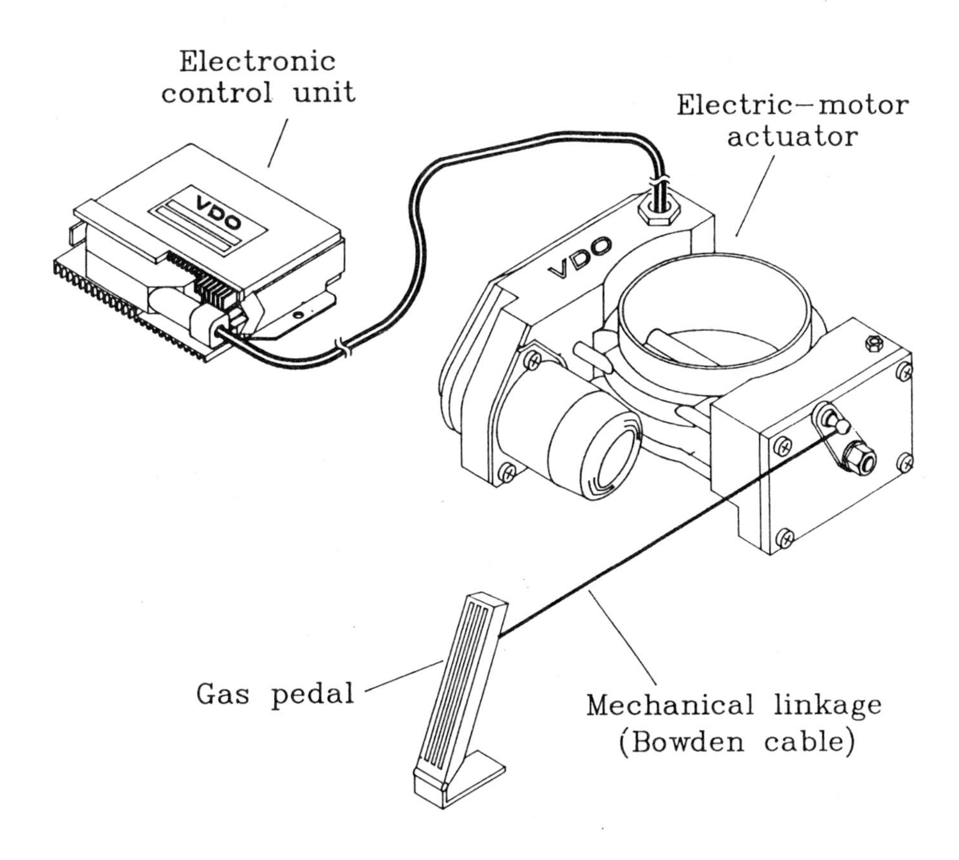

Fig 9 E-GAS third generation passenger car

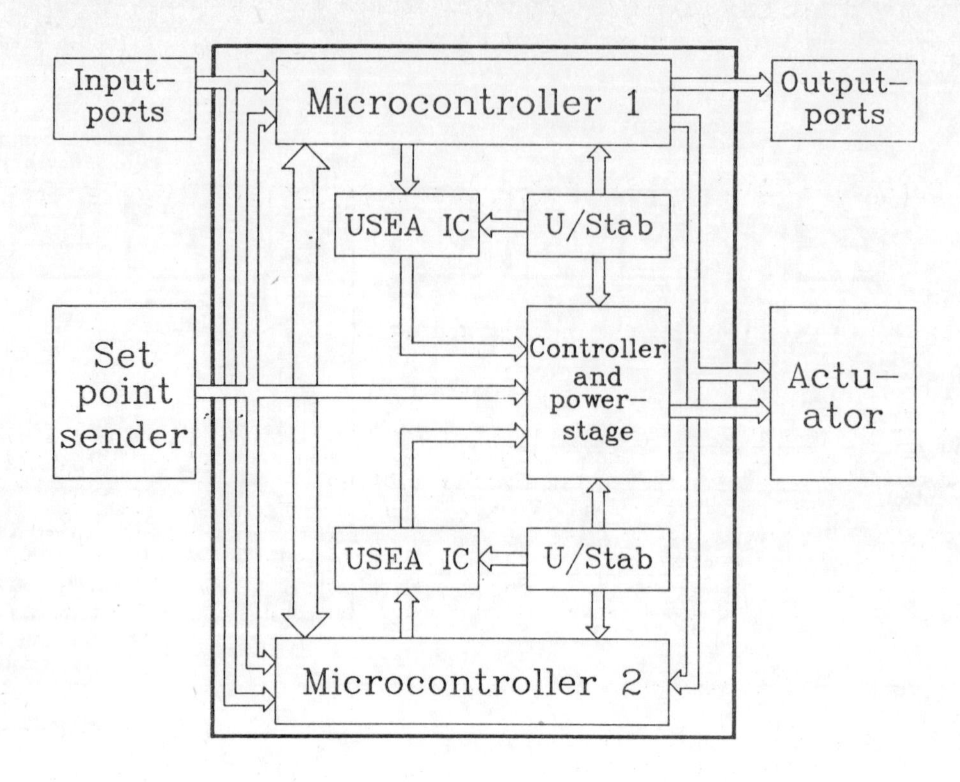

Fig 10 E-GAS third generation passenger car — system concept

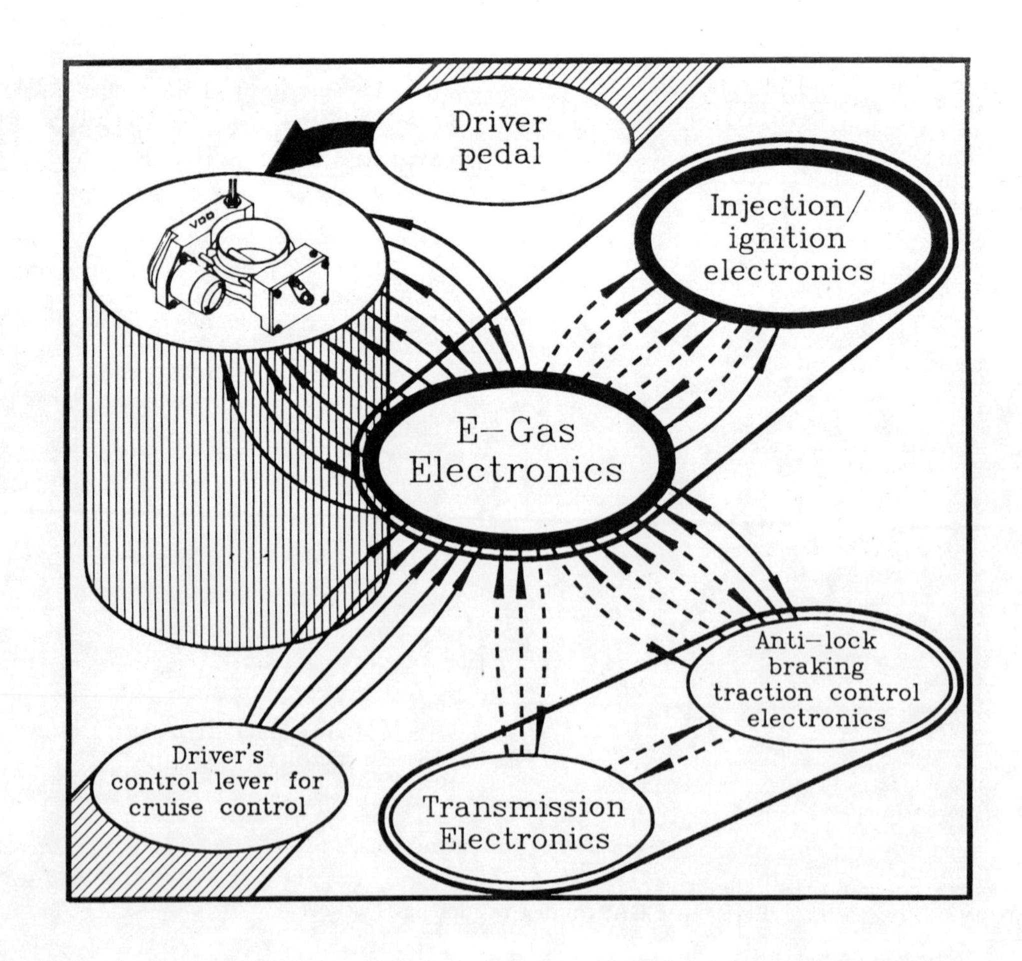

Fig 11 System connection with E-GAS third generation — management network

C391/075

A new high-performance tyre control system

J SCHUERMANN, IngFH, IEEE, VDE/ITG
Texas Instruments Deutschland GmbH, Freising, West Germany

Synopsis, A digital telemetric system for analog control of tyre pressure and temperature from -40..+125 deg C.The wheel modules are energized by a radio frequency burst and retransmit via the same frequency by FM actual measurement and sensor calibration data allowing the Central Unit (C.U.) to calculate absolute pressure and temperature from each measurement. The C.U. addresses the individual transponders by time division multiplex. Electronically calibrated silicon sensors are used. A single chip I/C provides all transponder functions including a novel high efficiency RF transmission.The system is fast,operates with low power and has a high resolution. Wheel transponder module measurement data and drive test results are given.

1 INTRODUCTION, TYRE CONTROL BENEFITS

Significant growth potential for semiconductor manufacturers is given in new automotive applications.Therefore system development is an essential tool to understand,define and provide cost/ performance optimized solid state solutions.

1.1 Introduction

One important new application is tyre control.The individual system functions require a combination of state-of-art silicon processing, packaging and telemetric circuit technology in order to provide an optimum solution.
The three essential functions of any system are the sensor H/W with signal conditioning, the data transmission and the system control.
As compared to threshold and other analog solutions, the new system provides:
- digital signal processing with Cyclic Redundancy Check coding (CRC)
- electronic sensor calibration techniques
- a novel RF data transmission/modulation circuit
- a detachable wheel module option
- high integration C-MOS chip to handle: Analog, RF, EEPROM, Digital Signal Processing and Power Supply functions
  EEPROM = Electronically Eraseable & Programmable Read Only Memory

Some of the advantages are:
- error and drift-free system operation
- high reliability by design, low cost potential
- reduction of power consumption
- high speed transmission of a complex yet redundant data telegram
- lower EMI susceptibility thru FM and a narrow RF receiver bandwidth.

Because of the high level of integration, a small wheel module can be realized which gives added flexibility for different mounting options and applications.

1.2 Tyre control benefits

Unlike other car features e.g. A B S, the need for tyre control is not recognized by the average driver. Sometimes benefits may not be considered as "worth the money". Nevertheless some facts may briefly underline the importance.
Tyres represent the critical interface between the car and the road. Improving safety on the roads means controlling the tyres with regard to the required pressure and to avoid critical temperatures.

Investigations have shown that a 20 per cent drop in pressure cuts life expectancy of the tyres by 30 to 40 per cent,it penalizes the maximum speed, increases the the fuel bill, reduces steering response as well as the directional control of the car.

The interaction with other automotive systems can be critical.High performance cars with suspension control rely on proper tyre pressure.On the other hand recognition of low tyre pressure by the driver is more difficult if the suspension control is in the "comfort" position as compared to a "sportive" driving state.
For the new "run- flat" wheels,a tyre control is almost imperative.

Table 1 Benefits of tyre control systems

Safety	- traction, directional control - steering response improved - max.safe cruising speed indication - early failure warning - support of suspension control
Comfort	- air filling level control - elimination of manual filling control
Economy	- higher tyre life expectancy - tyre replacement saving - lower fuel bill

Alarming is an analysis which says that 30% of all accidents are directly or indirectly caused by tyre malfunctions. They are ranging on second place following brake causes (1).

2 OVERVIEW OF SYSTEMS AND REQUIREMENTS

Tyre control can be categorized in analog/telemetric and threshold/passive type systems as summarized in Table 2.

2.1 Overview of systems

For threshold type systems, the warning point is preset by the manufacturing process. However mechanical means provide a compensation of the threshold point with regard to the temperature.

Table 2 System comparison

	Threshold	Digital/Analog (new system)
System Mode	passive	active,telemetr.
Sensoric's	pressure diaphragm,switch	press. & temper. silicon,analog
Parameter		
- pressure	preset level	1...5,+/-0.1 bar
- temperature	- - -	-40...+125 deg C +/-1 deg C
Transmission		
- powering	- - -	RF-burst
- data transm.	RF,absorption	RF,digit.FM,CRC.
- data rate	1 rotation/bit	60 microsec/bit
- distance	medium	medium / large
Sensor module		
- integr.level	- - -	high,single chip
- control	- - -	signal processor
- calibration	mechanical	EEPROM
Others		
- function	drive only	stop and drive
- power cons.	medium	low
- low-cost potential	N/A	yes

(2),(3)

With analog type measurement and telemetric data transmission systems, the warning level can be set at any point within the operating range of the system.
Enhanced safety features as well as optional trade-off's for comfort and economy can be realized via software (S/W) or controlled manually by the driver.
For example, a detection of the gradient of dP/dt or dT/dt allows an earlier warning and therefore control of a critical situation which in some cases may make a difference.

Threshold/passive type systems are in production, while analog/telemetric systems are in the development or pilot production phase.
The presented system is in the test phase using all the described H/W functions and with the S/W for the main functions in place.

2.2 Requirements

The specifications are determined by safety, environmental and functional needs. The service as well as construction limitations are also to be considered. Comfort items are either inherent to the system or can be realized with minor H/W or S/W additions.

Table 3 Requirements

SAFETY, FUNCTION, ENVIRONM.	- pressure range of 1...5 bar(absolute)
	- temperature range -40 ... +125 deg C
	- accuracy of 0.1 bar / 1 deg C
	- 10 year lifetime, redundancy, system failure warning, high reliability sealed wheel module
	- low power, low battery drain
	- fast, error-free data transmission
	- EMI, RFI & PTT compatibility
	- 1500 G linear acceleration,30 G shock
	- corrosion, icing avoidance
	- drive & stand-by functionality
COMFORT,	- indication of inflation level
SERVICE,	- simple replacement of tyre module
ECONOMY,	- small size, low mass of tyre module
CONSTRUCT.	- large distance bridging wheel to axle
	- added functions (wheel anti-theft)

3 CONCEPT AND FUNCTION OF THE NEW SYSTEM

The key system design guidelines are:
- Digitizing of the analog sensor signal for error-free signal processing and a Cyclic Redundancy Check (CRC) protected transmission
- Transponder & receiver with narrowband RF-FM and a high speed data transmission to minimize EMI which is primarily proportional to the bandwidth of the receiver
- Electronic calibration of all sensors
- Single chip I/C for all transponder functions
- Transponder with low power consumption

Furtheron the concept minimizes the functions to be handled by the transponder processor by placing the operations to calculate the absolute temperature and pressure in the Central Unit (C.U.) processor.This requires that each telegram transmits the actual sensor measurement data together with the calibration information of the sensors. These are permanently stored in the nonvolatile memory (EEPROM) of the transponder processor. The choosen work split between the C.U. and the transponder minimizes the amount of silicon and therefore the cost of the transponder units as there are five of these per system. On the other hand the processing power of the C.U. is available at no extra cost. This approach provides full compatibility between any transponder and any system. No adjustments or coding is needed.

3.1 System operation

The C.U. is designed to handle five wheels in a time division multiplex mode. It operates from the 12 volt supply and has a data interface to the main computer which controls or evaluates the information and provides warning or display of information. The H/W of the C.U. is comprised of a system processor, multiplexer (MUX) circuit, RF signal generator and a data receiver. The MUX addresses each wheel with a RF burst from the signal generator and allows the receiver to listen for the data telegram before switching to the next wheel.
Two RF interfaces link each sensor module to the wheel and to the body of the car.The first one provides a RF link from the car body to the wheel by using inductive coupling elements, later on refered to as stator and rotor antenna. The second interface acts as an inductive "wireless connector" to join the wheel-fixed rotor antenna

to the sensor module.This allows a contactless connection to the wheel and replacement is easy by just unscrewing the module when servicing is needed. The second benefit is that the module can be sealed.This enhances the reliability especially in such environmentally critical locations.

The wheel module transmits the following digital information to the C.U. :

- relative pressure of the tyre
- relative air temperature of the tyre
- relative temperature of the pressure sensor
- calibration data for sensitivity and offset parameters of all sensors
- CRC code

The CRC assures the proper reception of the telegram. If errors occur from the transmission or thru any electrical environmental influence like voltage spikes or EMI, the system recognizes the data to be erroneous. In this case the C.U. processor can either ignore, or in case of continued error indication, give the driver a system failure warning.

A low power or a low power programmable feature is very important. Especially for systems which can be set from the normal operating to a parking or stand-by mode. In the latter case one can use the system as a wheel-anti-theft alarm device.

The power consumption is mainly determined by the energy to operate the transponders. Variables which impact the energy level are the system accuracy, resolution, the distance between rotor and stator as well as the system speed respectively the measurement cycle time.
While the accuracy and resolution have to be defined by the design and therefore fixed, a simple trade-off and programmable power reduction can be implemented by S/W instructions at the cost of the cycle time. Another possibility to control the power level,is the reduction of the rotor-stator distance. A lower power level can also be set by the design or by providing a power level switch for instance controlling the drive level of the output circuit.

3.2 Wheel module function and operating mode

The most critical part with regard to power consumption, performance and size is the wheel sensor module.
Energizing the wheel module from the car body over a relative large distance is critical,so priority was given to the module efficiency.

The wheel module operates as a telemetric transponder by receiving RF energy and retransmitting on the same frequency.Generally transponders work in parallel mode with receiving and transmitting performed on different frequencies to provide a higher energy transfer thru continuous (or seriell) transmit/receive mode. However this either presents a component and complexity penalty besides adding some non-integrateable tuning circuit elements or untuned or wideband systems are more sensitive to EMI (4) (5).

The two most "power-hungry" functions of the transponder are:

a) the constant current powering of the sensors
b) the RF transmission

By using short current pulses combined with fast voltage sensing and digitizing techniques, the power consumption of one sensor function is in the order of 10 microwatts or less.
The RF transmission works with a high efficiency oscillator, combined with a new modulation circuit which cuts the transmission time, thus reducing the power requirement as well as the size of the storage capacitor.
With this technique, a short RF burst is sufficient to charge a small capacitor up to approx. five volts. At this value, all digital and analog I/C - functions operate at their maximum efficiency.

The sensor functions of the transponder are handled in a sequential mode, controlled by the sensor signal microprocessor thru the on-chip multiplexer.

3.3 Sensor approach and technology

Vacuum as reference is used to provide absolute pressure data under all conditions. Gas as a reference media would cause a further compensation because of the expansion with temperature rise.
Besides the three main sensors, the system can handle additional devices to provide redundancy. For instance, control of the hermeticity of the vacuum chamber can be provided by a second pressure sensor mounted inside the vacuum chamber.
The three sensors used in the module are:

a) A 6 bar pressure sensor, fabricated by ion implanting a resistor bridge onto silicon and thinning the backside by anisotropic etching, forming a thin diaphragm which acts as a carrier for the resistors.
b) A temperature sensor is provided to monitor the temperature of the pressure sensor. This allows the signal processor to cancel the inherent temperature dependancy of silicon pressure sensors.
This sensor is fabricated as "base-spreading" device using the silicon bulk material as the resistor element.
c) A temperature sensor to measure the air temperature of the tyre. This sensor is also a "base-spreading device" (6) (7)

During the module manufacturing process, each module is calibrated with regard to the individual offset voltages as well as the sensitivity factors. The set ofcalibration factors are stored in the memory (EEPROM) section of the sensor signal processor.

3.4 Transmission concept,speed,EMI considerations

Tyre control systems are, like other "open" transmission systems, more susceptible to EMI as compared to "wired" transmissions. The susceptibility is proportional to the bandwidth of the system or data receiver. Reduction of the transmission bandwidth is the most efficient tool to reduce potential EMI.
Inductive rotating transformer transmission systems have a large bandwidth of one or two octaves. The new data transmission system shrinks the bandwidth to 10 or 20 percent of the carrier frequency by using Narrow Band Frequency Modulation (NBFM). The EMI frequency sensitive range is about 5 to ten times reduced as compared to open inductive transformers.

The transponder RF oscillator operates with a

temperature compensated and electronically calibrated tuned circuit consisting of a ferrite inductor and a ceramic capacitor.
The inductor is used as part of the "wireless connector" arrangement. During the modulation process the frequency is shifted by a simple circuit which is directly controlled by the sequence of the binary "high - low" levels coming from the Random Access Memory (RAM).
The two methods for minimizing the power and enhancing the transmission efficiency are:

a) A new pulsed oscillator configuration with a digital, low duty cycle powering circuit yet providing a continuous RF carrier.
b) A high speed NBFM modulation circuit to cut the transmission time by a factor two compared to conventional NBFM without increasing the bandwidth.

The actual NBFM data transmission is done by Frequency Shift Keying (FSK) of the RF carrier. The data transmission speed can be as high as 8 or 16 k Baud depending on the modulation circuit and the receiver design.
Another by-product and advantage is the the high Signal-to-Noise ratio of the system, mainly due to the frequency selective data receiver.

3.5 Sensor processor chip configuration

A programmable constant current source is sequentially switched to the pressure and temperature measuring resistors of the sensors. The measured pressure and temperature proportional voltages are connected thru the multiplexer to a high resolution Analog/Digital converter (A/D) for digitizing and storage in the RAM.
All logic, analog and the RF-functions can be combined on a single C-MOS chip, together with the nonvolatile memory (EEPROM) function, needed for the module calibration. The processor requires only a few peripheral components to provide the full sensor module function.

4 FIELD TEST H/W AND SYSTEM MEASUREMENT DATA

4.1 Wheel components

The present H/W wheel components are preliminary but provided results from field tests over an operating temperature range from -20 to +100 deg C. The transponder modules have passed several temperature cycles from -40 to +125 deg C with pressure levels up to 5 bar. No destruction or noticeable deterioration of the parameters were recorded.
The "stator" antenna is the most critical component with regard to the operating temperature range, shock and corrosion. The upper temperature limit for the function is given by the Curie effect of the used ferrite material which was evaluated to be in the vicinity of 200 deg C. This temperature was not exceeded during the test.
The wheel-fixed rotor antenna is positioned near the center of the wheel. It is a simple one winding loop placed inside a high temperature resistant fixation ring. The single winding rotor with the stator antenna represents a rotating RF transformer and which maintains system function at any position of the wheel.
The wheel transponder module for the test was not fully miniaturized but uses C-MOS I/C components for all functions, which included a low power digital sensor signal processor (8).
The pressure sensor is contained in a low hysteresis housing. The transponder module is mounted inside the wheel with the inductive RF interface to "connect" it to the wheel "rotor" antenna.
Tests with a linear acceleration of 1500 G including a shock test of 30 G were carried out. No mechanical or electrical changes were noticed.

4.2 The Central Unit

The C.U. "talks & listens" thru the multiplexer to the individual wheels respective transponders. The multiplexer is designed as a bidirectional switching module to handle power and small signals.
The RF burst from the signal generator passes the multiplexer and goes thru the cables to the stator antenna and the wheels. The return path for the transponder signal provides a low loss, and a narrow bandwidth of only 10 per cent.
Power MOS devices are used in the signal generator for high efficiency. The oscillator operates with a low cost 2 Mhz type crystal combined with a frequency divider for long term stability. The signal generator works into the multiplexer by selective tuned circuits for EMI and RFI compatibility.
The data receiver was designed with a low cost RF I/C designed for the communications market as for instance cellular radio and cordless phone applications. These amplifier and NBFM demodulator circuits have a high inherent suppression for all kinds of amplitude type noise signals, for instance those generated by switching inductive or capacitive loads.
The processor provides the system control functions, assembles the decoded data telegram into the absolute temperature and pressure data for interfacing with a car-based data tracking computer.

4.3 System measurement data

Several systems were manufactured, they have accumulated considerable mileage. Although not optimized, the system functioned accurately and reliably.
The system accuracy is +/- 0.1 bar and +/-2 deg C from 0 to +80 deg C. The resolution of the pressure indication is approx. 0.03 bar.

Actual characterization data for pressure are shown as a normalized plot of the absolute pressure which are referenced to the different pressure levels between 2 and 5 bars. The X-axis gives the operating temperature range of -40 to + 100 deg C and at pressure levels of 2 to 5 bar. For each of the four pressure levels, two temperature cycles were run in order to determine the hysteresis.

The drive test shows a typical mixed motorway/ road test behaviour for all four wheels for the pressure and the associated tyre air temperature. The reason for the higher temperature rise for the rear-left wheel is due to the exhaust muffler heating the wheel by radiation.

The isochoric pressure, normalized to 20 deg C is rather flat which indicates good agreement between the actual average tyre air temperature and the temperature measured by the system.

RFI evaluation was carried out with regard to several European PTT specifications. All recorded

radiation levels were far below the allowable limits. Compatibility or interference tests with adjacent parked cars yield a safety factor of approximately 60 db. For this test the distance between the two (parallel) wheels or parked cars were as close as possible and in the range of 10 to 15 cm.
An investigation is underway to evaluate the temperature gradient between the tyre and the wheel for various conditions in order to determine a correction algorithm for a wheel mounted transponder module, not beeing perfectly isolated.

5 CONCLUSION

The new system can be considered as the second generation of analog tyre control systems because of the combination of State-of-Art digital signal processing, electronic sensor calibration, with a new carrier frequency modulation and fast data transmission system, to yield error-free, zero-drift communication between the low power wheel module and the central control unit.
High resolution and accuracy over a large operating temperature range was demonstrated.
A single and high integration C-MOS chip combines several process technologies or functions as digital and analog signal processing, RF signal generation and modulation as well as nonvolatile data storage in an EEPROM.
Part of the development program was the characterization of pressure and temperature behaviour of tyres or wheels under various conditions.
The system was installed in several test vehicles meeting functional and stress test conditions.

REFERENCES

(1) Schmid,H.-D. Sicher und wirtschaftlich fahren mit electronischer Reifendruckkontrolle. VDI Berichte 6/87

(2) Folger,J. Riedl,H. Wallentowitz,H. BMW-AG Electronic tire pressure control. EAEC Conference,Strasbourg,June 1989

(3) Hebert,M.J. Development of a tyre monitor. Michelin, C382/131,IMechE 1989

(4) Folger et al. Transmission method for variable measured values from vehicle wheels utilizing ambient temperature compensation. US Patent 4,567,459 BMW AG

(5) Smith. Temperature measuring system for rotating machines. US Patent 3,824,857

(6) Texas Instruments, Silicon Sensors Data book TDS 252

(7) Schuermann,J. Sensoren fuer die Automobil-Elektronik. Funkschau 14/1978, Texas Instruments

(8) Bierl,L. Brenninger,H. Diewald,H. TSS400 Application Report, EB 175E, SENSOR & SYSTEMS, Texas Instruments

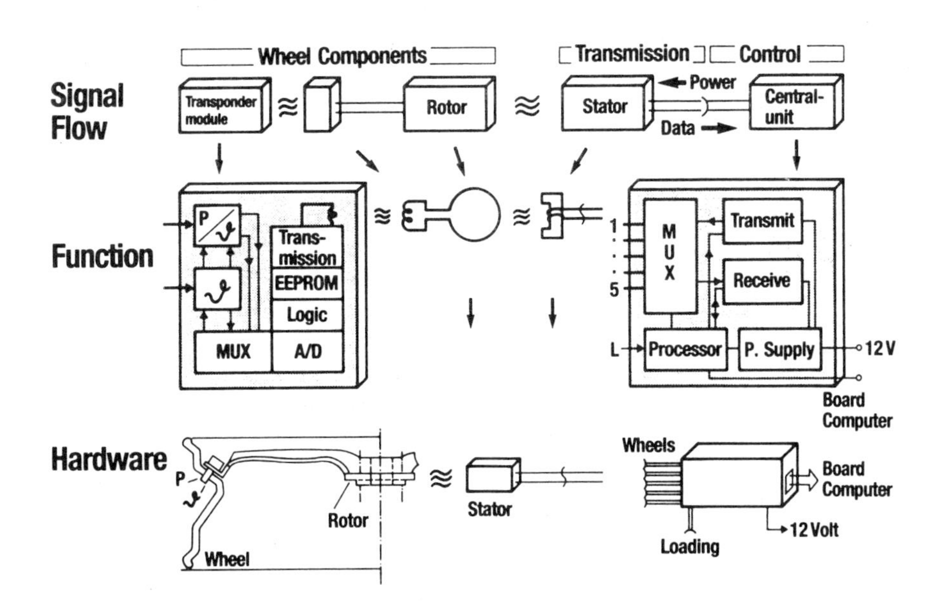

Fig 1 Tyre control system configuration

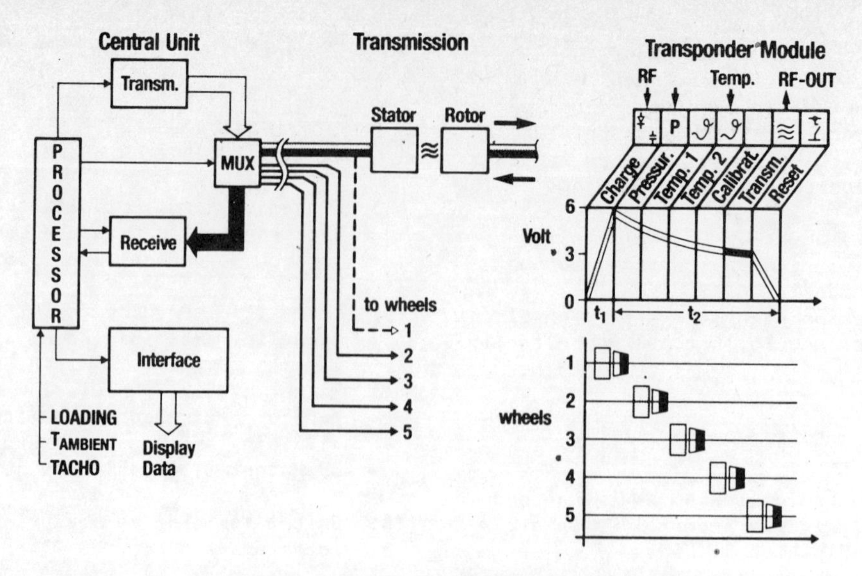

Fig 2 Functions and operation mode

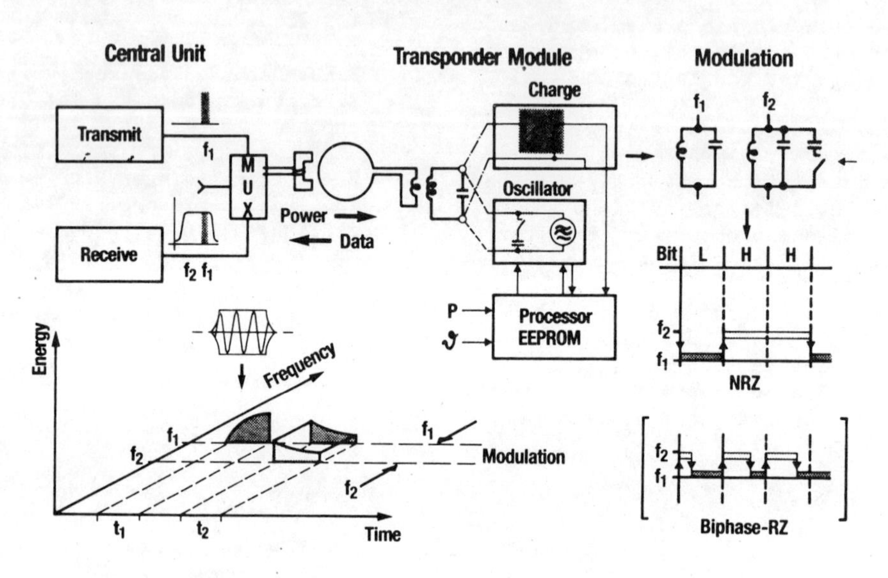

Fig 3 Transmission system

- **Mechanical part**
  - Temperature sensor integrated in wheel module to sense air temperature.
  - Pressure sensor uses vacuum as reference. conical aperture to avoid icing problems.
- **Electronic part**
  - All silicon sensors use short measurement cycle and low current pulses
  - Advantages
    - no self heating error
    - fast response time
    - low energy/small charge capacitor
    - small module size
  - **Example:**
    - Discharge Energy C = 3,2 Micro Amp. sec.
    - Power Consumption P = 4,2 Micro Watts
    - Measurement Cycle $t_2 \triangleq$ 4 % t,
  - **High accuracy**
    - Ratiometric,- Voltage independant measurement of pressure signal
    - Tracking of temperature gradient of pressure sensor for error correction

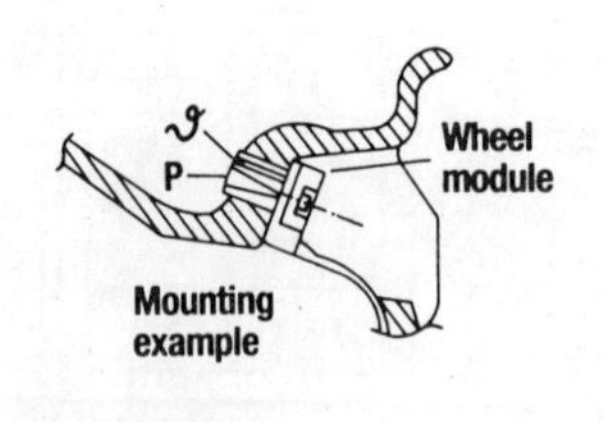

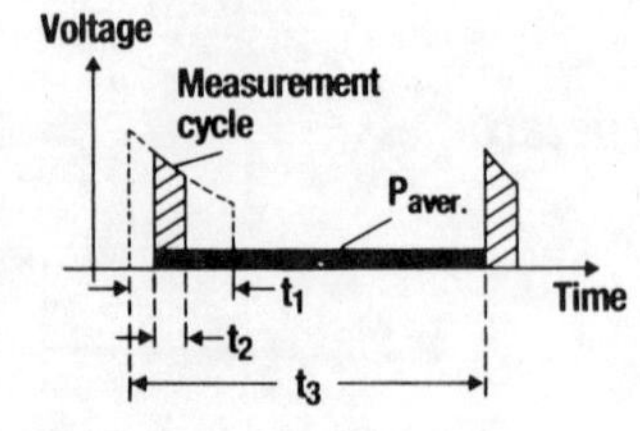

Fig 4 Sensor technology

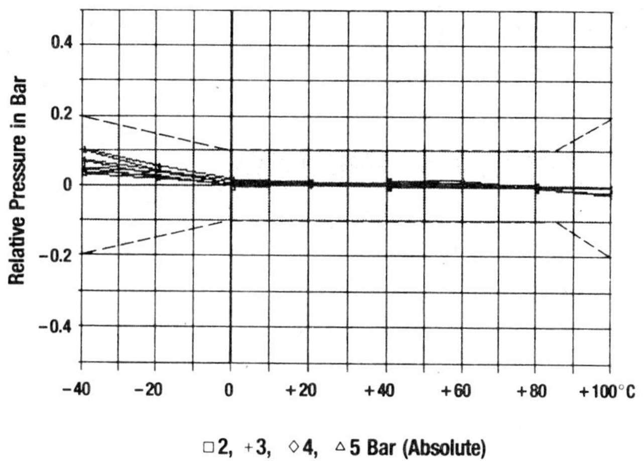

Fig 5 Pressure calibration data of a wheel transponder module

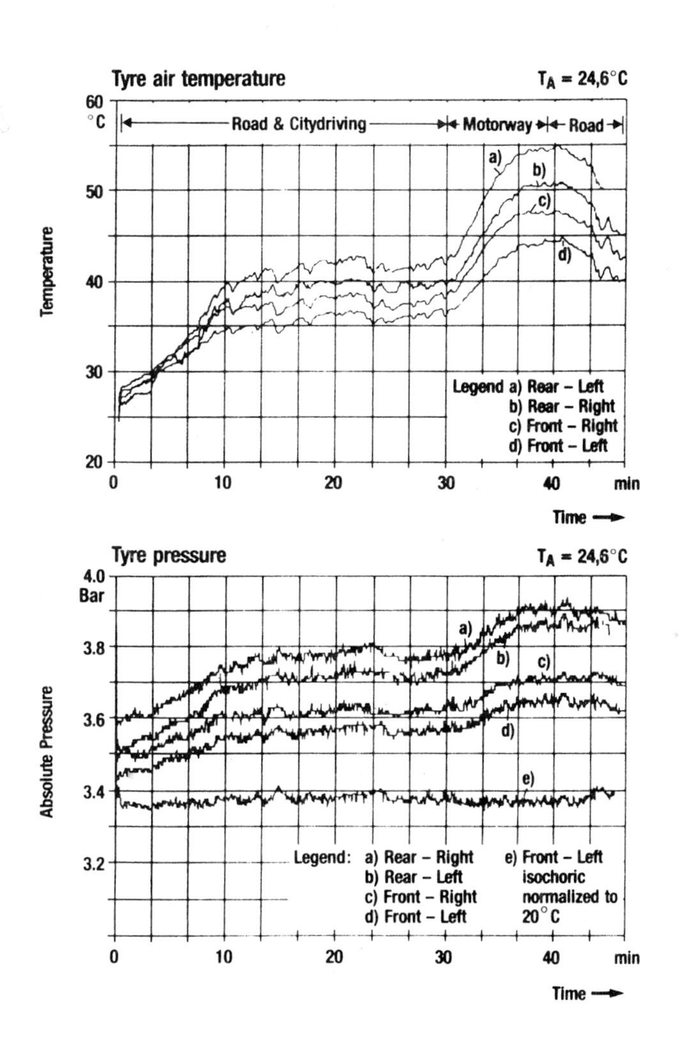

Fig 6 Test drive results

C391/026

Experience with a tyre monitoring system

D J MYATT, BSc, AMIEE
Manufacture des Pneumatiques Michelin, Cébazat, France

SYNOPSIS An outline of MICHELIN'S philosophy behind tyre monitoring systems is given with a subsequent description of the MTM (MICHELIN TYRE MONITOR) and our design objectives.

1 Introduction.

One could argue that punctures or other related problems with tyres are so infrequent that it is hardly worthwhile installing a monitoring system! In fact the same argument may apply to antilock brake systems, laminated windscreens or even seat-belts. This is not the view at Michelin where safety and customer-service are of the utmost importance. The air in the tyres is probably the last remaining fluid used by a vehicle that has not been fully instrumented and yet it is certainly by far the most important for road-handling, comfort and safety.

The fact is that there has always been a strong interest in tyre monitoring which is confirmed by the great number of patents that have been published over the years (more than 1000). One of Michelin's first patents on the subject in 1920, used an explosive cartridge mounted on the wheel, which detonated at low pressure!

Perhaps the biggest single problem hindering a viable solution being developed came from the fact that most of the proposals had to be implemented mechanically and the wheel/tyre location is fraught with problems of limited space, corrosion, temperature range, centrifugal force and non-contact transmission between wheel and vehicle.

The advent of low power electronics and especially micro-machined silicon pressure sensors has revived the subject, making tyre monitoring systems a reality at last .

2 Situation without tyre monitoring.

2.1 Some statistics.

A recent survey (1986) showed that in Europe, 19% of car drivers had at least one loss of pressure in the year and that car drivers could expect a mean time between pressure losses of about 5 years or 43000 miles.

Loss of pressure can be caused by one of the following:

Objects penetrating the tread area . . (49%)
Natural loss through the permeability of the rubber, valve leakage, porosity of rims, and other related causes. . . . (24%)
Sidewall damage . . . . (27%)

In 85% of these above first two cases, the driver could have avoided running on a flat tyre if only he had been warned early enough, the loss of air occurring after more than 30 seconds.

All tyres suffer from a very slow loss of pressure even when in perfect condition because of the porosity of rubber (less than 0.2 bars per year). Many people don't check their tyres very often and with today's cars it can be difficult to tell from driving the car whether the pressures are low, even dangerously low!

A recent study (1988) showed that in France the average pressure in all tyres tested was 15% lower than nominal, with a dispersion ranging from -70% to +40% ! In fact, 60% of all cars tested had under-inflated tyres.

2.2 Consequences of driving on under-inflated tyres.

2.2.1 Slightly under-inflated.

Increased running costs(rolling resistance and wear). Increased vulnerability to damage. The handling of the vehicle can be affected as too can be the performance of antilock brake systems or sophisticated suspensions.

2.2.2 Highly under-inflated.

The above mentioned points plus:

Possible internal damage to the tyre's casing which could lead to a failure later, even after the tyre has been correctly inflated.

An increase in internally generated temperatures which could lead to thermal runaway at high speeds or with heavy load.

3 Design objectives of the MTM system.

A possible tyre monitor in our view might well be one which measured tyre deflection, but this is difficult to measure and would not give good results in all circumstances. Pressure and temperature are however parameters of major importance in determining the global performance but must be considered indirectly by using a model of the tyre. We considered 12 points before designing the system :

1/ Accurate measurement and analysis of pressures and temperatures .
2/ No mechanical or galvanic contact between wheel and car.
3/ Fast measurements.
4/ Low power.

5/ Small size.
6/ No wearing parts and no maintenance.
7/ Low cost in mass production.
8/ Low sensitivity to centrifugal forces.
9/ Almost military temperature and vibration requirements.
10/ Low radio frequency interference generation (RFI) and immunity from high interfering fields (EMC) even using unshielded wiring.
11/ High reliability with fail-safe characteristics.
12/ A system that is operational as soon as the vehicle is in use.

Only the extensive use of integrated electronics and silicon pressure sensors on the wheels enabled all of these requirements to be met.

4 Principles used.

4.1 General outline.

The electrical equipment on each wheel acts as a passive transponder. It remains idle until it starts receiving a low frequency magnetic field that charges up a capacitor, which in turn supplies the DC power needed by the wheel electronics. By switching off this field, the central control unit orders the wheel electronics to start measuring and to transmit the resulting data back through the same antenna system. The whole cycle of Power-up/Receive only takes about 150 milliseconds to complete.

4.2 Wheel electronics.

From the beginning we made every effort to simplify the wheel electronics as much as possible in order to facilitate integration, reduce size and achieve low cost. The result is a standard silicon pressure sensor associated with a Michelin developed custom integrated circuit. An analog approach for part of the circuit gives us small size and low power whereas a microprocessor based system would have needed an analog-to-digital converter probably just as big as our circuit alone, not to mention the microcontroller.

After powering up the reservoir capacitor by the excitation signal, the pressure sensor is then switched on and begins delivering a voltage to the custom I.C.. Voltage ranges from 0 to 50 millivolts for full-scale pressure and so the first function of the I.C. is to amplify these low-level signals and to normalise them with respect to offset and sensitivity. Due to the high spread of these two parameters for standard 'catalogue' piezo-resistive sensors, each wheel-electronic-module must be laser-trimmed to ensure interchangeability between different modules.

Temperature information is obtained from a Band-gap device that is integrated within the same I.C.. Pressure and temperature signals are used to control a free-running oscillator from which are derived very short pulses at each change of state . Low-impedance switches connect the reservoir capacitor to the wheel antenna for the duration of the pulse.

The fixed antenna needs to receive only three pulses in order to correctly decode pressure and temperature. Since the pulses only last 5 microseconds, the energy and the time needed to complete a transmission are both very low (450 picojoules and 5 milliseconds).

Having the temperature measurement transmitted to the vehicle enabled the wheel electronics to be further simplified since the much needed temperature compensation for the pressure sensor was transferred to the software in the central control unit. This approach is more cost effective, allows more sophisticated compensation techniques to be used and avoids slowing down the otherwise simple trimming procedure.

The complete wheel electronics module which contains sensors, I.C., energy storing capacitor, calibrating resistors and protective packaging takes up less space than a sugar-cube (0.5in long by 0.6in dia.) and could probably be made even smaller.

It was built by the Californian company SENSYM which is now a subsidiary of Hawker Siddeley .

4.3 Electromagnetic coupling.

Transmission of both energy and signals between the antennas is achieved through near field magnetic coupling. Both antennas are like the primary and secondary coils of a very loosely coupled transformer. They are tuned to their resonant frequencies in order to compensate not only for the lack of coupling but also the shunting effect which the aluminium or steel wheels have on the rotating antenna. When the frequencies are kept low, the field strength decreases rapidly with this type of antenna so that problems with RFI and possible interference with other vehicles close by is avoided. Lots of natural shielding plus the low impedances used around the antennas efficiently ensure good immunity to high EMC field strengths.

Both antennas are mounted very close to the brake systems and so are subject to high thermal shocks. Only class-F or class-H insulated enamel wires plus one ceramic capacitor are needed, electronic circuits were avoided so that the only temperature restricting component is the type of plastic used. Highly resistant antennas can be produced through the use of modern materials like Kinel, Technyl or Noryl.

Proximity to the brakes is actually used to advantage in some applications by incorporating a temperature sensor in the fixed antenna. The MTM control unit can then use approximate brake temperatures in the driver warning strategy and this is of particular interest on lorries for example.

4.4 Central control unit.

Input signals from each fixed inspection antenna consists of 3 sharp pulses of about 150 millivolts and 5 microseconds duration. Amplitude can vary depending on climatic conditions and spacing between antennas but the timing between pulses does not alter. These low level input pulses are amplified by a differential low impedance front end before being converted into interruptions of an 8 bit microcontroller. Timer-counters contained within, are used to obtain the precise times separating the pulses, from which pressure and temperature values are derived .

The functions of this central control unit can quite easily be incorporated with other control systems including multiplexed bus structures which we are using at present in our version for lorries.

4.5 Driver warning strategy.

It is clear that few drivers will want their dashboards cluttered up with tyre pressure and temperature displays, although some may wish to see this data from time to time. Driver information should be simple, direct and easy to understand. In most cases, the way in which information will be displayed, is going to depend on the type of car and it's manufacturer.

Depending on speed, driving style, load, climatic conditions, tyre type and time, tyre pressures and temperatures vary enormously, even for a tyre that is correctly inflated. This means that alarm strategy is very situation-dependant and must be quite sophisticated in order to obtain the best results.

Without going as far as measuring all the external influences, good performance can be obtained by combining pressure and temperature signals. A 10% loss of air can be reliably detected without false alarms.

Measuring pressure and temperature provides close tyre monitoring. The temperature signal ensures compensation for the inevitable thermal drifts of the pressure sensor as already mentioned. There is also a third advantage in having two physically independent sensors because although they do not measure the same parameter, their signals are closely related by the thermodynamics of the tyre and so part of the fail-safe procedure is checking that temperature variations are accompanied by meaningful pressure variations.

Tyre data could be read out from a diagnostic socket. This also is something of particular interest on lorries right now.

5 Conclusions

The development of an effective tyre monitoring system needs a good understanding of tyres, wheels and vehicles. Only after close co-operation between tyre and car manufacturers is it possible to have a system that meets all the requirements.

The MTM system satisfies all these requirements and it's reliability and cost effectiveness deserve particular emphasis.

Pre-Series parts are now available. The whole system will be commercialised towards the end of 1990 on cars manufactured by BMW in Germany.

C391/009

Infra-red tyre condition monitor

M HUTCHINSON, BSc, MSc, DMS, CEng, MIMechE
A B Electronic Products Group, Newport, Gwent

SYNOPSIS

This paper presents : a practical tyre temperature monitor, which measures the surface temperature of vehicle tyres by measurement of radiation heat flux from the surface.

INFRA RED TYRE CONDITION MONITOR

1 INTRODUCTION

The physical condition of a tyre is intrinsic to the handling of commercial and passenger vehicles. Although modern suspension systems require very little maintenance to retain design handling, vehicles can become unstable if high 'G' manoeuvres are attempted with worn or damaged tyres. The condition of a tyre is at present monitored by the driver when the vehicle is stationary, by visual inspection and pressure measurement, but it would be much more useful to monitor the tyre condition when the vehicle is being driven.

Attempts have been made to develop a tyre condition monitor, but most have failed because of the relatively high price of the device and because most of these devices have been tyre pressure monitors which do not detect the general physical condition of a tyre.

It have been recognised (R. Smith (1), Herbert Kerner (2) and Avon Rubber (3) that a tyre in poor physical condition generates more heat than one in good condition, thus the surface temperature of a faulty tyre is higher.

This paper describes a device which measure the net infra red radiation from the side wall of the tyre and estimates the side wall temperature thus predicting the tyre condition.

2 GENERAL

2.1 Description of Tyre Sensor

2.1.1 The Preferred Tubular Design

The tyre condition monitor described in this paper is essentially a radiometer of the type described by Jacob (4) adapted for road vehicles.

This instrument consists of a window through which radiant heat enters and a sensing element which converts the radiation from the tyre into sensible heat. A temperature measuring device is bonded into the sensing element in order to measure the temperature and convert it to an electrical signal. This signal transmitted via the output leads to the controller inside the vehicle. Signals from other sensors are received by the controller. The sensing element is insulated from the body of the sensor in order to minimise conduction losses to the environment.

2.1.2 Radiation Receptors

Flat Plate Receptor

Figure No 1a shows an arrangement with a thin copper disc insulated from the outer shell of the instrument by means of low density 'plastic' foam disc. The temperature sensor is mounted on the copper disc bonded to give intimate contact.

Parabolic Reflector and Point Receptor

Figure 1b shows an alternative arrangement utilising a parabolic mirror to concentrate the radiation onto a very small diameter sensing element incorporating the temperature measuring device.

2.2 Design Criteria

2.2.1 General

The IR tyre sensors are suitable for light passenger and heavy goods vehicles. They have been designed to withstand the following operating and environmental condition:-

i) climatic temperature -40°C +105°C

ii) vehicle vibration (engine and road)

iii) water and dust penetration IP67

iv) diurnal, seasonal and geographical variation in climatic conditions expected by the current range of European cars and heavy goods vehicles.

2.2.2 Sensor Measurement

a) The temperature sensing element has been designed to detect changes of ±0.01°C in the receptor temperature, which corresponds with 0.05°C and 0.2°C change in the surface temperature of the tyre.

b) The thermal/electrical time constant 1.5 mins.

c) Compensation for variations in the following parameters will be incorporated into a micro processor as programme algorithms:

i) Environmental temperature

ii) Vehicle road speed

iii) Cross wind velocity

iv) Weather and road conditions

2.3 Electronic Controls, Measurement and Data Acquisition

2.3.1 General Considerations

The sensor circuit design was considered for commercial systems.

Design Criteria

The main objective was to provide cost effective electronic packages for automobiles and heavy goods vehicles. Tyre monitoring systems have to be designed for the following:

i) to fit to existing vehicles electrical systems

ii) to incorporate into new vehicles and utilise the automobile's computer management system

iii) to be incorporated as one of the sensing elements of an active suspension system

Circuit requirements are to:-

: be robust

: be cost effective

: require no further calibration after the initial factory calibration

: be suitable for expected environmental conditions

2.3.2 Experimental Programme

2.3.2.1 General

Thermocouples, thermistors and LM335 temperature measuring devices were used during the experimental programme, but only Thermistors and LM335 were considered for use in commercial sensors.

2.3.2.2 Thermocouples

Precision K type thermocouples were incorporated into the experimental prototype sensors because their very low thermal capacity and small size were suitable for surface temperature measurement. Thermocouples produced no self heating and heat losses by conduction through the leads was much less than for either of the other two devices. A COMARK 10 point temperature sensing instrument was used. However, thermocouples were not considered suitable for commercial sensors because they:

i) were less immune to noise than the other devices

ii) required a high gain amplifier for measurement

iii) were prone to corrosion at the interface between different metals, hence special connectors would be required

2.3.2.3 LM335 Temperature Measuring Devices

Although LM335 devices had the desirable property of immunity to noise because they use current loop, they were rejected for use in commercial sensors because of their large thermal capacity.

A BBC B computer was programmed for use as a data logger and a three term temperature controller to control the hot plate.

2.3.2.4 Thermistor Circuits

Thermistors replaced the LM335 devices because they had a thermal capacity less than a quarter and they were readily available in low cost precalibrated packages.

Smaller thermistor devices are available than those for the prototype sensors, they were more expensive and were not available in the calibrated form.

A temperature measuring circuit was designed suitable for laboratory and vehicle use. The thermistor was incorporated into a bridge circuit to linearise the output signal; BURKE (5).

Late in the programme the thermistor circuit was modified to include a differential term which measures the rate of change of temperature with respect to time so that by utilising equation 3.3 the tyre temperature was estimated during transient condition.

Figure 3 is a diagram which shows the basic elements of a simple tyre temperature monitor system. A 80C552 microprocessor is used to process data from each sensor and a warning lamp flashes to indicate when the condition of a tyre is suspect. It also shows the following additional parameters which should be taken into account when assessing tyre condition:-

i) Vehicle speed: used to compensate for the tangential air velocity across the sensor window.

ii) Wind speed: to compensate for an uneven distribution of air flow between different tyres.

iii) Vehicle loading and tyre pressure: to allow for tyre pressure and wheel load.

3 THERMAL MODEL FOR TYRE SENSOR

3.1 Radiation from Tyre

3.1.1 Stefan Boltzmann

All the heat energy received by the tyre sensor from the tyre will be radiant heat at a total emissive power of W_{TOT}, characterised by the modified Stefan Boltzmann equation:-

$$W_{TOT} = \sigma . E . T^4$$

$$W_{TOT} = 5.67 . \times 10^{-8} . T^4 . \quad (Wm^{-2})$$

where W_{TOT} = Heat Total Energy transmitted per unit area $(W\ m^{-2})$

σ = Stefan Boltzmann constant $(W\ m^{-2} . T^{-4})$

E = Emissivity (non dimensional)

T = Absolute Tyre Wall temperature (°K)

The environmental temperature varies from -40°C to 107°C hence the heat flux from a vehicle's tyres may vary from:

W_{TOT} = 167 $(W.M^2)$ at 233°K E=1

to W_{TOT} = 1,182 $(W.M^{-2})$ at 320°K E=1

(The emissivity for a tyre will be approx 0.94)

3.1.2 Spectral Energy Emitted from Tyre Wall

The monochromatic emissive power from the tyre wall will vary from 0 at a wave length $\lambda = 0$ reaching a maximum value of a = 20 microns (WIEN's displacement law) and back to 0 at $\lambda = \infty$. In practice most of the energy will be transmitted between wave length $\lambda = 0.8$ microns and wave length $\lambda = 33$ microns. (The visible spectrum is 0.4 to 0.8 microns).

Table No. 1 lists the maximum wave length λ at which 70%, 80% and 90% of the spectral activity is transmitted.

TABLE NO. 1

% Total Energy	λ at 300°K	λ at 370°K
70%	19 microns	15.4 microns
80%	23.3 microns	18.9 microns
90%	33.3 microns	27 microns

utilising the Planck, WIEN relationship to JACOB (4)

$$\frac{W_B}{T^5} = f(\lambda, T)$$

W_B = monochromatic emissive power (Wm^{-2})

T = absolute temperature (°K)

3.2 Heat Transfer through Window

3.2.1 Partition of Energy Transferred

The net radiant heat impinging on the outer surface of the window will be partitioned in accordance with KIRCHHOFFS Law (4) for black body conditions.

$$1 = \alpha + \tau + \rho$$

where α = fraction absorbed by body (absorptivity)

τ = fraction transmitted (transmissibility)

ρ = fraction reflected (reflectivity)

Figure 4a and 4b illustrate the way in which the radiant energy transmitted from the tyre is transferred through the sensor windows. A fraction 'p' of the incident radiant energy is reflected which leaves a maximum fraction $(\tau + \alpha)$ to be transmitted through the outer surface of the window.

All the infra red window material suitable for incorporating into the tyre sensor have poor transmission coefficient 'τ' hence most of the heat is transmitted through the window by molecular conduction.

Since the windows are poor transmitter of radiation a significant proportion is converted to sensible heat inside the window thus combining with the absorbed heat ' α ' which is conducted through this window. At the outer surface of the window an additional faction of the heat is lost by re-radiation and forced convection.

3.2.2 Window Materials

A survey of a range of window materials was carried out which demonstrated that all the suitable materials absorb or reflect most of the incident radiation from the tyre. Table No. 2 lists the percentage incident radiant energy transmitted through a range of different materials.

The percentage of the incident energy transmitted through the window was calculated by numerically integrating from the published spectral transmission data at 330 deg K and 370 deg K. It can be seen that glass which only transmits 0.001% of the incident radiant may be regarded as opaque to the radiation from a tyre. The incident energy striking the sensor window must be transmitted through the window by molecular conduction or be reflected from the surface.

Window materials such as Zinc Selinide and Polythene are relatively good transmitters of radiation since both transmit about 28% of the energy at 300°K. Experiments with polythene windows demonstrated that they transmit considerably more energy than glass and the sensor was more responsive to the rate of temperature change. Figure No. 5 shows how much of the total black body energy is transmitted by the Zince Selinide window (the best of the window materials). it was not possible to represent the percentage of the energy transmitted by glass because it was too small scale for the graph.

3.2.3 Conduction through windows

For radiometer mounted on a vehicle to measure the surface temperature of a tyre, part of the radiant energy absorbed at the outer surface of the window will be lost by the forced convective air flow across the window but a significant amount of this energy will be conducted through the window. A copper disc was substituted for the window and it gave a similar result to glass. At the inner surface of the window this conducted heat will be transmitted as radiant heat to the temperature sensing element inside the tyre sensor by:

a) Re-radiation from the window

b) Convection into the air space inside the sensor

With an LM335 temperature sensing element glued to the inner surface of the window, additional heat is transmitted to the LM335 by direct conduction through the component leads and the mounting socket.

Figure No. 6 demonstrates the effect of conduction through the window. It shows a temperature gradient across the window of about three degree C. for 2mm glass at a tyre temperature of 80°C and an overall temperature drop of 5°C.

When a window is fitted to a tyre sensor the thermal time constant is increased because:

i) the amount of radiant heat impinging upon the sensor is reduced,

and,

ii) the thermal capacity of the window acts as a capacitance which acts as a time delay element.

An increased time constant reduces the ability of the sensor to respond to changes in tyre temperature.

Experimental studies have showed that the window also has the effect of making the time constant longer for cooling than for heating. A computational allowance can be made for this effect.

3.2.4 Effect of Dirt on the Windows Performance

An early prototype sensor installed in an automobile continued to sense the tyre temperature with a thin layer of dirt on the sensor window, thus demonstrating that dirt applied to the window of their sensor had little effect on the general performance of the temperature sensor. It is believed that this tolerance to dirt was the result of using glass for the window because glass effectively transmits most the radiant energy by conduction. Thus, provided that the dirt layer is not too thick, it should not have any significant effect on the amount of heat conducted through the window. If however, a higher performance window (Polythene) were to be fitted instead of the glass, it would also give a measure of tolerance to dirt. Initially, when subjected to radiation, dirt would reduce the amount of radiant energy transmitted and the sensor temperature would fall. More of the incident radiant energy would then be absorbed by the dirty surface of the window and a significant proportion of it would be conducted through the window.

The final equilibrium temperature of the sensor will rise again after a time delay period. No experimental work has been carried out with fouled windows but it is possible to carry out this type of work on a bench scale with simulated dirty conditions. It may be concluded that polythene windows may also tolerate dirt but the expected disadvantage would be an increased time constant.

3.3 Dynamic Model of the Tyre and Sensor

3.3.1 General

A simplified dynamic model was developed in order to

i) assist in the understanding of the heat transfer processes.

ii) aid the analysis of experimental data.

iii) assist in the development of a real time algorithm for controlling a commercial monitor.

3.3.2 The Time Dependent Model

The following assumptions were made in order to derive the simplified model:

: the sensor is assumed to operate in still air

: near black body radiation is assumed because the sensor is held close to the tyre surface (20mm to 25mm)

: no window was fitted

The Heat Balance in Small Time Interval (δT)

Heat input in time δt

radiation from tyre assuming black body radiation from tyre

$$= (A_2 . \alpha_2 . E_1 . \zeta . \sigma . T_1^4) . \delta t$$

self heat from temperature sensing element

$$= Q_{SH} . \delta t$$

heat rejected from sensor in time (δt)

radiant heat from sensing disc

$$= (\sigma . A_2 . E_2 . \zeta . T_2^4(t)) . \delta t$$

heat rejected to ambient temperature T_3 in time

$$= A_c . U . (T_2 - T_3) . \delta t$$

through foam bonded to the copper sensing disc and convection from the outer surface to ambient

Heat Stored in the Copper Sensing in Time

$$= \omega . C_p . \frac{\partial T_2}{\partial t} . \delta t$$

Differential Equation for Tyre Sensor, Tyre System

$$\frac{dT_2}{dt} = \frac{E_2 . T_2(\infty)}{\alpha_2 . E_1} + \frac{A_s . U . (T_1(\infty) - T_3)}{\alpha_2 . E_1 . \zeta . A_2 . \sigma} - \frac{Q_{SH}}{\alpha_2 . E_2 . \zeta . A_2 . \sigma} \quad \text{—— 3.2}$$

The steady state equation 3.2 was derived by substituting the condition $\frac{dT_2}{dt} = 0$ in (3.1)

$$T_1 = \frac{E_2 . T_2(\infty)}{\alpha_2 . E_1} + \frac{A_s . U . (T_2(\infty) - T_3)}{\alpha_2 . E_1 . \zeta . A_2 . \sigma} - \frac{Q_{SH}}{\alpha_2 . E_2 . \zeta . A_2 . \sigma} \quad \text{—— 3.2}$$

The theoretical curve on Fig No. 7 was obtained by substituting numerical values into equation (3.3).

3.3.3 Stepwise Changes in Tyre Temperature

The differential equation (3.1) was solved for a stepwise change in tyre temperature T_1. It was considered an important solution because all the bench scale experiments were carried out with stepwise changes in the Hot Plate temperature. Equation 3.1 was linearised to make it amenable to solution by Laplace transforms (3.3).

$$\frac{dT}{dt} + m . T_2(t) = K_2 . T_1^4(t) + K_1 . T_3 + K_3 \quad \text{—— 3.3}$$

$$m = \left[\frac{A_s . U}{\omega . C_p} + 4 . \frac{\zeta . \sigma . A_2 . E_2 . T_1^3}{\omega . C_p} \right] \quad \text{—— 3.4}$$

$$\frac{1}{m} = \text{time constant (min)} \quad \text{—— 3.4a}$$

$$\alpha_2 = E_2$$

$$E_1 = \text{Emissivity of Tyre}$$

$$K_1 = \frac{U . A_s}{\omega . C_p}$$

$$K_3 = \frac{Q_{SH}}{\omega . C_p}$$

Q_{SH} : can be assumed zero for thermocouples but is not negligible for LM335 chips or thermistors.

The solution to equation (3.3) is given below:

Equation 3.5

$$m.(T_2(t) - T_2(0).\,e^{-m.t}) =$$

$$\left\{K_2.T_1^4(0)\left[1+4\,\frac{T_1(t)-T_1(0)}{T_1(0)}\right]+K_1.T_3+K_3\right\}.(1-e^{-m.t}) \quad \text{—3.5}$$

Tyre temperature T_1 (t) is the temperature required in a vehicle in order to establish the condition of a tyre. Hence an algorithm confirming experimental and theoretical models.

Time constants have been calculated for:-

Thermistor 25mm copper disc 0.5mm thick polythene window 1.975 - min

LM335 25mm copper disc 0.5 thick polythene window 4.854 - min

3.3.4 Modification of Differential Equation to include a correction factor fw for window

The correction factor has been calculated on the assumption that radiation energy is transferred through the window by:

a) Radiation transmission - (Radiation from tyre)

b) Calculation of the Energy absorbed by the surface.

$= \alpha . x$

where α = absorption coefficient (fraction)

x = fraction energy absorbed which is re-radiated from inner surface of window

$\therefore$ Total Heat Transferred

$\eta = (\tau + \alpha . x)$. (Radiation from tyre)

Hence η = fraction of incident radiation energy falling upon the sensing element.

$x = 0.5 \; ; \; \tau = 0.28 \; ; \; \rho = 0$

For a thin window for example 0.5mm polythene

hence η = 0.64

Nomenclature

A_1 = Area tyre exposed to the sensor (m^2)

A_2 = Area sensing element (m^2)

A_C = Area case exposed to environment

T_1 = Surface temperature tyre (°K)

T_2 = Surface temperature sensing element (°K)

T_3 = Ambient temperature (°K)

E_1 = Emissivity tyre (dimensionless)

E_2 = Emissivity sensing or element (dimensionless)

α_2 = Absorpivity sensing element (dimensionless)

($\alpha_2 = E_2$) near black body)

t = Time (sec)

σ = Stefan Boltzmann constant

U = Overall heat transfer coefficient

W = Mass sensing element (Kg)

C_P = Specific heat of sensing element

η = Coefficient giving fraction of radiation transmitted from tyre to sensor through window

4. EXPERIMENTAL PROGRAMME

4.1 Introduction

The experimental programme was divided into the following:

PHASE I Bench scale experiments to simulate the heating of the outer surface of a tyre.

PHASE II Rolling road experiments.

4.2 Phase I Bench Scale Laboratory Experiments

4.2.1 Objectives

The two main objectives were:-

i) Evaluate the main design parameters in order to determine the topology of the pre-production model.

ii) Study the performance of the different types of prototype designs when operating at the extreme limits of the expected environmental and operational conditions.

4.2.2 Apparatus

4.2.2.1 Test Equipment for Simulating Sensor Operating in Still Air

A vertical temperature controlled hot plate (300mm x 150mm) was used to simulate the effect of heating a tyre in still air. Heavy gauge 12mm thick aluminium plate was used to provide a uniform surface temperature and the surface was coated with matt black thermal paint of emissivity E=.95 similar to a tyre. The equipment was mounted in chamber to screen the experiment from draughts. A BBC B computer was used to control the hot plate temperature and to log the data. The computer was programmed to give stepwise or ramp changes in hot plate temperature.

4.2.2.2 Test Equipment to Provide Forced Convective Air Flow

A test rig was used to model the effects of cross wind and forward speed of a vehicle on the measurement accuracy of tyre sensors.

A 300mm x 150mm (temperature controlled) electrically heated plate was integrated into an enclosure so that the hot surface of the plate formed one surface of an air duct. A speed controlled centrefugal fan blew air through a duct of 75mm x 25mm at air velocities ranging from 0 to 30 m/s.

Six prototype tyre sensor were mounted in the air duct 25mm away from the hot plate surface with their axes normal to the surface of the plate.

The air duct made two passes across the surface of the plate and flow straighteners were used to generate a uniform air velocity profile over the hot plate.

The air flow circuit could be operated either as a closed or open loop to facilitate the control of air velocity and temperature.

4.2.2.3 Experimental Sensors

An experimental temperature sensor body was designed to accommodate changes in:-

window materials and thickness

body length

sensing elements

The experimental sensor body was made in 63mm, 36mm and 25mm diameters in plastic and aluminium.

The body was designed to be waterproof to the IP67 standard, so that samples were suitable for fitting onto vehicles.

SENSOR TYPE A	63 diameter body with parabolic reflector and LM335 temperature sensor at focus
SENSOR TYPE B	36mm diameter body in plastic and aluminium sensing element flat copper disc blackened with thermocouples, LM335 chips and thermistors
SENSOR TYPE C	25mm diameter bodies in plastic with copper disc and reflecting radiation collectors. LM335 and thermistors used for temperature measurement.

Window Materials

SENSOR A
2mm window glass
2mm optical glass
175 microns melinex
80 microns melinex
80 microns polypropylene
80 microns polythylene
IR camera filter
2 mm perspex

4.2.3 Test Procedure Phase I

4.2.3.1 Heating Trials in Still Air

As a control experiment one plastic bodied sensor of each body size was operated without any windows. The results were compared with theoretical calculations. Each experimental sensor was mounted 25mm distant from the electrically heated plate with the principal axis of the sensor normal to the hot plate. One control sensor and two sensors with different configurations were tested together. The temperature measuring devices were initially calibrated by retaining the rig at room temperature for more than 30 minutes. The hot plate temperature was then raised as rapidly as possible to simulate a stepwise temperature change. The temperature of the hot plate was controlled at a constant temperature for sufficient time to ensure equilibrium (1 to 2 hours). The hot plate temperature was raised in steps of 10 deg. C from 30 deg C to 90 deg C.

Data acquired :

T1 Hot plate temperature

T2 Ambient temperature

T3 Outer window surface temperature

T4 Temperature of the sensing element

4.2.3.2 Cooling Trials in Still Air

When the experiment with the hot plate temperature of 90 deg. C was completed, electrical power to the hot plate was switched off and the sensors and hot plate allowed to cool naturally. A full set of test data was taken in 10 deg. C steps from 90 deg. C to 30 deg. C.

4.2.3.3 Heating and Cooling with Forced Air Convection

Batches of six sensors were mounted simultaneously in the airflow rig. Four of the sensors had experimental configurations whilst the other two were used as controls. Air velocities of zero 6, 11 and 30 m/s were used during each test run. Point air velocities were measured across the cross section of the duct to obtain the average velocity. Sensor, ambient, plate and air duct temperatures were recorded.

4.3 Rolling Road Experiments

4.3.1 Introduction

Following the promising test results obtained in the laboratory, sensors and their associated electronics were used to monitor tyres undergoing tests at Avon Rubber plc, Melksham (Fig. 8).

4.3.2 Objectives

These trials were designed to find out how accurately the tyre sensor measured the surface temperature of tyres that were being run on a rolling road.

Specific objectives were:-

i) to determine whether or not the tyre sensor can distinguish between tyres operating at different loads and tyre pressures.

ii) to determine whether the tyre sensor can detect tyre surface temperature and temperature changes under controlled conditions leading up to tyre destruction.

4.3.3 Experimental Conditions

B. Lee (6) reported that tests were run with two identical types of tyre. Each test was run for time duration of 30 mins.

i) Loading and Inflation Pressure Test

Tyres were operated at the following conditions:-

CONDITION	LOAD (kg)	TYRE PRESSURE (bar)
Norm	320	1.8
Press Reduced	320	1.2
10% inc Load	425	1.8

The speed was maintained at a constant speed of 100km./hr. (70 mph).

ii) Increasing Speed Tests

The tyre was inflated to 40.6 p.s.i. (2.8 bar) and subjected to stepwise increases in speed whilst maintaining the load constant at 492kg. (80%).

SPEED (km/hr)	DURATION TEST (mins)
0 - 170	10
170	10
180	10
190	10
200	20
stop check tyre surface temperature	
210	10
220	10

These tests were repeated for a second tyre.

Two sensors were placed 20mm from the mid wall position on the tyre and a third one was used to record the ambient temperature. The tyre wall temperature was measured by a contact thermometer.

4.4 Discussion

4.4.1 General

In was concluded from analysis of the test data that the simple infra-red radiation sensor can detect relatively small charges in the surface temperature of a tyre which correspond to measured small changes in tyre conditions such as the 20% reduction in tyre pressure or 20% increase in wheel loading of Section 4.3.3. Temperature change, though indicative of a change in tyre condition, can only discriminate between causes if additional information such as tyre pressing loading etc are provided. However, the simplest form of sensor can provide a general warning which may prove useful to the driver, provided that he is willing to stop the vehicle to inspect the tyre. The effectiveness of the sensor would be significantly improved if additional information such as speed, loading and tyre pressures were also provided.

The main objective of the development programme was to establish the design specification and performance criteria for a commercial design.

As stated earlier, the experimental work was carried out in two stages:-

i) Laboratory experiments:

to optimise design,

and

to establish performance characteristics

ii) Rolling road trials

4.4.2 Evaluation of Design Parameters

4.4.2.1 Design Configuration

The main design parameters such as sensor diameter, length and the topology of the sensor were determined experimentally.

The final selection of diameter and length was a compromise between small dimensions to meet the requirements of the vehicle designer and the more effective larger sized radiation sensor.

A 25mm diameter sensor was finally selected because the performance was found by experiment to be satisfactory and suitable for mounting on either the front or rear suspension of an automobile. Fig 9 shows that the performance of a 25mm diameter final prototype sensor is much more sensitive than the original 36mm diameter prototype configurations.

4.4.2.2 Window Design

i) Window Material

Polythene HDPE was selected for the 25mm AB prototype although experiments were carried out with other materials. Figure 9 also compares different window materials for the equilibrium temperature condition in still air using a 36mm body shell. Selected materials were compared with curve A, the same 36mm sensor without a window. Significantly larger sensor temperatures 'T2' are obtained from the sensor without a window, but as expected the 17.5 micron melinex was better than all the other material tested with the exception of polythene on the 25mm sensor.

It may also be noted that the double glazed melinex window curve E performed very badly thus confirming the hypothesis that direct conduction was one of the most significant mechanisms for heat transfer through a sensor window. The air gap separating the two windows inhibited conduction through the windows.

ii) Window Thickness

Figure 10 summarises the results of two tests in which polyethylene windows of 0.5, 0.8, 1.5 and 3mm thickness were compared in still air and at an air velocity of 12m/s. In addition a sensor with a reflector and a .5mm window was also tested. The reflector proved superior with still air but surprisingly it was worse with high velocity air. The 0.5mm window with flat disc sensing element was consistently good, always better than the other window thicknesses and better than the reflector at high velocity.

4.4.3 Sensor Performance

4.4.3.1 General

The procedure adopted was to apply a stepwise change in the hot plate temperature and then observe the response of the sensor. A stepwise mode of operation was adopted because it was easy to reproduce tests and analysis was simplified. Each test run was continued until thermal equilibrium had been achieved.

Equation 3.5 was a simplified mathematical model. It characterises the exponential response of the sensor to stepwise tyre temperature by time constant (1/m) equations 3.4 and 3.4a and by the sensitivity ratio below.

where:

Sensitivity Ratio =

$$\frac{\text{(Temperature change of hot plate (Tyre))}}{\text{(Corresponding temperature change of Sensor)}}$$

This ration should be as small as possible for the most effective sensor.

4.4.3.2 Effect of Ambient Temperature

Graph Fig No. 7 shows the effect of changing the ambient temperature on the sensor which can be predicted by equation 3.3. Ambient temperature changes can be seen to have a very profound effect on the sensor temperature. It is essential to provide a method of compensating for ambient temperature changes in a commercial sensor.

4.4.3.3 Effect of Varying the Air Velocity

Both the equilibrium and dynamic performance of a sensor without a window could be predicted with reasonable accuracy by substituting numerical values into equation 3.5. However, with windows installed, the model was not accurate.

Fig No. 11 show the effect of air velocity on time constant and sensitivity ratio.

The high value of the 7 min for the equilibrium time constant (1/m) rapidly drops to less than 3 minutes with air flow. It appears to remain substantially constant with increasing velocity. Substantially constant with increasing velocity over the air velocities tested. The sensitivity also deteriorates with air flow and appears to reach a constant value of 4.1 at 12m/sec. These data must be extended to include air velocities of up to 60 m/s. It may be noted that the calculated time constant for still air condition was 4.854 min assumed no window.

This simple empirical relationship can be used to correct the predicted tyre temperatures for air velocity effects. As a first approximation the air velocity over a tyre will be equivalent to between twice the road speed at the point on the tyre remote from the road to zero where the tyre makes contact with the road. Hence a first order correction for air velocity could be related to the speed of the vehicle.

The AB sensors were fitted with LM335 devices which gave long time constants than the later thermistor designs.

4.4.4 Rolling Road Tests

Analysis of a short series of tests carried out on the tyre testing facility at Avon Rubber plc showed clearly that the 25mm prototype tyre condition monitor could detect small changes in tyre wall temperature as can be seen in Fig No.10. Test runs 2, 4 and 6 show a significant difference from the norm of runs 3 & 5. Whilst these rolling road tests were very promising a substantial experimental programme is required to characterise the sensor and tyre as a single operating system.

5. CONCLUSIONS

1. The infra-red tyre condition monitor was shown to be capable of responding to relatively small changes of tyre sidewall temperature when small changes in tyre conditions were introduced.

2. The optimised sensor design appears to operate satisfactorily the magnitudes of time constant and sensitivity were acceptable.

3. An infra-red Tyre Temperature Monitor has reached the stage of development where it can be integrated into vehicles to evaluate its use as either:-

   a) A simple vehicle tyre temperature comparator to give a warning when one tyre deviates from the average.

   or

   b) Part of a more comprehensive tyre condition monitor.

4. A micro processor control system as shown in Fig No. 3 could be developed to compensate for speed, ambient temperature etc.

ACKNOWLEDGEMENTS

The author wishes to acknowledge, with many thanks, the assistance and co-operation of his colleagues in the AB Technology Centre and to AB Electronic Products Group for permission to publish this paper.

REFERENCES

(1) SMITH, R. Internal communications, AB Electronic Products Group 1987.

(2) KERNER, H. Investigation of main landing gear tyre temperatures during taxying-out at heavy take-off masses.

(3) AVON RUBBER plc. Private communications 1988 (Dec).

(4) JACOB, M. Heat transfer Vol.1. John Wiley & Sons Inc. (1956).

(5) BURKE, A. Linearising thermistors with a single resistor. Burke Electronics, San Diego, California.

(6) LEE, B. Visit Report's Avon Rubber plc. 8th July 1988 and 30th September 1988 and internal communications 1987, 1988 & 1989.

TABLE 2

% TOTAL ENERGY TRANSMITTED BY WINDOW

Material of Window	Minimum Wave-Length (Microns)	Maximum Wave-Length (Microns)	300°K % Energy Transmitted	370°K Energy Transmitted
Optical Glass 2mm	0.3	2.8	0.001	0.055
Quartz W.F 2mm	0.15	3.5	0.047	0.475
Special Glass IRGZ	0.4	5.0	0.83	4.15
IRG 100	0.9	16.5	4.2	51
Zn. Se Zinc Selenide	0.45	12	28	40.7
Melinex		+16	8.0	
Polythene		+16	28	

TABLE No3

EXPERIMENTAL TYRE DESIGN PARAMETER				
1	BODY DESIGN PARAMETERS	SIZE	LENGTH 36mm	
			DIAMETER	63mm, 36mm, 24mm
		MATERIALS	ALUMINIUM	63mm & 36mm
			PLASTIC	63mm, 36mm, 24mm
2	WINDOW DESIGN	MATERIAL	POLYTHENE	
			GLASS	
			MELINEX	
			POLYPROPYLENE	
		THICKNESS	2mm GLASS	
			175 microns MELINEX	
			0.5, 0.8, 1.0, 1.5, 2, 3 POLYTHENE	
3	RADIATION SENSOR	PARABOLIC REFLECTOR	63mm DIA	
			24mm DIA	
		FLATE PLATE	TEMPERATURE MEASURING DEVICE	THERMOCOUPLE
				LM 335
				THERMISTER

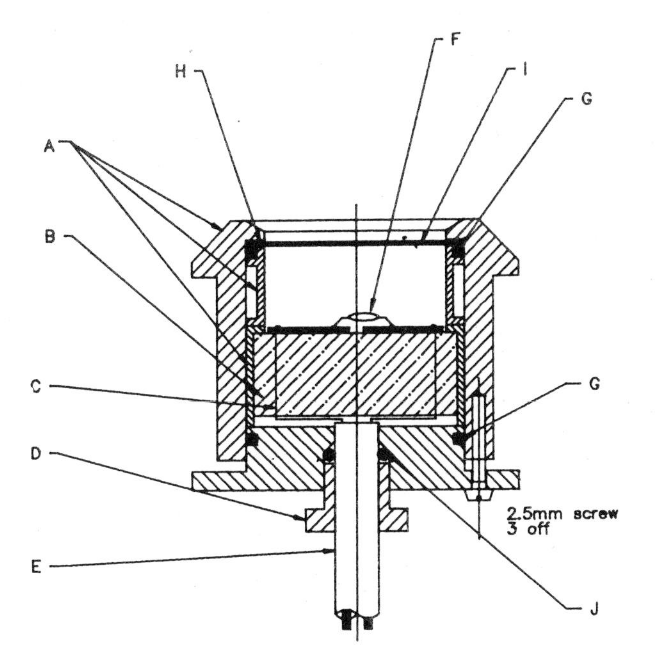

A. ACETAL (natural)

B. 6mm Foam white
P2101 24Kg/m (1.5LB/ft)

C. 0.08 Enamel Grade (F)
copper wire.

D. Enott gland nut 1/8 bsp or
equivalent metric size

E. 12mm O.D. 4core or 2 core
cable

F. Thermister type

G. 1 or 1.5 O ring
(order O ring kit from R.S.)

H. Silicone Sealant

I. 0.5 to 0.8mm (polythene)
(polyethelene) High Density

J. O ring sample available

Fig 1a Experimental tyre condition monitor with flat disc radiation

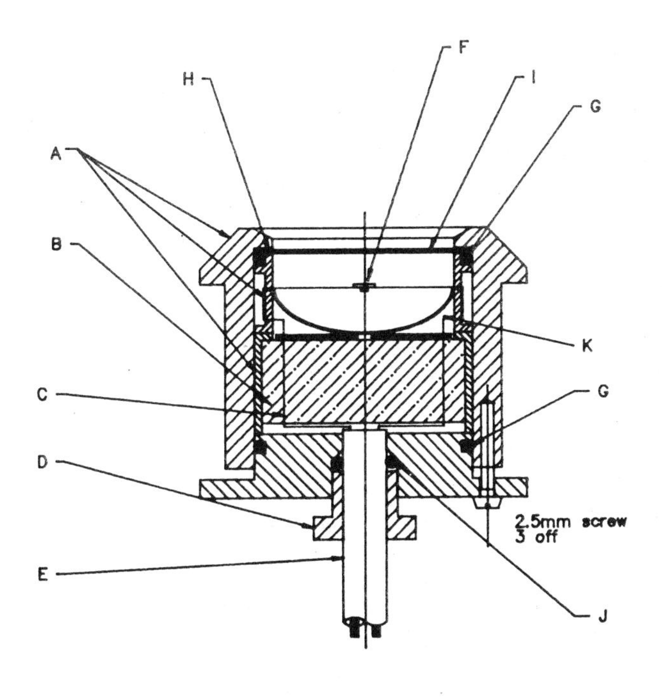

A. ACETAL (natural)

B. 6mm Foam white
P2101 24Kg/m (1.5LB/ft)

C. 0.08 Enamel Grade (F)
copper wire.

D. Enott gland nut 1/8 bsp or
equivalent metric size

E. 12mm O.D. 4core or 2 core
cable

F. Thermister type

G. 1 or 1.5 O ring
(order O ring kit from R.S.)

H. Silicone Sealant

I. 0.5 to 0.8mm (polythene)
(polyethelene) High Density

J. O ring sample available

K. PARABOLIC MIRROR

Fig 1b Experimental tyre condition monitor with parabolic reflector

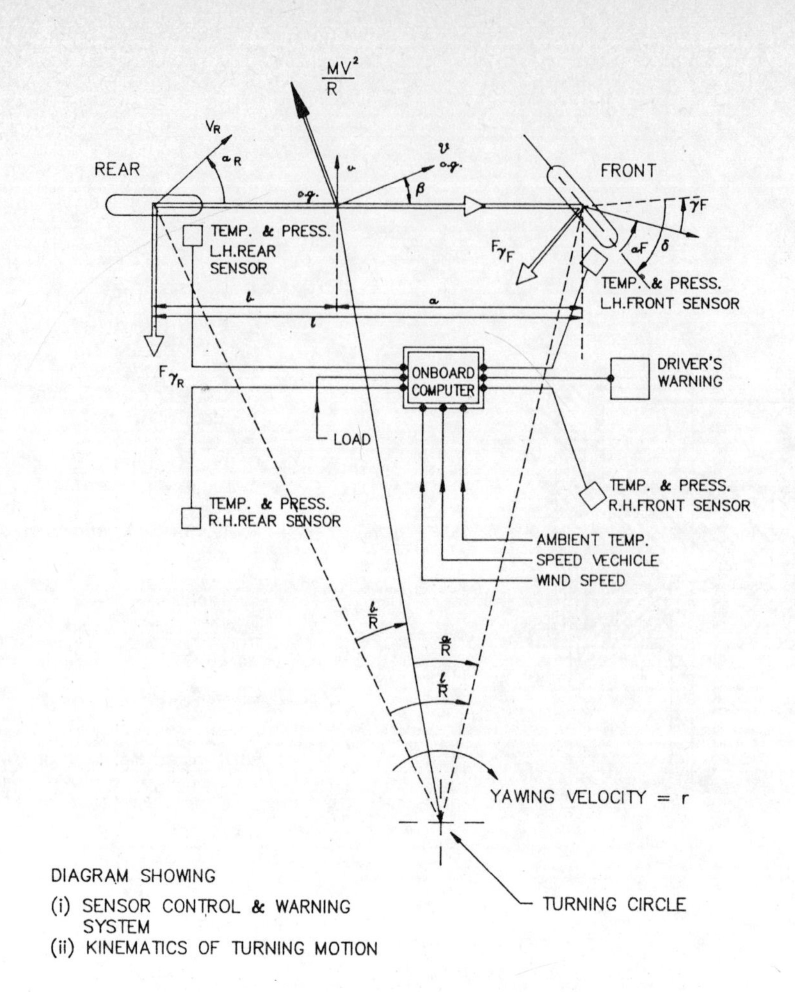

Fig 2 Diagram showing:
(i) sensor control and warning system and
(ii) kinematics of turning motion

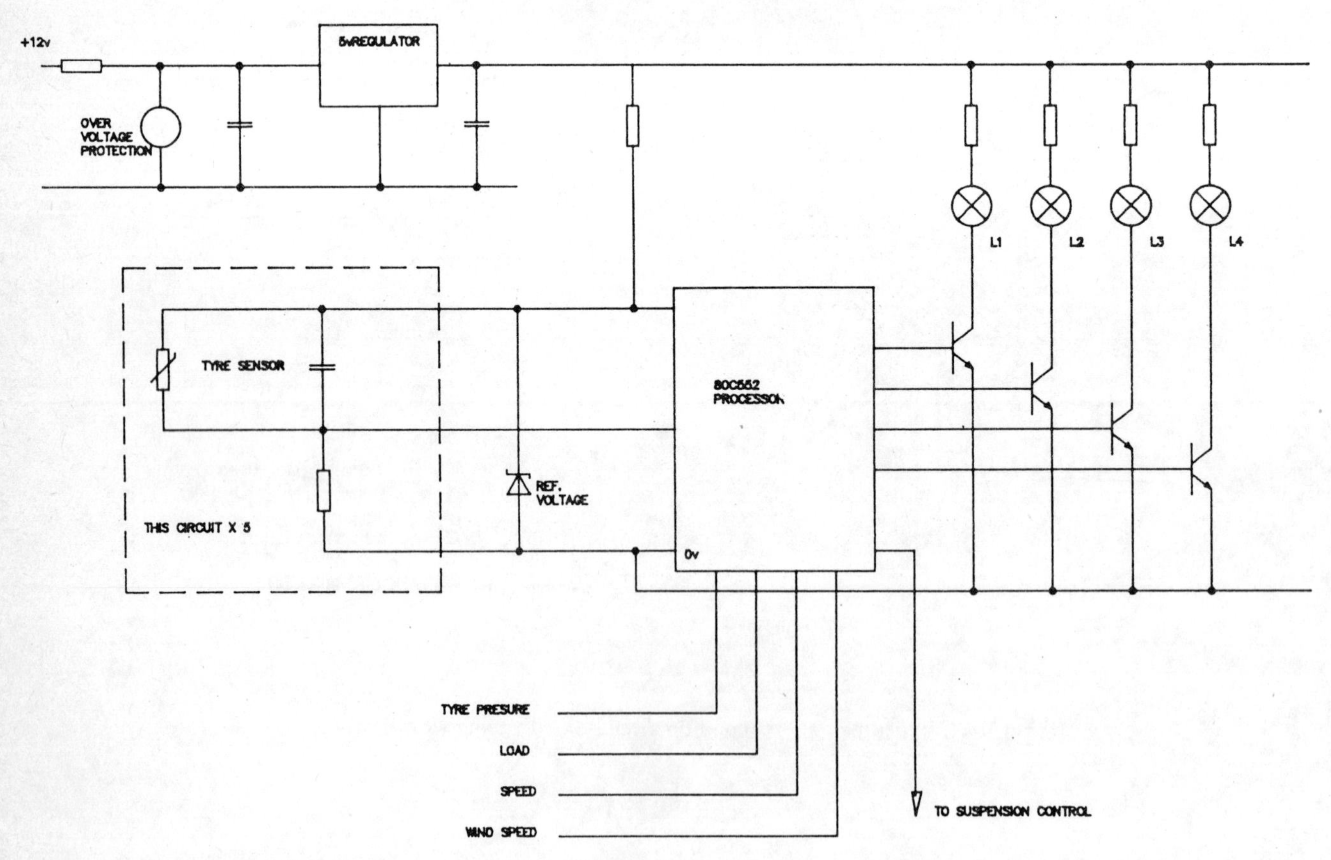

Fig 3 Tyre condition monitor

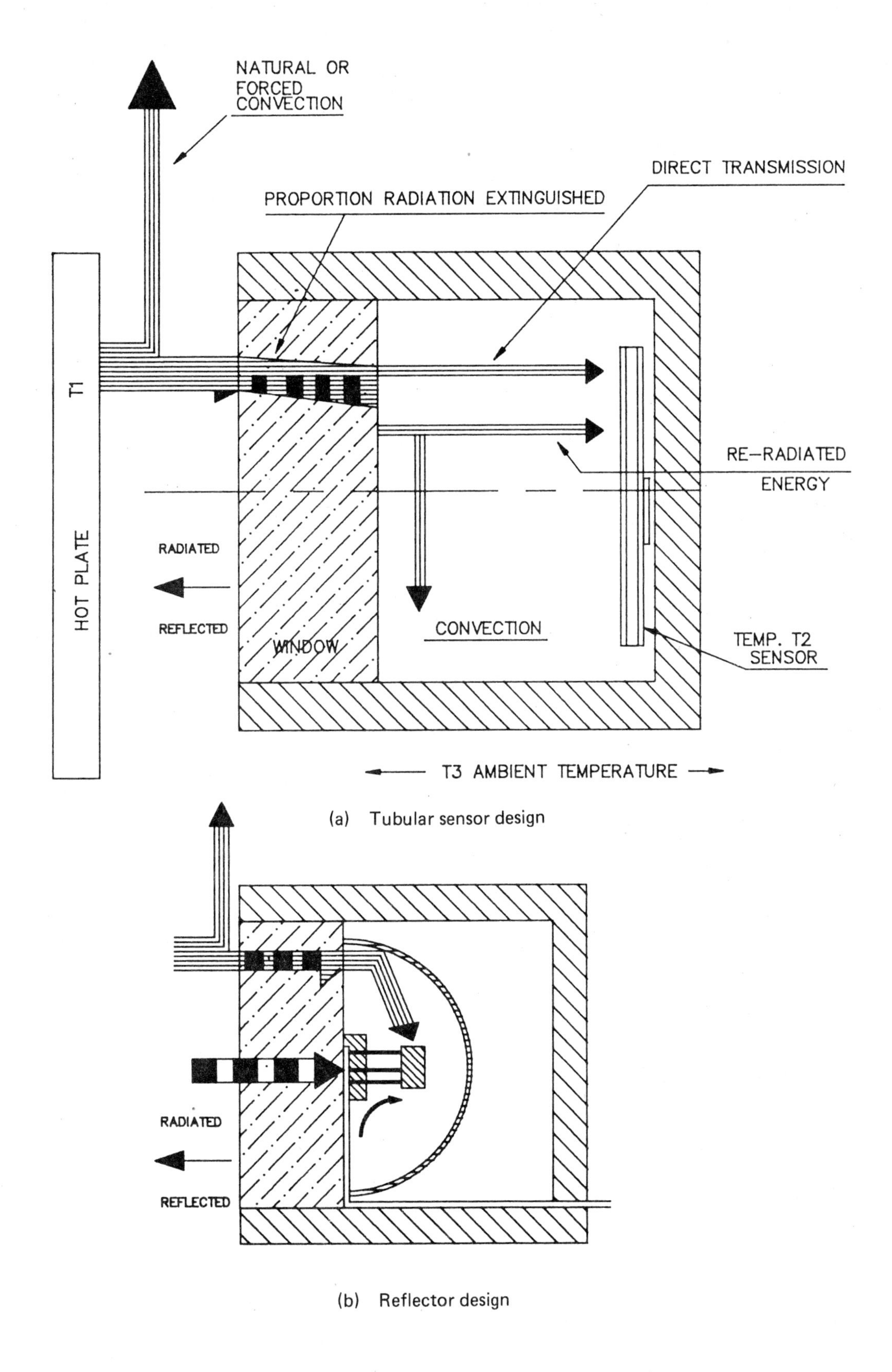

(a) Tubular sensor design

(b) Reflector design

Fig 4 Diagrammatic representation of heat distribution in tyre sensor

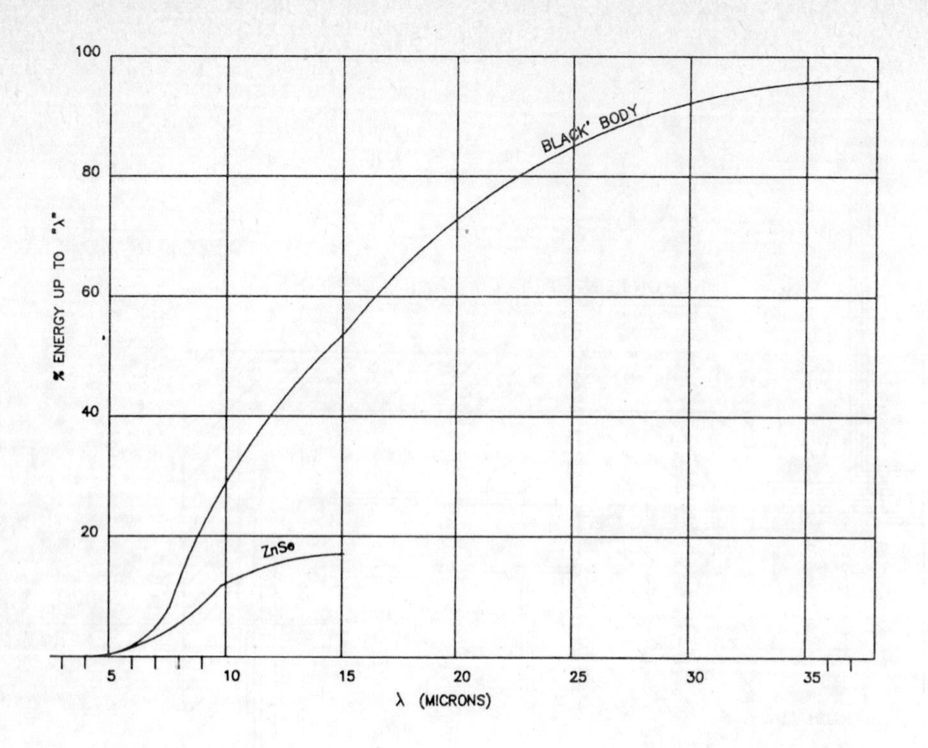

Fig 5 Percentage of total energy in spectrum transmitted by window

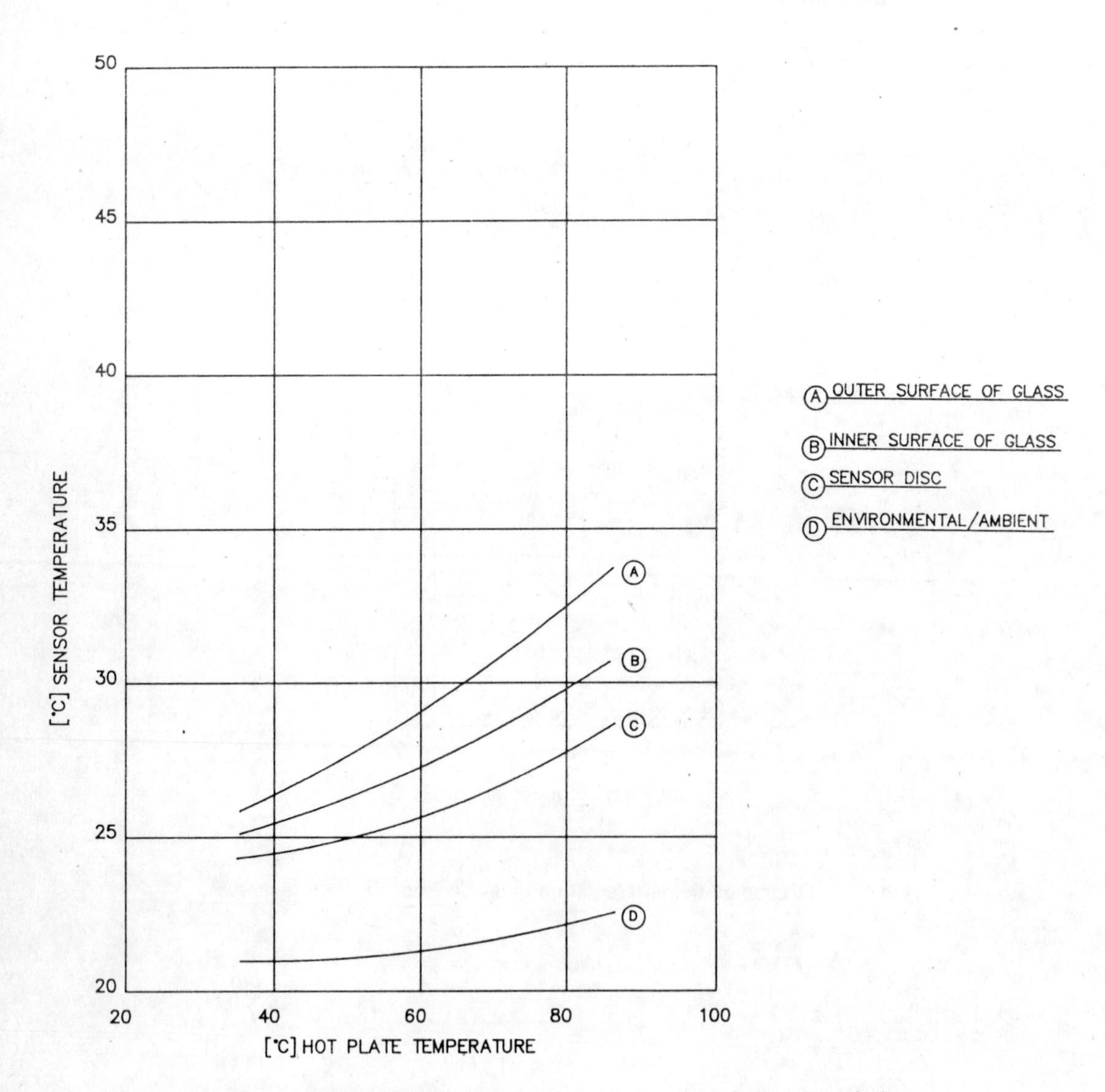

Fig 6 Temperature gradients in 36 mm sensor with 2 mm glass window

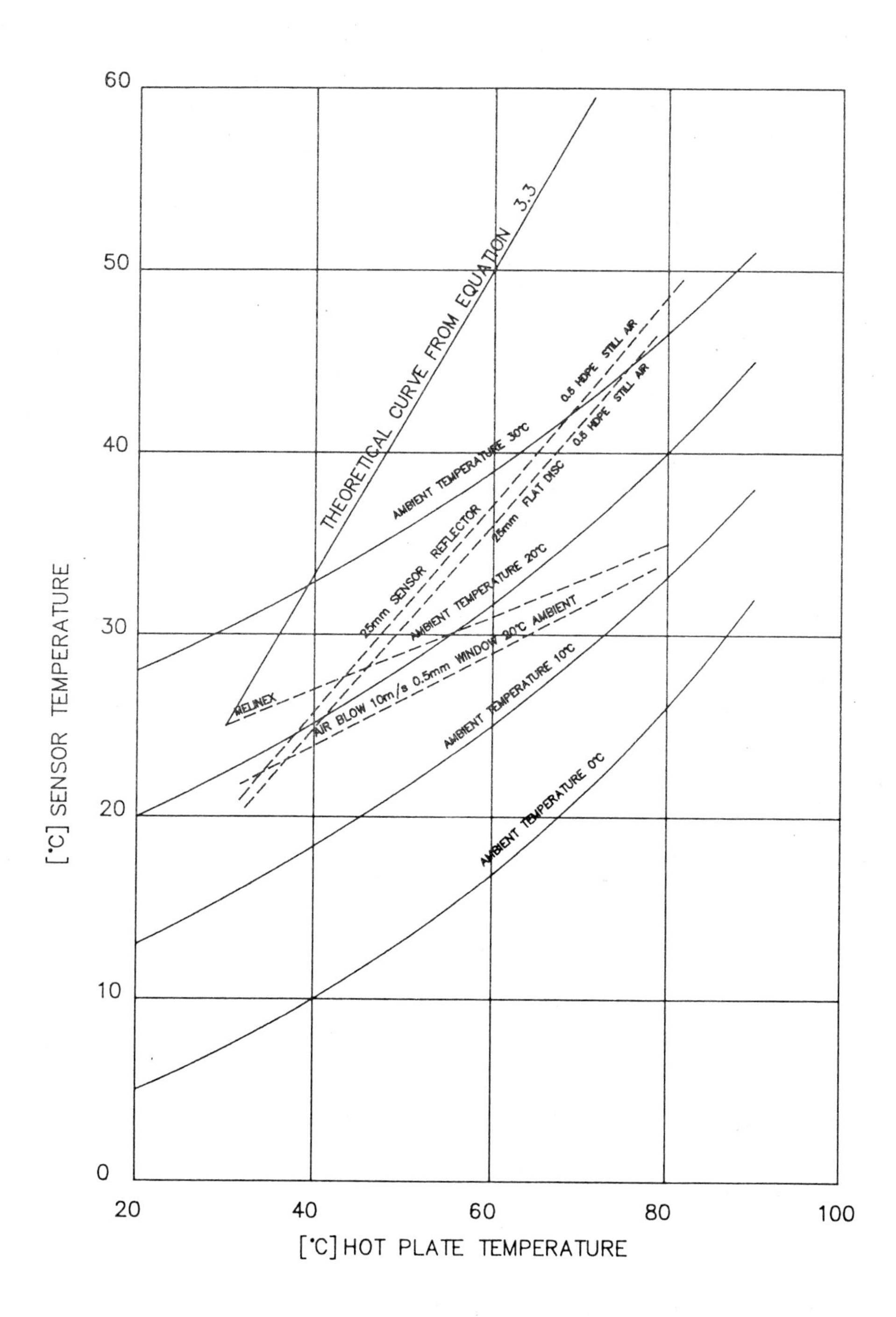

Fig 7 Effect of varying ambient temperature

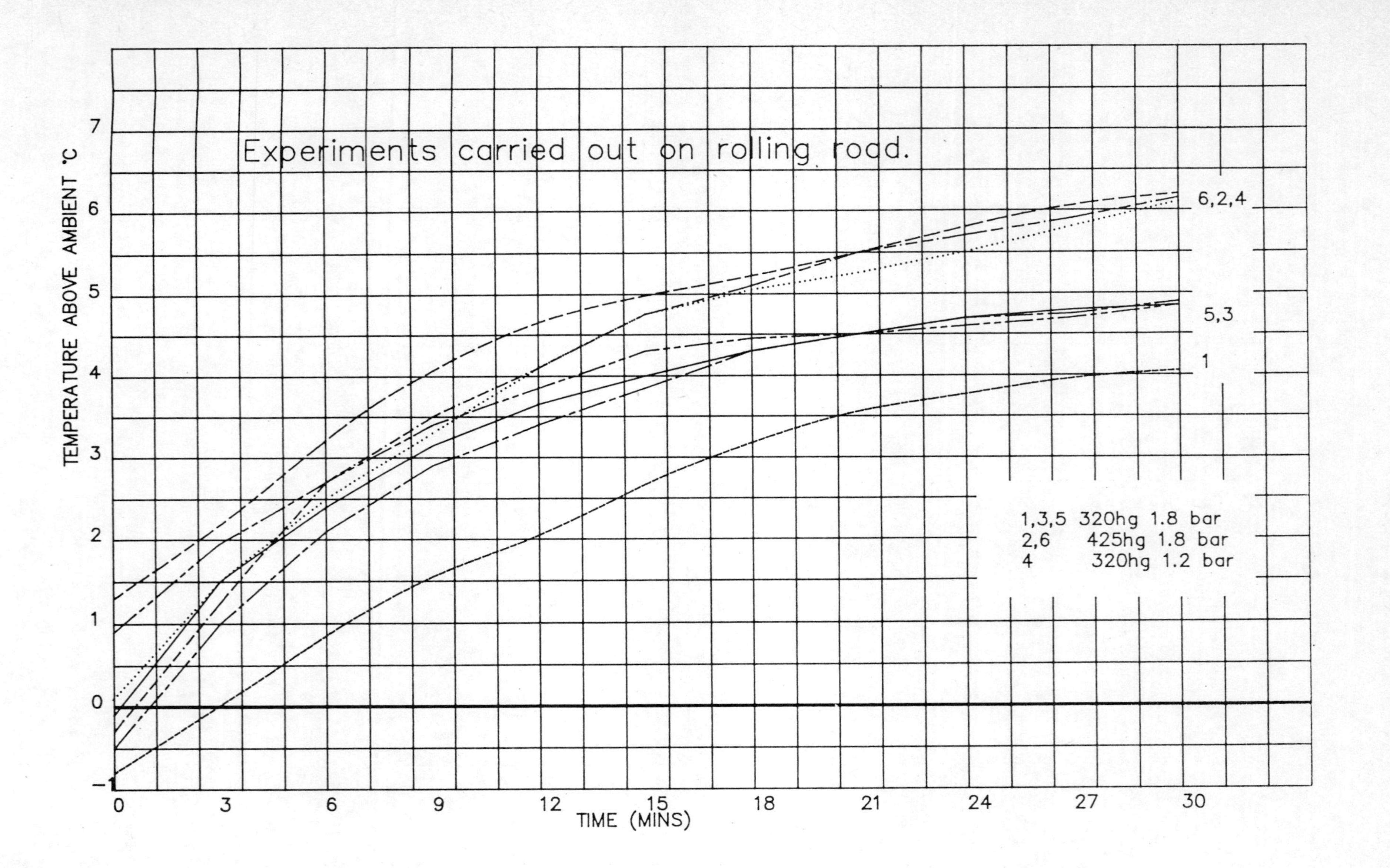

Fig 8 Tyre sensor—sensor temperature difference to ambient versus time

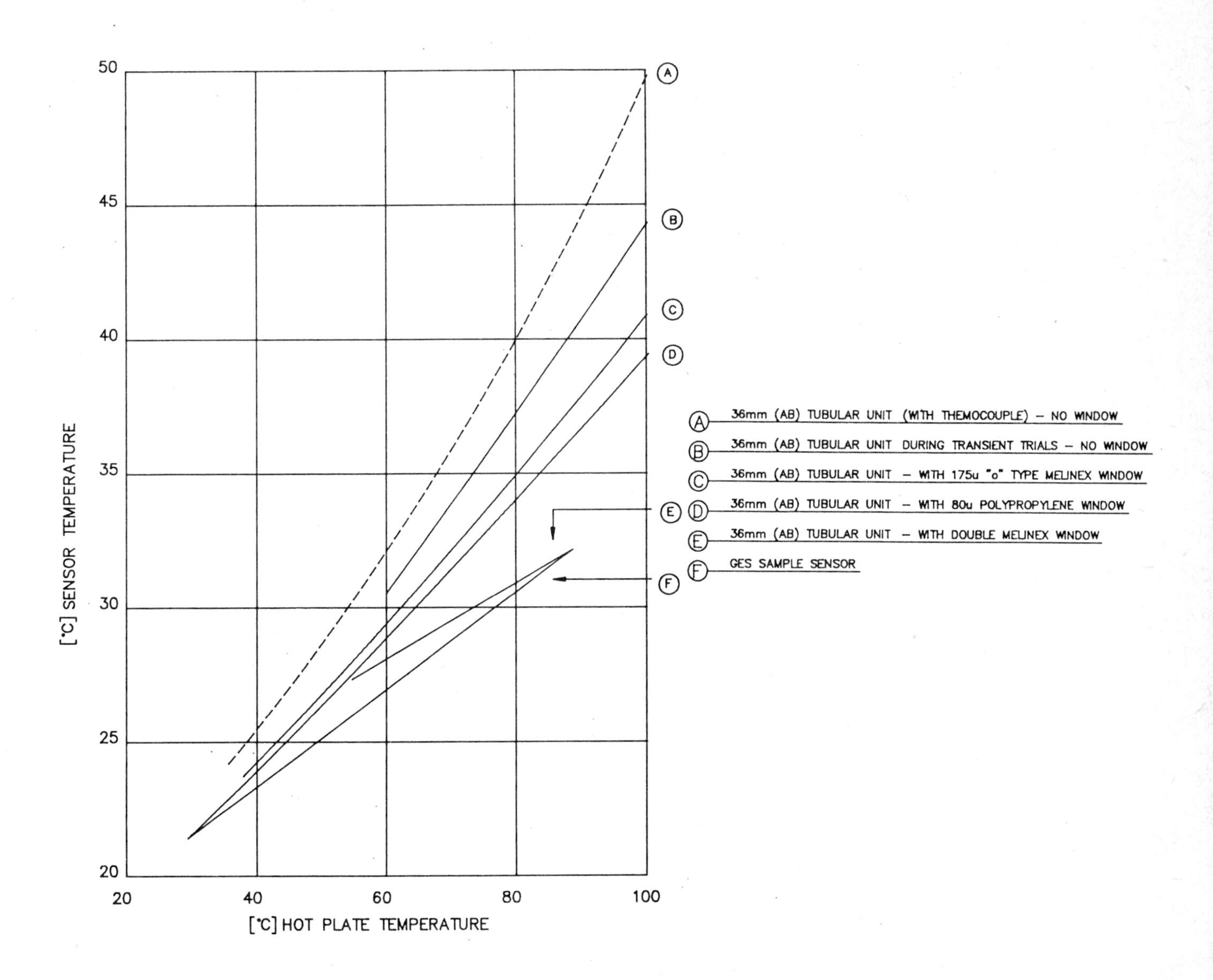

Fig 9 Performance of different window materials

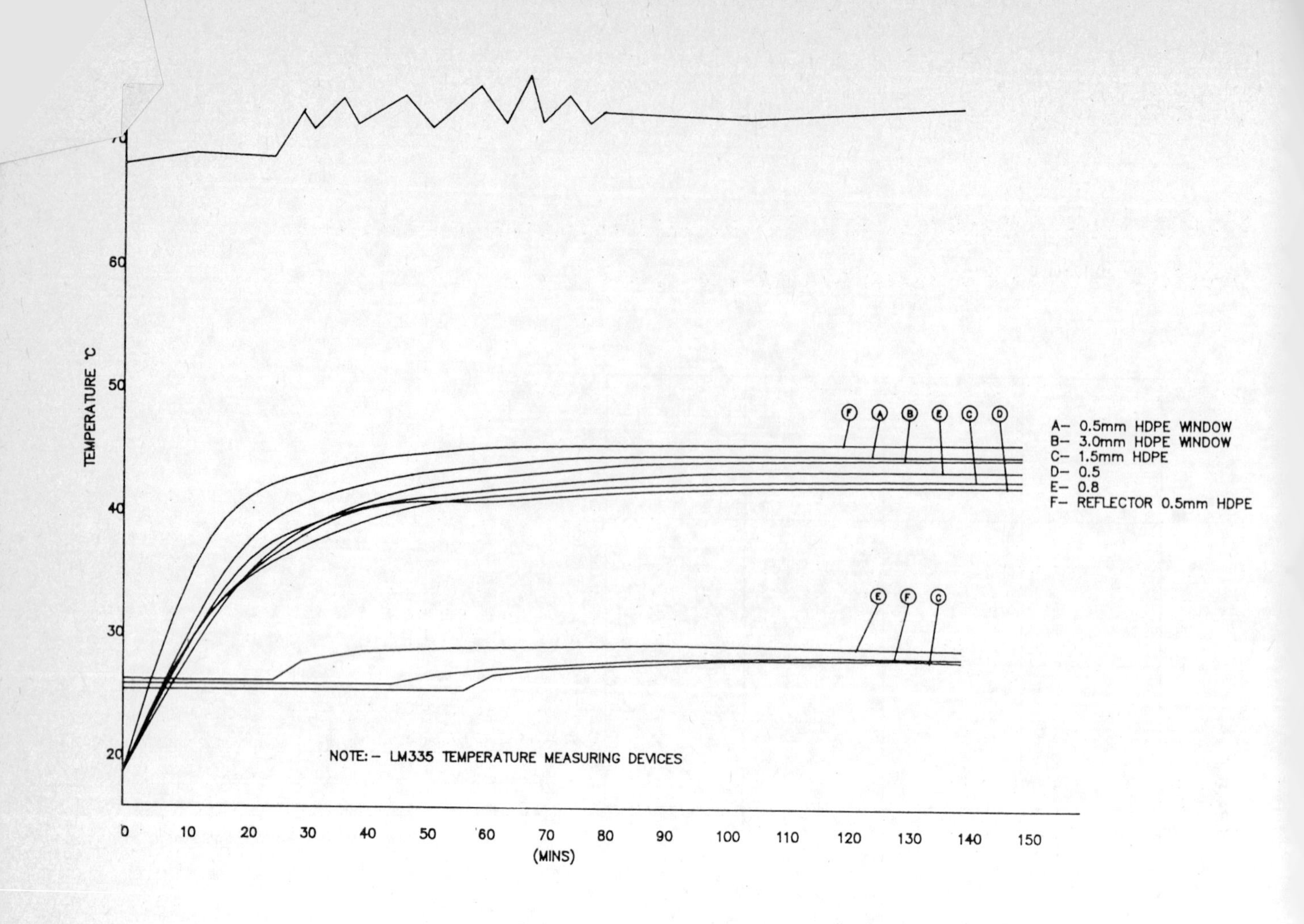

Fig 10 Graph of AB prototype sensors fitted with different thicknesses of polythene HDPE windows (with still air and forced convected air flow)

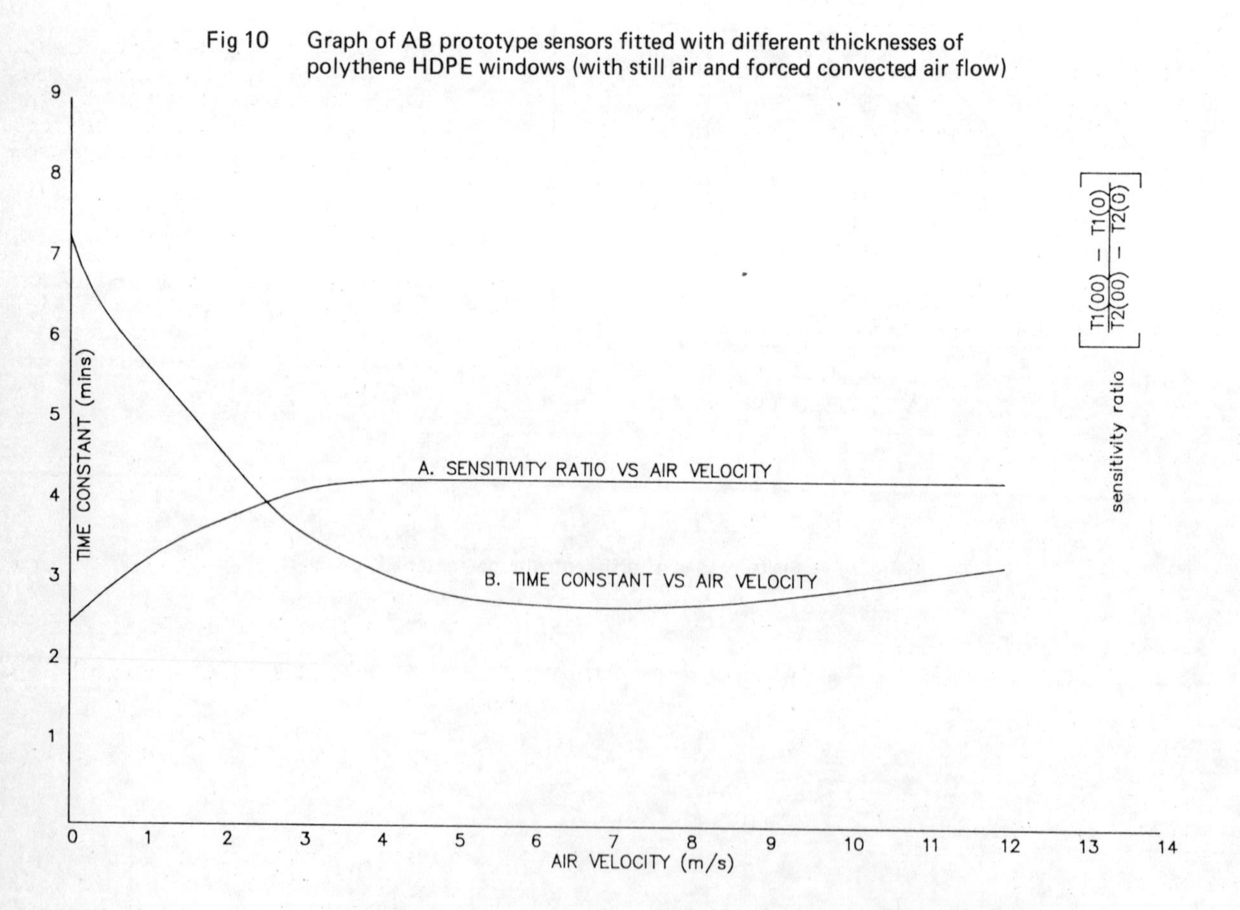

Fig 11 Graph showing the on change in the time constant of air velocity and sensitivity ratio for an AB prototype tyre sensor with LM335 temperature measuring device

C391/005

Steering control of an autonomous vehicle using a fuzzy logic controller

A HOSAKA, M TANIGUCHI, S UEKI, K KURAMI, A HATTORI and **K YAMADA**
Nissan Motor Company Limited, Yokosuka, Japan

SYNOPSIS We adapted fuzzy logic algorithms to the steering control of the autonomous vehicle. This vehicle has two types of moving mode, camera mode and ultra-sonic mode, originating from the differing methods used to detect the vehicle's position.
We performed computer simulations and showed the effectiveness of our control method. Moreover, at the test site, we verified that a test vehicle can be controlled by this fuzzy logic controller and move at 10km/h in spite of the uncertainty in the precision of the sensor. The steering control drove the vehicle as smoothly as a human being along both straight and curved roads.

1 INTRODUCTION

We envision a time when automatic drive control of cars is standard equipment on all new cars. In the future,the autonomous vehicle is expected to make a big contribution to the society composed largely of elderly people. It will be of great help for the physically handicapped, vastly improving their quality of life. Moreover, it also will be useful for people who are not used to driving cars. For these reasons, we are pursuing research in this area. Steering control is one of the most important technical systems for autonomous vehicles. Previous methods were primitive. For example,in factories, the steering angle of load carrying AGVs(Automated Guided Vehicles)is controlled by following an electromagnetic guide line buried in the ground. In this case, all that the control is required to be,is smooth enough so as not to shake the load from the vehicle.

On the other hand, in the case of a passenger carrying vehicle moving on the highway or the street,the steering control must be smoother. It must feel as if a human being were driving. It may seem that by improving the car's sensors and the algorithms that interpret the sensors' data the smoothness of ride can be improved. This is not,however,necessarily so. Higher precision sensing can result in a rougher ride because of its noisy signal.

In this case, the passengers may feel not only discomfort but also fear.
For this reason, we selected fuzzy logic control algorithms for the steering control of the vehicle. Fuzzy logic was chosen because of the following characteristics.

(1) It is easier to implement human skills using fuzzy logic than using other control algorithms.
(2) It filters even the uncertainty of the precision of the sensor.

Therefore,the vehicle may move as smoothly as a human being drives.

2 SIMULATION

We performed a computer simulation to evaluate the efficiency of the fuzzy logic control algorithms.

2.1 The map information

The simulator vehicle has a map of test site. The map information is classified into two parts, the nodes and the paths. The nodes include the names and the global coordinates of the nodes. Nodes represent specific points on the road.
The paths include the following items.

(1) The name of the starting node
(2) The name of the goal node
(3) The radius of the curved road
 This means the distance from the center of the road to the center of curvature of the curve in the road.
(4) The length of the paths
(5) The direction of the vehicle at the starting node
(6) The direction of the vehicle at the goal node
(7) The width of the road
(8) The number of lanes in the road
(9) The configuration of the road at the starting node
 (for example,whether it is a straight line or an intersection)
(10) The configuration of the road at the goal node

2.2 Global strategy of the speed and the steering

The global strategy of velocity and the steering commands consists of a set of node commands at each node and a set of path commands which are followed between nodes.

c of node commands is as follows.

The desired position of each node in the global coordinates
(2) The desired position of each node in the local coordinates
(3) The desired velocity at each node Vpo

The set of path commands is as follows.

(1) The name of the next node
(2) The distance to the next node
(3) The desired velocity Vpr
(4) The selection of the white line followed by TV camera
(5) The desired offset from the white line
(6) The desired orientation of the vehicle in the local coordinates

2.3 Moving Control

The vehicle's speed and steering angle are decided according to the path commands and node commands from the global strategy command unit.

2.3.1 Speed control

The speed variable V is calculated from the distance dr between global position calculated by wheel shaft encoders and the position of the node, the desired speed Vpr through the path, and the desired speed Vpo at the node.
In acceleration (Vpo>Vpr), V is given as follows,

$$V = \max(Vpo, Vpo - (Vpo - Vpr) \times \frac{dr}{dr1}) \qquad (1)$$

(dr1 is the turning point of acceleration)

In deceleration (Vpo<Vpr), V is given as follows,

$$V = \min(Vpr, Vpo + (Vpr - Vpo) \times \frac{dr}{dr2}) \qquad (2)$$

(dr2 is the turning point of deceleration)

2.3.2 Steering Control

The Steering command is defined as the steering derivation command ΔS. ΔS is derived from the deviation between the desired position and the current position. ΔS consists of the position deviation ΔX and the heading deviation $\Delta\theta$. The vehicle's position is shown in Fig.1. ΔX is derived from the difference, ΔXo_g, between the position of the next node(i+1), Xor(or Xol), and the position of the point at L meters ahead of the vehicle, Xgr(or Xgl), and the difference, ΔXr_g, between the desired position on the path, Xrr(or Xrl), and current position of the point at L meter ahead of the vehicle, Xgr(or Xgl).

In the same way, $\Delta\theta$ is derived from the difference, $\Delta\theta o\_g$, between the desired heading angle at the next node, Θos and the current heading angle at L meters ahead of the vehicle, θgr(or θgl),

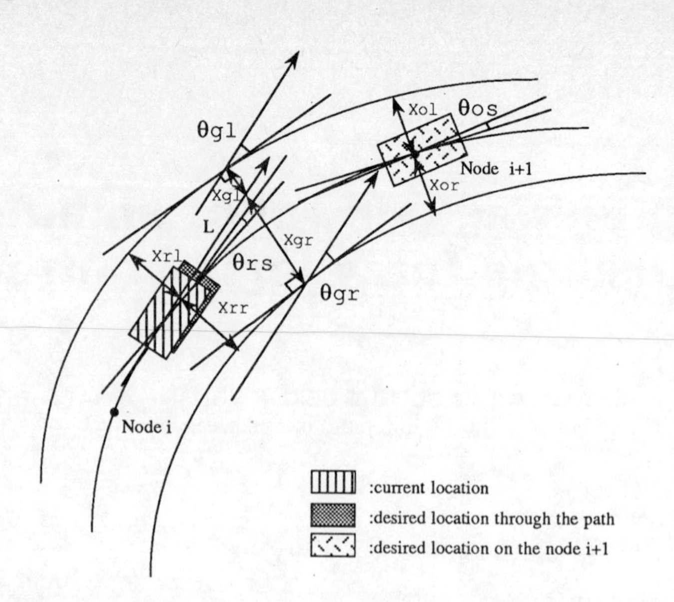

Fig 1 Vehicle position

and the difference, $\Delta\theta r\_g$, between the desired heading angle on the path, Θos, and current heading angle at L meters ahead of the vehicle, θgr(or θgl).
The distance L varies as follows,

$$L = 0.25v + 5 \qquad (3)$$

where v is the vehicle's speed(m/s).
Therefore, ΔXo_g, ΔXr_g, $\Delta\theta o\_g$, and $\Delta\theta r\_g$ are given as follows,

$$\Delta Xo\_g = Xor - Xgr \quad (\Theta os>0)$$
$$-|Xol| - (-|Xgl|) \quad (\Theta os<0) \qquad (4)$$
$$\Delta Xr\_g = Xrr - Xgr \quad (\text{for right line})$$
$$-|Xrl| - (-|Xgl|) \quad (\text{for left line}) \qquad (5)$$
$$\Delta\theta o\_g = \Theta os - \theta gr \quad (\Theta os>0)$$
$$\Theta os - \theta gl \quad (\Theta os<0) \qquad (6)$$
$$\Delta\theta r\_g = \Theta rs - \theta gr \quad (\text{for right line})$$
$$\Theta rs - \theta gl \quad (\text{for left line}) \qquad (7)$$

ΔX and $\Delta\theta$ are defined as the following,

$$\Delta X = w1 \times \Delta Xo\_g + w2 \times \Delta Xr\_g \qquad (8)$$

$$\Delta\theta = w1 \times \Delta\theta o\_g + w2 \times \Delta\theta r\_g \qquad (9)$$

Where, w1 and w2 are defined as

$$w1 = |1 - drf| \qquad (10)$$
$$w2 = drf \qquad (11)$$

Where,

$$drf = \min(1, \frac{dr}{dr0}) \qquad (12)$$

dr0: standard value(e.g. dr0=5m)

According to (4) and (5), ΔX and $\Delta\theta$ depend on path command when dr>dr0, or node command when dr<dr0.
ΔS is finally decided by fuzzy inference.

First of all, ΔX and $\Delta\theta$ have 5 membership function $\mu(\Delta x)$s, $\mu(\Delta\theta)$s (-Big, -Medium, Small, Medium, Big) respectively.

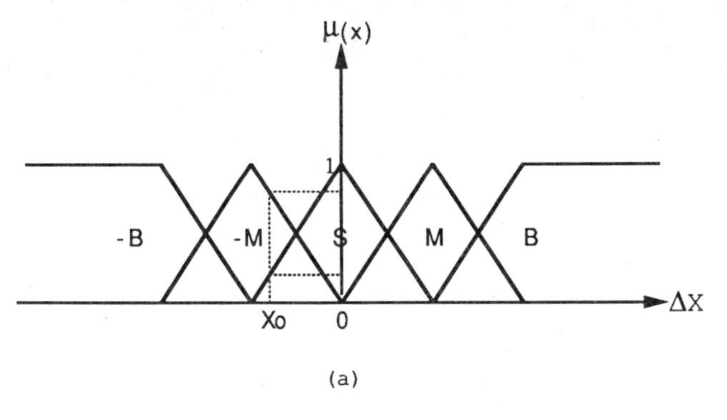

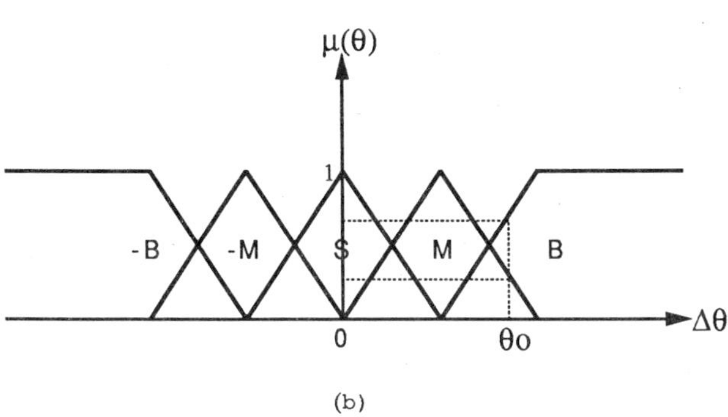

Fig 2 Membership function (a) $\mu(x)$ and (b) $\mu(\theta)$

Table 1 Control rule table $S(\Delta X, \Delta\theta)$

		ΔX				
		-B	-M	S	M	B
Δθ	-B	10	6	4	2	0
	-M	6	3	1	0	-1
	S	3	1	0	-1	-3
	M	1	-0	-1	-3	-6
	B	0	-2	-4	-6	-10

Here each set is composed of some linear functions. Each membership function represents the distribution of the probability (not exactly statistically) whose value varies between 0 and 1. Fig.2 shows these membership functions. In this figure, if ΔX is Xo, 2 memberships (-Medium, Small) have the non-zero values and the others have zeros.

Next, we define the control rules. for example,

(rule 1)
if Δx is -B(ig) and $\Delta\theta$ is S(mall),
then steering angle $S(\Delta x, \Delta\theta)=3$ (degree)

(rule 2)
if Δx is M(edium) and $\Delta\theta$ is B(ig),
then steering angle $S(\Delta x, \Delta\theta)=-6$ (degree)

These rules are shown in Table 1 which consists of 25 rules (Δx; 5 divisions, $\Delta\theta$; 5 divisions).

Finally steering angle ΔS is calculated as follows with $\mu(\Delta x)$, $\mu(\Delta\theta)$, and $S(\Delta x, \Delta\theta)$

$$\Delta S = \frac{\sum_{\Delta x}\sum_{\Delta\theta}\{\min(\mu(\Delta x), \mu(\Delta\theta)) \times S(\Delta x, \Delta\theta)\}}{\sum_{\Delta x}\sum_{\Delta\theta}\min(\mu(\Delta x), \mu(\Delta\theta))} \tag{13}$$

2.4 The simulation of moving

Fig.3 shows the simulation result of a vehicle turning a corner. In this figure, the square and the arrow show a vehicle and the moving direction of the vehicle, respectively. The width of the vehicle and the road is 2m and 4m wide, respectively. The velocity of the vehicle is 15km/h. The figure shows that the vehicle can turn the corner smoothly.

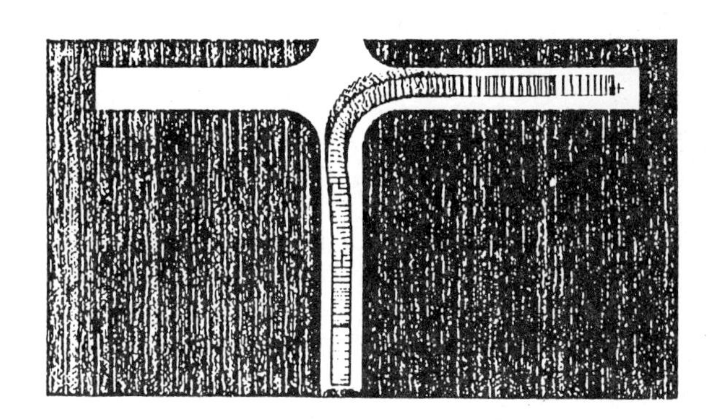

Fig 3 Simulation result of moving at a corner (15 km/h at a straight line, 10 km/h at a corner, road width 4 m, radius of the corner 4 m)

2.5 The simulation of obstacle avoidance

Fig.4 shows the simulation result of the obstacle avoidance. In this figure, the triangle, the square, and the arrow show an obstacle, the vehicle, and the moving direction of the vehicle, respectively. The width of the vehicle and the road is 2m and 8m, respectively. The velocity of the vehicle is 20km/h. When the vehicle recognizes the obstacle on the road 16m ahead, it decreases its velocity to 10km/h to avoid the obstacle. The figure shows that the vehicle can avoid the obstacle smoothly and return to the previous course with acceleration.

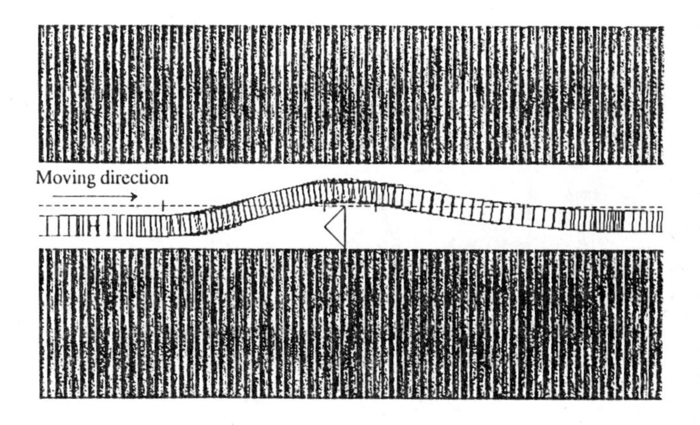

Fig 4 Simulation result of obstacle avoidance (20 km/h before obstacle detection, 10 km/h while avoiding, road width 8 m)

3 TEST VEHICLE

Using the results of the simulation,we manufactured the test vehicle(a 4.4 ton bus). Its control system is shown in Fig.5. It comprises 3 32bit-computers (main,sensor,and actuator computers) and an image processing computer (For a image processing unit, it has a FIVIS/VIP (Fujitsu Corp.)).

3.1 Sensing unit

The vehicle has the following types of sensors for recognizing its position and also the position of obstacles on the road.

(1) Cameras

The vehicle has 5 TV cameras. Three of them are on the roof of the car and are used for the detection of the white lines on both sides of the road. The other two are vertically aligned on the front panel of the car. They are used for the detection of obstacles on the road. The obstacles on the road must be non-moving objects whose width and height vary from 0.2m to 5m. The cameras can only detect one obstacle at a time.

(2) Wheel shaft encoders

The vehicle has two wheel shaft encoders on both sides of the rear wheel shaft. The vehicle calculates the direction that it is moving, its moving distance, and its speed from these encoders' information.

(3) Ultra-sonic sensors

The vehicle has four ultra-sonic sensors attached to its side panels. Two are at the front end and the other two are at the rear end. The sensors' measured distance ranges from 0.28m to 10m. The vehicle uses these sensors to measure the distance(with a resolution of 4mm) between the sensor and the guardrail along the edge of the road. Additionally they perceive the vehicle's orientation on the road and the distance to the white line on the edge of the road.

(4) Laser radar

On the vehicle's front panel is mounted a previously developed laser radar. This sensor's measured distance ranges from 1m to 120m. Since this radar does not scan horizontally, it can only measure the distance to the obstacle. It can be used for obstacle detection instead of the TV cameras.

3.2 Actuator unit

The vehicle has three actuator units for the throttle, brakes, and steering. The throttle is controlled by ASCD(Automatic Speed Control Device) which was previously developed by Nissan. The Braking actuator is controlled by a hydraulic and air servo controller. These actuators work cooperatively according to the velocity command from the PID servo control algorithms in the actuator control computer. The steering is driven by dc servo motor which is attached to the column shaft. The steering control system to which fuzzy logic control is applied is shown in Fig.6. The main computer sends the steering command ΔS to the actuator computer every 200msec, and using this information, the actuator computer updates the steering angle every 10msec.

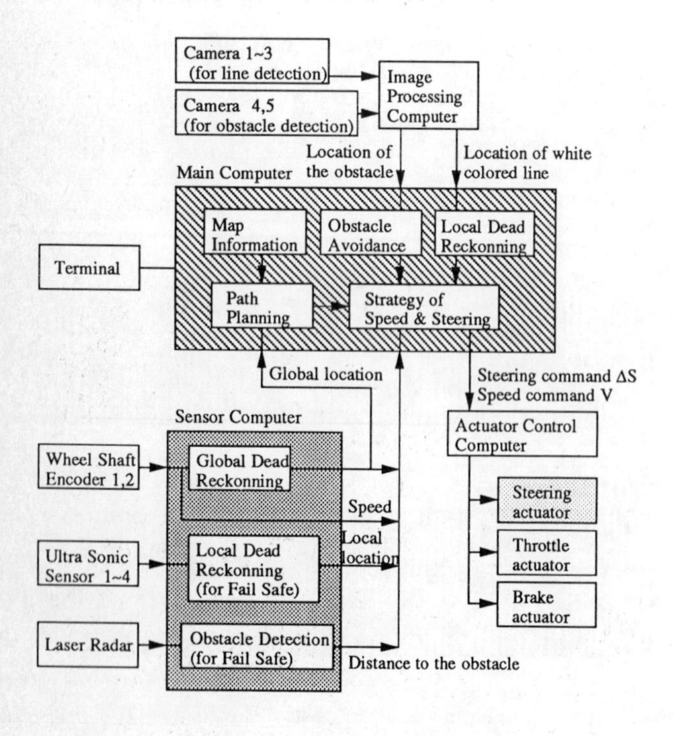

Fig 5 Control system

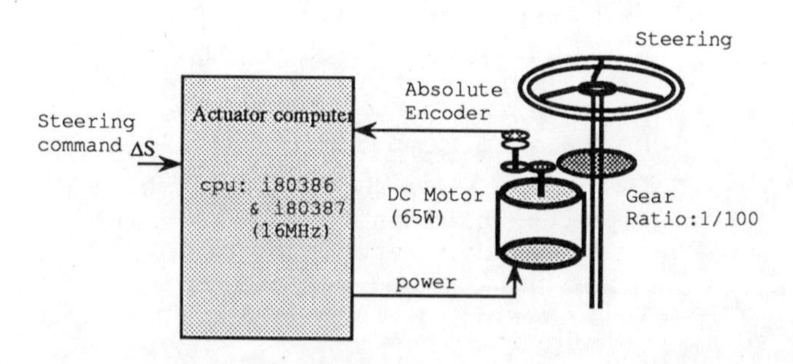

Fig 6 Steering control system

3.3 Moving Mode

As described before, this vehicle has two types of moving mode, ultra-sonic sensor mode, and camera mode.

3.3.1 Ultra-sonic sensor mode

In this mode,the vehicle uses ultra-sonic sensors to detect its position relative to the road or more accurately relative to the guardrail. The moving distance is determined using the vehicle's wheel shaft encoders.

On the other hand,it detects an obstacle on the road by the laser radar.

3.3.2 Camera mode

In this mode, the vehicle detects its position relative to the road and calculates the moving distance by using the TV cameras on the roof of the vehicle and wheel shaft encoders, respectively. The front panel TV cameras are also used for obstacle detection.

3.4 Control algorithms

The vehicle decides the most suitable path with the map in each mode, and generates commands for the vehicle's velocity and steering angle to maneuver the vehicle along the optimal path.

The control of the steering angle is decided by fuzzy logic algorithms. Since fuzzy logic compensates for the uncertainty in the precision of the sensor on the vehicle.

The order of the procedures is as follows.

(1) Accessing the sensors

The vehicle gains information on its position and the position of obstacles from the different sets of sensors. Exactly which set depends on the vehicle's mode of operation.

(2) Control of the field of vision of the TV camera

Depending on the road's curvature,the vehicle selects an optimal TV camera in order to ensure that the white line remains in the field of vision of the TV camera.

(3) Obstacle avoidance

As described before, when an obstacle on the road is detected, the vehicle executes the obstacle avoidance procedure. If the vehicle decides that the gap between the obstacle and the white line is adequate to pass through, it executes the procedure. If not, the vehicle stops before crashing into the obstacle.

(4) Command generation of the desired velocity and steering angle

If an obstacle on the road is detected, the vehicle generates the command of the desired velocity and steering angle for obstacle avoidance, and if no obstacle is detected, it generates a normal steering command based on its global strategy. Additionally,if the obstacle is removed, the vehicle starts executing the normal procedure again.

4 TEST SITE

The conditions at the test site were as follows.

(1) The width of the road :4(m)
(both on straight and curved pieces)
It is supposed that the vehicle moves on a standard lane of the roads in Japan.
(2) The minimum radius of the curvature :10(m)
This considered for the inner wheel difference of the vehicle.
(3) The white line width :15(cm)
(4) The geography :flat
(5) The pavement :black top asphalt
(6) The weather :fine
(3)~(6) are considered for the capability of the image processing unit.
(7) The guardrail :partly set up
This is used in the ultra-sonic sensor mode.

5 EXPERIMENT

Fig.7 shows the result of the vehicle moving at a test site. Here,the velocity was 10km/h.
The membership function was adjusted so that input/output characteristics had a parabolic function. The vehicle moved as smoothly as the simulation predicted that it would.

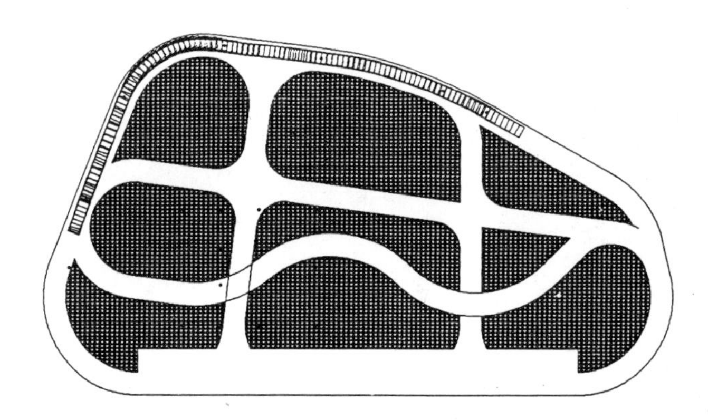

Fig 7 Result of moving at a test site (10 km/h)

6 Conclusion

We verified that
(1) The steering control using fuzzy logic algorithms is very suitable for an autonomous passenger vehicle because the vehicle can move very smoothly and also create the smooth ride and control that is normally only possible with human control.

(2) Using fuzzy logic,it is easier to adjust the control algorithms' parameters than in the other algorithms. Fuzzy control steering algorithms applied to the vehicle have only been tested at low speed.

In the future, it will be necessary to test the limit of the stability of the fuzzy control algorithms at higher speeds. This is because a vehicle is an inherently nonlinear system.

REFERENCES

(1) L.A.Zadeh,Fuzzy Sets. Inf.& Control 8, 338-358(1965)

(2) M.Sugeno and M.Nishida, Fuzzy Control of Model Car. Fuzzy Sets and Systems, 16, 103-113(1985)

(3) M.Sugeno and K.Murakami, Fuzzy Parking Control of Model Car. 23rd IEEE Conference on Decision and Control, Las Vegas(1984)

(4) Volker Graefe, A highway Autopilot for Motor Vehicles.International Symposium "Computer World '88". Kobe, October (1988)

(5) Masao Sakata and Shigeru Okabayashi,et al.,Laser Radar Rear-end Collision Warning System for Heavy-duty Trucks. Associazione Tecnica Dell' Automobile International Symposium,471-487 (1988)

C391/081

Car obstacle avoidance radar at 94GHz

P MALLINSON, BSc, AMIEE and **A G STOVE**, MA, DPhil, CEng, MIEE
Philips Research Laboratories, Redhill, Surrey

This paper will describe work carried out by Philips (primarily at the Research Laboratories in Redhill) on Car Obstacle Avoidance Radar. I shall be describing a realisable radar system using millimetre wave FMCW (Frequency Modulated Continuous Wave) techniques at 94 GHZ. Millimetric radar is still relatively novel at 94 GHz, but progress is fast and mass production is now possible (e.g. using plastic microwave assemblies with a high degree of cost effective microwave integration techniques). The system which will be described in detail, consists of a 94 GHz Front End (the scanning antenna, microwave feed, RF circuitry and IF downconversion), a powerful data processor/funnel and a variety of display options.

The advantages of the millimetre wave solution will be explored in detail and will include such discussion points as:
* The dimensions of a system are inherently small due to the short wavelengths involved.
* Accurate and narrow beam patterns are easy to form with a suitable antenna.
* Beams are easy to steer using pseudo optical techniques.
* Complexity per unit volume is favourable due to progress in the millimetre wave components.
* The front end can now be moulded in metallised plastics (including the scanning antenna) and assembled using mass production techniques.

The results of a number of live field trials will be presented, as will our ideas on the forms that a practical, car obstacle avoidance radar might take.

Introduction

Radar has been used since the mid 1930's to detect such "targets" as aircraft, ships and missiles. It can also be used for mapping ground movements such as those at an air or sea port. Other applications include systems for presence/intruder detection and proximity fuses.

For a number of years it has been hoped that a form of radar could be employed in a road vehicle to warn the driver of obstacles and other dangerous objects in his path. A number of radars have been proposed for use on cars and other vehicles but there has been no successful outcome to date.

This paper describes a possible millimetric wave (94GHz) radar solution which could meet all of the demanding constraints for vehicle use.

The Task

The physical constraints which exist when discussing a vehicular radar (particularly for a compact car) means that only a small number of types of radar are suitable. The main areas of general concern include:-

1. Space - quite limited (also weight must be low)
2. Cost - must be low enough to make the system attractive and it must be mass producible
3. Reliability - must require (almost) no maintenance
4. Integrity - Minimum false alarms or else the user will distrust the system
5. Safety - There are two areas of concern here. Firstly the radar power transmitted must be very low and secondly the display/warning presented should not burden the driver with any significant workload

These are general requirements to which must be added the technical requirements. Following our earlier work on obstacle detection radar we are able to say the following:-

1. The system must be able to detect objects of $0.1m^2$ radar cross section (typical minimum)
2. The range must be 300m min in bad weather (giving the driver approx 7 secs warning at 100mph)
3. A range resolution of 5m (at max range) is required
4. The data update rate must be better than 1Hz
5. The coverage should be typically $\pm$ 15° (1° resolution) in azimuth and in elevation
6. The system should warn the driver of a potential threat to his progress and not give him any other information unless he requests it
7. The display to warn the driver must not intrude on the drivers concentration

The Experimental Data Gathering System

Figure 1 shows a block diagram of the experimental hardware and a photograph of the lab demonstrator.

Front End

The technology is based on that developed for use in Smart Weapons for terminal guidance. The requirement here is for a radar seeker that is lightweight, disposable and therefore low cost, small and accurate - the ideal general basis for an automative radar. 94GHz has been chosen as the operating frequency for the system - resulting in a compact antenna. The target weight for the microwave transceiver is under 200 grams, with an overall weight of about 2kg for the whole microwave front end including the antenna. This is achieved using plastic injection moulding techniques including a state-of-the art plastic microwave subsystem. Figure 2 shows one such subsystem manufactured by Philips Microwave Hazel Grove.

The system radiates approx 10mW of RF power in the form of an FMCW waveform which is transformed into range information using a Fast Fourier Transform based processor.

The proposed system will be well protected against interference from other radars of the same type. There are three features protecting the radar:-

1. The natural frequency drifts of the radars make it unlikely that their transmission spectra will coincide. This is allowable with the wide spectra available at 90GHz
2. The high time-bandwidth product of the radar gives it about 50dB rejection of other radars, as the sweeps will not be synchronised
3. Because the radar is a low power, narrow beam, line of sight system there will in any case be relatively few potential sources of co-channel interference, especially if the radar is mounted relatively low down on the car.

Principles of FMCW

Very briefly, an FMCW radar operates as follows:- The carrier radio frequency (in this case 94GHz) is modulated with a linear frequency sweep of a few MHz as shown in Figure 3. This is transmitted continuously from the antenna. The return from a target (eg an obstacle) will be received at a time t later, t being proportional to the distance to the obstacle from the radar. After suitable amplification the resultant difference Δf can be extracted - where Δf is proportional to the range.

For scenarios with several targets (the normal case) several different frequencies will be present simultaneously. In order to separate these a Fast Fourier Transform needs to be performed. The output of this transform is used to obtain the range information for the scene - hence giving us the range to each obstacle in this case.

Processing and Display

Having built a radar head to detect obstacles considerable signal processing is required to support it - not least in the area of presenting only relevant information to the driver in a simple error free form. The final processor should occupy minimum space (suggesting the highest practical level of integration) and consume only a small amount of power (eg 20W). Simulation of this processor and display is currently taking place on a workstation and forms part of our ongoing research programme. The algorithms necessary to extract the relevant information to warn the driver only of obstacles are being investigated as are possible output format for the data.

Trials and Results

Trials of the proposed system are currently underway. Various hardware configurations are being examined for a number of different scenarios. The data gathered thus far includes the returns from road surfaces, parked cars, selected obstacles, people and moving vehicles.

The detection results are summarised in Figure 4 which shows the typical radar cross sections of some common articles. This table shows that the 94GHz radar system is able to detect the obstacles which could potentially cause a hazard to road users.

The quality of the data from the front end is very good using the 94GHz scanner and even the raw data is intelligible to the untrained eye for a number of everyday roadscenes.

As an example, Figure 5 shows a plot of some raw data. The amplitude of the return is plotted against the range cell information (where cell 128 represents approximately 500m in this case). The man presents a target of approx $1m^2$ radar cross section at 88m, the car (a Fiesta) presents a target of $10m^2$ at 160m. Also visible are trees, a fence and a distant hedge. The targets shown were verified visually.

Figure 6 shows the road as the radar sees it and illustrates how the picture may be processed to give useful information to the car driver.

Figure 6a shows the scenario, a small country road and 6b is the raw radar map obtained by the experimental radar standing at the right hand side of the road. The map is scaled to indicate perspective with azimuth ($\pm 10.5^o$) as abscissa and a distorted range scale as ordinate. Maximum range is 300m and half range about 120m. The white areas are cells where the radar detected a target. The raw data clearly shows the line of the road, but is obviously too complicated to be presented to the driver as it stands.

Figure 6c shows the same data after some processing to identify the targets lining the roadside (hatched cells) and the edges of the roadway, deduced from the road edges (white lines). The white cells at the right hand edges of the roadway are targets within the roadway, which therefore represent potential hazards. These hazards are artefacts which were deliberately introduced for the illustration by defining the roadway to be slightly too wide.

Note that the algorithm correctly follows the curve of the road and does not classify the hedge at the top left as a hazard.

This picture is obviously getting closer to what is needed, but is still too crude to show the driver and much more work is still required on the man-machine interface. Figure 6d is a conceptual sketch for a possible display format, based on the experimental data. The picture could well be presented to the driver via a low cost Head Up Display.

Conclusions

The research so far has demonstrated that using a 94GHz radar system we can reliably detect obstacles in an automobile scenario.

Even a simple display of this data, with no manipulation, clearly suggests that a millimetre-wave based sensor could form the basis of an obstacle detection system.

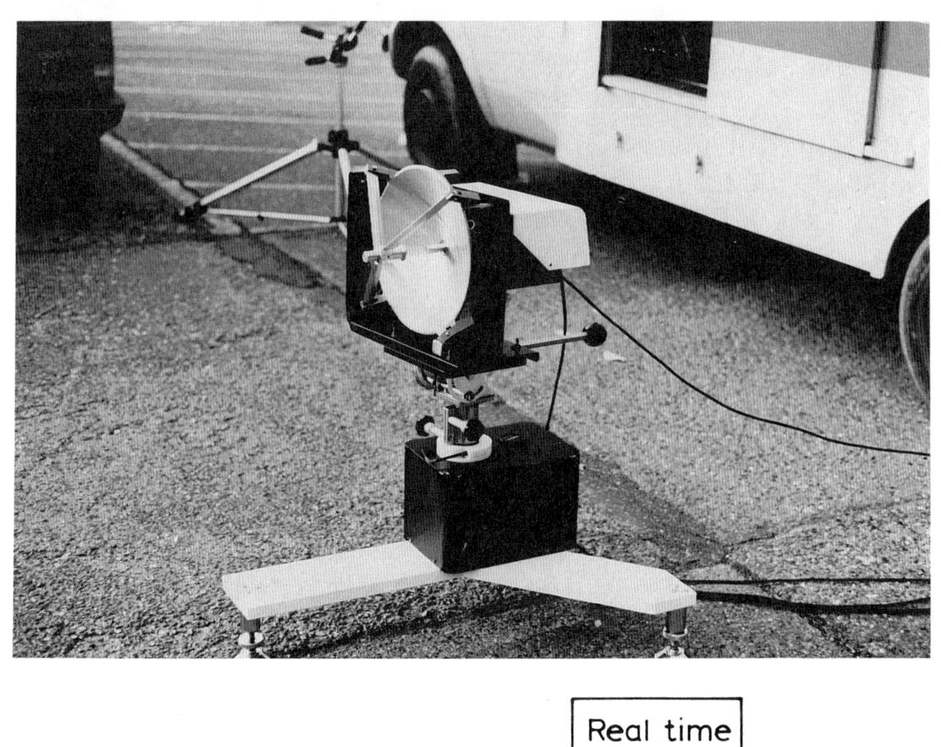

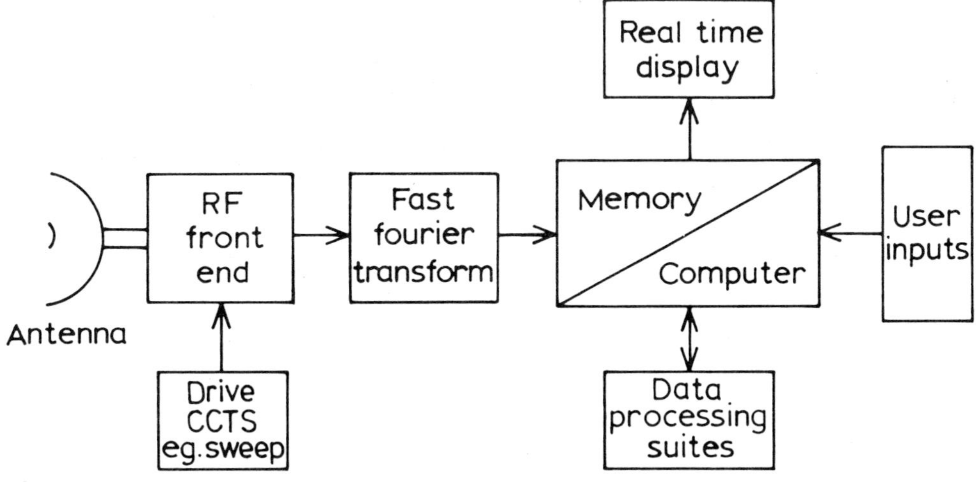

Fig 1 The experimental system

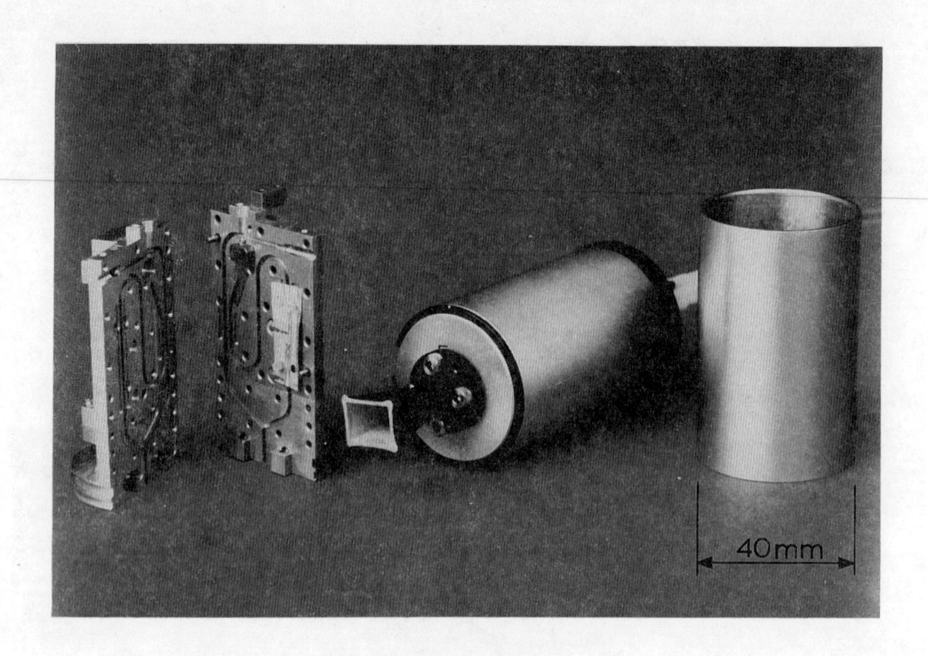

Fig 2 A 90 GHz radar front end from Philips Microwave Hazel Grove

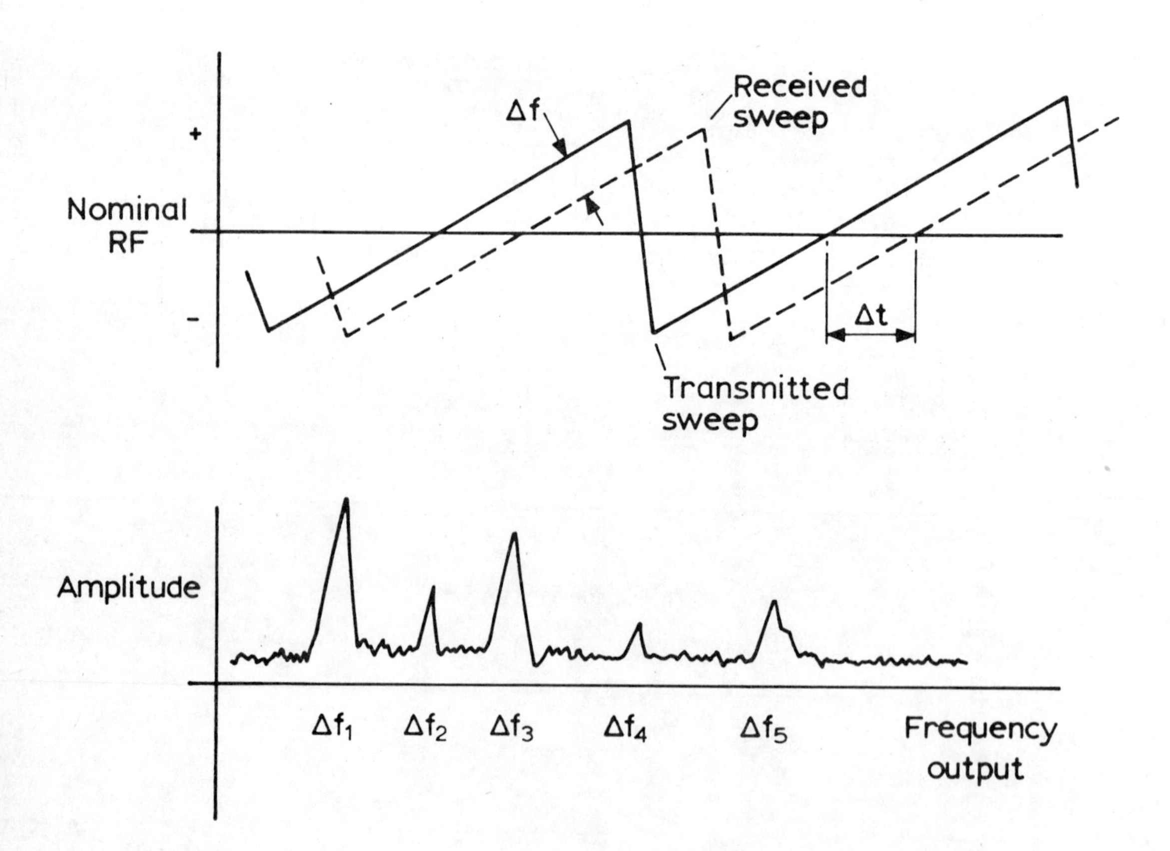

Fig 3 The principle of FMCW

Radar cross sections at 94 GHz
(Horizontal polorisation)

Target	RCS (dBm^2)
Small automobile	10
Van	15
Man	0
Debris	-10 to +10
Road surface (5° incidence)	≈ -40 dB/m^2

Fig 4 Typical radar cross-sections

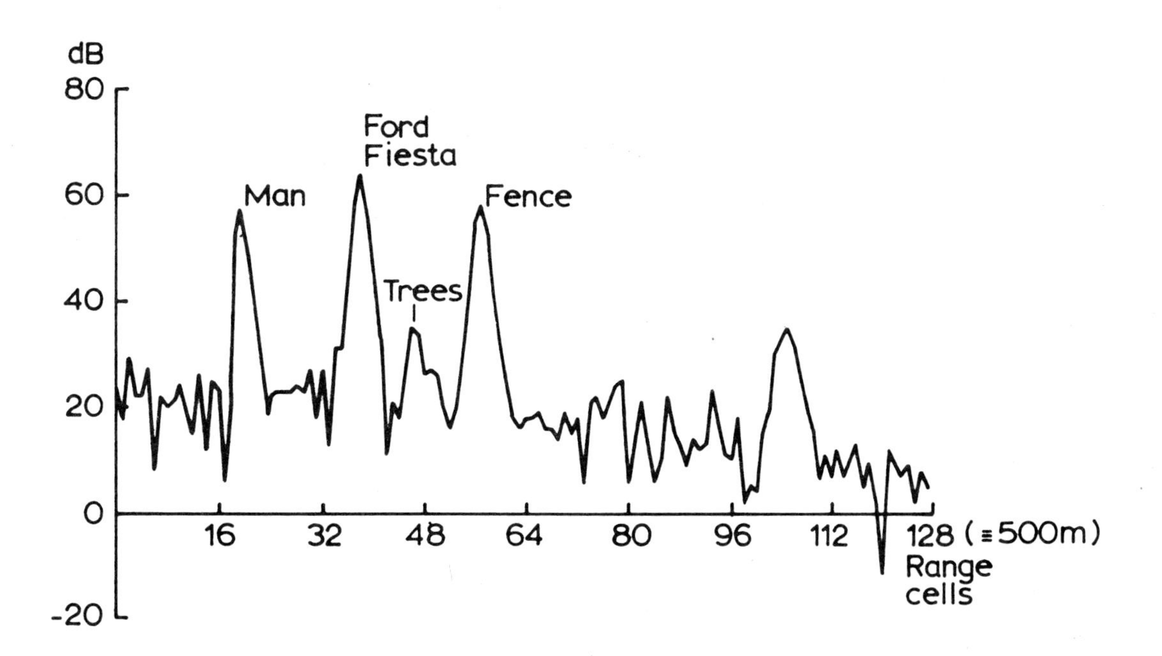

Fig 5 Raw data from the system

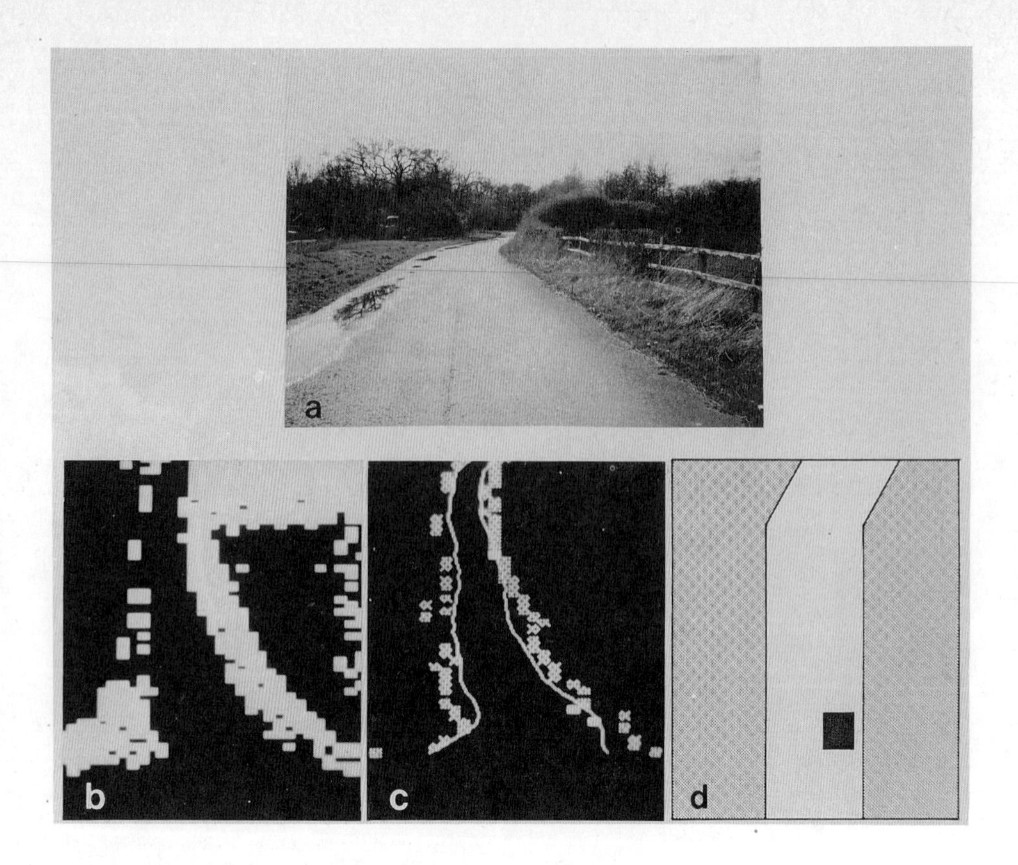

Fig 6 Results from a typical scene

C391/061

PROMETHEUS research—a Pro-Road segment survey

W H ZIMDAHL, Dr-Ing, IEEE
Volkswagen AG, Wolfsburg, West Germany

SYNOPSIS PRO-ROAD is that part of the pan-european research programme PROMETHEUS which comprises industrial research work concerning the whole road traffic information and control system. It should include global functions like data acquisition, communication, information processing, road network description and on-board elements. This survey combines the reports of three European lead researchers about the topics of research. Via the infrastructure elements PRO-ROAD is linked to the traffic engineering research in the PRO-GEN segment.

1 INTRODUCTION

The situation on Europe's roads is slowly reaching catastrophic proportions. Private transport is being forced into an impasse as a result of the continuous rise in car registrations and the limited space available for extending the road network. In the European Community alone, 50,000 deaths and over 1.7 million casualties are recorded from road accidents every year. Exhaust gas and noise pollution paint a black picture for the future of private transport.

Individual solutions to detailed problems are not longer sufficient. What is needed is a new fundamental global approach to solve the problem of the "System Road Transport". "Systemic Thinking" will provide the key to move away from an individual towards a collective approach. Global utility is therefore the prime target.

Based on these prerequisites, the European Research Programme PROMETHEUS was started in October 1986. It will supply the technical foundations for advanced developments of the road traffic system. Its objectives are to create the concepts and solutions to make road transport significantly safer, less harmful to the environment, more economical and more efficient.

The objectives and complexity of this immense undertaking are extraordinary and demand a new form of cooperation between information technology and automobile and traffic engineering and will also involve the authorities responsible for transport and telecommunications. Common interests and joint responsibilities for the problems and tasks can only be solved by joint actions.

What is special about PROMETHEUS is the cooperation between **basic scientific research** conducted at universities and research institutes and **application-specific research** carried out by industry. Basic research is composed of the subprogrammes **PRO-ART, PRO-CHIP, PRO-COM and PRO-GENeral.**
The subprogramme **PRO-ART** will elaborate methods using artificial intelligence for processing the complex mass of information which will occur in the vehicle and in traffic control networks.
The subprogramme **PRO-CHIP** will design the hardware components to enable intelligent real-time data processing in the vehicle.
The subprogramme **PRO-COM** will develop new communications methods which will be compatible and standardised throughout Europe.
The subprogramme **PRO-GENeral** will concentrate on appropriate introduction strategies and prepare a global model of road transport in Europe. This will ensure the optimum technological link between vehicle and the infrastructure.

In the sphere of industrial research, there are three subprogrammes: **PRO-CAR, PRO-NET** and **PRO-ROAD.** The main goal of industrial research is to design, develop and test demonstrators.
The objective of the subprogramme **PRO-CAR** is the development of computer-aided systems in the vehicle to assist and relieve the driver. Special sensor systems and actuators will be integrated into an electronic and hydraulic on-board architecture which will be totally redefined. A central theme of development of PRO-CAR is matching technology to human needs. Special fail-safe strategies ensure perfect vehicle operation even if systems fail or the information flow is interrupted.

In the subprogramme **PRO-NET,** a communications network between on-board computers of different vehicles is being developed. The driver will be presented with information from areas which he cannot perceive. Through its forward-seeing strategy, PRO-NET can inform the driver in good time of the critical manoeuvres of other drivers. This will allow him to make a better assessment of the situation and take safer action.

The **PRO-ROAD** subprogramme will create roadside communications and information devices for a superordinated traffic management system. Drivers will then be provided with the information they require for their individual decisions and actions in order to reach their destination by the best and safest route.

2 PRO-ROAD SYSTEM ENGINEERING

System engineering is like playing a puzzle. Several PRO-ROAD thematic projects will deliver parts of the whole PRO-ROAD traffic information and control system, for instance communication systems, data acquisition systems, architecture for information processing, road network information, on-board elements and not to forget PRO-GEN will develop suitable control strategies. It is the task of system engineering to ensure the correct implementation of all parts into a global approach towards the use of a common infrastructure in an integrated road transport environment.

The framework or the strategy according to that global approach of the PRO-ROAD traffic management system which will be realised, was defined during the definition phase as follows:
Autonomous vehicle route guidance systems with on-board digital road maps limited to the main road network of Europe will be combined with roadside infrastructure in dense traffic areas. This infrastructure consists of a beacon network for traffic information collection, connected to central computer systems for generation of routing information. Links to the public communication network will provide auxiliary driver and passenger information.

Systems according to this concept can be introduced immediately via the autonomous components and then be upgraded step by step by installing roadside infrastructure and communication links.

To study the implementation of the information part of traffic management systems, we have foreseen the following activities:

- participation in the RDS field test taking place in a corridor along the river Rhine from Rotterdam to Basle. RDS allows the transmission of digitally coded traffic information via the ordinary radio broadcast service.

- satellite communication systems will be tested in european fleet management applications. Currently two system approaches, called PRODAT and INMARSAT Standard C are at hand.

- an electronic road book, a kind of enhanced electronic traffic sign, will allow the transmission of very detailed information concerning a specific stretch of road, such as advisory speeds, warnings, gear shift recommendations and so on.

To study the implementation of the control part of traffic management systems we participate in several European field trials in order to obtain experience for setting up a near to perfect global PROMETHEUS approach for traffic management systems. We participate in the AUTOGUIDE trial in London, in the LISB-trial in Berlin and in the CARMINAT trial in the Paris area. These systems collect data centrally for calculating optimum routes and for giving the driver up-to-the-minute recommendations about traffic conditions and recommended routes. Besides this, new ideas such as improving the efficiency of junction traffic by using electronically coupled vehicle clusters will be pushed forward. We also hope to establish similar field trials in the northern, the scandinavian, part and in the southern, the italian, part of Europe to learn more about the traffic peculiarities of these countries.

The DRIVE project "The total traffic management environment" concludes all these activities and develops recommendations how to build up the puzzle play in a functional and strategically suitable way. The project is treated by a consortium of road administrations, traffic engineers, electronic companies and car manufacturers and will deliver agreed sets of functional requirements and specifications, as well as requirements for international standards related to the integrated road transport environment.
This leads to the next topic area which is standardization of interfaces for data communication on different levels and protocolls for the organisation and handling of huge data bases.

3 STANDARDIZATION

Concerning standardization in general, the standardization of the European digital road map is emphasized at the moment, since only few components are in a similar development stage. The Task Force "European Digital Road Map" running as a DRIVE project within a 3 year time span represents the European projects CARMINAT, DEMETER, TELEATLAS and PROMETHEUS as well as national surveying agencies and road administrations. Main concern is to prepare the business of digital road maps, in order to allow a rapid commercial introduction of digital maps on the market.

One of the major problems in this respect is the development of organization concepts for a European Digital Road Data Base. This comprises of the whole logistic process from road data recording to distribution of digital maps for various applications. The technical organization of the data base as well as the updating problem will be defined in order to cope with the tremendous amount of digital road data expected for Europe.

To provide exact values for calculation of costs and time needed to digitize the whole of Europe a benchmark test "European Digital Road Map" will be performed in 1989. 12 small areas all over Europe were selected to study specific data recording problems by using different methods. Precise estimation of the data quality delivered by these methods will be obtained. During this test the first release of a draft common standard for European digital road maps will be checked for necessary changes. All activities of the Task Force are aiming at a common European standardization procedure as a basis for a digital road map business.

Another area where standardization is realy necessary for a pan-European approach to improve road traffic is data communication. A limited number of projects for different levels and applications are on the way already towards standardization, but they need a stronger support from the concerned actors and consensus in order to become standards within a few years from now.

For the purpose of long range, one-way communication (broadcasting) the already standardized radio data system RDS (from EBU) is very probable to incorporate a so called traffic message channel TMC. Digitally coded information about traffic incidents due to roadworks, accidents or just congestion will be transmitted to on-board units in oder to improve drivers information or even route planning and route guidance.

As a means for medium range, two-way communication a special service within the already standardized pan-European mobile telephone, GSM, will be used in order to get access to a large variety of commercial data bases as explained later.

The already mentioned beacon network needs standardization for a very short range, two-way communication link between on-board and infrastructure elements. A first release was already drafted on the level of U.K. and German road administration with the assistance of industrial companies. As a physical means infrared light (IR) is recommended, and the technical elements as well as the protocolls are already under test within the mentioned field trials AUTOGUIDE (in London) and LISB (in Berlin).

A totally new concept of short range, one- and two-way commmunication link between vehicles and between on-board and infrastructure elements originated from several new demands for PROMETHEUS-functions like convoy-driving (road trains), automatic merging, intelligent intersection control or "looking around the corner".

This task is a real challenge for industrial and basic research to come up with a mobile communication net dynamic in space and time between all road traffic participants (including pedestrians, bicyclists and motorcyclists). For the long run we have the dream, that vehicle clusters might behave like birds or fish swarms in cooperative maneouvering. But some preliminary specifications have been agreed already, and the common efforts of industrial and basic research will provide first real test and simulation tools in 1990.

4 INFRASTRUCTURE INFORMATION PROCESSING AND DATA ACQUISITION

The research work within this area will be done mainly by the electronic companies who committed to be associates to the PROMETHEUS programme. The following ideas about architecture, databases and data acquisition should be regarded as guidelines from the requirements which the car manufacturers assume to represent the drivers demands.

4.1 Architecture

The introduction of new digital communication channels offers new possibilities and services to the road users.

The radio data system RDS, will be employed for broadcasting of various kinds of information.

The pan-European mobile telephone, GSM, will be introduced in 1991 and will cover the main roads and all major cities before 1995. GSM supports digital communication which will be used for a broad range of applications.

A concept for a beacon network via infrared (IR) is tested intensively within LISB (in Berlin) and Autoguide (in London).

Europe has gained a lead over the US and Japan, concerning the important communication infrastructure. This lead can be turned into a market advantage. A prerequisite is that external supporting databases will be developed to the same extent as the applications in the vehicles. The design of this information processing architecture is one of the major challenges in PRO-ROAD.

4.2 Database Supported Features

An example is new possible features for cellular telephones and route guidance systems, such as:

- Extensive up to date telephone directory available from the car.

- Yellow pages with geographical information. The three most important keys in the usage of the yellow pages are product, company name and location. The yellow page systems will be improved and will be made available to the drivers supporting them in their choice of destination, but also to access the telephone numbers.

- Route planning support, permits the driver, not only to optimize route, but also to optimize the destination, time of departure, avoid queues and thereby probable helping to reduce the total level of congestion.

The project with partners from the telecommunication companies is aiming at defining the protocol and demonstrating the possibilities.

4.3 Traffic Data Acquisition

Equipment for data acquisition of traffic data is widely used.
The present systems are mainly traffic flow sensing loops and there is a strong need for more sophisticated detectors.

There is for example a lack of sensors sensing traffic density, travelling time, individual vehicles and pedestrians. Video based and road user detectors, can be used in intersections but the technique still need a substantial development performed by industrial and basic research.

The floating car concept is one of the more interesting possibilities. The navigation system in the cars transmits spontaneously information to the traffic management computer. The transmitted information is mainly the time spent for the last part of the trip. A statistical acceptable value is achieved even with very few equipped cars and the accuracy will be higher in dense traffic. The LISB trial in Berlin has proven some very promising results already.

4.4 Data Acquisition of Environment Data

Weather monitoring stations are introduced on large scale in many european countries. The information is mainly used to optimize snow clearance but different ways to provide the drivers with this information is under investigation. The most imminent are TV, radio and RDS.

The present acquisition methods for environment data are not trustworthy enough and the research field comprises topics for:

- Forecast of road surface condition
- Visibility factors
- Water film detection
- Direct measurement of ice
- Measurement of pollution in tunnels and at strategic points in urban areas
- Detection of wild animals and
- Roadwork which is one of the major causes to traffic disturbances.

One of the ongoing projects will improve the detectors so they may distinguish between different types of ice and snow. Several companies and research institutes are participating in projects in this field.

A demonstrator project investigated by one of the automotive companies, comprises a weather monitoring station and a FM transmitter. Traffic relevant information is received by the RDS radio and presented to the driver on a display and by a synthetic voice. The car also contains a small computer allowing to search for a hotel and make reservations. The computer is connected to an external database via the Mobitex radio network.

5 ON-BOARD-ELEMENTS

The last but not least research area in PRO-ROAD is concerning the on-board elements, mainly for interactive dynamic route guidance. The task is to provide the driver with recommendations for routing while driving which comply with his individual trip criteria (short, inexpensive, comfortable a.s.o.) and takes into account the real traffic situation in the concerned road network.

A system which is able to fulfill this tasks needs to have several modules to perform different functions interacting with each other:

- retrieving static information from a large data base: a digital description of the road network which includes not only topological data but also logical attributes like: connection of roads, oneways, speed limits, average travelling speed and many others;

- sensing and collecting information: on one hand sensors for travelled distance and heading from which an estimated position can be calculated by a dead reckoning algorithm; on the other hand receivers for exact on-board localization and for dynamic data (e.g. roadworks, congestion a.s.o.);

- processing these available information: to update the stored static data of the road network, expecially the traffic dependent data; to compare and correct estimated position in order to find exact location values; and last but not least to calculate the "best route" from all this information according to the actual demands of the driver and to provide him with suitable recommendations (in the long run via a centralized man-machine-interface).

Although this general architecture was already realized in different prototypes from electronic suppliers intensive testing has shown the need for further research.

The first one deals with the route planner: the first generation which calculate the nearest or the shortest route from a static database has to be improved. Apart from the problem of computation time which is not sufficient today to ensure real guidance on every network and in a dynamic environment, the main point is the criteria. What is the "best route" for the driver? It is neither the shortest nor the cheapest but in most cases the most comfortable one. What does this mean in terms of algorithms? Two conclusions have been made to answer these questions: First it is very likely to use artificial intelligence techniques to explore this issue and prepare a new generation of route planners; second it is obvious that a tool is needed to do that job by which existing algorithms can be evaluated as far as performance and quality are concerned and new concepts can be specified and designed. The development of this tool has been started already.

Another research topic is that of localization. Even if recent research and development has focussed on autonomous systems (i.e. dead reckoning plus map matching), it seems valuable to look carefully to other localization systems, that is to say infrastructure supported systems. One reason is that map matching will be extremely difficult where only limited digitized road network data is available.

Some of these already well known systems are promising for road traffic application also but are not totally available now as space as well as time is concerned. The most interesting are GPS (satellite based), LORAN C, or systems based on cellular phone base stations or based on synchronized RDS-signals.
There is no clear preference now, therefore development and test work is necessary on different systems.

6 PRO-GEN TRAFFIC ENGINEERING

PRO-GEN has two main work areas:
(1) Design and optimization of an integrated traffic control system based on higher-order framework conditions and extensive driver information.
(2) Impact determination of the system solutions investigated and evaluation of the effects on the basis of overall requirements made on a road traffic system of the future.

In order to prevent from misunderstanding it should be pointed that the task of impact determination and effects evaluation concerns all different system ideas: not only the above mentioned PRO-ROAD sytems but also those being investigated under PRO-NET for cooperative driving and PRO-CAR for intellignt dynamic vehicle control and driver information.

But a very close liaison to PRO-ROAD industrial research has been established to the work area of integrated traffic control system design. The main tasks of this are optimization of existing systems and specification and advanced development of interfaces in order to create an overall system architecture; and development and testing of integrated control strategies taking into account the possibilities of intelligent vehicles in the future.

The following list of main strategical functions serves to give some ideas about the modules a future integrated traffic management will comprise:

- Trip planning
- Route guidance
- Traffic demand management
- Traffic flow control
- Network control (urban and motorway traffic)
- Intelligent intersection control
- Parking management/control
- Improved interface to public transport

But at the end it should be mentioned again that integration is the final goal of the PROMETHEUS approach.

7 CONCLUSION AND ACKNOWLEDGEMENTS

After the analysis of deficiencies and definition of research topics automotive industry, electronic companies and suppliers and basic research institutions have started the research work: PROMETHEUS is rolling. I want to acknowledge the subtantial help in preparing this survey coming from the European lead researches Mrs. P. Fast, P. Häussermann and Y. Gueguen and from intensive discussion with representatives of PRO-GEN.

C391/045

Prototyping a navigation database of road network attributes (PANDORA)

A B SMITH, MA, MAppSci, ARICS
Ordnance Survey, Maybush, Southampton

SYNOPSIS

The Automobile Association, Ordnance Survey, Philips International BV and Robert Bosch GmbH are collaborating in a project to create and test a prototype navigation database. The project, which is being managed by consultants MVA Systematica of Woking, has been accepted for support under the European Community's DRIVE initiative.

The data requirements of future vehicle navigation systems such as CARIN from Philips, EVA from Bosch and the Autoguide pilot scheme for London are being examined. Digital street networks, extracted from Ordnance Survey's large scale digital mapping, and the necessary road and traffic attributes, will be gathered and integrated in a prototype database for parts of London and Birmingham and the motorway connections in between. This database will then be tested using the Autoguide, CARIN and Travelpilot/EVA systems. The data set will also be available for testing as benchmark number 12 within the Eureka project PROMETHEUS.

The PANDORA project is due to last from January 1989 to June 1990. This paper covers the objectives and work programme of the project, its relationship to other European initiatives on vehicle navigation, and a review of the progress achieved up to the date of submission.

1. INTRODUCTION

In October 1988, a proposal for a project entitled 'Prototyping a Navigation Database of Road Network Attributes (PANDORA)' was submitted to Directorate General XIII of the Commission of the European Communities (CEC) based in Brussels. The proposal was one of 189 received in response to an invitation to tender under the Community's DRIVE Programme, formally initiated by the Council of Ministers on 29 June 1988. DRIVE, or 'Dedicated Road Infrastructure for Vehicle Safety in Europe' to give its full title, is a partnership research programme between the Community and European industry, each of which will contribute a total of ECU 60 million (about £40 million) over a 3-year period, beginning on 1 January 1989.

The objectives of the DRIVE programme are:

- to improve road safety;
- to improve transport efficiency;
- to reduce the negative environmental impact of road transport.

Following a detailed technical evaluation, approval was given in December 1988 for work to begin on PANDORA, subject to completion of contractual formalities. It was one of almost 60 projects accepted at this time, representing 12 000 man months of effort by 450 participants from throughout the Community countries and from a wide variety of academic, industrial and governmental backgrounds.

2. THE PANDORA PARTNERS

The 4 partners in the PANDORA project are:

- the Automobile Association;
- Ordnance Survey;
- Philips International BV;
- Robert Bosch GmbH.

The Automobile Association (AA), based in Basingstoke, is the largest motoring organisation in the United Kingdom. It offers a full range of roadside breakdown services, technical, touring and legal advice to its over 6.9 million members. AA Roadwatch is the largest road traffic broadcasting operation in Europe, providing live radio and television broadcasts to more than 40 stations in the UK as well as traffic information to the Oracle and Prestel teletext services. The AA's nationwide staff of touring information officers, patrols and Roadwatch units provide a unique source of information about roadworks, traffic congestion and weather-related hazards in support of AA's services.

Ordnance Survey (OS) is the national governmental mapping agency for Great Britain, responsible for maintaining and supplying maps at scales which range from 1:1250 to 1:625 000. Its headquarters are in Southampton, but most of the survey work is conducted from small offices spread throughout the country. Although perhaps best known for the 204 maps in its 1:50 000 scale Landranger series, OS's main task is to keep up to date over 220 000 maps at the 'basic scales' of 1:1250, 1:2500 and 1:10 000. In 1972 it embarked on a programme of digitising these basic scale maps, in order both to automate map production and to supply data to an increasing number of digital customers.

Philips and Bosch are both major European forces in the field of in-car entertainment and information. Philips, based in Eindhoven in the Netherlands, has a long history of participation in the development of vehicle navigation systems, including CARIN (Car Information and Navigation System) their autonomous vehicle guidance product. Philips and Bosch have been collaborating in the Eureka DEMETER project which has created the Geographical Data File (GDF) format specification, the most detailed and comprehensive statement, in Europe at least, of the data requirements for vehicle navigation systems. Robert Bosch GmbH, and its subsidiary Blaupunkt are also familiar names in the automotive electronics industry. Bosch are active in the LISB experiment in Berlin, which uses a system similar to the UK Autoguide. They created considerable interest at the Birmingham Motor Show in October 1988 when they launched their Travelpilot system in UK, using a sample of Ordnance Survey street network data. Bosch's successor to Travelpilot, known as EVA, is already under development. It is planned that both Travelpilot/EVA and Philips' CARIN will be used in the field trials of PANDORA.

The other major contributor to PANDORA is MVA Systematica of Woking. Although strictly a sub-contractor rather than a partner, MVA is providing the project manager and making a substantial input to the work. MVA Systematica is an information technology consultant and software house, specialising in the transport and travel sector. It has been involved in several projects for the European Commission including the DRIVE Planning Exercise Contract Management and in 1988 conducted research for Ordnance Survey into the market for digital mapping data in support of vehicle navigation.

3. PANDORA OBJECTIVES

The PANDORA project is directed at 3 of the tasks which were described in the DRIVE Workplan, issued early in 1988:

T304 Integrated autonomous and infrastructure-supported route guidance systems.

T328 Digitised Road Maps

T329 Study of a Road Database Management Structure.

PANDORA covers the entire requirements of T328 and T329 and also those aspects of T304 which primarily depend on the supply and update of digital road maps. To meet these requirements it has 6 main objectives:

1. To develop a comprehensive and re-usable data model for a digital road network database for vehicle navigation, location and other traffic and transportation purposes in Europe.

2. To provide the British element (Task 12) of the PROMETHEUS Benchmark Test 'European Digital Road Map'.

3. To develop, prototype and demonstrate the methodology, systems and software for extracting road network data from a digital map database, and integrating traffic-related attribute data, taking account of the GDF standard and the requirements of the Autoguide pilot system.

4. To demonstrate the correctness of the integrated data in field trials using prototype autonomous and infrastructure-based navigation systems.

5. To apply, develop and publish standards for digital road databases.

6. To determine legal measures for protecting data providers.

These objectives all contribute to the development of standards in digital road mapping which match user-requirements, which are cost-effective to work to, and which have the support of the major industrial parties. The fundamental component of the project is the data model, in which the basis for the standards and the integration work is stated in clear and unambiguous terms. In addition, the project is verifying the integrity of this data model and the test data itself through a series of field trials.

Task T304 of the DRIVE Workplan seeks to "ensure that European developments in autonomous route guidance systems and infrastructure-supported systems are able to take best advantage of each others' resources and to recommend integrated systems ..."

In order that autonomous and infrastructure-based systems can achieve an effective level of integration, one of the primary pre-requisites is that the digital road network data and exchange formats for transferring the data are compatible. This requirement also extends elsewhere to include the location system proposed for use in the RDS-TMC message codes and receiver units. This compatibility of data and exchange formats will not come about by accident, but will have to be designed in the data model which describes the digital road mapping application in fine detail.

Most of the PANDORA project objectives are designed to meet these requirements from T304. The development of the data model, development of integration systems and methodology, demonstrations and trials of both autonomous and infrastructure-based systems, development of standards and attention to legal issues are all required if mapping compatibility is to be assured.

The objectives of the PANDORA project are wholly aligned with those of T328 and T329. T328 has the objective of "developing a data format for a digital map of the roads of Europe" "to support the selection of a general purpose data exchange format." T329 is concerned with how such a digital map system would be operated, including "data collection, control, updating, editing, distribution", "setting up the legal measures to guarantee the interests of data furnishers", and the "realisation of a local database so as to validate the coding norms and test the various configurations of car navigation systems".

4. PANDORA WORK PROGRAMME

A major deliverable from the PANDORA project is a set of agreed standards for a road network database for vehicle navigation and traffic engineering purposes. The technical approach required for specifying and accepting such standards involves:

- reviewing current standards - an examination of GDF and the technical specification for Autoguide together with the UK National Transfer Format (NTF) to compare existing data content, structure and exchange format specifications for digital data with navigation database requirements.

- establishing sources for geometric and attribute data - research into, and evaluation of methods of data capture and existing sources of digital road data;

- designing a data model which can best relate data and attributes to serve the needs of vehicle guidance and navigation systems;

- integrating geometric and attribute data under agreed standards - the physical construction of a relational database in which navigation data are stored;

- supplying integrated data to vehicle navigation systems, followed by field trials - the data are to be tested in both autonomous and infrastructure-based navigation systems;

- a critical review and amendment of data supply, integration system and standards following monitoring of field trials;

- the production of proposed standards - these will embrace the existing GDF Release 1.0 but will also take other system requirements into consideration;

- determining legal requirements for protecting the data providers.

5. PROVISION OF THE DIGITAL ROAD NETWORK

The PANDORA trial area covers just under 7000 sq km, between the eastern side of Birmingham and West London (fig 1), ranging from heavily built-up urban and suburban areas through to rural countryside.

The digital road network for the urban areas of Birmingham and London (fig 2) consists of street centre-lines extracted from Ordnance Survey's large scale digital data. These centre-lines were originally digitised from 1:1250 scale maps in order to create automatically derived mapping at 1:10 000 and 1:25 000 scales. This process was used by OS for a few years, but it proved less economical than conventional methods of updating maps and a decision to abandon it was taken towards the end of 1985. Since then, centre-lines have continued to be digitised because of their value as digital data rather than merely as a means to derived map production.

Because they are taken from the largest scale of mapping which OS produces, the centre-lines are very accurate and detailed and are continuously updated as part of normal day-to-day OS survey operations. New Digital Field Update Systems (DFUS) are being installed in some OS field offices so that the digital data can be updated as soon as the fieldwork is completed. Elsewhere, digital updates are done at the OS headquarters in Southampton when sufficient work has accumulated on a particular sheet.

A small production unit was set up in mid-1988 in order to:

- extract centre-lines from the mass of OS large-scale data and check them against the road-edge pecks;

- classify and structure them into a cohesive network without mismatches at sheet edges;

- attach road names and route numbers;

- store the data in a relational database;

- output data in National Transfer Format.

Their first target, which was achieved on schedule by March 1989, was to complete London, out as far as the M25 orbital motorway, the West Midlands conurbation, and 6 other major urban areas of Britain: Greater Manchester, Liverpool, Leeds, Sheffield, Edinburgh and Glasgow. The London block alone covers 2474 sq km, requiring almost 10 000 sheets of 1:1250 digital mapping, each map unit covering only a quarter of a square kilometre.

Only a small wedge of this London data will be used in PANDORA, plus 25 sq km of Birmingham. The roads for a suburban/rural area to the West of London are being specially digitised and for the rest of the test block small scale digital data, probably taken from OS 1:50 000 scale mapping, will be used. To provide rapid route selection, a simplified topological roads network is needed, without all the detail of complex junctions and intermediate co-ordinates. This may be provided from still smaller scale data, probably that digitised from the OS 1:250 000 scale Routemaster series.

6. ATTRIBUTE INFORMATION

Creating the digital road network is only the first stage of producing a database which can be used by all manner of vehicle navigation and traffic management systems. The far more difficult part of the operation is to collect and integrate all the necessary attributes.

The GDF specification lists in great detail which attributes are required. Some, such as road names and route numbers, are already part of the OS digital roads file. Others may be derived from existing map sources. These include measures of road width, hilliness and bendiness, house number ranges and administrative area names.

A further group are collected by other organisations for a variety of purposes. Fortunately the Automobile Association already gathers and verifies a vast amount of attribute data from these organisations for its existing products and services. Whereas the provision of the digital road network will be OS's part in PANDORA, AA's main role will be the provision of attribute data.

But then there will inevitably be a further body of data for which there is no existing reliable source. In the PANDORA project we shall investigate how these items can be economically collected and managed, but we may have to conclude that some items cannot be economically gathered and maintained.

Creation of specifications such as the GDF have to be seen as an iterative process. The first edition of GDF represents a valuable statement of requirement, but its authors understand that it will need to be modified in the light of the experience gained in trying to meet it.

7. DATABASE MANAGEMENT AND UPDATING

MVA's main task in PANDORA, apart from project management, is to design a database scheme which can handle the road network and attribute data, taking it in from suppliers, storing it and then outputting it in appropriate forms to system developers. A commercial relational database will almost certainly provide the basis of this work, but it will need considerable tuning and development.

No attempt is being made at this stage to set up a complete commercial organisation to manage roads data, although it is hoped that working together in PANDORA will help to develop an understanding which will eventually make this easier. It remains to be seen whether the role of database administration should eventually be undertaken by one of the PANDORA partners or by another player altogether.

Although the initial stages of PANDORA are aimed at creating a suitable data set, maintenance has not been forgotten. Each item of data will have its own update requirement. Renewed versions of the basic road network data may not need to be supplied to users more than once or twice per year. At the other extreme, information on short-term delays in the route ahead must be provided on an up-to-the-minute basis if it is to be useful. This is where infrastructure-based systems such as Autoguide, feeding back information from users via the central computer to vehicles not yet ensnared in a hold-up, have a considerable advantage over purely autonomous systems.

8. LIAISON WITH OTHER EUROPEAN GROUPS

So far in this paper, only a British test area has been mentioned. But PANDORA is very much a European project. Close liaison is being maintained with other DRIVE projects concerned with digital road databases and with earlier projects such as PROMETHEUS and DEMETER under the Eureka initiative. DRIVE places great emphasis on liaison, or 'concertation', between projects.

The DEMETER project (Digital Electronic Mapping of the European Territory) created the GDF specification through the efforts of Philips and Bosch. This specification was adopted for its benchmark tests by the Task Force European Digital Road Map which started life as part of the enormous PROMETHEUS project (Programme for European Traffic with Highest Efficiency and Unprecedented Safety!). These tests, which cover 11 tasks in 6 countries, are designed to evaluate the costs and methods of creating data for a European digital road database. Data from the PANDORA test area could be made available for assessment alongside the other 11 tasks if required, and this has been referred to as benchmark test number 12.

A DRIVE project, also entitled Task Force European Digital Road Map, now includes the benchmark tests. The partners in this project are Daimler-Benz and Robert Bosch from Germany, Graphic Science from Belgium, Philips and TeleAtlas from the Netherlands and Renault of France. The objectives of this project as a whole run parallel to those of PANDORA and we shall need to work closely together to ensure that the results are in sympathy.

Another organisation which has been involved in this work from a very early stage and has helped to stimulate good co-operation between countries is the Comite Europeen des Responsables de la Cartographie Officielle (CERCO). The Director-General of Ordnance Survey is a member of CERCO and OS is represented on several of its working groups, including Working Group VII, concerned with the European Roads Database. Participation at the meetings of this group by representatives of Eureka, PROMETHEUS, DEMETER and more recently DRIVE has ensured that the national mapping agencies' potential for contributing to databases for vehicle navigation has been fully appreciated. Members of CERCO Working Group VII and representatives of national road administrations are playing an important role on steering groups in each country involved in the benchmark tests. Other CERCO working groups on databases and copyright have also made contributions.

9. CONCLUSION

By the time this paper is being presented, PANDORA will be just past half-way through its 18 months' duration and 5 of the work packages should be completed. It remains to be seen how long it will take for the concepts which are being prototyped in PANDORA and the other projects which have been mentioned to be realised in commercially available systems. There has been great excitement and activity for several years over the development of systems for vehicle location and navigation, but not enough consideration of where the data will come from. But now there is real work being done in a genuine spirit of international cooperation to consider how the databases which will turn these exciting systems into reality can be created.

ACKNOWLEDGEMENT
The author wishes to thank the DRIVE Central Office and the other 4 organisations participating in this project for their permission to publish this paper. The contributions of Mr F S Hoffman (Automobile Association), Dr W Zechnall (Robert Bosch GmbH), Mr H F van Leiden and Dr C H de Voogd (Philips), and Mr D G McCallum and Dr C Queree (MVA Systematica) in the preparation of this paper and within the project are gratefully acknowledged.

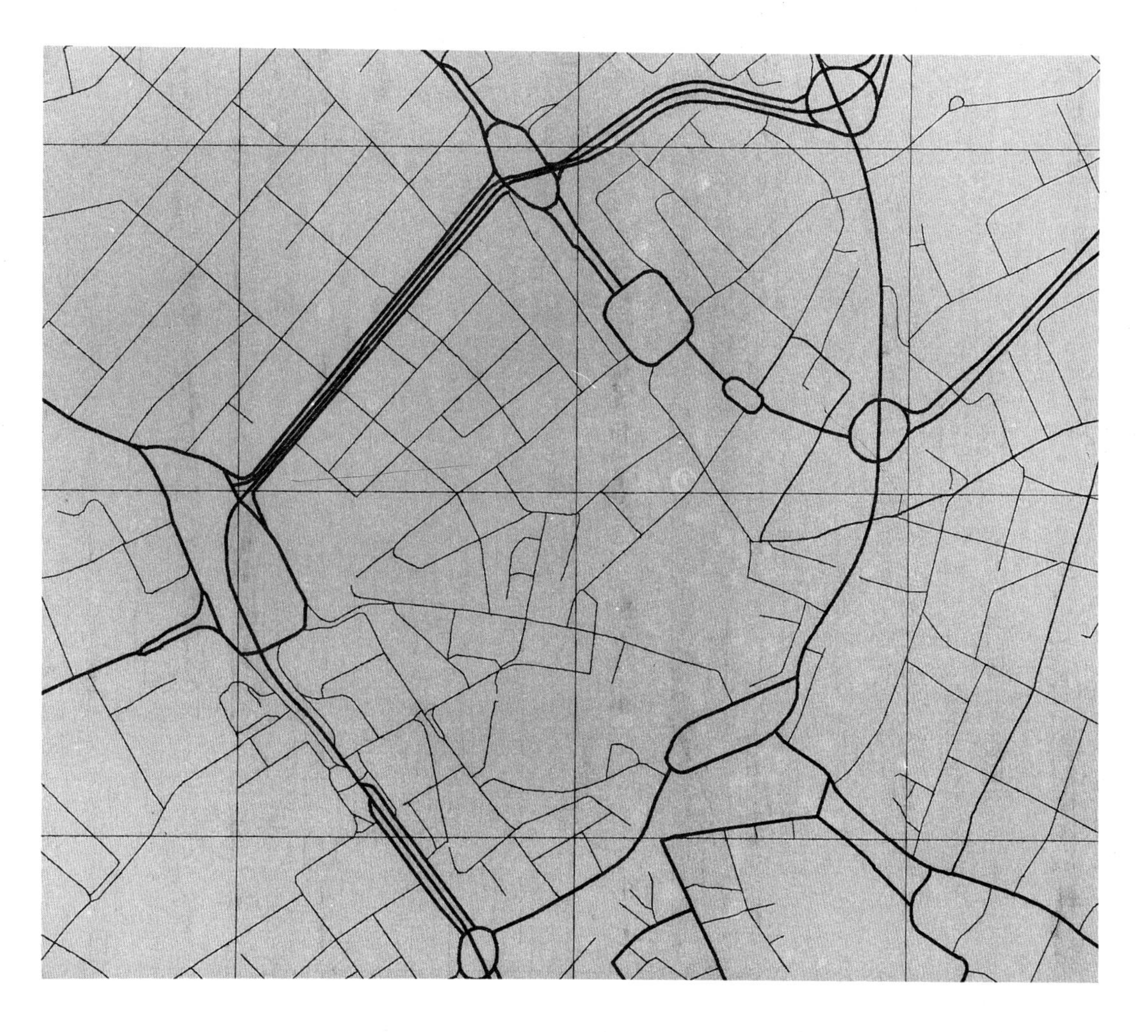

Fig 2 An extract from an electrostatic plot drawn using OS digital street network data

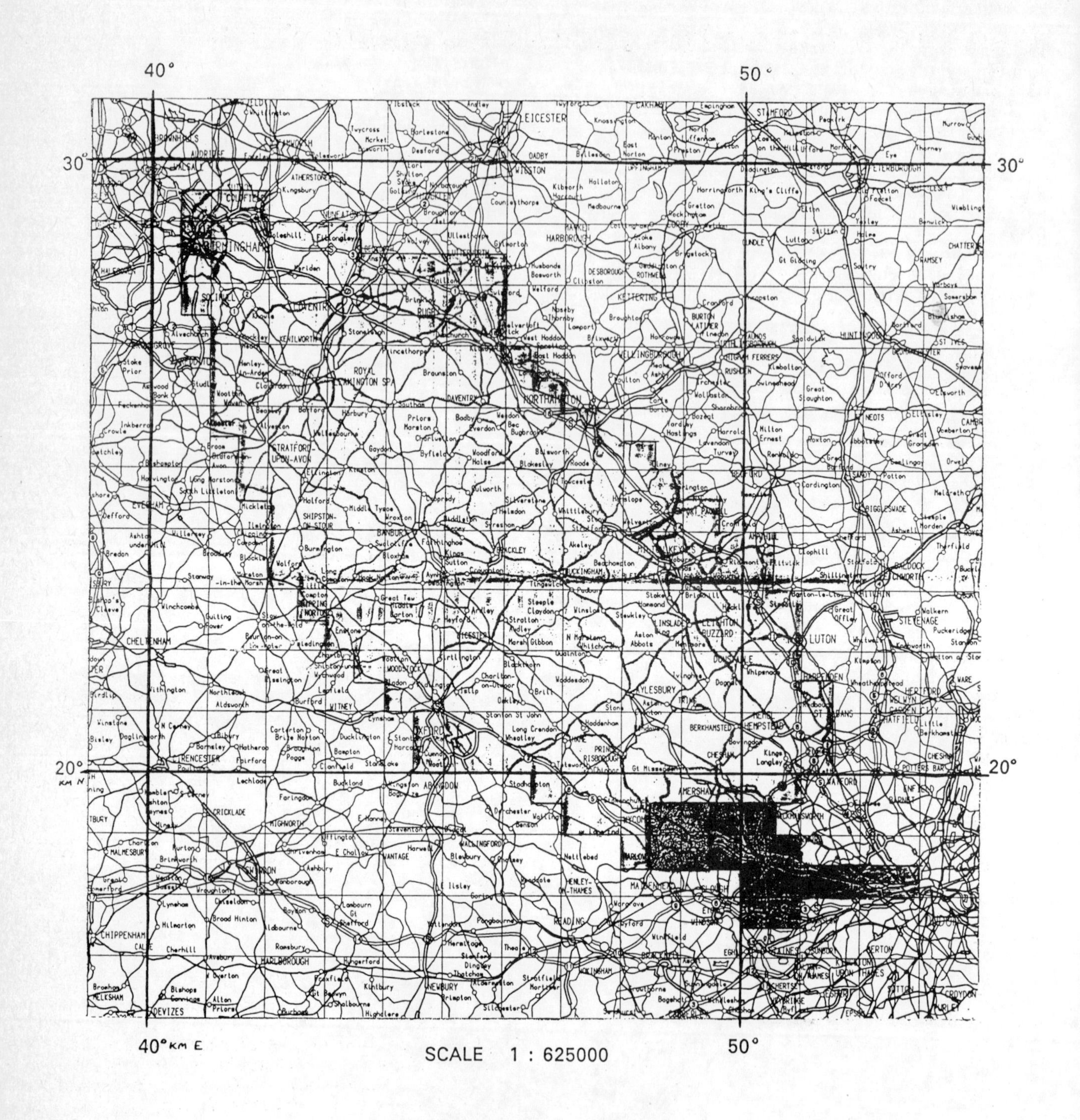

Fig 1 The PANDORA trial area

C391/053

Standards for traffic messages using RDS-ALERT

P DAVIES, PhD, CEng, **G A KLEIN**, BEng and **N W BALDING**, BSc, PhD
Castle Rock Consultants, Nottingham

SYNOPSIS The Radio Data System (RDS) consists of a silent data channel already being broadcast by most VHF-FM radio stations in the UK. One of the additional features currently being developed for RDS is the Traffic Message Channel (TMC). This will provide motorists with a constant stream of traffic information, either displayed on an in-vehicle receiver or by speech synthesis. Significant work has already been carried out on messages to be broadcast, location codes and message management issues for RDS-TMC, which is reported in this paper. Through the European DRIVE Road Transport Informatics project called RDS-ALERT (Advice and Problem Location for European Road Traffic), this work is now continuing towards standards for RDS-TMC throughout Europe.

1 INTRODUCTION

The Radio Data System (RDS) is currently being introduced throughout Europe. RDS is a silent data channel defined by European Broadcasting Union (EBU) specification, which can be thought of as the radio equivalent of teletext. Its primary purposes are to identify radio broadcasters and support the new self-tuning receivers, which automatically select the strongest signal carrying any particular programme. However, many secondary and additional features are also proposed, as outlined below.

The three kinds of RDS applications are as follows:

(a) Primary Features

These comprise programme identification codes; the programme service name; the alternative frequencies list and the traffic programme/traffic announcement flags. Several of these features are already implemented on most UK FM transmitters, and RDS car radios are now available from several manufacturers.

(b) Secondary Features

These include codes for other networks data; clock time and date; programme type and alarm code; radiotext and a transparent data channel.

(c) Additional Features

The additional features currently proposed for RDS use capacity in the RDS channel as a convenient data transport mechanism. At present, they comprise radio paging and the traffic message channel.

The traffic message channel (TMC) will provide continuous information to motorists using digitally-encoded silent messages which can trigger a speech synthesizer or display in the receiver. The messages will be language-independent, and will be stored in receiver memory to be selected by the user according to region or road number.

This paper outlines recent work carried out by Castle Rock Consultants (CRC) on behalf of the Commission of the European Communities in developing proposals for a European standard on RDS traffic message coding and management. It begins by defining RDS terminology and then goes on to look at issues relating to TMC location coding, message coding and message management. The paper concludes by summarizing recent proposals for structuring traffic message codes within the RDS multiplex.

2 BACKGROUND

The radio data system (RDS) provides for a silent data stream superimposed on normal FM audio broadcasts. The data rate is 1187.5 baud, divided into groups or sequences of 104 bits. Each sequence is made up of four blocks of 26 bits. Each block of 26 bits comprises 16 data bits and ten error correction bits. The structure of the RDS sequences is shown in Figure 1.

An early proposal for defining the allocation of RDS-TMC data fields was made to the European Conference of Ministers of Transport (ECMT) in Madrid, in 1987. This proposal was based on the use of at least two sequences for each message. Within these two sequences, only 32 bits per sequence would be used for actual traffic data, giving a total message length of 64 bits.

Another proposal was made by Blaupunkt and BAST (the German Federal Road Authority) for encoding 80% to 90% of all messages within a single sequence. ECMT supported the one-sequence concept as offering about twice the channel capacity as that of earlier proposals, with an enhanced probability of successful message reception. The only question raised within the group was whether the single sequences could offer sufficient flexibility to deal with the full range of situations in various countries.

More recently, a report was prepared by Castle Rock Consultants addressing the above question. It sets out to define an allocation of bits within RDS-TMC which retains the flexibility needed for covering a full range of situations, while retaining the potential benefits of single-sequence messages for a majority of actual events. The report suggests that flexibility can be retained within a highly efficient coding structure.

3 LOCATION CODING

One objective of recent work has been to resolve coding issues relating to location, leading to a detailed set of coding rules. Since every message will contain some kind of problem site location, the coding needs to be

* comprehensive, to ensure that all locations can be covered;
* flexible, dealing with current situations in different countries and allowing the system to adapt to future changes; and
* efficient, to enable the limited capacity of RDS-TMC to be used to best advantage.

The first question addressed by the study was the total number of locations to be coded. Initial estimates suggested that a maximum of 65 535 inter-urban locations might be needed within the Federal Republic of Germany. An efficient coding system would require only 16 bits to code these by simply numbering the locations from 1 to 65 535.

Calculations for France and Britain suggested that around 30 000 to 40 000 locations should be enough for a national system. CRC therefore recommended a standard 16-bit location code for inter-urban networks. An extended location set bit would also be reserved for defining local places within cities.

The Madrid proposal of 1987, by comparison, required 33 bits to code problem location, with separate fields for road number, road class, area of the country, etc. These 33 bits gave a theoretical total of 8.5 billion location codes, most of which could never be used.

One reason for the greater efficiency of the CRC location coding system over the Madrid proposal is due to the omission of seven-bit area code within the broadcast message. These were needed to indicate the administrative department in France, where many roads are numbered by Department; and to give a 'user-selectable area' facility for motorists.

Neither of these functions actually requires the area code to be broadcast, however. Instead, the first seven bits of the 16-bit problem site location code could indicate the area of that site's location. This would allow the broadcasting of area codes to be dropped.

A second increase in efficiency was caused by the omission of road class and number from the CRC location codes. In practice, there are anomalies in several countries which would create difficulties in uniquely identifying a road by its class and number alone. In Britain, for example, we must distinguish between the A1, M1 and A1(M). Spur connectors off motorways are not always clearly numbered. Similarly in France, we have the A6a and A6b entering Paris, and the suffix 'bis' commonly added to numbers to distinguish alternative routes. Some autoroutes may be better known by name than by number, while the Peripherique around central Paris is not numbered at all. In Germany, secondary roads are not numbered on traffic signs, while this is generally true in all countries for tertiary roads. Considerations of efficiency, as well as flexibility, therefore, supported the proposal that road number and/or name be stored in memory within the TMC decoder.

Simple site numbering has the further advantage that codes can be allocated to highways flexibly, according to their numbers of problem locations. Some location codes will represent 'all roads in the Black Forest'; 'all roads leaving Amsterdam'; 'all bridges at high altitudes'; etc. Others will represent ferry terminals, the Channel Tunnel, and Alpine rail tunnels.

In all approaches, the extent of the problem will be indicated by an offset system of 'steps' (numbers of problem sites) along the highway. CRC recommended a sign convention to indicate the direction of queue formation, in which the location is given as the problem site immediately downstream of the accident or bottleneck. This avoids the need to broadcast a changed location for the problem each time the queue gets longer.

The problem of accommodating alterations to the highway network in the location coding system is more difficult to overcome without destroying the benefit of the offset system. Sequential numbering along highways recommended in the Madrid proposal can be eliminated by storing, for each problem site, the 'address' (problem site number) of the adjacent points. Such a network could grow indefinitely, up to the capacity of 65 535 problem sites, without having to guess in advance where new construction or traffic management will occur.

A further advantage of simple site numbering is that each problem site location can be pre-assigned to a major highway segment. These can be used in site descriptions, as well as to select highway segments of interest to individual drivers.

A final problem with location coding deals with the case where a decoder from one country receives broadcasts from another. The TMC decoder has to 'know' that the broadcast codes relate to a different country. A partial solution is provided by a four-bit country code already defined in RDS which divides 16 codes between 51 territories of Europe in such a way as to avoid duplication in adjacent territories. To deal with the increasingly common situation where a broadcast in one country deals with conditions in another, however, CRC recommended that each country include selected locations in border areas of adjacent countries in its national location coding look-up table.

4 MESSAGE CODING

As in the case of problem location, a draft standard for coding traffic messages had been recommended as a basis for further work in Madrid. Requirements for a flexible and efficient basic message coding structure were further addressed in the recent work carried out by CRC. Message category, basic message text and advice messages are three areas of message coding which need to be considered. These are dealt with in turn.

The case for separate categories of message relates to operator convenience. National traffic administrations and their respective broadcasting authorities need to have messages divided into broad categories such as weather, traffic, alarm and 'other' (Figure 2) so that the operator can quickly find any particular item in the list. In order to make efficient use of the channel space available, CRC recommended preparing lists of messages which are categorized for the convenience of the system operators, but are numbered on a simple, sequential basis. The advantage of this approach is that the numbers of messages in each category need not be pre-determined or equal, as they had been in the Madrid protocol. New messages can be added any time retaining the convenience of message categorization, while gaining the benefits of increased flexibility.

Earlier proposals for basic message texts divided message elements into causes and effects. In reality, however, there are problems arising from this method. Traffic congestion, floods, or snow on the road can each appear as effects or causes. More seriously, many combinations are implausible. Defining rigid cause/effect fields therefore tends to create an inflexible and inefficient framework for message coding. This problem is overcome by using a single sequence of basic message texts using 11 bits, giving over 2000 messages. Some texts contain explicitly-linked ideas in the form of causes and effects, while others convey a single concept. Memory storage can be reduced by techniques which permit long lists of messages to be reconstructed from much shorter lists of message sub-texts.

Advice messages fall into two main categories. The first constitutes common-sense, well-meant advice. The second type contains specific instructions for diverting around a problem site. Common-sense advice messages was considered best coded in a second RDS sequence, so it will only be used occasionally. Predetermined diversion routes, as used in Germany and Switzerland, can be programmed into the TMC decoder and activated by a single bit in the first sequence. More complex diversion coding will require additional message sequence(s). Only a small proportion of all broadcast messages contain explicit diversion instructions, however, and these can be dealt with using multi-sequence messages upwardly compatible with the basic message codes.

5 MESSAGE MANAGEMENT

Detailed message content development required careful consideration of the ways in which messages will be handled by broadcasters and receivers. TMC receivers will process a constant, cyclic stream of incoming messages. Of these, some will be repeats of previous messages, and some will be more urgent than others. All messages will eventually become out of date. So the main problems for the receiver are as follows:

* to ensure the validity of each received message;
* to check if it has been received before;
* to determine the urgency; and
* to delete messages which have become out of date.

Message repetition is needed to help verify the validity of the received messages. Each message transmission may consist of several repeat transmissions, which the receiver can verify by comparing the message content. Messages need only be accepted as valid after they have been identically received several times. Including repeat transmissions of urgent messages within the overall cycle will also increase the probability of their being received quickly. Adapting a fixed cycle time for the full set of messages and repeats, with fixed offsets between different radio networks, offers the potential for intelligent receivers to switch automatically between channels without loss of data. An appropriate cycle time has yet to be determined, but may be several minutes.

It is likely that the broadcaster will want to transmit urgent messages as soon as they come to hand. If the fixed cycle is in progress this could be accomplished by reducing the number of repeats of 'old' messages in order to accommodate the new. The new message would then take its place at the head of the message list within the next cycle.

Each message received by the TMC decoder has to be checked with all current messages to see whether it has previously been received or is new information. In addition to simple repeat messages, there will be message updates which supersede the contents of an earlier message. Normally these will be messages relating to the same location as an earlier broadcast, but with a different severity of traffic disruption. Occasionally the location may change, indicating a problem spreading or a mobile hazard such as transport of an exceptional load.

The CRC report recommended verification of message validity by full comparison of the message code. Messages will be accepted as valid only after they have been received at least twice, identically. New messages, repeats and updates can then be determined by comparison of location and message type. This results in a flexible system and avoids the need for message numbering.

Messages must be cancelled when they become out-of-date so that motorists are not misinformed. This also helps to ensure that TMC decoders do not fill up with irrelevant or superseded messages. There are two ways in which messages can be cancelled; automatically by the receiver, or by broadcasting cancellation messages. Both options are allowed for in the CRC protocol.

Under current proposals, messages will be valid for a specified period after receipt. This valid period could be fixed; predefined, relating the message type; explicitly broadcast, as a stop-time; or related to the cycle time of the messages. The recommended method of assigning a valid period is to make all messages valid for one of two alternative fixed periods, known as short or long persistance periods. Long persistance messages would be valid for several cycle lengths. The short persistance period would be only slightly more than the longest anticipated cycle time.

Under this system, traffic authorities would initially classify incidents into 'transient' and 'long-lasting' categories. Transient incidents would be broadcast as short-persistance messages. They would quickly clear from TMC receivers after broadcasting stopped. Longer-lasting incidents would initially be broadcast as long-persistance messages, retained by the receiver during significant periods of poor reception.

This approach to automatic message cancellation provides reasonable flexibility to broadcasters, without requiring unrealistically accurate forecasts of incident durations from the start. It will also facilitate reasonably rapid, automatic removal of old messages.

However efficient the automatic message cancellation method, it will still be necessary to manually override the system on some occasions. Explicit cancellation messages would generate their own text correction, drawing the driver's attention to the amendment. They can be used to automatically delete an earlier message of the same type for the relevant location. Implicit cancellation messages would quietly delete outdated messages. They could be used to cover exceptional situations where routine message management procedures for automatic cancelling or superseding of messages might be unsuitable.

It may be advantageous for the receiver to treat messages differently according to their relative importance and urgency. Messages such as 'driver on the wrong side of the motorway' might be so urgent that broadcasters should put the message out immediately and repeat it very frequently. RDS-TMC receivers should also inform drivers of such messages immediately, interrupting the radio or cassette player. On the other hand, messages such as 'rain forecast' might only be output on request.

Conceptually, there are three levels of urgency at a receiver-functional level:

1. Extremely urgent, which will interrupt the radio or cassette on all equipped vehicles in the broadcasting region;

2. Urgent, which will interrupt the radio or tape on vehicles which have selected the relevant route or area; and

3. Normal, which will be made available to drivers on request.

Each message can have an associated, default level of urgency, which the receiver will assume unless told otherwise. Provision is also made for overriding the default urgency in exceptional cases, using a second sequence.

6 MULTI-SEQUENCE MESSAGES

Under the message coding proposal developed by CRC, the great majority of traffic information messages could be coded in a single sequence. A few messages would require additional information, however, which could be contained within a second or subsequent sequence.

The precise content and extent of the additional information will vary considerably between different multi-sequence messages. In many cases, only one kind of additional information will be necessary. In other cases, multiple codes may be required - for example, to code complex diversion routes.

This substantial variability in multi-sequence message content can be accommodated using a 'free format' approach for programming the second and subsequent sequences. This will allow the additional sequences to be used for any combination of kinds of information. Flexibility can be maintained by specifying a coded 'label' before each additional data field.

An early proposal suggested using five bits of the two-sequence message for information on broadcast incident duration. This duration information would be provided to assist the driver, and is quite separate from the persistence period in the automatic cancellation system. Two, optional, second-sequence codes are now proposed for this driver information function, one for start time and one for stop time of the problem, each with an eight-bit data field. It is likely that only a small minority of RDS-TMC messages will contain details of expected problem start and finish times, however, because in practice these are very difficult to predict.

Another recommended use of second sequence data fields is to carry optional, detailed message quantifiers. Advisory speed, weight and width limits are examples of TMC message quantifiers. A few common values can be included in the basic message codes, but others will require the transmission of a second sequence. In the case of weight limits, for example, to cover all axle and gross limits of up to (say) 60 tonnes in 0.5 tonne increments would require 120 codes (seven bits). A similar approach is recommended for other quantitative data.

CRC has proposed that a specific label be allocated to advisory speed limits, which might accompany a wide range of basic message types. A further label can be used to indicate 'other quantitative data' used to supplement basic messages according to type. The receiver would interpret these as weights or other values according to the message context.

In many countries where diversions cannot currently be pre-programmed into the receiver, a second or subsequent sequence can be used to list points on the highway network along the diversion. These points can be specified in terms of location codes already programmed into the TMC decoder. Any reasonable number of locations can be listed to specify a diversion. These can be separated by labels interpreted as 'Diversion recommended by ....' followed by the place name and road number. Subsequent labels would be interpreted as '.... and then by ....', again followed by another place name and road number.

Each point along the diversion will require a four-bit label and 16-bit location code. A diversion via three points would therefore require 60 bits, contained in two additional sequences. Fortunately, diversions of this complexity are expected to be rare. The system is not limited to this number, however, and may in reality be constrained more by drivers' ability to understand directions than by limitations of the proposed coding technique.

Receiver control codes can also be specified using a second sequence. At present, two codes are required for non-default urgencies, and a further one or two codes for unusual device destinations. The remainder can be reserved for future system expansions, particularly involving on-board navigation systems or other intelligent equipment on the vehicle.

7 OVERALL MESSAGE STRUCTURE

This final section brings together all of the earlier discussion. It shows how the CRC recommendations can be 'packaged' within the context of the existing RDS specification. A new RDS 'type 8' group would be defined for the TMC, with the information content illustrated in Figure 3.

Block 1 of type 8 groups would contain the programmed identification code of the broadcaster, plus the checkword. Block 2 would contain the group type code indicating that this is a type 8 (TMC) group; the traffic programme code, indicating whether and when spoken traffic announcements are carried on this station; the programme type code; 5 'spare' bits; plus the checkword and offset. Up to this point the format of blocks 1 and 2 is already defined by EBU. The five spare bits of block 2 and the whole of blocks 3 and 4 would be assigned for the TMC.

Most TMC messages will comprise a single sequence. The 5 spare bits in block 2 can be used for message management and system extensions as follows:

(a) An extension bit, to indicate when the sequence is a one-sequence message. Otherwise, the sequence is part of a multi-sequence message.

(b) An undefined bit, which could be used for a future system expansion. This bit is needed within the first sequence of multi-sequence messages and therefore should not be given over to any incompatible application.

(c) A diversion bit, to cause the receiver to give drivers pre-programmed diversion instructions.

(d) The persistence bit, to indicate when a message is to be retained by the receiver for a long period. Otherwise, the message will be kept for only a short period after cessation of broadcasting before being automatically cleared from the receiver's memory.

(e) The extended location set, which can double the number of locations from 65 535 to 131 072. In most countries, this will be used to distinguish detailed urban locations from national inter-urban route locations.

The whole of block 3 (except the checkword) would be given over to the 16-bit location code. When read in conjunction with the final bit of block 2, sufficient locations should be available for all national and local location-coding needs. The whole of block 4 (except the checkword) would be given over to the five-bit location offset and 11-bit basic message.

To preserve compatibility between one-sequence and multi-sequence messages, CRC has recommended that exactly the same bit allocation be used in the first sequence of multi-sequence messages as that for one-sequence messages. One potential difficulty arises, however; how to tell that a first sequence is indeed the first of a new message, should it happen to follow a previous multi-sequence message, or a period of poor reception.

This problem was addressed by Philips/Blaupunkt using the 'border bit' concept. CRC has recommended a similar but slightly different concept of a first-sequence bit. This would be set only when a sequence is the first of a multi-sequence message. The first-sequence bit would take the place of the undefined bit in one-sequence messages.

All subsequent sequences of multi-sequence messages would be identified with a synchro-word in block 2 and would comprise wholly free-format coding space in blocks 3 and 4. The synchro-word for the subsequent sequences of multi-sequence messages would comprise an extension bit, a first sequence bit and three sequence identifier bits. As before, the extension bit will show this is part of a multi-sequence message. The first sequence bit will show this is not the first sequence, so that receivers coming in part-way through a message will not be confused.

The sequence identifier counts down the subsequent sequences. If, for example, a message needs three additional sequences, the SI for the first additional sequence will be two, for the second will be one, and for the last will be zero. Blocks 3 and 4 will be used for free-format messages as outlined earlier in this paper.

8 CONCLUSIONS

The recommendation of Castle Rock Consultants, after detailed consultations with ECMT, was to replace the existing message category, cause, effect and advice by a single 11-bit basic message code. This permits up to 2 048 basic messages to be broadcast, which should cover all known requirements and allow a reasonable safety margin.

The CRC proposal significantly increases the efficiency of message coding, shortening the basic message content from 18 to 11 bits. In conjunction with the revised location codes which saved 17 of the 33 bits previously needed, this will allow the great majority of messages to be broadcast using a single sequence. The few multi-sequence messages containing additional, detailed information would be wholly compatible with the recommended one-sequence structure, through the use of virtually identical fields in the first sequence.

The work on setting standards for the traffic message channel is now entering its final phase, through the RDS-ALERT project of the DRIVE Road Transport Informatics programme. Through further consultations with EBU and ECMT, backed up by extensive practical fieldwork in three countries, it is proposed to finalize an ALERT protocol covering advice and problem location for European road traffic. The ALERT project seeks to build a real consensus among broadcasters, industry and government. It offers exciting prospect of a practical application of information technology to highway transportation in the 1990s.

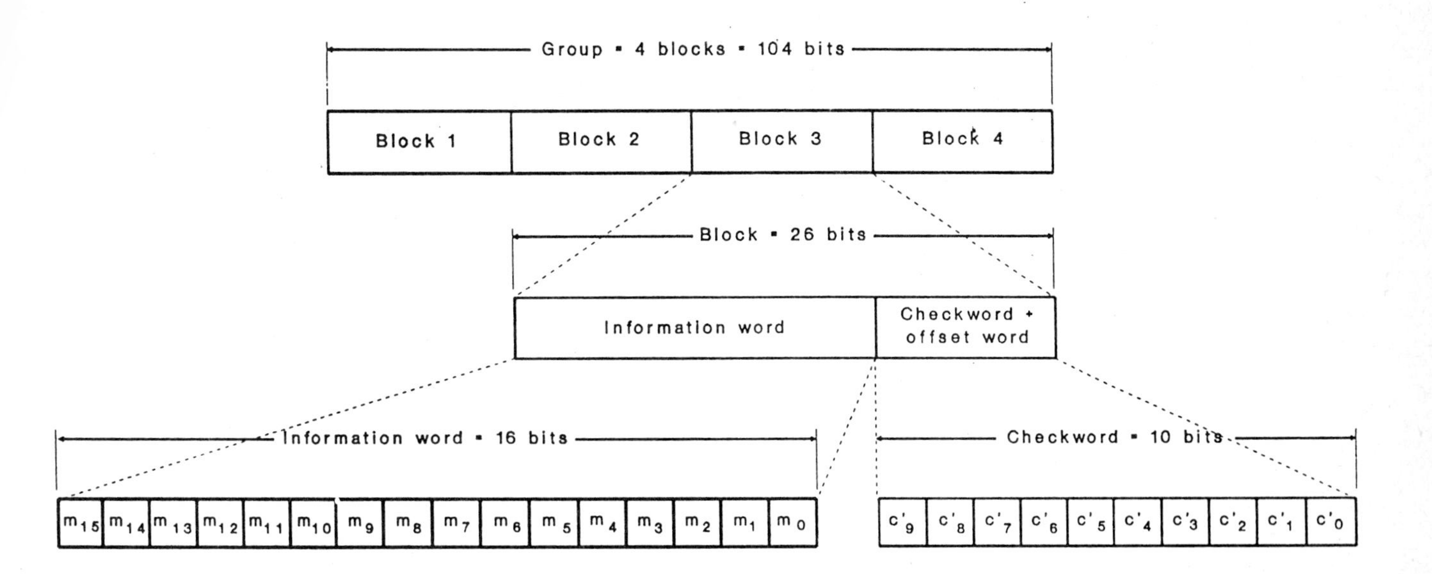

Fig 1 Structure of the RDS sequences

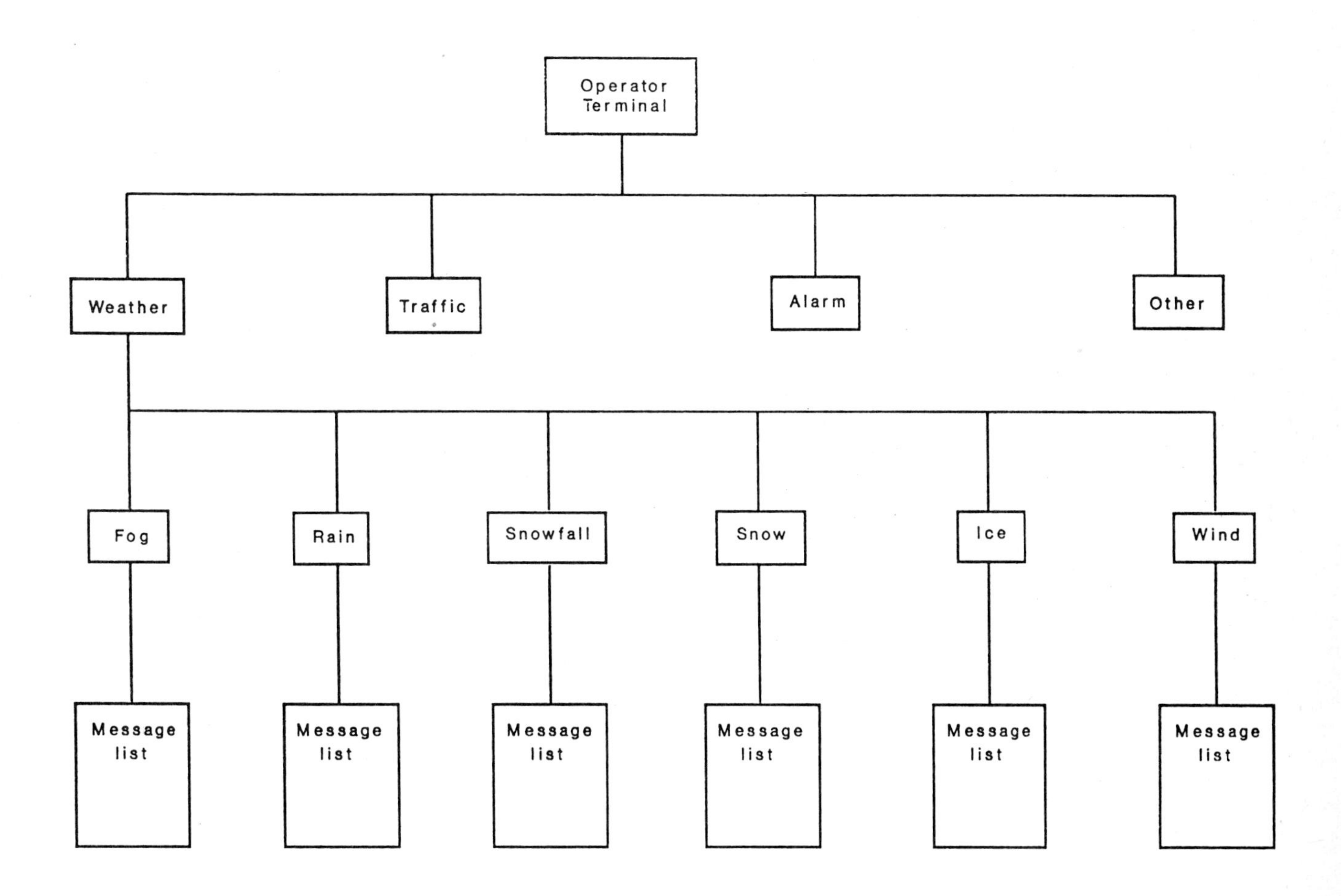

Fig 2 Typical message categories for operator terminal

First Sequence

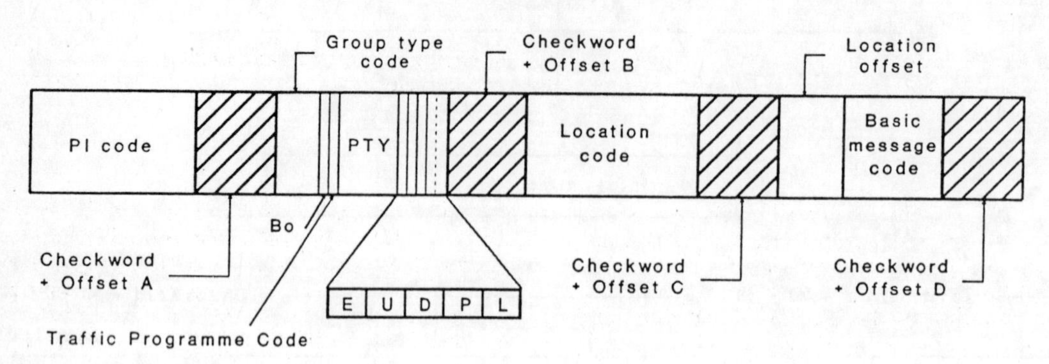

Subsequent Sequence

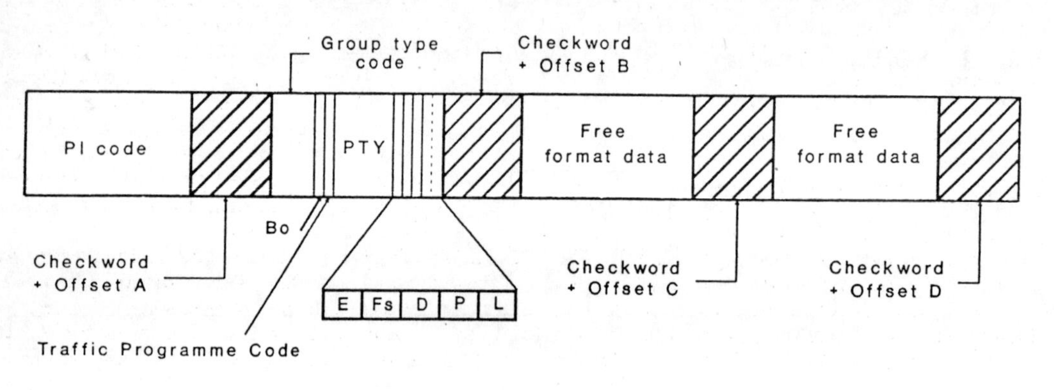

Notes :
E = extension bit
Fs = first sequence bit
D = diversion bit
P = persistance bit
L = extended location set
U = * undefined bit
* or first sequence bit
PI = programme Identification code
PTY = programme type code
TP = traffic programme code

Fig 3 Proposed RDS Type 8 group content

C391/060

The LISB field trial, forerunner of AUTOGUIDE

R von TOMKEWITSCH
Siemens AG, Munich, West Germany

SYNOPSIS

700 vehicles measure journey and congestion times in accordance with the floating car method.
A guidance computer processes the measured data and guides the vehicles to their destinations over the fastest routes possible.

A large-scale field trial is currently taking place in Berlin with the aim of testing the technical operability, acceptance and traffic engineering benefit of a traffic guidance and information system known in Great Britain under the name of AUTOGUIDE. Following completion of a development and realization phase lasting approximately three years, practical trials have begun in Berlin. This large-scale field trial is being sponsored by the Federal Ministry of Research and Technology and by Berlin's Senate. It is being conducted by the Bosch and Siemens companies and involves decisive cooperation by the Studiengesellschaft Nahverkehr (SNV) and Berlin's Technical University. The trial is being supported by the motor vehicle manufacturers BMW, Daimler-Benz, Opel and Volkswagen and also by Mannesmann Kienzle.

What task does the system have to fulfill?

Traffic information systems of the kind shown in Figure 1, i.e. where the traffic situation is registered by inductive loops, already exist in all European countries. The data preprocessed by an electronic system along the route is forwarded to a traffic control computer which, depending on the traffic situation, displays warnings on variable message signs, declares speed limits or recommends alternative routes. No matter how helpful they may be at restricted traffic focal points, such systems have two fundamental disadvantages: on the one hand, above all the equipment is expensive – we only have to think of the large traffic sign gantries above motorways. On the other hand, the amount of information which can be conveyed by variable message signs to motorists travelling past at high speeds is very limited. This is why there is no point in envisaging full area coverage either nationally or even all over Europe. Above all in traffic concentration centres, i.e. in our cities, traffic guidance by this method is unimaginable. Especially, where it would be particularly beneficial, there is a lack of a communication link element between road users and traffic strategists. Thus, even the most efficient traffic guidance computers we are capable of constructing today are of no use in attempting to justly distribute a commodity in short supply, which is what space for traffic is today.

LISB (Berlin guidance and information system) is aimed at creating this communicative link. The functions of all detection and display devices shown in Figure 1 are taken on by an in-vehicle unit which exchanges information with a central traffic guidance

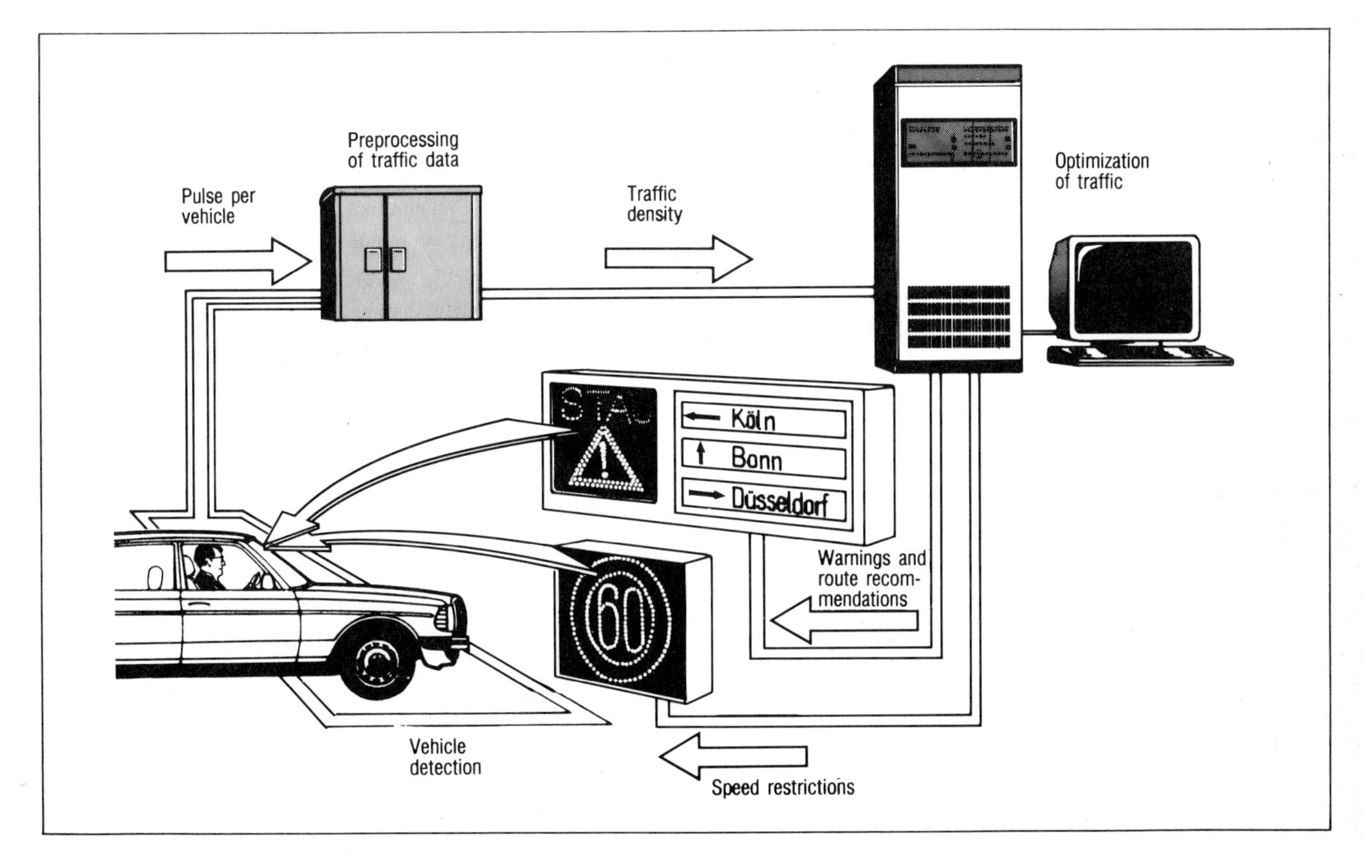

Fig 1 Conventional dynamic collective traffic information system

computer through infrared beacons installed at selected traffic signals (Figure 2). The nature and quantity of the exchanged data differ fundamentally from those of conventional traffic guidance systems; that is to say, between 2 intersections with beacons, the in-vehicle units measure the individual journey times per road link and the queueing times ahead of all traffic signals. In traffic engineering, this process is known as the "floating car" method. In LISB, all appropriately equipped vehicles (hundreds are involved) perform this function. Conversely, the in-vehicle units receive a detailed description of the main road network surrounding the respective beacon – all described "route trees" combined together are capable of comprising a route length of 100 km and more. They do, however, only display to their driver the information relevant to the individual journey by means of easily understandable symbols wherever changes in direction, for instance, are recommended.

How does LISB function in principle?

The operational principle of LISB (or of ALI-SCOUT, as the system was named in the Federal Republic of Germany) has already been described several times. This is why it will only be summarized briefly here:

The core of the in-vehicle equipment (Figure 3) is the navigation computer N. At the start of a journey, it receives a message from the destination memory M concerning the destination itself. The position finding system P continuously determines the vehicle's current position by means of dead reckoning with the aid of wheel pulses and a magnetic field sensor F.

When a beacon is passed, an infrared signal receiver receives the description of the surrounding main road network in the form of so-called route trees. Guidance recommendations depending on the actual traffic situation are given for destination areas by allocating the most favorable routes.

During the journey, the journey time meter T measures the journey times required per section of the route and the queueing times ahead of all traffic signals which have been passed since the last beacon and reports this information at the next beacon to the traffic guidance computer. "Direct vision contact" lasting one second suffices for the exchange of data between the vehicle and beacon. A display device conveys the guidance recommendations to the drivers both visually and audibly.

Characteristic system quantities of LISB

The most important characteristic system quantities pertaining to the LISB field trial are summarized in Table 1: the trial area lies within a square of approximately 25 km x 25 km. Up to 700 trial participants are guided within a network of arterial roads with a total length of approximately 1500 km. This road network branches at approximately 4500 intersections. Since various different traffic flows (straight ahead traffic, left- or right-hand turners) frequently require different times for passing intersections, those intersections in the network which are subjected to extreme loads have been split up.

Accordingly, the LISB network comprises approximately 7500 so-called (traffic) connections.

The LISB guidance centre is housed next to the conventional traffic control centre, from where the police is capable of monitoring and controlling all of the city's 1300 traffic signal installations. Infrared beacons have been added to approximately 250 of these 1300 traffic signal installations. This suffices for a guidance system which covers the entire area of West Berlin. All 250 intersections with beacons are connected to the traffic guidance computer by means of one wire pair each in the cable network for the centralized traffic light control system. In total, the radial transmission network through which data exchange takes place at a rate of 9.6 kbits/s has a length of around 1200 km. If we take a closer look at these figures, we realize that LISB is a very large remote control system in which a central station communicates with hundreds of computer terminals (in-vehicle units). Experienced computer specialists will be able to comprehend the degree of

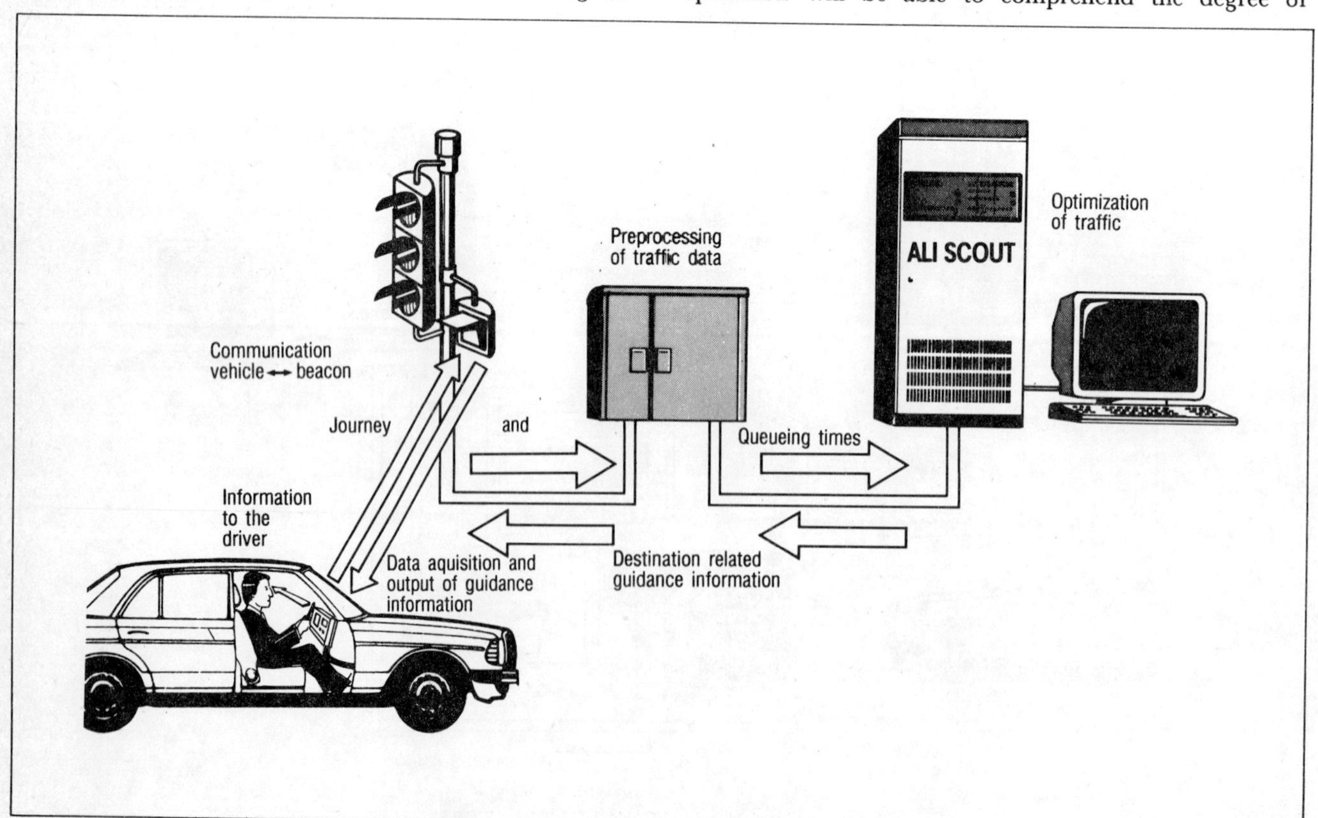

Fig 2 Dynamic individual traffic information with ALI SCOUT

complexity involved if they take a look at the additional information given in Table 1 about the traffic guidance computer software and the software in the in-vehicle units.

Standard journey time profiles as a result of the first learning phase

As already stated, the essential feature of LISB is the acquisition of data in the vehicle. Reliable route recommendations are possible only if enough information is available in the traffic guidance centre to be able to issue realistic forecasts concerning journey times in all links of the road network for a period of up to two hours. To put it in a nutshell, LISB is capable of learning; that is to say, during an initial phase the vehicles collect data only, thus enabling the traffic guidance computer to establish a historical knowledge. Output of route recommendations is still inhibited during this phase. Journey time profiles for each of the 4500 links or each of the 7500 connections in the main road network are the result of this initial learning phase. Figures 4 and 5 show how these standard journey time profiles are produced. The initial basis consists of theoretically calculated profiles resulting from the length of the link concerned and the assumed travelling speed during the four traffic phases consisting of rush hour traffic in the mornings, midday traffic, rush hour traffic in the evenings and nighttime traffic. The journey times assumed in the example in Figure 4 were obviously too high because first journey times actually measured by test vehicles are all lower. A certain amount of leveling can be witnessed over a different route, as shown in the example in Figure 5, where journey times during rush hour traffic in the morning and during the midday period scarcely seem to differ. As far as rush hour traffic in the evening is concerned, journey times sway around the theoretical starting value. Each of the spikes depicted shows the average value of all measured journey times within a time interval of 5 minutes. The result

table 1

LISB

System Characteristics

Road Network Configuration	
area of Berlin (west):	appr. 25 km x 25 km
length of digitized streets:	appr. 1500 km
number of nodes:	appr. 4500
number of connections:	appr. 7500
System Configuration	
number of traffic lights:	appr. 1300
number of beacon junctions:	appr. 250
length of transmission lines:	appr. 1200 km
Central Computer Software	
program:	appr. 140.000 lines of code
network data:	appr. 150 Mbyte
number of route recommandations:	appr. 1.000.000 / 5 min
vehicle messages:	now 1 Mbyte / workday
IVU Software	
program:	appr. 64 Kbyte
received data:	appr. 8 Kbyte / beacon
transmitted data:	appr. 100 byte / beacon

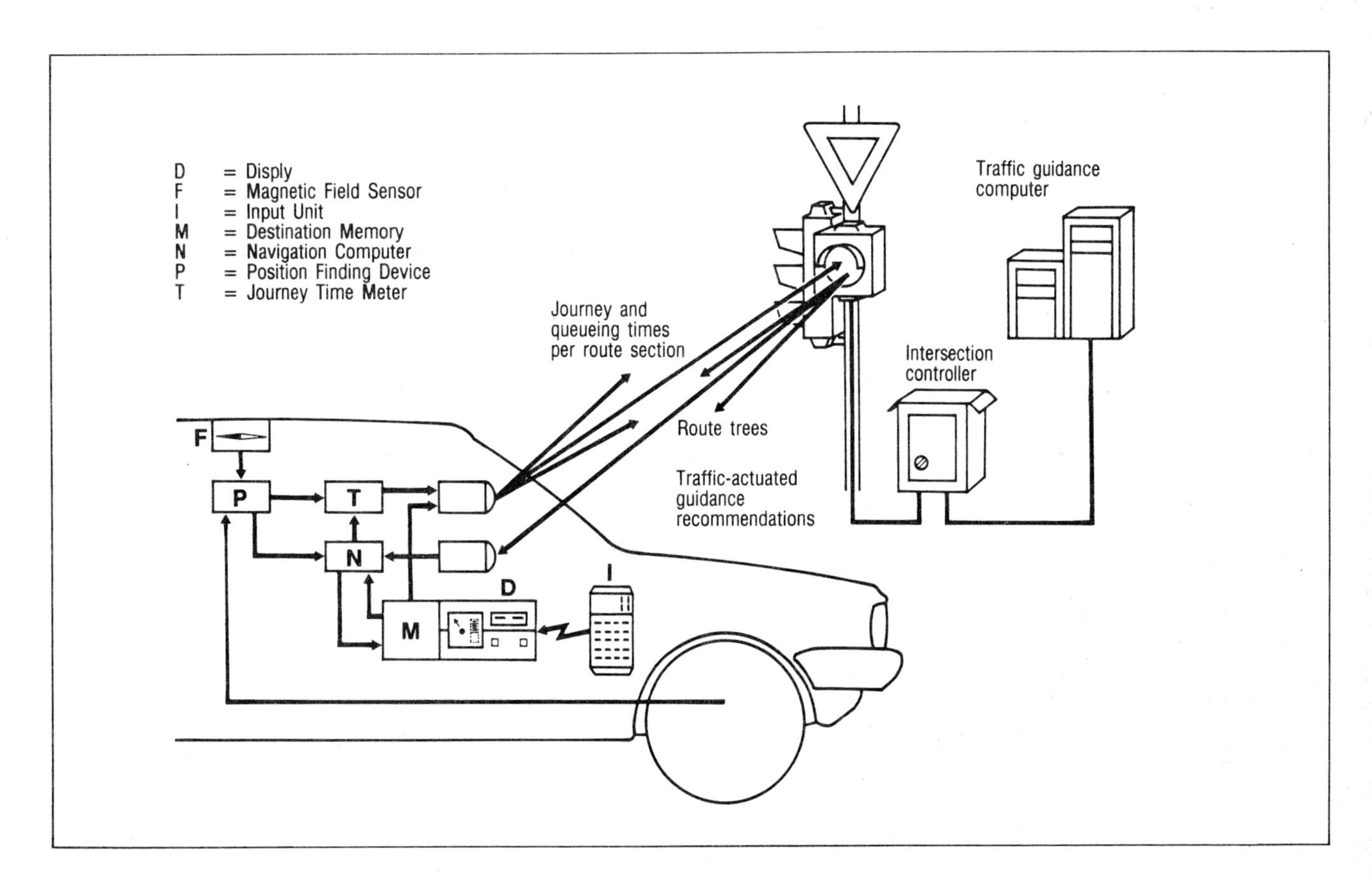

Fig 3 The components of LISB

will be the standard journey time profile in the form of a smoothed curve between the spikes. After a few weeks, the original journey time profile will have changed to a continuous line.

The phase of static guidance

The phase of static guidance begins as soon as the traffic guidance computer has adequately assured standard journey time profiles. Statistically, this means that the registered standard journey time profiles serve as the basis for calculation of the route recommendations, but are not yet updated by newly arriving journey time measurements. The traffic guidance computer therefore takes into account the rising and falling of the traffic during the course of day, naturally also including road works, but does not yet react to daily deviations from the standard journey time profiles determined in the long term.

For the first time ever, the vehicles are now receiving route recommendations to enable the drivers to get accustomed to the system's principle of operation. The drivers participating in the trial are also requested to participate in testing the system. They are provided with the opportunity to submit suggestions for correction and they are given corresponding forms to do this.

The reason for this is that the few test drivers in the development team have no possibility of testing the approximate total of 100 000 guidance recommendations enabling them to change lanes in good time, to take the correct turnings and to choose suitable lanes. It is especially the correct output position ahead of the intersection concerned and the clarity of this information which is particularly important. Information provided too late or in a confusing manner is capable of paying a negative contribution towards traffic safety, while route recommendations issued in good time and which are clearly understandable will increase traffic safety.

It goes without saying that the data for output of the route information was defined in accordance with specific rules. Since, however, this large-scale field trial is the first of its kind, further experience has to be gathered. It would also be unrealistic to assume that the data quantity of 150 MBytes stated in Table 1 which is required to describe Berlin's main road network with all attributes and journey time forecasts was collected flawlessly. Table 2 provides information about what kind of data is involved.

table 2

Geographic Data
- beacon data
- destination zones
- links
- guidance vectors

Graphnet Data
- connecting links
- destination zones/departure points
- subdiveded junctions
- time limited connections
- journey times for all links
  * daily journey time profile
  * standard journey time profile
  * predicted journey time profile

System Data
- beakon datas
- junction codes
- link codes
- destination zone codes
- line listings

Traffic Control Data
- position of traffic ligths
- queueing times per traffic light
  * daily queueing time profile
  * standard queueing time profile

Display Data
- bar display length
- angle indication
- orientation bar
- lane information
- text information

Mapping Data
- confidence angle
- capture zone
- special mapping infos

The phase of dynamic guidance

It is not intended for the phase of dynamic guidance to begin until static guidance has established itself. During this phase, approximately one million routes are recomputed during each interval of five minutes. These are based on journey time forecast profiles resulting from continuously updated standard journey time profiles: i.e. as long as no measurements are yet available during the early hours of the morning, the fastest routes are determined on the basis of the standard journey time profiles. If the first journey times of the day to arrive should deviate from the standard journey time profiles, for instance because of different weather or visibility conditions, forecast profiles are computed and extrapolated in a floating manner for the next 120 minutes. This phase will commence in Summer 1989.

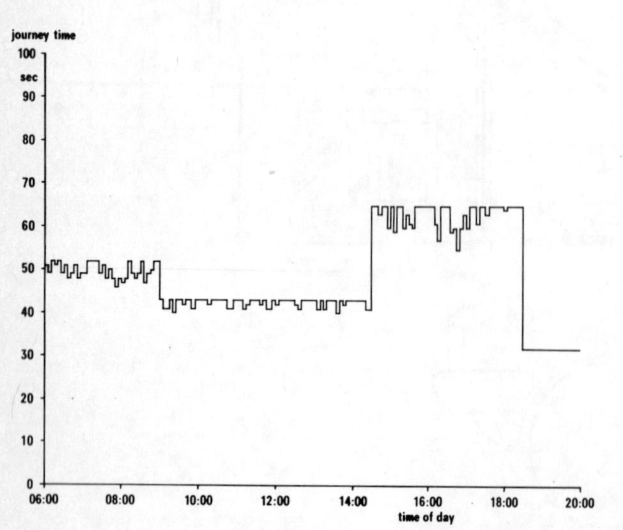

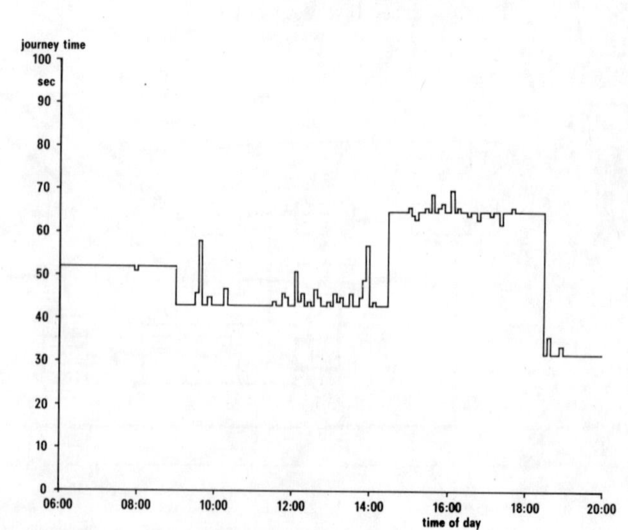

Fig 4 and Fig 5 Floating tracking of journey times for two road links of Berlin's road network. Based on theoretically assumed journey time patterns, the curves show the variation trends after a measurement period of two days. With the chosen smoothing algorithm, averaging involves 25% of the most up-to-date measurements in each case

Queueing time measurements

A traffic guidance system of the type described here should not only provide advantages for drivers whose vehicles are equipped with an LISB in-vehicle unit. Although these drivers will receive good route recommendations, they will not be able to drive faster than any of the other road users if all roads are congested. LISB should help individuals by helping everybody!

Traffic congestions in urban areas mostly originate at intersections. This is found to be particularly unpleasant when traffic signal plans are not adapted optimally to the traffic volume. Up to now, it has not been possible to measure queueing times, that is to say, times needed by vehicles to approach an intersection by "stop and go" before they can finally cross it. In any case, this has not been possible up to now using permanently installed units for online optimization of signal plans. Within the scope of LISB, for the first time ever these measurements will be carried out systematically by the in-vehicle units and will be transmitted to the traffic guidance computer by way of the beacons. Although online coupling of the LISB guidance computer to the signal controllers is initially not intended, depictions of queueing time patterns for intersections subjected to high loads are intended to discover periods during which there are traffic imbalances and hence congestions in the approaches.

Queueing times are the result of the addition of all intervals during which vehicles move slower than specified by a defined limit speed. This speed threshold is currently defined as 20 km per hour, but can be varied during the trial by means of parameter variations.

To distinguish between queueing times caused by the traffic and queueing times caused by other circumstances such as when a driver stops his car and leaves the engine running to buy cigarettes or newspapers, the in-vehicle units additionally transmit a weighting factor. This weighting factor describes the number of speed changes characteristic of stop and go traffic. High weighting factors are an indication of the fact that long measured queueing times are very probably the result of the traffic situation. The Senate's traffic experts are hoping to achieve crucial traffic flow improvements from such a green time optimization within the entire urban area of Berlin. This will benefit all drivers and consequently justifies the LISB infrastructure's expense.

Completion of the field trial

The field trial is to continue until September 1990 and will be completed with a final report which will describe its technical feasibility, acceptance and traffic benefit for the individual driver as well as for the general public and which will propose strategies for introduction of such a system in the Federal Republic of Germany.

C391/065

A review of developments in vehicle navigation systems worldwide

I CATLING, BSc and **D SACKER**, BSc
Ian Catling Consultancy, Carshalton Beeches, Surrey

SYNOPSIS The background to vehicle navigation systems is presented and three categories of navigation aid are identified: radio navigation systems, autonomous navigation systems and infrastructure-based systems. These systems are described together with examples. The paper concludes with a discussion of the likely future implementation of dynamic route guidance systems.

1 INTRODUCTION

Road vehicle navigation systems are designed to help the driver get to his destination efficiently and safely. Their functions can include keeping track of the vehicle's location, giving the driver information about the road network and about traffic conditions, finding the best route to the driver's destination and giving recommendations to the driver to enable him to follow it.

Approaches to vehicle navigation are not new. In the early 1900s devices such as the "Jones Live Map" shown in figure 1 advertised that "you leave all the guide books, maps and folders behind." But in recent years increasingly sophisticated techniques have meant that genuinely useful systems have been developed which are now close to becoming widely available.

Navigation systems encompass aids such as electronic map displays through to genuine route guidance systems. Route guidance can be either static - i.e. giving routeing recommendations based on typical traffic conditions - or dynamic. Dynamic systems offer the greatest potential benefits because they react to actual traffic conditions and can make recommendations to re-route equipped vehicles away from the worst congestion spots.

A variety of technologies need to be integrated for succesful guidance. These include dead reckoning, digital cartography, map-matching, radiolocation, roadside beacons, traffic monitoring, mobile data communications such as cellular radio and RDS (Radio Data System) and advanced "man-machine interface" techniques within the vehicle.

Considerable research in the UK (1), the US and Japan has indicated that besides being a convenience to individual drivers, there are significant benefits for the community to be gained by efficient route finding.

This paper reviews the various technologies that exist now or are currently under development, and gives some consideration to the likely developments in the future.

Approaches to vehicle navigation systems can be categorised into three main types, although most systems contain elements of more than one type. The three are radio navigation systems, autonomous navigation systems and infrastructure-based systems. These systems together with examples are described in the following sections.

2 RADIO NAVIGATION SYSTEMS

Radio navigation, which is commonly used in aerospace and marine navigation, has in recent years been considered for use by land based vehicles. It has the advantage of being able to provide the absolute position of the vehicle, although this can often be difficult to relate to standard map coordinates. The disadvantages are that signals can be blocked by high structures or mountainous terrain and that radio receivers are subject to interference. The present high cost of satellite receivers are likely to be reduced with higher volume production.

Radio signals have until now found more use in Automatic Vehicle Location Monitoring (AVLM) systems than in navigational aids, but the trend towards navigation aids could increase in the next decade.

These systems can be divided into satellite systems such as the NAVSTAR Global Positioning System (GPS), the Transit system and the Geostar (Locstar) system, and land-based systems such as Loran-C, OMEGA/VLF and Datatrak.

2.1 Satellite Systems

The Transit system is operated by the US Navy and is scheduled to be phased out in 1996. It has a number of satellites in polar orbit which provide worldwide although intermittent coverage. Each satellite transmits information which can be analysed using the Doppler effect to provide precise location. However due to the intermittent coverage, any system using Transit must be backed up by dead reckoning. Ford demonstrated the feasibilty of using Transit in its Concept 100 car in 1983.

The NAVSTAR GPS system (2) is scheduled to be completed in 1996 when 24 satellites will be in orbit. Then any point on earth will be in range of at least 4 satellites. A GPS receiver analyses the data from the 4 satellites and will be able to pinpoint location to within 100 metres for general (non-government) users. Combined with broadcasts from a local receiver, location to within a few metres will be possible. Both General Motors and Chrysler have used GPS in demonstration vehicles.

Any vehicle system using GPS would need auxiliary dead reckoning for those times when radio signals were obscured by shadowing. A car system could use GPS either as the primary system backed up by dead-reckoning or alternatively could use GPS as a backup to a primary dead-reckoning system. A Nissan system uses both systems and switches to whichever system appears to be currently the most accurate.

The Geostar (or Locstar) system , when completed in 1989, will provide AVLM and two way digital communications from the same equipment. Triangulation will be achieved by using 2 Geostar satellites in conjunction with a groundstation at a control centre. The control centre's computer then translates the absolute location into map coordinates for use by the fleet operator.

2.2 Land-based systems

Loran-C (3) is a system of land-based transmitters operated by the US coast guard in the US and to a limited extent in the Medittéranean and France. The European transmitters may be taken over by a European consortium when the US Navy ceases support after the completion of GPS.

The Loran-C chain transmits synchronized pulses at 100KHz. The time difference of the arrival of pulses are used to compute the location, normally by re-transmission of the time differences to a central monitoring station.

High performance low cost receivers are now commonplace in the US and many types of AVLM sytsems are now in use by bus, police, utilities and fleet management organizations. Loran-C has yet to be used seriously for navigation but when it is, it will need to be backed up by dead-reckoning as for the satellite systems.

OMEGA is a Very Low Frequency (VLF) radio system (4) available worldwide which has been experimented with successfully for vehicle location in the UK.

Datatrak (5) is an AVLM system available in the UK based on low frequency radio transmitters which cover the country so that generally a vehicle is within range of at least 4 synchronised transmissions. Each pair of transmitters creates a hyperbola of points with constant phase difference, and hence distance, from the tranmitters; the vehicle's position is determined from the intersection of at least two hyperbolae.

3 AUTONOMOUS NAVIGATION SYSTEMS

An autonomous navigation system is self-contained within the vehicle and is capable of useful performance in the absence of any communication with external signals such as radio or roadside beacons.

Three separate types of autonomous system are identifiable - direction guides, electronic maps and route guidance systems.

3.1 Direction Guides

The information provided by a direction guide is simply a crow-fly direction possibly combined with a distance to destination indication. The simplest direction guide is of course a compass, but as early as the 2nd century AD the Chinese (figure 2) made use of the principal of the differential odometer which maintains a constant direction indication by measuring the differences in rotation of opposite wheels on the same axle.

The components of a modern direction guide system would include a key-pad or some other means to input data; a display unit to display the direction to destination; a micro processor; a distance sensor driven by the speedometer cable or by the measurement of wheel pulses; and a direction sensor for example a solid state flux gate or magneto-resistance effect magnetic compass.

The driver keys in the coordinates of his destination, for example in terms of the Ordnance Survey Grid Reference system, and the microprocessor computes the distance and bearing of the vector connecting his present position to the destination. If necessary the driver can correct the current position of his vehicle if it is known to be in error.

As the car travels, the distance and direction sensors are used to update continuously the present position of the car and hence the direction and distance to the destination are calculated and displayed on the screen.

The system thus works solely with the use of dead reckoning. Dead reckoning is a method of determining position by summing the small changes in distance and direction from a known starting point and for centuries has been a principle of marine navigation.

Basic components of the system are the direction sensor and the distance sensor. A magnetic compass would commonly be used for the direction sensor, although gyroscopic techniques can give greater accuracy at greater cost (laser gyroscopes are becoming available - they are extremely accurate, but as yet too expensive for widespread use in vehicles), and an odometer for the distance sensor. The analysis of data from a differential odometer recording the differences in movements between the rear wheels when cornering can be used to provide both distance and direction changes.

A number of simple direction guides have been developed for demonstration purposes, including DRIVEGUIDE from Nissan in Japan and CITYPILOT from the German company VDD. The beacon-based systems Ali-Scout and Autoguide function as simple direction guides when not supported by beacon information.

One of the main problems of this type of system is the inherent inaccuracy of dead reckoning. Errors accumulate as the journey progresses, typically of the order of 2-5% of the distance travelled, and if no other external data are available for correcting the position, the system rapidly becomes unreliable.

A second significant problem is that the guidance provided is unrelated to the road network and is therefore liable to direct drivers illogically, for example across a river estuary where no physical crossing point was provided. However the dead reckoning system is an inexpensive option.

These problems are partially overcome in the second type of autonomous system which includes an electronic map.

3.2 Electronic maps

An early form of this type of system was the Honda Electro Gyrocator in 1981, which used transparent map overlays to show the road network.

However this has now been superceded by the use of digital map displays. Road maps are stored as digital data and different parts of the map can be displayed at different scales depending on the car's location and its proximity to its destination. The display can either be on a CRT (Cathode Ray Tube) screen or on a flat panel LCD (Liquid Crystal Display).

The vehicle location can be displayed together with the destination and different classes of roads can be displayed in different colours or line thickness.

The digital data for the map can now be stored on CD-ROM (Compact Disk - Read Only Memory). The typical capacity for a CD-ROM is 550 megabytes, enough to store the details of most of the roads in Europe.

A significant advantage of having an on-board digital map is that the capability is provided for eliminating the dead reckoning errors of the Direction Guide systems. This is achieved by a process known as map-matching.

In map-matching, the changes in position of the car are continuously compared with the digital map data. When a turn is made whose magnitude and position can be confidently identified on the map, the vehicle is assumed to be at that location and the coordinates of the position of the car can be appropriately corrected.

Equipment of this type is available in the Electro Multivision system available as an option in Japan on the Toyota Crown (figure 3). Another commercially available system is the ETAK Navigator (6) introduced in California in the mid 1980's. This system originally used a magnetic tape cassette, with a capacity of approximately 3.5 megabytes, to hold the map data; the map is displayed on a CRT screen and can be shown at different scales depending on the distance of the vehicle from the destination. The vehicle location is displayed as a fixed symbol pointing to the top of the screen and as the car moves, the map rotates and moves around the position of the car. The destination, which is input by address or road junction, is also displayed on the screen.

A version of the ETAK system has been launched in Europe by Bosch-Blaupunkt (figure 4). It includes map data storage on compact disc and is called Travelpilot. A version is being marketed in Japan by Clarion.

Electronic map displays are considerably more useful to the driver than simple direction guides, although they are sometimes criticised as potentially distracting of the driver's attention. It remains the driver's responsibilty to choose a route through the network which may well not be a "good" one.

Navigation systems are taken one step further by the use of a route selection algorithm and the provision of turning recommendations which are available in the third type of autonomous system, Route Guidance Systems.

3.3 Route guidance systems

Autonomous route guidance systems use route choice algorithms to devise a route through the road network, digitally stored on board the vehicle, and provide turn by turn instructions to the driver along the route to his destination.

The instructions can be provided either on a display and/or backed up by a synthesised voice.

This type of system frees the driver from making route decisions, allowing him to concentrate on driving. Experience of demonstration systems suggests that this can lead to considerable relief of driver stress in busy city driving conditions.

One of the earliest prototype systems of this type was developed by Wootton Jeffreys in the early 1980s and involved the use of an Apple Microcomputer linked only to the vehicle odomoter for distance measuring. At each junction at which a turn was needed, a synthesised voice gave appropriate instructions. The Transport and Road Research Laboratory (TRRL) in the UK developed in conjunction with Lucas a similar system called Navigator, shown in figure 5.

More recent examples include the Philips CARIN system, Bosch-Blaupunkt's EVA and the Nissan NAV system.

The CARIN system (7) uses CD-ROM for storage of map data and uses dead reckoning (differential odometer and magneto-resistance magnetic compass) together with map-matching. Route instructions are displayed on an LCD screen and announced by a synthesised voice. A CRT map display may be provided as an option.

The EVA system from Bosch originally used map data "hard-coded" into memory chips (EPROMs) but is now developing CD-ROM technology. It also has map-matching and uses a differential odometer. Junction displays are provided on an LCD screen indicating the path to be followed through the junction and synthesised voice is also available.

The NAV system also uses dead reckoning combined with map-matching. Routes are calculated and displayed on a CRT screen and special displays guide the driver through junctions.

None of these systems are yet widely commercially available. One of the main reasons for this is the lack of common standards and data for digital maps and road network representations. Philips and Bosch in particular are participating in European projects designed to produce this data and it is likely that very soon there will be the widespread availablity of digital map data to a common standard suitable for use in on-board navigation systems.

4 AUTONOMOUS SYSTEMS SUPPORTED BY TRAFFIC INFORMATION

By definition, autonomous systems are independent of information from external sources. However the benefit of autonomous navigation aids can be considerably enhanced if access is available to relevant traffic and weather conditions and also to local data such as parking availability. For route guidance systems, research at TRRL has led to estimates that the benefits will be roughly doubled by building routes taking into account actual traffic information, compared with purely autonomous systems.

Assuming that data is available in a suitable form, there are a number of different methods of communicating the information to the vehicle. These methods include Radio Data Systems (RDS), teleterminals and cellular telephone.

RDS (8) is the system gaining popularity in Europe which uses side-band capacity of existing FM radio transmission. It has been developed primarily to give programme and frequency information to intelligent radio receivers, but has been designed to include a Traffic Message Channel (TMC) to improve driver information.

RDS is relatively cheap to implement and car radio receivers will readily be modified to accept RDS signals. RDS signals can automatically interrupt radio broadcasts or the radio/cassette player to provide relevant local data. The DRIVE programme of research in Europe (see section 6 below) includes the investigation of RDS-TMC for improved driver information, and there are plans for using RDS-TMC to enable systems such as CARIN to react to changes in traffic conditions.

Teleterminals are being used in the AMTICS (Advanced Mobile Information and Communication System) project in Japan (9). These are radio transmitters arranged in a mesh at approximately 3 km intervals and are used to transmit traffic information to equipped vehicles. The system uses a method of packet data transmission and operates at 800MHz with data rates of 4800 or 9600 baud. Vehicles will be equipped with compact disc systems for autonomous map displays, and the dynamic traffic information will be superimposed. The Pathfinder Project in California (10), a collaborative venture between the California Department of Transport, the Federal Highway Administration and General Motors is planning to equip 25 vehicles with the ETAK equipment previously described. Traffic data will be transmitted to the cars by radio and displayed on the ETAK map screen; link travel times will be transmitted back to the Operations Centre. The project will take place in the SMART corridor in Los Angeles where state-of-the-art traffic management techniques are being tested with the objective of combatting congestion.

5 INFRASTRUCTURE-BASED SYSTEMS

There is widespread support for the view that the most effective form of vehicle navigation - dynamic route guidance - will be most successfully achieved using a control centre and a communications infrastructure to transfer information between the control centre and equipped vehicles.

A control centre would cover a complete conurbation and possibly a large area surrounding it, and would be the central point at which traffic conditions were monitored directly. There would be direct links between the control centre and conventional traffic control systems. Routes would be calculated at the control centre based on up-to-the-minute traffic data and would be transmitted via the infrastructure to equipped vehicles; alternatively, or additionally, dynamic traffic data might be transmitted directly to vehicles with more sophisticated on-board equipment capable of calculating routes directly.

The communications infrastructure is likely to be based on roadside beacons, although work is also in progress in the DRIVE SOCRATES project to investigate the use of cellular radio.

5.1 Beacon-based systems

Roadside beacons are strategically located short range transceivers (transmitter/receivers) which provide two-way communication with vehicles. The transmission could be via inductive loops using radio frequencies or by means of microwave or infra-red transceivers mounted on roadside equipment. The roadside beacon approach has traditionally been used for AVLM for buses, but more recently large scale route guidance systems are being implemented.

For route guidance, systems could work in a number of different ways. The roadside equipment might receive from the passing vehicle a destination code, based on the information keyed in by the driver. The roadside unit could then determine and instantly transmit driving instructions to the car for subsequent display to the driver. This is the basis of the original approach adopted for Autoguide by TRRL and used in the CACS project in Japan and the ERGS project in the US during the 1970s and early 1980s; these systems were based on inductive loops buried in the road surface and have now been superceded.

The approach adopted in the Autoguide and ALI-SCOUT systems being installed in the UK and Germany is based on the transmission of the same large data set to each equipped vehicle passing a particular beacon at a particular time. The transmission medium is infra-red pulses giving a very high data rate and the data set includes location, map and routeing data. The in-vehicle unit (IVU) of each vehicle is able to derive the details of the most appropriate route for the driver's current destination, together with the appropriate routeing instructions to be given to the driver.

A bilateral working party was set up by the British and German governments in 1987 and it has produced a draft standard (11) for the all-important communication link between roadside and vehicle.

ALI-SCOUT is a development of the earlier ALI loop-based and AUTOSCOUT infra-red systems. A full scale trial of ALI-SCOUT is currently in operation in the LISB (Leit und Information System Berlin) project in Berlin (12). A demonstration Autoguide system has been operational in London since early 1988, and the Government introduced legislation in the 1988-89 session to enable the commercial operation of Autoguide. During 1989 work is expected to begin on a large scale pilot system covering a major part of London. If successful, it is expected that the pilot system could be expanded to a fully commercially available system by 1992 or 1993 (13).

The Autoguide/ALI-SCOUT system combines the use of roadside beacons with dead-reckoning and map-matching. The driver begins his journey by keying in his destination by means of a place-name and/or coordinates. This destination can be held in the on-board computer's memory for future reference if required. In future systems the ability might be included to specify a post code as a destination.

The journey begins in autonomous mode with the driver being given the direction and distance to his destination. When the car passes a beacon (figure 6), up to 32 kilo-bytes of map and route data are transmitted into the car, and the IVU selects the appropriate route to follow to the destination. At each junction where a turn is to be made, the driver is informed of the manoeuvre both on the LCD display (figure 7) and by synthesised voice instructions.

Other beacons will be passed en route which will update the information in the car. As the vehicle nears its destination it will leave the main road network temporarily stored in the vehicle and revert to autonomous mode for the last part of the journey. Whilst in guidance mode, dead-reckoning and map-matching are used to update the vehicle's coordinates.

The communication with the beacons is two-way. Travel time data from fitted vehicles is transmitted to the beacons and hence to the control centre. This will provide an unprecedented source of traffic data, both real time and historic, which, together with data from other traffic management control systems, will enable journey times to be continually kept up to date. The optimum routes provided to drivers can be dynamically updated to take account of prevailing traffic conditions.

5.2 Cellular radio

An alternative approach to roadside beacons is being investigated in the SOCRATES (System of Cellular RAdio for Traffic Efficiency and Safety) project (see 15) in which major partners are Philips and Bosch. The use of cellular radio for widespread two-way communication as a genuine dialogue will always face capacity restraints. However, the plan in SOCRATES is to use specific frequencies from the pan-European GSM system (which will replace current cellular radio systems during the 1990s) to broadcast, in a similar way to Autoguide/ALI-SCOUT, the same data set to all equipped vehicles in a particular cell. A multiple-access protocol will provide for transmission back to the control centre from equipped vehicles.

5.3 Progress towards infrastructure-based systems

The UK and West Germany have established a firm basis for international progress towards compatible infrastructure-supported route guidance systems; there is considerable interest from many other European countries in investigating the possibilities for implementing similar systems to Autoguide and ALI-SCOUT.

The British TRRL has been carrying out various research projects to investigate the potential benefits to the user and the community resulting from the introduction of Autoguide (14).

In the US, the FHWA (Federal Highway Administration) is currently carrying out a 30 month project comparing the potential benefits of static ETAK, dynamic ETAK (i.e. Pathfinder), Autoguide and an 'Advanced Experimental System'.

In Japan, the RACS project (Road Automobile Communication System) is using a method of roadside radio beacons (eg 247KHz, 9600 baud) to transmit location information as well as for communication. The location information can be used by on-board dead reckoning and map matching systems.

In 1986, the Public Works Road Institute in Japan began a 3 year cooperative project with a number of private companies to look at various types of transmission including inductive radio and microwave.

In systems of this type, there is clearly a balance to be struck in the distribution of intelligence between the roadside equipment and the car. The overall system cost will be kept down if the cost of the in-car equipment is kept to a minimum; the objective of the Autoguide/ALI-SCOUT approach is to produce the basic vehicle unit as inexpensively as possible in order to maximise the market penetration. The corollary however is the requirement for sugnificant investment in the infrastructure. In the UK, the system is likely to be funded privately with costs being recovered from user charges, possibly in the form of annual licence fees. SOCRATES is particularly intended to provide dynamic traffic information to autonomous systems such as CARIN and EVA.

6 THE FUTURE

The main use of radio navigation is likely to remain with AVLM for use in fleet management and control, although there could be some emergence of the use of radio or satellite-based techniques to support autonomous navigation systems, particularly for long distance high value commercial traffic.

During the next decade the most far-reaching developments are likely to be found in infrastructure-based route guidance systems provided that suitable mechanisms for funding for the infrastructure can be found. In Europe there is likely to be a significant investment in beacon-based systems; these may later be integrated with cellular radio capabilities as the new European GSM system becomes widely installed and the results from SOCRATES become available.

While the infrastructures are being developed and installed, simultaneously the marketplace is likely to see the genuine emergence of autonomous systems which will include digital maps to support the combination of dead reckoning and map matching. These will take time to become common whilst there is a lack of available quality digital map data and whilst standards for digital map data are still being finalised.

However, one of the most active areas of international collaboration is the development of standards for the data content, data acquisition and exchange formats for digital maps. European standards should have been agreed during 1989 and by 1994 there should be the first generation of common European digital road maps. NavTech in the US has set out to develop digital data road maps (ie including one-way streets and banned turns) of the 100 largest cities in the US.

These developments will therefore ease the way for the introduction of automous navigation systems, which are then likely to be extended by the use of radio communications such as RDS to provide congestion, local traffic, weather and parking information.

Later, the trend should be towards the integration of autonomous and infrastructure-supported systems. Autonomous systems should start to make full use of the dynamic information available from the infrastructure whilst the autonomous capabilities of purely infrastructure-based systems become more sophisticated.

A communications infrastructure designed initially for route guidance will have numerous other applications, many of them almost unavoidably implemented as direct extensions of the basic route guidance system. These applications include parking management and information, vehicle location and fleet management, tourist information, public transport management and information, trip planning, hazard warning, enhanced traffic control, automatic incident detection, data collection for real time traffic monitoring and for long term transport planning, and automatic toll collection and other forms of automatic debiting.

The role of international research and development programmes is likely to be increasingly important in promoting the progress towards integrated systems. In particular the European Community programme DRIVE and the Eureka programme PROMETHEUS are especially concerned with integrated systems of "Road Transport Informatics" (RTI).

DRIVE (Dedicated Road Infrastructure for Vehicle safety in Europe) (15) is an £80m programme designed to bring Europe closer towards an "Integrated Road Transport Environment" (IRTE) in order to improve traffic efficiency and safety and to reduce the environmental impact of the motor vehicle. It consists of some 60 multinational projects funded jointly by industry and the Community research budget; plans are already being formulated for a "DRIVE II" programme beginning in 1992. DRIVE II will be intended to help bring about the implementation of the integrated systems developed during the initial three-year programme.

Eureka is the framework for innovative research agreed in 1986 by 19 European countries under which some 200 projects are currently active. One of these is the PROMETHEUS programme (16) run by a consortium of most of the leading European automotive manufacturers; PROMETHEUS has similar objectives to DRIVE but its approach is more from the point of view of the vehicle than of the infrastructure. Other relevant Eureka projects include DEMETER, EUROPOLIS and CARMINAT.

Progress is therefore being made rapidly in three continents. Whilst there will inevitably be duplication of effort and the discarding of some good ideas in the move towards international compatibility, there is no doubt that during the 1990s the motor vehicle and the infrastructure which supports it will become increasingly intelligent. Fundamental to the implementation of the Integrated Road Transport Environment will be systmes for improved vehicle navigation and route guidance. It is quite probable that the implementation of the guidance display in the car will parallel the introduction of the car radio - in the early 1960s a radio was a luxury additional item to a car, as the route guidance system will be in the early 1990s. But very few cars are purchased today without a radio, and by the year 2000 the average driver may well feel genuinely lost without his on board route guidance computer to help him to his destination.

REFERENCES

(1) Jeffery, D.J., Russam, K, and Robertson, D.I. Electronic route guidance by Autoguide: the research background. Traffic Engineering and Control, Vol 28 No. 10, October 1987

(2) Johannessen, R. GPS as a component in land mobile navigation. IEE digest no 1987/21, February 1987

(3) Herbault, P. PSA Report on Localisation Methods based on LORAN C., PROMETHEUS PRO-ROAD document 51 110, July 1987

(4) Stratton, A. Land vehicle location and navigation using OMEGA/VLF radiolocation. IEE digest no 1987/21, February 1987

(5) Banks, K.M. Initial trials and evaluation of the Datatrak location and position reporting system. IEE digest no 1987/21, February 1987

(6) Honey, S.K. and Zavoli, W.B. A novel approach to automotive navigation and map display. Proceedings, RIN conference on land navigation and location for mobile applications, September 1985

(7) Thoone, M.L.G. et al. The car information and navigation system CARIN and the use of compact disc interactive. SAE Technical Paper no 870139, 1987

(8) What the Radio Data System can do for you. BBC Engineering Information, Information sheet 1505(2)8610, 1986

(9) Okamoto, H. Advanced Mobile Information and Communication System (AMTICS). SAE Technical Paper Series no. 881176, August 1988

(10) Blackburn, L., Takasaki, G.M., Wasielewski, P.F. and Mammano, F.J. PATHFINDER - moving ahead with motorist information and in-vehicle road navigation. California Department of Transportation internal document, December 1988

(11) Route guidance information systems - draft standard for the road-vehicle communication link. Bundesministerium fuer Verkehr, Germany, and Department of Transport (TCC Division), UK, May 1988

(12) von Tomkewitsch, R. LISB: Large-scale test 'Navigation and Information System Berlin'. Proceedings of ITTT seminar, volume P302, PTRC Education and Research Services Ltd, September 1987

(13) Catling, I. and Belcher, P.L. Autoguide - electronic route guidance for London and the UK. IEE conference publication number 299, February 1989

(14) TRRL contractor reports:
CR128 Study to show the benefits of Autoguide in London (JMP Consultants Ltd), March 1989
CR129 Estimates of Autoguide Traffic Effects in London (London Transportation Studies Unit), June 1988
CR130 Recommended Methodology for Evaluation of Route Guidance in London and the Potential for Comparability with Berlin (Institute for Transport Studies, University of Leeds), October 1988
CR131 A study to show patterns of vehicle use in the London area, and CR132 Supplementary report (Wootton Jeffreys Consultants Ltd), November 1987

(15) Commission of the European Communities, DGXIII, "The DRIVE programme in 1989", DRI 200, March 1989

(16) Karlsson, T. PROMETHEUS: the European research programme. TRB annual meeting proceedings, January 1988

Fig 1 Jones live map

Fig 2   Chinese south-pointing carriage

Fig 3   Toyota electro-multivision

Fig 4   Travelpilot

Fig 5 TRRL/Lucas navigator

Fig 6 ALI-SCOUT beacon

Fig 7 Autoguide display

C391/087

Concepts for vehicle route guidance

B G MARCHENT, BSc, PhD, CEng, MIEE
Plessey Research and Technology, Romsey, Hampshire

SYNOPSIS - System architectures are considered for vehicle route guidance. The aspects considered include requirements, vehicle equipment, roadside beacons, system control, traffic management and system evolution. The recommended architecture consists of an infrastructure of infra red roadside beacons which communicate guidance information to passing vehicles with a simple compass/odometer navigation system in the vehicle to provide navigation between beacons. Routing is provided to post code resolution with automatic recovery in case of driver error or traffic congestion. Integration with other forms of navigation systems and vehicle information systems would provide enhanced vehicle route guidance facilities.

1. INTRODUCTION

Developments in information technology have progressed to a point where vehicles can communicate with a complex infrastructure to provide benefits to the vehicle drivers and road traffic authorities (reference 1). One of the first applications will be the introduction of vehicle route guidance. A number of countries within Europe have recognised this opportunity and are in the process of developing and evaluating prototype systems (reference 2 and 3).

A common European system architecture is required which will encompass the needs of various countries, drivers and vehicle types. This would provide an architecture on which other information and safety systems could be developed. This paper describes the results of a research study into possible system architectures for vehicle route guidance.

2. REQUIREMENT

2.1 Main Issues

If vehicle route guidance is to achieve wide acceptance then it should :

1) Provide guidance for the entire vehicle journey including inter-urban, urban, local, rural and international guidance.
2) Take into account current and predicted traffic conditions and update routes as appropriate to avoid congestion.
3) Achieve low vehicle equipment cost combined with operational reliability.
4) Be provided as a public service or private operation funded by users' licence fees.

2.2 Ergonomics

The vehicle driver should be able to:

1) Enter intended destination in a simple form.
2) Receive manoeuvre instructions in a clear manner at an appropriate distance before each road intersection.
3) Receive a new route if he has ignored a manoeuvre instruction.
4) Select a route to suit his requirements.

2.3 Traffic Management

The system architecture must include means by which traffic congestion can be detected or predicted in critical parts of the road network. As the number of vehicles using the system increases then the route guidance could affect the distribution of traffic flow on the road network. The resulting distribution of traffic flow must be stable and acceptable to the local authorities.

3. SYSTEM ARCHITECTURE

A number of possible system architectures have been studied and the main architecture options are described below.

3.1 Map Storage in Vehicle

The road network data could be contained entirely within the vehicle. This could be provided either on compact disc for the whole country, as for the Philips CARIN system (reference 4), or as plug in modules for specific parts of the country. This static data could be supplemented by an infrastructure to broadcast information on major traffic congestion, e.g using RDS (reference 5).

This approach can lead to expensive vehicle equipment but the investment in the infrastructure would be low.

3.2 Information from Roadside Beacons

As an alternative to map storage in the vehicle routeing information could be provided from roadside beacons.

The vehicle could transmit its intended destination to roadside equipment and the manoeuvre instruction for that road intersection transmitted back to the vehicle, e.g. using an inductive loop in the road. This leads to low cost vehicle equipment but the disadvantages are that the communication time and data rate available are limited and roadside beacons would be required at every road intersection.

Roadside beacons could broadcast the same information to all passing vehicles (reference 6). The vehicle equipment would select the information it requires to follow a route to its destination. Three possible levels of information transfer could be considered:

1) Road links on which severe congestion has occurred to update the map data stored in the vehicle.
2) Preferred routeing information for specific destination regions as for the Siemens ALI-SCOUT system (reference 2).
3) Extensive road network information so the vehicle can select its own route.

3.3 Recommended Architecture (Figure 1)

A system based on broadcast from roadside beacons of extensive road network information was accepted as the most suitable architecture. This structure allows for a considerable distance, including many road intersections, between beacons. The roadside beacons would monitor road link times from passing vehicles and collect traffic flow information from other sensors on the road network.

The cost of the vehicle units would be low but the system is dependant on an infrastructure of roadside beacons. An initial hybrid solution may be necessary with road network data stored in the vehicle to provide route guidance in areas where beacons have not as yet been installed.

4. ROADSIDE BEACON INFORMATION

4.1 Beacon/Vehicle Data Link

A wide range of techniques were considered for the communication of information between roadside beacons and passing vehicles including :

1) inductive loops
2) UHF and 60GHz micro cellular radio
3) acoustic
4) infra-red

Only an infra-red or UHF/microwave link could meet the requirements. The only areas where radio is significantly better are its ability not to be blocked by obstructions, its immunity to fog, and its capability of operating at data rates significantly higher than those supported by the low cost infra-red option. Overall infra-red is the preferred option with several infra-red beacons deployed at each intersection to avoid obstructions.

This confirms the suitability of the infra-red data link proposed by Siemens for the ALI-SCOUT system. The beacon to vehicle data rate could be increased to take account of technology developments to enable extensive road network information to be broadcast to passing vehicles.

4.2 Link & Node Data Structure

A Link & Node data structure is recommended, figure 2, for the road network information broadcast from roadside beacons consisting of nodes (i.e. road intersections) joined by links (i.e. roads).

In the beacon region, up to the next beacon, the beacon data would describe all or most links and nodes including local roads. The information on each road link would be sufficient to provide vehicle tracking along any road link in either direction and manoeuvre instructions for all possible turns at each node. Beyond the next beacon the Link & Node data would include all main roads but not necessarily local roads. This would enable partial routeing beyond the next beacon in the case of communication failure at that beacon.

The exit points into the distant region would provide a list of the destinations that can be accessed by that exit or alternatively a coarse route tree for main destination centres. In the latter case a rough route preview and estimated journey time to main destination centres along the route could be provided.

The detail of the road network in the beacon data can be adjusted as further beacons are deployed. Initially for a low beacon density only the main road network structure would be included. When the full beacon density is deployed then route guidance would be available to UK postcode resolution.

4.3 Destination Specification

A postcode could be appended to each Link, Node or destination centre although it would be sufficient to identify a central node for a group of destinations. For distant destinations one could use only the first two characters of the code (e.g. SO for Southampton area), intermediate destinations the outward code (e.g. SO51 for Romsey) and destinations just outside the beacon region the sector code (e.g. SO51 0). Destinations within the immediate vicinity would be described with their full postcode (e.g. SO51 0ZN).

Map reference information could also be added to each node to provide an alternative means of destination input. In addition specific destinations could be added to road links (e.g. street names, railway station, etc).

4.4 Route Choice Information

Distance, delay time and cost would be added to the road links and vehicle turning movements at nodes. Initially these would be preset to defined values depending on the road type but would change as the system control becomes aware of the distribution of traffic flow on the network. Attributes would be added to specific links to allow routeing for a wide range of vehicle types, e.g. low bridge with a height constraint of 4.5m, taxis only, etc.

4.5 Route Recovery

If the driver deviates from the recommended route then the vehicle has sufficient information to compute a new route to the destination. This could be required if the driver ignores a turn instruction or drives through a car park or along a back street. Also immediate route guidance is available on entering a new destination from the information received from the last beacon.

5. VEHICLE EQUIPMENT (Figure 3)

5.1 Data From Beacons

As the vehicle passes a roadside beacon, route guidance information is collected and stored in the vehicle unit. The roadside beacons are designed for short range directional communications so only vehicles approaching the beacon within 100m are able to collect information from that beacon. The point at which the beacon signal is lost enables the vehicle to confirm or reset its position on the road network.

5.2 Vehicle Location Techniques

To enable the vehicle to track its progress along the recommended route the vehicle must be able to locate its position on the road network provided from the last beacon. Techniques for vehicle location, reference 7, can be divided into absolute position systems and dead reckoning navigation.

5.2.1 Absolute position navigation

Absolute position systems are usually based on the reception of satellite or ground based radio beacon signals. The radio signal propagation from such a system, particularly in urban areas, is often distorted or severely attenuated so that accurate and continuous vehicle location becomes difficult.

5.2.2 Dead reckoning navigation

Dead reckoning navigation systems require vehicle sensors to measure heading and distance. A compass and odometer dead reckoning navigation system leads to a low cost vehicle unit. Alternatively a differential odometer with wheel sensors on both front wheels could be used to measure heading but this is only accurate over relatively short distances.

5.2.3 Compass/Odometer Performance

A compass and odometer dead reckoning navigation system was recommended, such as the Plessey Adaptive Compass Equipment (PACE) (reference 8). This system operating on reasonably flat ground after calibration provides an accuracy of 0.25 degree on heading and 0.2% on distance. Calibration techniques have been developed for the automatic compensation of the vehicle interference fields.

Errors can accumulate due to road cambers/hills, and road network limitations/errors. A slope of 5 degrees could cause a worst case error in heading of 11.5 degrees in the UK.

Map matching of the vehicle's progress to the known road network would be used to reduce these errors. Position confirmation beacons could be used if ambiguity could arise at certain intersections.

5.3 Route Selection, Tracking and Display

The vehicle equipment would locate its intended destination in the beacon data, provide a route to that destination, track the vehicle along that route and detect if the vehicle deviates from the route.

The display to the vehicle driver would be manoeuvre instructions at each road intersection in the form of a simple display, the lane he should take on approaching the intersection and the exit from that intersection. This could be supported by voice output to avoid distraction to the driver.

Other information could be extracted from the Link & Node data structure including road distance to destination, time to destination, intermediate destinations and road numbers. A complete map display of the vicinity could be generated from the beacon data with the current position of the vehicle marked but, for safety reasons, this could only be used by a passenger.

5.4 Detection of Link Delays

As the vehicle proceeds along its route between beacons it accumulates journey time and stopped time along each road link. As the vehicle passes the next beacon this information is passed to that beacon and thence to the system control.

As the vehicle proceeds it could also collect system test data to enable the system control to monitor total system operation and detect faults. For example if a large number of vehicles do not travel along a road then that road may be blocked.

5.5 Encryption

To ensure that only licenced subscribers' vehicles can use the route guidance information then some of the data transmitted from beacons would be encrypted. The vehicle user would enter a Private Access Number (PAN) into his vehicle equipment. This would be combined with a code stored in the vehicle equipment, i.e. the Unit Identification Code (UIC), to enable beacon data to be decrypted.

6. ROADSIDE BEACONS (Figure 4)

6.1 Beacon Data

The information to be broadcast from roadside beacons would be derived in a computer centre for the total system (or part of that system). The details of the local road network could be stored in the central computer, in the beacon or down loaded to the beacon say once per day. The beacon data could be assembled in the computer centre, or in the beacons (possibly as transmission of changes only) to enable a considerable reduction in data communications from the computer centre.

6.2 Beacon Deployment

Where possible beacons would be deployed at major road intersections. The attachment of beacons to traffic lights provides a low cost solution as beacon mounting points, power supply and often communications are available. In some cases new routeing information must be available before a road intersection is reached in which case beacons must be deployed some distance before that intersection.

The spacing between beacons will be chosen to ensure that alternative vehicle routeing could be provided within an acceptable time delay after detection of traffic congestion. The maximum beacon density will also be constrained by the amount of information that can be broadcast from beacons. For full route guidance a typical spacing of 0.5km is expected in urban areas and 5km in rural areas.

7. SYSTEM CONTROL (Figure 5)

The system control would include the task of controlling the total system whether this implies a Pilot, London, National or International system. The full system control is expected to be distributed in terms of processing power with distribution depending on how the system should grow and the needs of various operating authorities.

7.1 Road Network Database

The road network database will form the core of the system control. Software would be used to convert the database into a suitable format and to distribute the road network information over the total distributed system and, if appropriate, for the generation of vehicle plug-in map modules.

7.2 Data Processing

Each computer control centre will generate routeing information to be communicated to vehicles via roadside beacons. The computer centre will also monitor traffic flow in its own and other areas to update the road network information, control total system operation, detect faults and provide alternative processing and/or vehicle routeing in the case of failure of system components.

7.3 Encryption

The keyword used to encrypt the beacon data would be changed at the end of each month. New Private Access Numbers would be distributed to vehicle owners probably as a complete set for one year. A computer centre will be required to derive these Private Access Numbers from the vehicle unit identification codes.

8. TRAFFIC MANAGEMENT

8.1 Detection of Traffic Congestion

There would be a time delay before the system control would be aware of changing traffic conditions due to the delay before vehicles pass the next beacon and the delay to average sufficient link time samples. To provide route guidance to avoid congestion this information would be supplemented with other traffic flow data acquired from traffic flow sensors on critical parts of the network plus manual input, e.g. from the police.

8.2 Historic Traffic Flow Data

Congestion normally occurs at the same time each day. Thus historic link time profiles would be developed throughout each day and used on similar following days. Information from equipped vehicles and traffic flow sensors would be used to detect and respond to unexpected congestion.

8.3 Beacon Update

At one time a beacon update every 5 minutes was suggested but a priority system would be more appropriate. The computer centre would detect traffic congestion and control the priority in which beacons should be updated.

8.4 Guidance Stability

Once a certain proportion of vehicles are equipped then routeing of these vehicles could cause a significant change in distribution of traffic on the road network. Instability could result for example with the recommended vehicle routeing oscillating between two parallel roads as congestion builds up on each road in turn. Proportional routeing may be required, i.e. to provide different routeing recommendations for consecutive vehicles passing a beacon.

9. SYSTEM EVOLUTION

The eventual intention is to provide route guidance throughout Europe but this will not be possible in one step. Guidance could be restricted to areas where the infrastructure has been developed or supported by other forms of route guidance.

9.1 Static Route Guidance

Initially a static system could be developed with communications from roadside beacons to the system control being added at a later date. Beacon data would be prepared in a computer centre and stored in PROM in the beacons including variations in link attributes, e.g. for peak and off peak periods.

9.2 Main Routes only

Initially a low beacon density would be deployed with only main routes included in the beacon data structure. The vehicle equipment would be able to provide guidance up to the nearest known road intersection to the required destination with autonomous mode, i.e. compass heading and distance, supplied for the final part of the journey. If the vehicle travels along a road that is not known in the beacon data structure then the vehicle equipment can locate the vehicle position as it rejoins the known road network.

9.3 Map Storage in Vehicle

Map storage in the vehicle via plug in modules, with expected road link times for peak and off peak periods, could supplement information provided from beacons for long distance routeing or provide a complete route guidance system in areas where an infrastructure has not been deployed. Additional facilities can be provided, e.g. detailed route preview and user route selection. The roadside beacons would then provide routeing to reach each milestone in turn avoiding traffic congestion as appropriate. If routeing is required in areas where beacons have not been deployed then the vehicle driver would need means by which he can occasionally reset or confirm his position on the known road network, e.g. by entering a road name.

9.4 Radio Data Systems

Digital data transmissions are being added to VHF/FM transmitters, reference 5, to provide 1187.5bit/s. Most of this data has been allocated but a small amount could be available to provide an immediate indication of major blockages on the road network if the vehicle is equipped with an RDS receiver for other purposes.

9.5 Cellular Radio

If cellular radio is fitted to the vehicle then this could form a link with the system control to enable the driver to select a route to a distant destination. Beacons then provide routeing via milestones on that route and indicate if the vehicle should change from that route or contact the system control, via cellular radio, for a new route.

9.6 Additional System Facilities

The route guidance system would provide an architecture on which other vehicle information systems could be developed. These include :-
a) Information to vehicles, e.g tourist information
b) Fleet location systems, e.g. for taxis, police, etc
c) Integrated transport, e.g. with ferries, railway.
d) Simple message transmission to/from vehicles

10. REFERENCES

1) Queree, C., "Application of Information and Telecommunication Technology to Road Safety", 6th International Conference on Automotive Electronics, IEE London, UK, 12-15 October 1987

2) Von Tomkewitsch, R., "ALI-SCOUT : A Universal Guidance and Information System for Road Traffic", 6th International Conference on Automotive Electronics, IEE London, UK, 12-15 October 1987

3) "Autoguide Pilot Stage Proposals : A Consultation Document", Department of Transport, UK, 1988

4) Fernhout, H.C., " The CARIN Information and Navigation System and the Extension to CARMINAT", 6th International Conference on Automotive Electronics, IEE London, UK, 12-15 October 1987

5) "Specification of the Radio Data System RDS for VHF/FM Sound Broadcasting", Tech. 3244-E, European Broadcasting Union, Technical Centre Brussels, March 1984

6) Pawley, A.J.R., "In Vehicle Information Systems using Roadside Beacons", 6th International Conference on Automotive Electronics, IEE London, UK, 12-15 October 1987

7) Foster, M.R., "Vehicle Location and Navigation in the Urban Environment", 6th International Conference on Automotive Electronics, IEE London, UK, 12-15 October 1987

8) Foster, M.R., "Vehicle Navigation Using Adaptive Compass", Journal of Navigation, Vol 39, No 2, May 1986

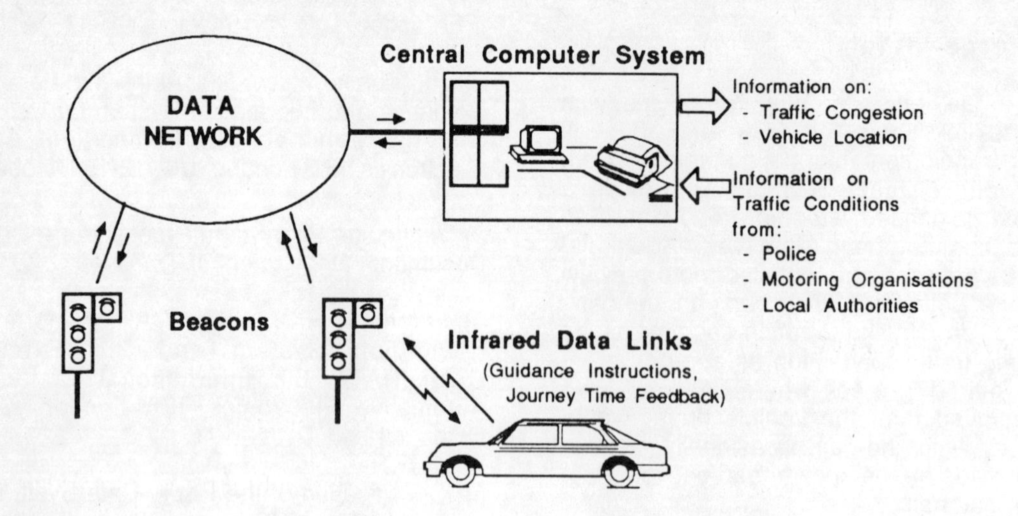

Fig 1 Recommended system

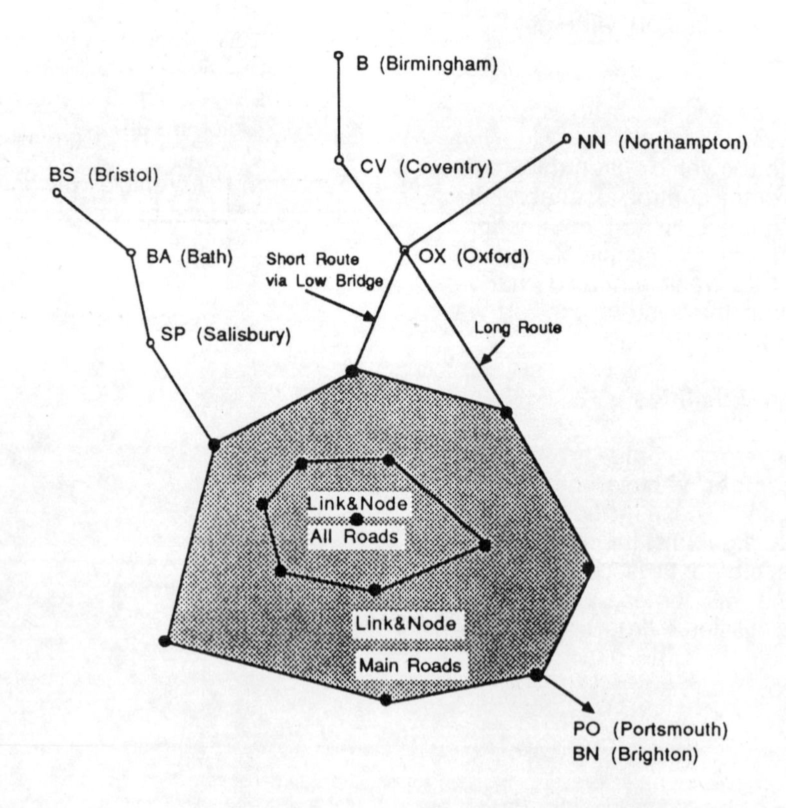

Fig 2 Link and node data structure

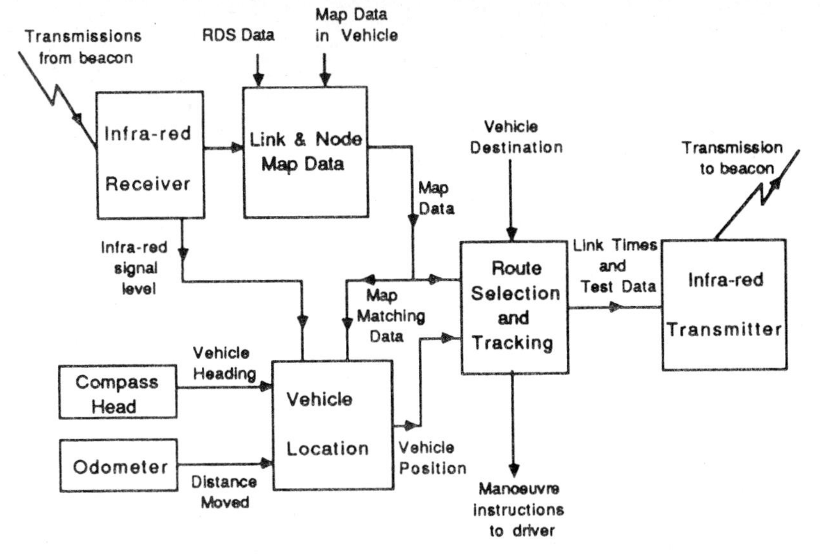

Fig 3 Vehicle equipment functions

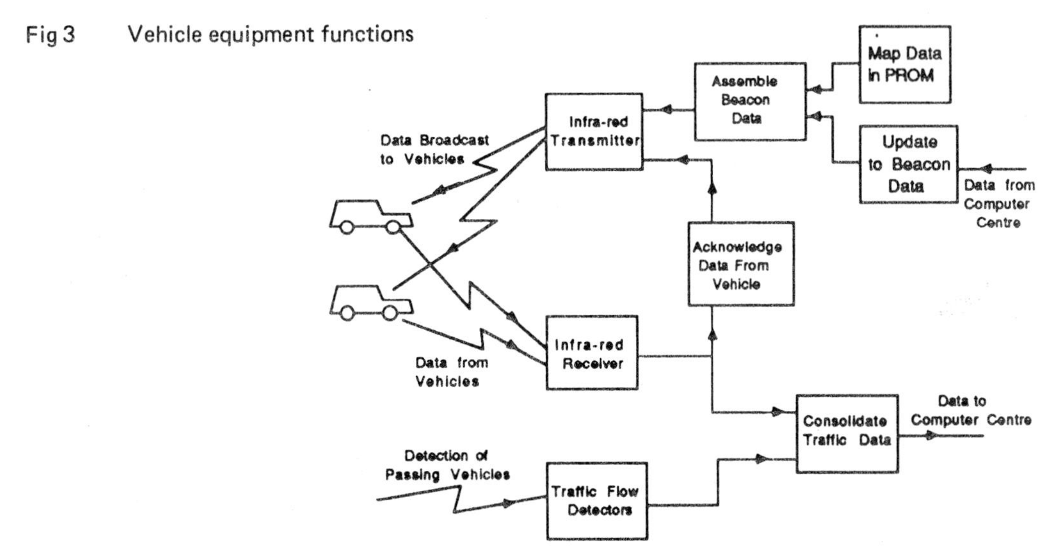

Fig 4 Beacon functions

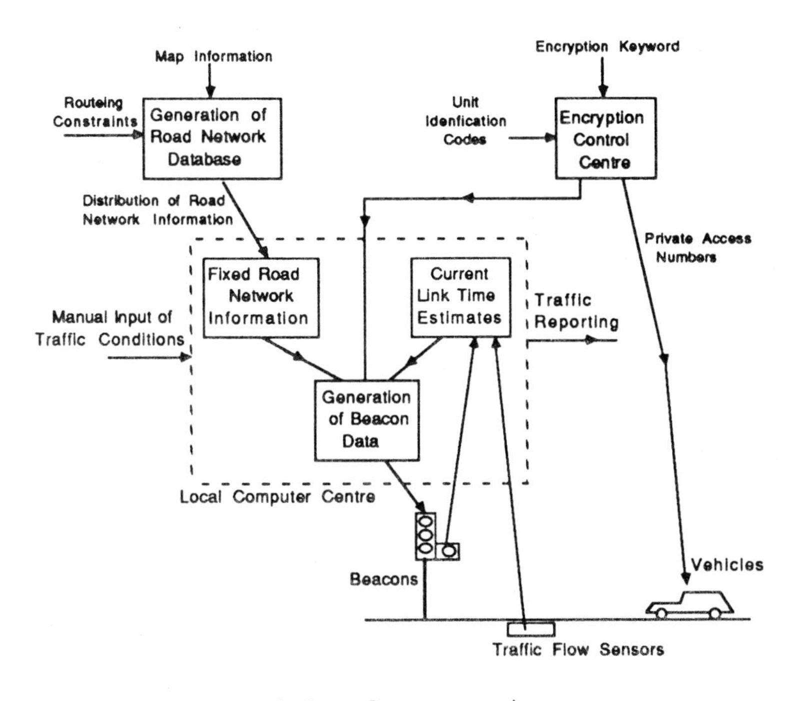

Fig 5 System control